BIODEGRADABLE POLYMER BLENDS AND COMPOSITES FROM RENEWABLE RESOURCES

BIODEGRADABLE POLYMER BLENDS AND COMPOSITES FROM RENEWABLE RESOURCES

Edited by

Long Yu

A JOHN WILEY & SONS, INC., PUBLICATION

Published by John Wiley & Sons, Inc., Hoboken, New Jersey
Published simultaneously in Canada

For general information on our other products and services or for technical support, please contact our Customer Care Department within the United States at 877-762-2974, outside the United States at 317-572-3993 or fax 317-572-4002.

Wiley also publishes its books in variety of electronic formats. Some content that appears in print may not be available in electronic format. For more information about Wiley products, visit our web site at www.wiley.com.

Library of Congress Cataloging-in-Publication Data:

Yu, Long.
   Biodegradable polymer blends and composites from renewable resources/Long Yu.
      p. cm.
   Includes index.
   ISBN 978-0-470-14683-5 (cloth)
1. Biodegradable plastics. 2. Polymeric composites. 3. Renewable natural resources.
I. Title.
   TP1180.B55Y8 2009
   620.1′92323--dc22   2008009441

Printed in the United States of America

10 9 8 7 6 5 4 3 2 1

CONTENTS

The history of composites from renewable resources is far longer than that of conventional polymers. For example, in the biblical Book of Exodus, Moses' mother built a basket from rushes, pitch, and slime—a kind of fiber-reinforced composite, according to the modern classification of materials. Also, during the opium wars in the middle of the nineteenth century, the Chinese built their defenses using a kind of mineral particle-reinforced composite made from gluten rice, sugar, calcium carbonate, and sand.

There are now many well-developed techniques that are used to produce conventional polymer blends and composites, and various products have been widely commercialized. There are also numerous papers, patents, books, and handbooks that introduce and discuss the development and application of various polymeric blends and composites.

Over the past two decades, however, polymers from renewable resources (PFRR) have been attracting increasing attention, primarily for two major reasons: environmental concerns, and the realization that our petroleum resources are finite. A third reason for the growing interest in polymers from renewable resources relates to adding value to agricultural products, which is economically important for many countries.

Generally, polymers from renewable resources can be classified into three groups: (1) natural polymers, such as starch, protein, and cellulose; (2) synthetic polymers from natural monomers, such as poly(lactic acid); and (3) polymers from microbial fermentation, such as poly(hydroxybutyrate). A major advantage of all these materials is that they are biodegradable, and that their final products of degradation are environmentally friendly.

As with numerous petroleum-based polymers, many properties of PFRR can be improved through appropriate blending and composite formulation. These new blends and composites are extending the utilization of PFRR into new value-added products.

The book comprises six sections that highlight recent developments in biodegradable polymer blends and composites from renewable resources, and discusses their potential markets. Following an overview of polymers from renewable resources (Chapter 1), the rest of Part I (Chapters 2–6) emphasizes blends from natural polymers, involving both melting and aqueous blending, as well as reactive blending. Part II (Chapters 7 and 8) focuses on aliphatic polymer blends, in particular improving the thermal properties of these systems. Part III (Chapters 9 and 10) discusses various hydrophobic and hydrophilic blends, in particular polyesters and natural

vii

polymers. Part IV (Chapters 11–14) discusses composites reinforced with natural fibers, while Part V (Chapters 15–17) introduces the development of nanoclay-reinforced composites, including novel techniques of delaminating clay for use in natural polymers. Part VI (Chapter 18) introduces multilayered systems from renewable resources.

Throughout the book, attention is given to the relationship between microstructure and properties, in particular interfacial compatibility, and mechanical and thermal characteristics. Written primarily for materials and polymer scientists and technologists, this book will also be of value to those in the commercial market concerned about the environmental impact of plastics.

I wish to take this opportunity to thank each of the authors, and their many collaborators, for their splendid contributions to the development of biodegradable polymers. I also wish to thank my postgraduate students both in CSIRO Materials Science and Engineering and the Centre for Polymers from Renewable Resources at SCUT for their hard work in this area. Lastly, I thank Cathy Bowditch for editing my publications over the years.

LONG YU

CSIRO Materials Science and Engineering
Melbourne, Victoria, Australia

Centre for Polymers from Renewable Resources
SCUT, Guangzhou, China
June 2008

CONTRIBUTORS (CORRESPONDING AUTHORS*)

Vera Alejandra Alvarez, Research Institute of Material Science and Technology, Engineering Faculty, Mar del Plata University, Juan B. Justo 4302 (7600), Mar del Plata, Argentina

Ioannis S. Arvanitoyannis*, Department of Agriculture, Animal Production and Aquatic Environment, School of Agricultural Sciences, University of Thessaly, Fytokou Str., 38446 Nea. Ionia Magnesias, Volos, Greece, Tel: +30-4210-93104; Fax: +30-4210-93144; Email: parmenion@uth.gr

Luc Avérous*, ECPM-LIPHT (UMR CNRS 7165), University Louis Pasteur, 25 rue Becquerel, 67087 Strasbourg Cedex 2, France, Tel: +33-3-90-24-27-07; Fax: +33-3-90-24-27-16; Email: AvérousL@ecpm.u-strasbg.fr

Cheng Chen, State Key Laboratory of Polymer Physics and Chemistry, Changchun Institute of Applied Chemistry, Chinese Academy of Science, Changchun, 130022, P. R. China

Guo-Qiang Chen*, Multidisciplinary Research Center, Shantou University, Shantou, Guangdong 515063, P. R. China, Tel: +86-754-2901186; Fax: +86-754-2901175; Email: chengq@stu.edu.cn

Ling Chen, Centre for Polymers from Renewable Resources, ERSPSP, South China University of Technology, Guangzhou, China

J. J. de Vlieger, TNO Science and Technology, Materials Performance, P.O. Box 6235, 5600 HE Eindhoven, The Netherlands

Lisong Dong*, State Key Laboratory of Polymer Physics and Chemistry, Changchun Institute of Applied Chemistry, Chinese Academy of Science, Changchun, 130022, P. R. China, Tel: +86-431-5262076; Fax: +86-431-5685653; Email: dongls@ciac.jl.cn

Hartmut Fischer*, TNO Science and Technology, Materials Performance, P.O. Box 6235, 5600 HE Eindhoven, The Netherlands, Tel: +31-40-2650151; Fax: +31-40-2650850; Email: hartmut.fischer@tno.nl

Baochun Guo, School of Materials Science and Engineering, South China University of Technology, Wushan, Guangzhou, 510641, P. R. China

Milford A. Hanna*, Department of Biological System Engineering, University of Nebraska, 208 L.W. Chase Hall, Lincoln, NE 68583-0730, USA, Tel: +1-402-472-1634; Fax: +1-402-472-6338; Email: mhanna@unl.edu

Yoshito Ikada*, Department of Environmental Medicine, Faculty of Medicine, Nara Medical University, Shijo-cho, Kashihara, Nara 634-8521, Japan, Email: ikada@naramed-u.ac.jp

Demin Jia*, School of Materials Science and Engineering, South China University of Technology, Wushan, Guangzhou, 510641, P. R. China, Tel: +86-020-87113374; Fax: +86-020-87110273; Email: psdmjia@scut.edu.cn

A. Kassaveti, Department of Agriculture, Crop Production and Agricultural Environment, School of Agricultural Sciences, University of Thessaly, 384 46 N. Ionia Magnesias, Volos, Greece

Lan Liu, School of Materials Science and Engineering, South China University of Technology, Wushan, Guangzhou, 510641, P. R. China

Yongshang Lu, Department of Chemistry, Iowa State University, Ames IA 50011, USA

Rong-Cong Luo, Department of Biological Sciences and Biotechnology, Tsinghua University, Beijing 100084, P. R. China

Yuanfang Luo, School of Materials Science and Engineering, South China University of Technology, Wushan, Guangzhou, 510641, P. R. China

James Ramontja, National Centre for Nano-Structured Materials, CSIR Materials Science and Manufacturing, P.O. Box 395, Pretoria 0001, Republic of South Africa

Robert A. Shanks*, Co-operative Research Centre for Polymers, School of Applied Sciences, Science, Engineering and Technology, RMIT University, GPO Box 2476V, Melbourne, Victoria, 3001 Australia, Tel: +61-3-9925-2122; Fax: +61-3-9639-1321; Email: robert.shanks@rmit.edu.au

Mitsuhiro Shibata*, Department of Life and Environmental Sciences, Faculty of Engineering, Chiba Institute of Technology, 2-17-1 Tsudanuma, Narashino, Chiba 275-0016, Japan, Tel: +81-47-478-0423; Fax: +81-47-478-0423; Email: shibata@sky.it-chiba.ac.jp

Randal L. Shogren*, Plant Polymer Research Unit, National Centre for Agriculture Utilization Research, USDA/ARS, 1815 N. University St., Peoria, IL 61604, USA, Tel: +1-309-681-6354; Fax: +1-309-681-6691; Email: Randy.Shogren@ars.usda.gov

Suprakas Sinha Ray*, National Centre for Nano-Structured Materials, CSIR Materials Science and Manufacturing, P.O. Box 395, Pretoria 0001, Republic of South Africa, Tel: +27-12-841-2388; Fax: +27-12-841-2135; Email: rsuprakas@csir.co.za

Nilda Soares*, Departamento de Tecnologia de Alimentos, Universidade Federal de Viçosa, 36570-000 Viçosa-MG, Brazil 36570-000, Brazil, Tel: +55-31-3899-2208; Fax: +55-31-3899-1624; Email: nfsoares@ufv.br

Lan Tighzert*, Laboratoire d'Etudes des Matériaux Polymères d'Emballage (LEMPE), Ecole Supérieure d'Ingénieurs en Emballage et Conditionnement (ESIEC), Université de Reims Champagne-Ardenne (URCA), Esplanade Roland Garros—Pôle Henri Farman, BP 1029, 51686 Reims Cedex 2, France, Tel: +33-(0)3-26913764; Fax: +33-(0)3-26913764; Email: lan.tighzert@univ-reims.fr

Persefoni Tserkezou, Department of Agriculture, Icthyology and Aquatic Environment, School of Agricultural Sciences, University of Thessaly, Fytokou Str., 38446 Nea Ionia Magnesias, Volos, Greece

Hideto Tsuji*, Department of Ecological Engineering, Faculty of Engineering, Toyohashi University of Technology, Tempaku-cho, Toyohashi, Aichi 441-8580, Japan, Email: tsuji@eco.tut.ac.jp

Analía Vázquez*, Research Institute of Material Science and Technology, Engineering Faculty, Mar del Plata University, Juan B. Justo 4302 (7600), Mar del Plata, Argentina, Email: anvazque@fi.mdp.edu.ar

Xiaoping Wang, School of Materials Science and Engineering, South China University of Technology, Wushan, Guangzhou, 510641, P. R. China

Yixiang Wang, Department of Chemistry, Wuhan University, Wuhan 430072, P. R. China

Susan Wong, Co-operative Research Centre for Polymers, School of Applied Sciences, Science, Engineering and Technology, RMIT University, GPO Box 2476V, Melbourne, Victoria, 3001 Australia

Yixiang Xu, Industrial Agricultural Products Center, 15A L.W. Chase Hall, University of Nebraska, Lincoln, NE 68583–0730, USA

Long Yu*, Commonwealth Scientific and Industrial Research Organization, Materials Science and Engineering, Clayton South, Melbourne, Vic. 3168, Australia, Tel: +61-3-9545-2797; Fax: +61-3-9544-1128; Email: long.yu@csiro.au

Lina Zhang*, Department of Chemistry, Wuhan University, Wuhan 430072, P. R. China, Tel: +86-27-87219274; Fax: +86-27-68754067; Email: lnzhang@public.wh.hb.cn

■■■■ **CHAPTER 1**

Polymeric Materials from Renewable Resources

LONG YU

CSIRO, Materials Science and Engineering, Melbourne, Australia

LING CHEN

Centre for Polymer from Renewable Resources, ERSPSP, SCUT, Guangzhou, China

1.1 INTRODUCTION

Polymers from renewable resources have been attracting ever-increasing attention over the past two decades, predominantly for two reasons: the first being environmental concerns and the second being the realization that our petroleum resources are finite. In addition, this kind of material will provide additional income to those

Biodegradable Polymer Blends and Composites from Renewable Resources. Edited by Long Yu
Copyright © 2009 John Wiley & Sons, Inc.

involved in agriculture. Generally, polymers from renewable resources can be classified into three groups: (1) natural polymers such as starch, protein, and cellulose; (2) synthetic polymers from bioderived monomers such as poly(lactic acid) (PLA); and (3) polymers from microbial fermentation such as polyhydroxybutyrate.

In this short chapter, various polymers from renewable resources, in particular those that have been used as or have been shown to have potential for use as polymeric materials, are briefly reviewed. The review focuses on the microstructure, general properties and some of the applications for these materials. Some comparisons are also made between natural and conventional synthetic polymers.

1.2   NATURAL POLYMERS

The study and utilization of natural polymers is an ancient science. Typical examples, such as paper, silk, skin and bone artifacts can be found in museums around the world. These natural polymers perform a diverse set of functions in their native setting. For example, polysaccharides function in membranes and intracellular communication; proteins function as structural materials and catalysts; and lipids function as energy stores and so on. Nature can provide an impressive array of polymers that can be used in fibers, adhesives, coatings, gels, foams, films, thermoplastics, and thermoset resins. However, the availability of petroleum at a lower cost and the biochemical inertness of petroleum-based products proved disastrous for the natural polymers market. It is only after a lapse of almost 50 years that the significance of eco-friendly materials has once again been realized. These ancient materials have evolved rapidly over the past decade, primarily due to environmental issues and the increasing shortage of oil. Modern technologies provide powerful tools to elucidate microstructures at different levels and to understand the relationships between structures and properties. These new levels of understanding bring opportunities to develop materials for new applications.

Wide ranges of naturally occurring polymers that are derived from renewable resources are available for various materials applications (Charles et al., 1983; Fuller et al., 1996; Kaplan, 1998; Scholz and Gross, 2000; Gross & Scholz, 2001). Some of them (e.g., starch, cellulose and rubber), are actively used in products today whereas many others remain underutilized. Natural polymers can sometimes be classified according to their physical character. For example, starch granules and cellulose fibers are classified into different groups, but they both belong to polysaccharides according to chemical classification. Table 1-1 lists some natural polymers (Kaplan 1998).

1.2.1   Natural Rubber

One of the best known mature materials is natural rubber, which has been and is widely used in modern life. Although rubber's usefulness was known about in the early seventeenth century, it was not until the early nineteenth century that the rubber manufacturing industry became established. Rubber biosynthesis can occur

TABLE 1-1   List of Natural Polymers (Kaplan, 1998)

Polysaccharides
- from plant/algal: starch, cellulose, pectin, konjac, alginate, caragreenan, gums
- from animal: hyluronic acid,
- from fungal: pulluan, elsinan, scleroglucan
- from bacterial: chitin, chitosan, levan, xanthan, polygalactosamine, curdlan, gellan, dextran

Protein
soy, zein, wheat gluten, casein, serum, albumin, collagen/gelatine
silks, resilin, polylysine, adhesives, polyamino acids, poly(γ-glutamic acid), elastin, polyarginyl-polyaspartic acid

Lipids/Surfactants
acetoglycerides, waxes, surfactants, emulsan

Speciality Polymers
lignin, shellac, natural rubber

---

in either of two different types of plant cells: specialized latex vessels or parenchyma cells. Latex vessels are the more common route. A number of research groups have investigated ways of transferring the genes responsible for rubber biosynthesis into other species (Backhaus, 1998).

Natural rubber is a *cis*-polyisoprene that occurs as natural latex or a submicroscopic dispersion of the rubber in saplike materials. All such latexes appear milk white and the polyisoprene has the chemical structure shown in Fig. 1-1. Natural rubber can be recovered by coagulation processes. It has a glass transition temperature of about $-70°C$ and a molecular weight of 3×10^6 g/mole (Tanaka, 1991). The central portion of the rubber molecules is marked by extensive *cis*-polymerizations, which denote the structural hallmark of rubber. Rubber from different sources varies most with respect to the number of these condensations, which results in different molecular weight and molecular weight distribution.

Natural rubber is a very reactive polymer because of the presence of olefinic double bonds at every fifth carbon atom. The extensive degree of unsaturation and the close spacing of these double bonds provide a highly vulnerable target for free-radical attack and oxidation. The molecule can also undergo numerous chemical reactions such as hydrogenation, addition, and substitution. Rubber can be made inert by fully chlorinating the double bonds to afford up to 65 wt% chlorine. Natural rubber is soluble in most aromatic, aliphatic, and chlorinated solvents, but its high molecular weight makes it difficult to dissolve.

In practice, natural rubber has limited utility and must be compounded with other ingredients such as carbon black fillers, antioxidants, plasticizers, pigments, and vulcanizing agents to improve its chemical and physical properties. It should be

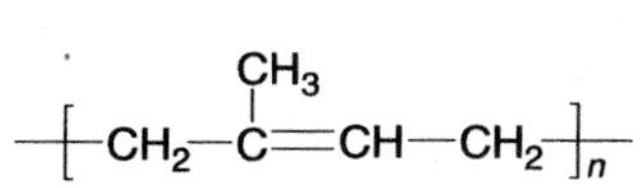

$$-\left[CH_2 - \underset{\underset{CH_3}{|}}{C} = CH - CH_2 \right]_n-$$

Fig. 1-1   The chemical structure of natural rubber.

realized that since vulcanized rubber is not biodegradable, natural rubber-based products are not frequently mentioned in the literature on biodegradables even though natural rubber is one of the earliest natural materials to have been used.

1.2.2 Starch

Starch is a polysaccharide produced by most higher plants as a means of storing energy. It is stored intracellularly in the form of spherical granules that are 2–100 μm in diameter (Whistler et al., 1984). Most commercially available starches are isolated from grains such as corn, rice, wheat, and from tubers such as potato and tapioca. The starch granule is a heterogeneous material: chemically, it contains both linear (amylose) and branched (amylopectin) structures; physically, it has both amorphous and crystalline regions (French, 1984). The ratio of amylose to amylopectin in starch varies as a function of the source, age, etc.

1.2.2.1 Chemical Structures
Starch is the principal carbohydrate reserve of plants. It is a polymeric carbohydrate consisting of anhydroglucose units linked together primarily through α-D-(1 → 4) glucosidic bonds. Although the detailed microstructures of starch are still being elucidated, it has been generally established that starch is a heterogeneous material containing at the extremes two microstructures according to their chain structure: amylose and amylopectin. Amylose is essentially a linear structure of α-1,4-linked glucose units and amylopectin is a highly branched structure of short α-1,4 chains linked by α-1,6 bonds. Figure 1-2 shows the structure of amylose and amylopectin.

The linear structure of amylose makes its behavior closer to that of conventional synthetic polymers. The molecular weight of amylose is about 10^6 (200–2000 anhydroglucose units) depending on the source and processing conditions employed in extracting the starch, which is 10 times larger than conventional synthetic polymers. Amylopectin, on the other hand, is a branched polymer. The molecular weight of amylopectin is much larger than that of amylose. Light-scattering measurements indicate a molecular weight in millions. The large size and branched structure of amylopectin reduce the mobility of the polymer chains and interfere with any tendency for them to become oriented closely enough to permit significant levels of hydrogen

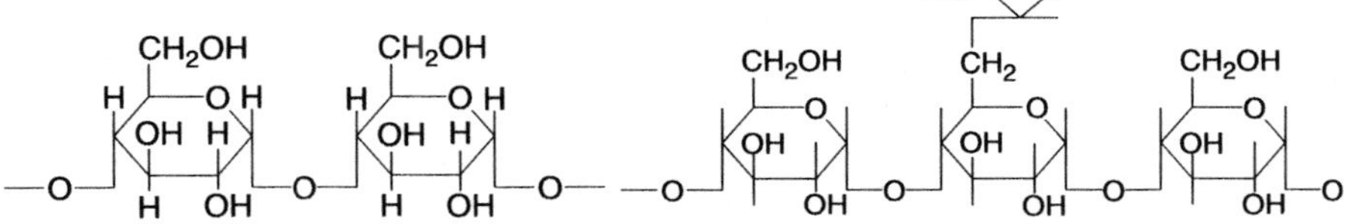

Fig. 1-2   The structures of amylose (left) and amylopectin (right).

bonding. Except the linear amylose and the short-branched amylopectin, starch of a long-branched structure has been detected (e.g., tapioca starch).

1.2.2.2 *Physical Structures*   Most native starches are semicrystalline with a crystallinity of ~20–45%. Amylose and the branching points of amylopectin form the amorphous regions. The short branching chains in the amylopectin are the main crystalline component in granular starch. The crystalline regions are present in the form of double helices with a length of approximately 5 nm. The amylopectin segments in the crystalline regions are all parallel to the axis of the large helix. The amylose/amylopectin ratio depends upon the source of the starch but can also be controlled by extraction processing. Starch granules also contain small amounts of lipids and proteins.

Figure 1-3 shows the wide angle x-ray scattering (WAXS) patterns of cornstarch with different amylose/amylopectin content. The amylose/amylopectin ratios in waxy, maize, G50 and G80 starches are 0/100, 23/77, 50/50 and 80/20, respectively (Chen et al., 2006). It is seen that waxy and maize starches show a typical A-type pattern with strong reflections at 2θ of about $13°$ and $21°$ and an unresolved large doublet between them. G50 and G80 give the strongest diffraction peak at around $2\theta = 16°$ and a few small peaks at 2θ values of $18°$, $20°$ and $22°$. An additional peak appears at about $2\theta = 4°$. These latter spectra are basically the same as the characteristic B-type. From Fig. 1-3, it is also seen that the crystalline area of amylopectin-rich starches is higher than that of amylose-rich starches, which is to be expected since it is well known that amylopectin in starch granules is considered to be responsible for the crystalline structure.

1.2.2.3 *Morphologies and Phase Transition During Processing*   Starch granules are a mixture of rounded granules from the floury endosperm, and angular granules from the horny endosperm (French, 1984; Chen et al., 2006). In their native state, starch granules do not have membranes; their surfaces consist simply of tightly

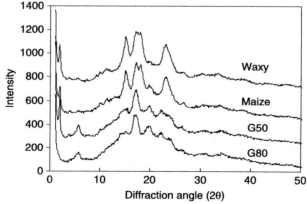

Fig. 1-3   WAXS patterns of cornstarch with different amylose/amylopectin content.

packed chain ends resembling the bottom of a broom with the straws pressed tightly together. Undamaged starch granules are not soluble in cold water but can reversibly imbibe water and swell slightly.

One of the unique characteristics of starch-based materials is their phase transition during processing, which encompasses various chemical and physical reactions including starch swelling, gelatinization, melting, crystallization, decomposition and so on (Lelievre, 1974, 1976; Whistler et al., 1984; Yu and Christie, 2001). It is well known that an order–disorder phase transition occurs when starch granules are heated in the presence of water. When sufficient water is present, this transition— referred to as "gelatinization"—results in near-solubilization of the starch (Lelievre, 1974; Donovan, 1979; Whistler, 1984). The well-accepted concept of gelatinization means destroying the crystalline structure in starch granules. Starch gelatinization is an irreversible process and includes granular swelling, native crystalline melting, loss of birefringence and starch solubilization. The concomitant changes of measurable properties such as viscosity, heat uptake, crystallinity, and size variation of starch granules have been used to detect the extent of starch gelatinization (Yu and Christie, 2001; Liu et al., 2006; Chen et al., 2007; Xie et al., 2008; Xue et al., 2008).

It has been shown that under shearless conditions, full gelatinization of starch requires about 70% water content (Wang et al. 1989; Liu 2006), while gelatinization under shear conditions requires less water since shear stress enhances processing. Extrusion cooking or processing of starch-based materials relies on the proper conversion of starch within the raw materials. In an extrusion environment, gelatinization is typically achieved with low amounts of water under high-shear and high-pressure conditions.

1.2.2.4 Starch Modification In practice, the raw materials of starch are not straightforwardly suitable for any specific nonfood application. Various modified starches have been developed for nonfood applications (e.g., starch graft copolymers, glycosides, cationic starch, or oxidized starch), to meet different specific requirements (Wurzburg, 2000). For example, sodium hypochlorite-oxidized starch which has the advantages of bright white color, easy gelatinization, and high solubility, is more suitable than native starches for applications in the papermaking and textile industries. Cationic starch, which has positive charges, can more easily be attracted by negative charges on fibers and hence is also more effective than native starches in applications in the papermaking and textile industries. Superabsorbent polymers, prepared by grafting acrylonitrile onto starch, can be used in various fields such as hygiene, cosmetics, and agriculture.

Recently, the technology of reactive extrusion has been used for starch modification (Xie et al., 2006). It has been shown that reactive extrusion is a feasible and efficient way to modify starches and to produce more applicable products. The use of an extruder as a chemical reactor allows high-viscosity polymers to be handled in the absence of solvents. It also affords large operational flexibility as a result of the broad range of processing conditions in pressure (0–500 atm [0–50 MPa]) and temperature (70–500°C), the possibility of multiple injections, the controlled residence time (distribution) and degree of mixing.

1.2.3 Protein

Proteins are one of three essential macromolecules in biological systems and can easily be isolated from natural resources. Proteins have been studied for decades for their ability to spontaneously form primary, secondary, and higher-order structures that can exhibit biological function and supramolecular protein organization in tissues and organs.

1.2.3.1 Microstructures and Properties of Proteins

Proteins are constructed mainly of α-amino acids. These amino acids can be neutral such as glycine, basic (containing one or more additional amines) such as lysine, or acidic (containing one or more additional acid groups) such as aspartic acid. Also, they may contain alcohol or thio functional groups, each representing a chemical "handle" with which chemists (synthetic, inorganic and organic) are able to play their trade on either proteins containing these "available" functional groups or on the amino acids themselves. Figure 1-4 shows the structure of the amide bond linking amino acids.

Proteins have four levels of structural organization: *Primary* structure refers to the sequence of amino acids in the polypeptide chain. Proteins or polypeptides are polymers of amino acids linked by amide linkages (peptide bonds); *Secondary* structure refers to the extended or helically coiled conformation of the polypeptide chains; *Tertiary* structure refers to the manner in which polypeptide chains are folded to form a tightly compact structure of globular protein; *Quaternary* structure refers to how subunit polypeptides are spatially organized. Several interactions and linkages are known to contribute to the formation of secondary, tertiary, and quaternary structures, such as steric strain, van der Waals interactions, electrostatic interactions, hydrogen bonding, hydrophobic interactions, and disulfide cross-links (Cheftel and Cuq, 1985).

Proteins interact with water through their peptide bonds or through their amino acid side-chains. The water solubility of proteins is a function of numerous parameters such as thermodynamic standpoint. Solubilization corresponds to separating the molecules of a solvent/protein molecules and dispersing the latter in the solvent for maximum interaction between the protein and solvent. The solubility of a protein depends mainly on pH value, ionic strength, the type of solvent, and temperature (Cheftel and Cuq, 1985). Protein denaturation is any modification in conformation (secondary, tertiary, or quaternary) not accompanied by the rupture of peptide bonds involved in primary structure. Denaturation is an elaborate phenomenon during which new conformations appear, although often intermediate and short-lived. The sensitivity of a protein to denaturation is related to the readiness with which a denaturing agent breaks the interactions or linkages that stabilize the protein's secondary, tertiary, or quaternary structure. Denaturation agents can be classified as physical

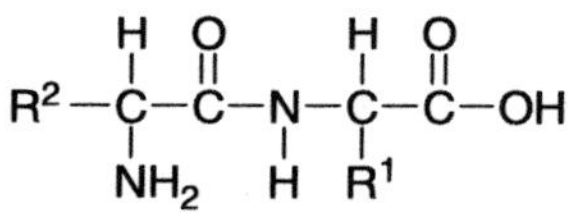

$$R^2 - \underset{\underset{NH_2}{|}}{\overset{\overset{H}{|}}{C}} - \overset{\overset{O}{\|}}{C} - \underset{\underset{H}{|}}{\overset{\overset{H}{|}}{N}} - \underset{\underset{R^1}{|}}{\overset{\overset{H}{|}}{C}} - \overset{\overset{O}{\|}}{C} - OH$$

Fig. 1-4   Structure of the amide bond linking amino acids.

agents (heat, cold, mechanical treatment, hydrostatic pressure, irradiation, interfaces, etc.) and chemical agents (acids and alkalis, metals, organic solvents, etc.).

The most widely available and highly refined plant proteins are from soybeans. Biotechnology offers the opportunity to modify all four levels of soy protein structure and therefore improve its potential and usefulness for applications as a biomaterials or food protein (Utsumi, 1992). Soybeans have a particularly high protein content, typically ∼40%. Processing of soybeans into protein ingredients for food and industrial products involves two major industries: soybean crushing (oil extraction) and protein ingredient processing. The soy proteins are globular, reactive, and often water soluble. The preponderance of polar and hydrophilic amino acids confers unusually good water solubility among plant proteins. Chemically reactive groups in amino acids are the carboxylic, primary and secondary amine, aliphatic and aromatic hydroxyl, and sulfhydryl groups.

1.2.3.2 *Processing of Protein-Based Materials* Similar to the processing of starch, the processing of protein-based materials (in particular thermal extrusion), is much more complex than that of conventional polymers. Some primary studies on the extrusion of soy protein have mainly focused on food applications. Irreversible and complex changes in the physicochemical interactions of protein molecules such as unfolding and disulfide–disulfide interactions, make it difficult to analyze the melt rheology for biopolymers. Plasticizers are widely used in protein-based materials not only for reducing brightness but also for improving processing properties. Several processes, traditional and new, are currently being developed for protein-based materials. A better understanding of the physical and chemical modifications underlying the various processes will lead to a further improvement in the materials generated by these means.

1.2.3.3 *Development of Protein-Based Materials* Early research on soy protein-based plastics was conducted in the 1940s. At that time, soy protein was mainly used as a filler or extender to decrease the cost of petroleum-based plastics. The predominant industrial application of soy protein is for coating paper since soy protein has unusual adhesive properties. Recent research work has been aimed at developing soy protein-based biodegradable plastics. The experimental results show that soy protein (alone or mixed with starch) can be molded into plastics items such as disposable containers, utensils, toys, and sporting goods. Soy protein can also be extrusion-blown or cast into films. Soy protein-based films have been shown to be good oxygen barriers and UV-blockers that are useful as packaging materials. With proper processing, soy protein can also be made into foam products. With its biodegradable, nonflammable, and nonelectrostatic properties, soy protein-based plastics provide unique and attractive features.

1.2.4 Cellulose

Natural fibers can generally be classified into several groups according to their resources: (1) wood fibers (such as soft and hard woods); (2) vegetable fibers

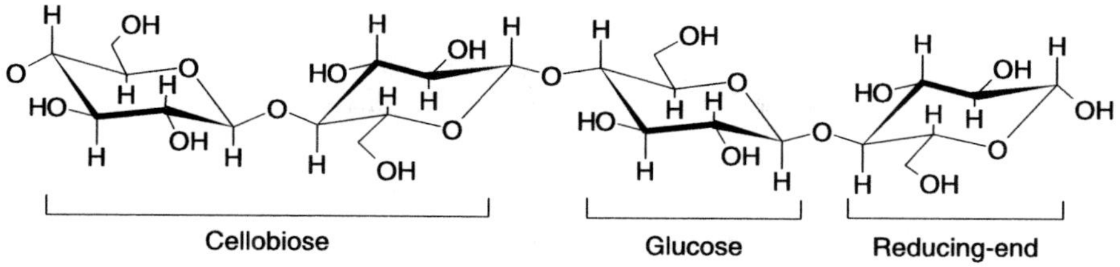

Fig. 1-5   The chemical structure of cellulose.

(such as cotton, hemp, jute, ramie, kenaf); (3) animal fibers (such as wool, silk, spider silk, feather, down); (4) mineral fibers (such as asbestos, inorganic whiskers). Biodegradable fibers usually means wood and vegetable fibers, which consist mainly of cellulose. Cellulose is the most abundant and renewable polymeric resource, but only a small proportion of sources are utilized. Cellulose occurs in all plants as the principal structural component of cell walls. The principal sources of cellulose for industrial processes are wood, cotton fiber, and cotton linters.

Cellulose is a homoglucan composed of linear chains of $(1\rightarrow4)$-β-D-glucopyranosyl units (see Fig. 1-5). The essential linearity of cellulose makes it easy for molecules to associate strongly in a side-by-side manner, as occurs extensively in native plant cellulose, especially in trees and woody parts of other pants. Cellulose has both amorphous and crystalline structure. The amorphous regions can be easily attacked by solvents and chemical reagents. However, cellulose is generally insoluble and highly crystalline, and the strong glucosidic bonds ensure the stability of cellulose in such as water, organic solvents etc.

Cellulose has been used since ancient times in such applications as rope, canvas, and sacking made from kenaf. However, because of its crystalline structure, cellulose will not melt upon heating. It is impossible to thermally process cellulose as simply as other conventional polymeric materials. Chemical reactions such as etherification and esterification are conducted on the free hydroxyl group to improve its thermoplastic behavior. Numerous cellulose derivatives have been commercialized including cellulose acetate, ethyl cellulose, hydroxyethyl cellulose, hydroxypropyl cellulose, hydroxyalkyl cellulose, carboxymethyl cellulose, fatty acid esters of cellulose, and others (Chiellini et al., 2002; Blackburn, 2005). Various solvent systems have been developed for aqueous processing (Zhang, 2001; Cai et al., 2006). In practice, cellulose has been widely used as a reinforcing agent in various polymer composites, especially for green composites (Baillie, 2004).

1.2.5   Chitin and Chitosan

Chitin is the second most abundant natural polymer after cellulose. Chitin is found widely in nature and is the main structural polysaccharide in many invertebrate animals such as insects and crustaceans. Marine crustaceans such as crabs and shrimps are the source of the most easily isolated chitin. Chitin plays the role of a structural material in many life forms. It is to be found in the cell walls of fungi where it

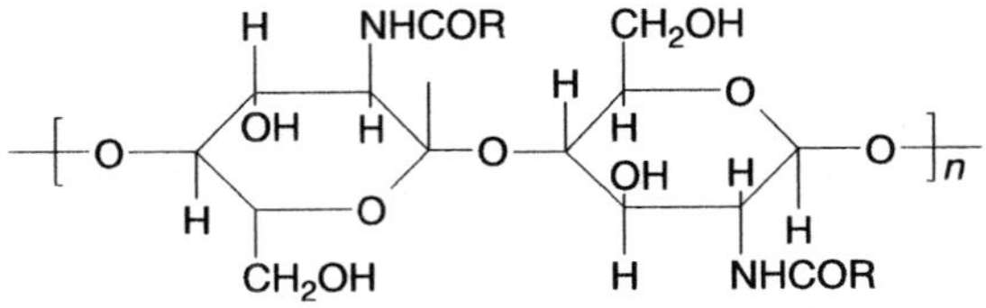

Fig. 1-6 Representative structure of chitin.

contributes stabilizing strength. The biosynthesis of chitin is a very old activity of cells and has survived from the earliest periods of life on this planet. Parrisher has provided a bibliographic listing of specific sources for chitin (Parrisher, 1989).

Chitin is essentially a homopolymer of 2-acetamido-2-deoxy-β-ᴅ-glucopyranose, although some of the glucopyranose residues are in the deacetylated form as 2-amino-2-deoxy-β-ᴅ-glucopyranose. When chitin is further deacetylated to ∼50%, it becomes soluble in dilute acids and is referred to as chitosan. A representative structure is shown in Fig. 1-6. Chitosan is commonly obtained by the alkaline hydrolysis of the amid group in chitin. The chemistry of chitosan is similar to that of cellulose but also reflects the presence of a primary aliphatic amine. The molecular weight of chitosan is $\sim 0.1-4 \times 10^6$ (Wu et al., 1976). Chitosan reacts readily with carbonyl compounds (i.e., by acylation with acid anhydrides), to form a wide range of ester and amide products. Chitosan's primary usefulness is a result of its ability to act as a cationic polyelectrolyte, its bioactivity and biocompatibility, its use as a thickening agent in water, and its selective chelation properties. Chitosan is also readily converted into fibers, films, coatings, and beads as well as into powders and solutions, further enhancing its usefulness (Hudson, 1998).

1.3 SYNTHETIC POLYMERS FROM BIODERIVED MONOMERS

The development of synthetic polymers using bioderived monomers provides a new direction for the production of biodegradable polymers from renewable resources. One of the most promising polymers in this regard is poly(lactic acid) (PLA) because it is made from agricultural products and is readily biodegradable. Lactide is a cyclic dimer prepared by the controlled depolymerization of lactic acid, which in turn can be obtained by the fermentation of corn, sugar cane, or sugar beat. Although PLA is not a new polymer, better manufacturing practices have improved the economics of producing monomers from agricultural feedstocks and, as such, PLA is at the forefront of the emerging biodegradable plastics industries.

1.3.1 Poly(lactic acid)

PLA belongs to the family of aliphatic polyesters commonly made from α-hydroxy acids, which include polyglycolic acid and polymandelic acid (Garlotta, 2001). PLA is commercially interesting because of its good strength properties, film transparency, biodegradability, and availability. PLA is also manufactured by biotechnological

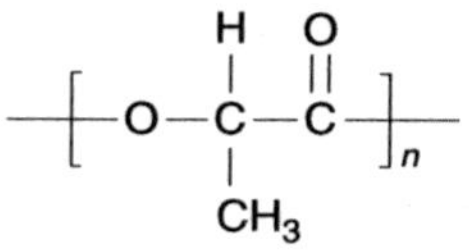

$$\left[\begin{array}{c} \overset{\displaystyle H}{\underset{\displaystyle CH_3}{\mathrm{O-C}}} \overset{\displaystyle O}{\mathrm{-C}} \end{array}\right]_n$$

Fig. 1-7   Basic structure of PLA.

processes from renewable resources. Although many sources of biomass can be used, corn has the advantage of providing the required high-purity lactic acid. PLA can be synthesized from lactic acid in two ways: a direct polycondensation reaction or ring-opening polymerization of a lactide monomer. The technique of ring-opening polymerization has the advantage of providing a product with a higher molecular weight.

The basic building block for PLA is lactic acid (2-hydroxypropionic acid) and its chemical structure is shown in Fig. 1-7. The stereochemistry of PLA is complex because of the chiral nature of lactic acid monomers. The stereoisomeric L/D ratio of the lactate units influences the properties of PLA. Poly(D,L-lactic acid), or poly(mesolactic acid), a racemic polymer obtained from a mixture of D- and L-lactic acid, is amorphous with weak mechanical properties.

PLA homopolymers have a glass transition and melt temperatures of about 60°C and 180°C, respectively. PLA is well suited to many conventional thermoplastics processing techniques such as extrusion and injection. In order to avoid degradation, PLA must not be exposed to high temperatures and moisture. PLA undergoes thermal degradation at temperatures $>200°C$ by hydrolysis, lactide reformation, oxidative main-chain scission, and inter- or intramolecular transesterification reactions (Garlotta, 2001). Mineral fillers have been used as nucleation sites for injection products (Bleach et al., 2002), and plasticizers may be employed for films and to improve toughness (Avérous and Martin, 2001; Piorkowska et al., 2006). PLA has been successfully utilized in surgical-implant materials and drug-delivery systems. The application of biodegradable plastics are now attracting much attention with regard to thermoformed products and biaxially-oriented film.

1.3.2   Propanediol

Under aerobic conditions, glycerol can be used as a substrate for the growth of many microorganisms. The glycerol can then be converted by facultative anaerobic bacteria to propanediol, ethanol, butanediol, acetic and lactic acid (Mickelson and Werkman, 1940). Among the fermentation products, propanediol (PDO) is of particular interest as it can be used as a monomer for producing plastics such as polyesters, polyethers, and polyurethanes (Deckwer, 1995).

1.4   POLYMERS FROM MICROBIAL FERMENTATION

In nature, a special group of polyesters is produced by a wide variety of microorganisms for internal carbon and energy storage as part of their survival mechanism

(Hocking and Marchessault, 1994; Scholz and Gross, 2000; Suriyamongkol et al., 2007). Poly(β-hydroxybutyrate) (PHB) was first mentioned in the scientific literature as early as 1901 and detailed studies began in 1925. Over the next 30 years, PHB inclusion bodies were studied primarily as an academic curiosity. The energy crisis of the 1970s was an incentive to seek naturally occurring substitutes for synthetic plastics, which then sped-up the research and commercialization of PHB. This biopolymer has received much research attention in recent years, with a large number of publications concerned with biosynthesis, microstructure, mechanical and thermal properties, and biodegradation through to genetic engineering.

1.4.1   Polyhydroxyalkanoates

Figure 1-8 shows the generic structure of polyhydroxyalkanoates (PHAs). The simplest of the family of PHAs is poly(β-hydroxybutyrate) (PHB). Analogous to starch, the reserve material in plants, these biopolyesters occur as submicrometer inclusions inside the cell. The present commercial development is based on the fermentation technology, but genetic engineers are already cloning the genes that have shown some potential.

1.4.2   Copolymers of the PHA Family

Pure PHB is brittle and has a low extension to breaking. The brittleness of PHB is largely due to the presence of large crystals in the form of spherulites. Interest in copolymers, in particular in copolymers of 3-hydroxybutyrate (3HB) and 3-hydroxyvalerate (3HV) (i.e., poly(3-hydroxybutyrate-co-3-hydroxyvalerate) (PHBV), stemmed from the fact that they have much lower melting temperatures and are less crystalline, more ductile, easier to mold, and tougher than PHB (Braunegg, 2002; Suriyamongkol et al., 2007). The biosynthetic origin of poly(β-hydroxyvalerate) (PHV) copolymers adds a number of interesting properties, such as full biodegradability and optical activity. Some research shows that the mechanical properties of PHBVs are sometimes similar to those of polypropylene since they have similar morphologies. The tensile strength was reported to be ∼45 MPa and elongation was about 10%. PHB is susceptible to thermal degradation at temperatures well below its melting point. Careful control of temperature is required during thermal processing.

Several applications for PHB and PHBVs can be envisaged due to their wide range of specialized properties. One of the simplest applications for PHBVs is as a biodegradable alternative to existing materials. Its gas barrier properties could lead to applications in food packaging. More specialized applications of PHBVs include controlled release and surgical swabs (Hocking and Marchessault, 1994).

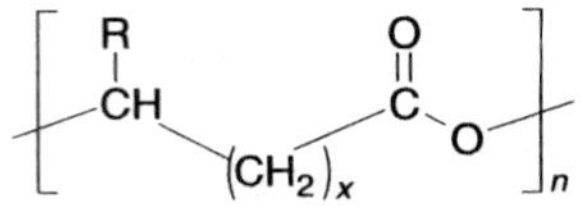

Fig. 1-8   Generic structure of PHAs.

1.5 SUMMARY

Polymers from renewable resources have world-wide interest and are attracting an increasing amount of attention for predominantly two major reasons: environmental concerns and the fact that petroleum resources are finite. Generally, polymers from renewable resources can be classified into three groups: natural polymers, synthetic polymers from bioderived monomers, and polymers from microbial fermentation.

There are some natural polymers that are actively used in everyday products while many others remain underutilized. New and more efficient isolation and purification methods along with chemical and physical modification are future research directions for natural polymers. PLA is a typical synthetic polymer from bioderived monomers. Although it can be processed by many conventional thermoplastics processing techniques, high price is still the major stumbling block to its general application. Biodegradable polymeric materials are now a major research focus and some products are becoming available in the marketplace. However, when it comes to replacing most of the conventional polymers, even in the most promising applications such as packaging and mulch films, it is either poor performance or high cost that is currently preventing the changeover.

Like most polymers from petroleum, polymers from renewable resources are rarely used by themselves. Blending and compositing are common and mature technologies for improving the performance and reducing price. In fact, the history of composites from renewable resources is far longer than that of conventional polymers. In the biblical Book of Exodus, Moses' mother built a basket from rushes, pitch, and slime—a kind of fiber-reinforced composite according to the modern classification of materials. During the opium wars in the mid-nineteenth century, the Chinese built castles to defend against invaders using a kind of mineral particle-reinforced composite made from gluten, rice, sugar, calcium carbonate, and sand.

Fibers are widely used in polymeric materials to improve mechanical properties. Cellulose fibers are the major substance obtained from plant life, and applications for cellulose fiber-reinforced polymers have again come to the forefront with the focus now on renewable raw materials. Hydrophilic cellulose fibers are very compatible with most natural polymers. The reinforcement with filler is particularly important for polymers from renewable resources since most of them have the disadvantage of lower softening temperatures and lower modulus. Furthermore, the hydrophilic behavior of most natural polymers offers a significant advantage since it provides a compatible interface with the mineral filler.

Many natural polymers are hydrophilic and some of them are water-soluble. Water solubility increases degradability and the speed of degradation, and this moisture sensitivity limits application. Blends and multilayers of natural polymers with other kinds of polymers from renewable resources can be used to improve their properties. Blends can also aid in the development of new low-cost products with better performance.

These new blends and composites are extending the utilization of polymers from renewable resources into new value-added products. Various polymeric blends and composites are discussed in the following chapters.

REFERENCES

Avérous L, Martin O. 2001. Poly(lactic acid): plasticization and properties of biodegradable multiphase systems. *Polymer* 42:6209–6219.

Backhaus RA. 1998. Natural rubber from plants. In: Kaplan DL, editor. *Biopolymers from Renewable Resources*. Berlin: Springer-Verlag. Chapter 13.

Baillie C. 2004. *Green Composites: Polymer Composites and the Environment*. Cambridge UK: Woodhead Publishing.

Blackburn RS. 2005. *Biodegradable and Sustainable Fibres*. Cambridge, UK: Woodhead Publishing.

Blanshard JMV. 1987. Starch granule structure and function: a physicochemical approach. In: Galliard T, editor. *Starch: Properties and Potential*. New York: John Wiley & Sons. p. 16.

Bleach CN, Nazhat NS, Tanner EK, Kellomäki M, Tormälä P. 2002. Effect of filler content on mechanical and dynamic mechanical properties of particular biphasic calcium phosphate-polylactide composites. *Biomaterials* 23:1579–1585.

Braunegg G. 2002. Sustainable poly(hydroxyalkanoate) (PHA) production. In: Scott G, editor. *Degradable Polymers*, 2nd ed. Dordrecht: Kluwer Academic. p. 235–293.

Cai J, Liu Y, Zhang L. 2006. Dilute solution properties of cellulose in LiOH/urea aqueous system. *J Polym Sci: Part B Polym Phys* 44:3093–3101.

Charles E, Carraher Jr, Sperling HL. 1983. *Polymer Applications of Renewable–Resource Materials*. New York: Plenum Press.

Cheftel CJ, Cuq JL. 1985. Amino acids, peptides and proteins. In: Fennema RO, editor. *Food Chemistry*, 2nd ed. New York: Marcel Dekker. p. 254–265.

Chen P, Yu L, Chen L, Li X. 2006. Morphologies and microstructure of cornstarch with different amylose/amylopectin content. *Starch* 58:611–615.

Chen P, Yu L, Kealy T, Chen L, Li L. 2007. Phase transition of cornstarch under shearless and shear stress conditions. *Carbohydr Polym* 68:495–501.

Chiellini E, Chiellini F, Cinelli P. 2002. Polymers from renewable resources. In: Scott G, editor. *Degradable Polymers Principle and Application*, 2nd ed. Dordrecht: Kluwer Academic. p. 163–233.

Deckwer WD. 1995. *FEMS Microbiology Review* 16:143–149.

Donovan J. 1979. Phase transitions of the starch–water system. *Biopolymers* 18:263–275.

Fuller G, McKeon AT, Bills DD. 1996. *Agricultural Materials as Renewable Resources*. Washington DC: American Chemical Society.

French D. 1984. Organization of starch granules. In: Whistler LR, Bemiller NJ, Paschall FE, editors. *Starch: Chemistry and Technology*, 2nd ed. New York: Academic Press. p. 184–248.

Garlotta DA. 2001. Literature review of poly(lactic acid), *J Polym Environ* 9(2):63–84.

Gross RA, Scholz C. 2001. *Biopolymers from Polysaccharides and Agroproteins*. Washington DC: American Chemical Society.

Hocking JP, Marchessault HR. 1994. Biopolyesters. In: Griffin G, editor. *Chemistry and Technology of Biodegradable Polymers*. London: Chapman & Hall. p. 48–96.

Hudson MS. 1998. Polysaccharides: chitin and chitosan: chemistry and technology of their use as structure materials. In: Kaplan DL, editor. *Biopolymers from Renewable Resources*. Berlin: Springer-Verlag. p. 97–118.

Kaplan DL. 1998. *Biopolymers from Renewable Resources*. Berlin: Springer-Verlag.

Lelievre J. 1974. Starch gelatinization. *J Polym Sci.* 18:293–296.

Lelievre J. 1976. Theory of gelatinization in a starch water solute system. *Polymer* 17:854–858.

Liu H, Yu L, Xie FW, Chen L. 2006. Gelatinization of cornstarch with different amylase/amylopectin content. *Carbohydr Polym* 65:357–363.

Mickelson MN, Werkman CH. 1940. The dissimilation of glycerol by coli-aerogenes intermediates. *J Bacteriol* 39:709–715.

Parrisher E. 1989. *The Chitin Source Book.* New York: Wiley Interscience.

Piorkowska E, Kulinski Z, Galeski A, Masirek R. 2006. Plasticization of semicrystalline poly(L-lactide) with poly(propylene glycol). *Polymer* 47:7178–7188.

Scholz C, Gross AR. 2000. *Polymer from Renewable Resources: Biopolyesters and Biocatalysis.* ACS Symposium Series 764. Washington DC: American Chemical Society.

Suriyamongkol P, Weselake R, Narine S, Moloney M, Shah S. 2007. *Biotechnol Adv* 25:148–175.

Tanaka Y. 1991. Rubber and related polyprenols. In: Charlwood BV, Banthorpe DV, editors. *Methods in Plant Biochemistry: Terpenoids.* London: Academic Press. p. 519.

Utsumi S. 1992. Plant food protein engineering. In: Kinsella JE, editor. *Advances in Food and Nutrition Research,* vol 36. London: Academic Press.

Whistler LR, Bemiller NJ, Paschall FE. 1984. *Starch: Chemistry and Technology,* 2nd ed. New York: Academic Press. p. 1–9; 153–242.

Wang SS, Chiang C-W, Ye A-I, Zhao B, Kim I-H. 1989. Kinetics of phase transition of waxy corn starch at extrusion temperatures and moisture contents. *J Food Sci* 54(5):1298–1301.

Wu AC, Bough WA, Conrad EC, Alden KE. 1976. Determination of molecular-weight distribution of chitosan by high-performance liquid chromatography. *Chromatography* 128(1):87–99.

Wurzburg OB. 2000. *Modified Starches: Properties and Uses.* Boca Raton, FL: CRC Press.

Xue T, Yu L, Xie F, Chen L, Li L. 2008. Rheological properties and phase transition of starch under shear stress. *Food Hydrocolloids* 22(6):973–978.

Xie F, Yu L, Chen L, Li L. 2008. A new study of starch gelatinization under shear stress using dynamic mechanical analysis. *Carbohydr Polym* 72(2):229–234.

Xie F, Yu L, Liu H, Chen L. 2006. Starch modification using reactive extrusion. *Starch* 548:131–139.

Yu L, Christie G. 2001. Measurement of starch thermal transitions using differential scanning calorimetry. *Carbohydr Polym* 46(2):179–184.

Zhang LM. 2001. New water-soluble cellulosic polymers: a review. *Macromol Mater Eng* 286(5):267–275.

PART I

NATURAL POLYMER BLENDS AND COMPOSITES

CHAPTER 2

Starch–Cellulose Blends

IOANNIS S. ARVANITOYANNIS

Department of Agriculture, Animal Production and Aquatic Environment, School of Agricultural Sciences, University of Thessaly, Fytokou Str., 38446 Nea Ionia Magnesias, Volos, Greece

A. KASSAVETI

Department of Agriculture, Crop Production and Agricultural Environment, School of Agricultural Sciences, University of Thessaly, Fytokou Str., 38446 Nea Ionia Magnesias, Volos, Greece

2.1 INTRODUCTION

The first polymers used were natural products, especially cotton, starch, proteins, and wool. Early in the twentieth century, synthetic polymers began to be produced. The first polymers of importance, Bakelite and Nylon, showed the tremendous possibilities of the new materials (Sperling, 2006). Global consumption of biodegradable

---

Biodegradable Polymer Blends and Composites from Renewable Resources. Edited by Long Yu
Copyright © 2009 John Wiley & Sons, Inc.

polymers increased from 14 million kg in 1996 to 68 million kg in 2001 (URL1). The advantages of natural polymers include the renewable resources from which they originate, their biodegradability, and the environmentally friendly products of degradation that are produced (Yu et al., 2007).

Ecological concerns have resulted in a renewed interest in natural and compostable materials and issues such as biodegradability and environmental safety are becoming correspondingly important. Tailoring of new products within a perspective of sustainable development or eco-design is a philosophy that is being applied to more and more materials. This is why components such as natural fibers, and biodegradable polymers are considered as "interesting"—environmentally safe—alternatives for the development of new biodegradable composites (Avérous and Boquillon, 2004).

Among the many kinds of candidates for biodegradable polymer (see Table 2-1), starch is one of the most promising materials for biodegradable plastics because it is a

TABLE 2-1   Overview of Natural Biodegradable Polymers (Biopolymers)[a]

Polysaccharides (vegetable, animal, microbial, algae)	Starch and derivatives
	Cellulose and derivatives
	Chitin and chitosan
	Heparin
	Hyaluronic acid
	Gellan
	Alginic acid
	Xanthan
	Dextran
	Pullulan
	Pectin
	Agar
	Gums
Proteins	Albumin
	Casein
	Collagen/gelatin
	Fibrinogen/fibrin
	Wheat gluten, soy protein
	Zein (corn protein)
	Silk
Polyesters	Polylactic acid
	Polyhydroxyalkanoates
Polyphenols	Lignin
	Tannin
	Humic acid
Lipids	Waxes
	Surfactants
Specialty polymers	Shellac
	Natural rubber
	Nylon (from castor oil)

[a]URL1; URL2; Briassoulis (2004).

versatile biopolymer with immense potential and low price for use in the nonfood industries (Choi et al., 1999; Mohanty et al., 2000). Starch constitutes more than 60% of cereal kernels and is relatively easy to separate from the other chemical components. These other components (fibers, protein, and fat) have a market value of their own as food or feed (Rexen et al., 1988). Starch is derived from renewable resources and is suitable for industrial applications. Every year nearly 20% of the starch produced in Europe is consumed by nonfood industries (Poutanen and Forsell, 1994). Moreover, the use of starch in plastic materials would reduce dependence on synthetic polymers made from imported oil and offers socioeconomic benefits because it generates rural jobs and a nonfood agricultural-based economy (Dufresne et al., 2000). The bioplastics found in the market are made mainly from starch. Starch-based bioplastics represent 85–90% of market's bioplastics (Bastioli, 2000). Among starch bioplastics are those manufactured with native or slightly modified starches, either isolated or blended with natural or synthetic molecules (Vilpoux and Avérous, 2004).

On the other hand, cellulose is the most abundant natural polymer, with wood as the greatest source. It is the main component (40–50% or more) of lignocellulose fibers, the other two being hemicellulose (20–30%) and lignin (10–30%). The purest form of cellulose sources is cotton linters (85–95%). Others are jute, kenaf, and chanvre (60–75%), as well as woods (40–50%) (URL2). Cellulose, starch, and their constituents are the two most important raw materials for the preparation of films (Engelhardt, 1995). Blends of starch and certain synthetic polymers or cellulose derivatives hold the lead among some new materials that have been used successfully (Ramkumar et al., 1996).

2.2 STARCH AND STARCH DERIVATIVES

Starch is a highly hydrophilic material that contains anhydroglucose units linked by α-D-1,4-glycosidic bonds. The water absorbed by starch granules is mainly bound in the amorphous phase. At certain temperatures, shear stress, and pressure in an extruder the gelatinization processing disrupts the crystalline and ordered structure in starch granules to produce an amorphous phase (Yu et al., 2007). Starch granules vary in size from about 2 mm to 150 mm (Kerr, 1950). Pure starch, as distinct from commercial starch, is a white, odorless, tasteless, neutral powder, insoluble in cold water or organic solvents (Radley, 1953). Most starch granules are composed of a mixture of two polymers (Ahmad et al., 1999), namely, amylose and amylopectin. The amylose content can vary over a broad range, from 0% to about 75%, but typically is 20–25% (w/w) (Orford et al., 1987; Parker and Ring, 1996). In addition, starch is easily hydrolyzed to fermentable substrates (Rexen et al., 1988).

Starch is mostly water soluble, difficult to process, and brittle when used without the addition of a plasticizer. In addition, its mechanical properties are very sensitive to moisture content, which is difficult to control. In principle, some properties of starch can be significantly improved by blending it with synthetic polymers (Dufresne et al., 2000). Starch alone cannot form films with satisfactory mechanical properties (high percentage elongation and tensile and flexural strength) unless it is plasticized, blended with other materials, chemically modified, or modified with a combination of these treatments (Liu, 2006).

Starch derivatives include dextrins, starch esters, and starch ethers. The starch esters most commonly used are esters with phosphoric acid (phosphate starches) and acetic acid (acetyl starches). The three most important starch ethers are the hydroxyethyl, hydroxypropyl, and carboxymethyl starches. Starch derivatives based on esters can be biologically reduced, while starches based on ethers are more difficult to biodegrade. However, the great variety of possible chemical modifications makes it possible to produce good slashing agents, which can be nearly completely biodegraded (URL3).

2.2.1 Thermoplastic-like Starch (or Destructurized Starch)

Thermoplastic-like starch (TPS) is a relatively new concept and, today, is one of the main research avenues for the manufacturing of biodegradable materials (Curvelo et al., 2001). The starch is not a true thermoplastic, but, in the presence of a plasticizer (water, glycerin, sorbitol, etc.), high temperature (90–180°C), and shearing, it melts and fluidizes, enabling its use in injection, extraction, and blowing equipment, such as that used for synthetic plastics (Lourdin et al., 1999). Unfortunately, TPS shows some drawbacks such as a strong hydrophilic character (water sensitivity), rather poor mechanical properties compared with conventional polymers, and an important postprocessing variation of its properties (Avérous and Boquillon, 2004).

The main use of destructurized starch alone is in soluble compostable foams such as loose fills, expanded trays, shape-molded parts, and expanded sheets and as a replacement for expanded polystyrene (Fang and Fowler, 2003).

2.2.2 Mechanical Properties

Avérous and Boquillon (2004) investigated the thermal and mechanical behavior of two different plasticized starch matrixes (TPS$_1$: dried wheat starch [70 wt%], glycerol [18 wt%] and water [12 wt%]; TPS$_2$: dried wheat starch [65 wt%] and glycerol [35 wt%]). The elongation at break was 124% for TPS$_1$ and 60% for TPS$_2$. The maximum tensile stress was 3.6 MPa and 1.4 MPa for TPS$_1$ and TPS$_2$, respectively, and the tensile modulus was 87 MPa for TPS$_1$, and 12 MPa for TPS$_2$. It can be concluded that the less plasticized matrix shows higher mechanical behavior. Avérous et al. (2001b) reported that the tensile modulus and strength of plasticized wheat starch (PWS) were 52 MPa and 3 MPa, respectively, and the elongation at break was 126%. According to Avérous et al. (2001a), the tensile strength and modulus of TPS were 3.6 MPa and 87 MPa, respectively, and the elongation at break was 124%.

Starch acetate was blended with corn stalk fibers at different concentrations (0%, 2%, 6%, 10%, and 14% w/w) for the production of biodegradable extruded foams. The results showed that the force required to shear the extrudates increased with increase in fiber content from 1% to 10%. There was a drastic increase in the force (0.06–0.075 N/mm^2) required to shear the extrudates with 14% fiber content (Ganjyal et al., 2004).

Dufresne et al. (2000) examined the mechanical behavior of an unfilled starch–glycerol matrix at room temperature as a function of glycerol content and relative humidity (RH). With increasing glycerol content, a decrease in the modulus was observed regardless of RH. This decrease was linear up to ∼15% glycerol. At higher glycerol content the drop in modulus was strongly increased.

The tensile modulus of unfilled starch–glycerol matrix films ranged between 0.24 and 20 MPa depending on the water content. The 35% RH conditioned sample displayed a typical rubberlike behavior with a high strain at break (∼140%). As the water content increased, starch showed a tendency to crystallize, and the elongation at break decreased strongly (by up to 18% at 43% RH). No significant difference was reported for more moist specimens. The strength of the material was almost moisture independent and ranged between 0.25 and 1 MPa (Anglés and Dufresne, 2001).

The mechanical properties of starch are presented in Table 2-2.

2.2.3 Thermal Properties

In the experiment conducted by Avérous and Boquillon (2004) (see above), the glass transition temperature (T_g) (by differential scanning calorimetry [DSC] analysis) was 87°C for TPS_1 and 12°C for TPS_2. In contrast, Avérous and Fringant (2001) found that the T_g of various TPS formulations was below ambient temperature except for the 74% starch/10% glycerol/16% water blend.

Native starch (NS) was extracted from natural corn grain or from alkaline temperature-treated corn (ATS) (95°C/40 min in a 10% lime solution). The NS and the ATS starches were stored (30°C) at different relative humidities (%RH) (i.e., 11–84%). The T_g and melting temperature (T_m) (by DSC) of the starches were determined at different storage times. The T_g of the starches in 80% water dispersion was between 60 and 65°C (see Table 2-3).

2.3 CELLULOSE AND CELLULOSE DERIVATIVES

Cellulose is made up of glucopyranose units in β-(1–4) linkages (Kennedy et al., 1985). Sodium carboxymethyl cellulose (CMC) is the only water-soluble cellulose derivative used as a sizing agent (URL3). It is known that native celluloses, when subjected to strong acid hydrolysis, can readily break down into "microcrystalline cellulose" (MCC) with almost no weight loss (Battista, 1975; Ebeling et al., 1999).

Cellulose derivatives can be made by etherification, esterification, crosslinking, or graft-copolymerization reactions. Cellulose acetate is an important organic ester of cellulose. Cellulose ethers comprise methylcellulose, ethyl cellulose, hydroxyethyl cellulose, hydroxypropyl cellulose, and their derivatives (Hon, 2006). CMC is the largest industrial cellulose ether in terms of annual production (Richardson and Gorton, 2003). CMC is produced by reacting cellulose with sodium hydroxide and sodium chloroacetate, whereby the cellulose polymer is also depolymerized. CMC

TABLE 2-2 Mechanical Properties of Starch–Cellulose Blends

Components	% Elongation	Tensile Strength (MPa)	Tensile Modulus (MPa)	References
100% Plasticized starch films/ 0% cellulose crystallites	110	2.6	40	Lu et al. (2005)
90% Plasticized starch films/ 10% cellulose crystallites	85	5.7	130	
80% Plasticized starch films/ 20% cellulose crystallites	73	6.5	260	
70% Plasticized starch films/ 30% cellulose crystallites	47	7.7	310	
47.5% Corn starch/47.5% MCC/ 0% glycerol/5% water	2.3	90	158.9	Psomiadou et al. (1996)
42.5% Corn starch/42.5% MCC/ 10% glycerol/5% water	4.4	66.3	124.2	
34% Corn starch/34% MCC/ 27% glycerol/5% water	8.5	55.0	53.4	
45% Corn starch/45% MCC/ 5% sorbitol/5% water	4.1	73.3	131.5	
39.5% Corn starch/39.5% MCC/ 16% sorbitol/5% water	6.0	48.9	94.2	
33.5% Corn starch/33.5% MCC/ 28% sorbitol/5% water	9.3	18.3	40.0	
44.5% Corn starch/44.5% MCC/ 6% sucrose/5% water	2.9	65.0	120.5	
40% Corn starch/40% MCC/ 15% sucrose/5% water	3.5	43.4	82.7	
34.5% Corn starch/34.5% MCC/ 26% sucrose/5% water	3.1	21.0	33.5	

Material				Reference
45% Corn starch/45% MCC/ 5% xylose/5% water	3.6	63.9	119.6	Lee et al. (2006)
39.5% Corn starch/39.5% MCC/ 16% xylose/5% water	5.6	41.7	81.2	
35% Corn starch/35% MCC/ 25% xylose/5% water	8.8	20.5	34.7	
Cellulose diacetate	4 6	22 49.0	6254 1727	
10% starch/90% plasticized cellulose diacetate	10	44.0	1551	
20% starch/80% plasticized cellulose diacetate	12	26.0	1064	
30% starch/70% plasticized cellulose diacetate	15	23.0	953	
40% starch/60% plasticized cellulose diacetate	18	16.0	858	
50% starch/50% plasticized cellulose diacetate				
CP (clear-pure) cellulose acetate propionate	50	41.3	1241	URL5
Cellulose acetate plastic nanocomposites (Process A, 0% clay)	24	26	1300	Wibowo et al. (2006)
Cellulose acetate plastic nanocomposites (Process A, 5% clay)	20	27	1500	
Cellulose acetate plastic nanocomposites (Process B, 0% clay)	14	32	2100	
Cellulose acetate plastic nanocomposites (Process B, 5% clay)	15	44	2800	
100% Plasticized starch films/0% ramie cellulose nanocrystallites	95	2.8	60	Lu et al. (2006)

(Continued)

TABLE 2-2 *Continued*

Components	% Elongation	Tensile Strength (MPa)	Tensile Modulus (MPa)	References
90% Plasticized starch films/10% ramie cellulose nanocrystallites	70	5.3	160	Cao et al. (2006)
80% Plasticized starch films/20% ramie cellulose nanocrystallites	60	5.5	240	
70% Plasticized starch films/30% ramie cellulose nanocrystallites	30	6.1	450	
60% Plasticized starch films/40% ramie cellulose nanocrystallites	15	7.0	510	
0% Benzyl starch/100% regenerated cellulose	8	66	–	
10% Benzyl starch/90% regenerated cellulose	9	72	–	
20% Benzyl starch/80% regenerated cellulose	6	76	–	
30% Benzyl starch/70% regenerated cellulose	7	78	–	
40% Benzyl starch/60% regenerated cellulose	6.5	80	–	
50% Benzyl starch/50% regenerated cellulose	9.5	93	–	
60% Benzyl starch/40% regenerated cellulose	14	102	–	
70% Benzyl starch/30% regenerated cellulose	12	103	–	
100% Plasticized starch/0% tunicin[a] whiskers	0 (35% RH)	0 (35% RH)	1.4 (35% RH)	Anglés and Dufresne (2001)
	20 (75% RH)	1 (75% RH)	0.1 (75% RH)	

95% Plasticized starch/5% tunicin whiskers	40 (35% RH)	4 (35% RH)	0.7 (35% RH)	Avérous et al. (2001b)
	20 (75% RH)	1 (75% RH)	0.2 (75% RH)	
90% Plasticized starch/10% tunicin whiskers	50 (35% RH)	4.5 (35% RH)	0.6 (35% RH)	
	30 (75% RH)	1 (75% RH)	0.15 (75% RH)	
85% Plasticized starch/15% tunicin whiskers	60 (35% RH)	5 (35% RH)	0.55 (35% RH)	
	30 (75% RH)	1.5 (75% RH)	0.1 (75% RH)	
80% Plasticized starch/20% tunicin whiskers	120 (35% RH)	9 (35% RH)	0.5 (35% RH)	
	40 (75% RH)	1.5 (75% RH)	0.1 (75% RH)	
75% Plasticized starch/25% tunicin whiskers	320 (35% RH)	15 (35% RH)	0.4 (35% RH)	
	50 (75% RH)	1 (75% RH)	0.1 (75% RH)	
Plasticized wheat starch (PWS)	126 (50% RH)	3 (50% RH)	52 (50% RH)	
85% PWS/15% B400	31 (50% RH)	13 (50% RH)	430 (50% RH)	
70% PWS/30% B400	19 (50% RH)	22 (50% RH)	670 (50% RH)	
85% PWS/15% BC200	33 (50% RH)	10 (50% RH)	350 (50% RH)	
70% PWS/30% BC200	19 (50% RH)	15 (50% RH)	630 (50% RH)	
85% PWS/15% B600	47 (50% RH)	7 (50% RH)	296 (50% RH)	
70% PWS/30% B600	10 (50% RH)	13 (50% RH)	757 (50% RH)	
PWS 1 (reference)	3 (50% RH)	25 (50% RH)	1110 (50% RH)	
PWS 1 (reference)	60 (75% RH)	3 (75% RH)	100 (75% RH)	
PWS 2/20% cellulose fibers	8 (50% RH)	26 (50% RH)	1150 (50% RH)	
PWS 2/20% cellulose fibers	n.d. (75% RH)	n.d. (75% RH)	n.d. (75% RH)	
PWS 3/30% cellulose fibers	6 (50% RH)	24 (50% RH)	1280 (50% RH)	
PWS 3/30% cellulose fibers	15 (75% RH)	7 (75% RH)	200 (75% RH)	
90% Soluble starch/0% methyl cellulose/10% water	4.2	50.5	–	Arvanitoyannis and Biliaderis (1999)
70% Soluble starch/20% methyl cellulose/10% water	8	33.2	–	
50% Soluble starch/40% methyl cellulose/10% water	10.6	12.6	–	

(Continued)

TABLE 2-2 *Continued*

Components	% Elongation	Tensile Strength (MPa)	Tensile Modulus (MPa)	References
45% Soluble starch/45% methyl cellulose/10% water	8.9	61.2	–	
40% Soluble starch/40% methyl cellulose/20% water	16.7	49.5	–	
35% Soluble starch/35% methyl cellulose/30% water	24.5	38.7	–	
45% Corn starch/45% methyl cellulose/5% glycerol/5% water	11.9	59.4	–	
40% Corn starch/40% methyl cellulose/15% glycerol/5% water	25.3	43.3	–	
32.5% Corn starch/32.5% methyl cellulose/30% glycerol/5% water	34.2	33.5	–	
45% Corn starch/45% methyl cellulose/5% sorbitol/5% water	14.3	54.6	–	
40% Corn starch/40% methyl cellulose/15% sorbitol/5% water	28.7	35.2	–	
32.5% Corn starch/32.5% methyl cellulose/30% sorbitol/5% water	39.4	28.4	–	
45% Corn starch/45% methyl cellulose/5% xylose/5% water	8.8	48.4	–	
40% Corn starch/40% methyl cellulose/15% xylose/5% water	21.5	32.5	–	
32.5% Corn starch/32.5% methyl cellulose/30% xylose/5% water	28.4	25.6	–	
TPS₁ (dried wheat starch [70 wt%], glycerol [18 wt%] and water [12 wt%])	124	3.6	87	Avérous and Boquillon (2004)

TPS₂ (dried wheat starch [65 wt%] and glycerol [35 wt%])	1.4	60	12	Avérous et al. (2001a)
TPS	3.6	124	87	Dufresne et al. (1999)
100% Starch/0% glycerol	–	–	2000 (25% RH) 1700 (58% RH) 1600 (75% RH)	
90% Starch/10% glycerol	–	–	1900 (25% RH) 1600 (58% RH) 1500 (75% RH)	
80% Starch/20% glycerol	–	–	1250 (25% RH) 750 (58% RH) 500 (75% RH)	
70% Starch/30% glycerol	–	–	250 (25% RH) 250 (58% RH) 250 (75% RH)	
100% Starch/0% cellulose microfibrils/0% glycerol	–	–	2200 (25% RH) 1900 (58% RH) 1800 (75% RH)	
90% Starch/10% cellulose microfibrils/0% glycerol	–	–	3500 (25% RH) 3000 (58% RH) 2800 (75% RH)	
80% Starch/20% cellulose microfibrils/0% glycerol	–	–	4500 (25% RH) 3500 (58% RH) 3500 (75% RH)	
70% Starch/30% cellulose microfibrils/0% glycerol	–	–	5400 (25% RH) 4500 (58% RH) 4500 (75% RH)	
85% Starch/0% cellulose microfibrils/15% glycerol	–	–	2000 (25% RH) 1800 (58% RH)	

(Continued)

29

TABLE 2-2 *Continued*

Components	% Elongation	Tensile Strength (MPa)	Tensile Modulus (MPa)	References
75% Starch/10% cellulose microfibrils/ 15% glycerol	—	—	1000 (75% RH) 1700 (25% RH) 2100 (58% RH) 2000 (75% RH)	
65% Starch/20% cellulose microfibrils/ 15% glycerol	—	—	4000 (25% RH) 2600 (58% RH) 2100 (75% RH)	
55% Starch/30% cellulose microfibrils/ 15% glycerol	—	—	4000 (25% RH) 3000 (58% RH) 2500 (75% RH)	
70% Starch/0% cellulose microfibrils/ 30% glycerol	—	—	400 (25% RH) 100 (58% RH) 0 (75% RH)	
60% Starch/10% cellulose microfibrils/ 30% glycerol	—	—	1000 (25% RH) 700 (58% RH) 500 (75% RH)	
50% Starch/20% cellulose microfibrils/ 30% glycerol	—	—	2000 (25% RH) 1700 (58% RH) 1000 (75% RH)	
40% Starch/30% cellulose microfibrils/ 30% glycerol	—	—	2500 (25% RH) 2000 (58% RH) 1100 (75% RH)	

[a]Tunicin: animal cellulose.

TABLE 2-3 Glass Transition Temperatures (T_g) of Starch–Cellulose Blends

Components	T_g (°C)	Method of T_g Determination	References
100% Plasticized starch films/0% cellulose nanocrystallites	23	DSC	Lu et al. (2005)
90% Plasticized starch films/10% cellulose nanocrystallites	30	DSC	
80% Plasticized starch films/20% cellulose nanocrystallites	40	DSC	
70% Plasticized starch films/30% cellulose nanocrystallites	48	DSC	
100% Plasticized starch films/0% cellulose nanocrystallites	25	DMTA	
90% Plasticized starch films/10% cellulose nanocrystallites	45	DMTA	
80% Plasticized starch films/20% cellulose nanocrystallites	40	DMTA	
70% Plasticized starch films/30% cellulose nanocrystallites	50	DMTA	
47% Corn starch/47% methyl cellulose/6% water	110.9	DTA (2nd run)	Arvanitoyannis (1999); Psomiadou et al. (1996)
45% Corn starch/45% methyl cellulose/10% water	92.2	DTA (2nd run)	
42% Corn starch/42% methyl cellulose/16% water	67.3	DTA (2nd run)	
39% Corn starch/39% methyl cellulose/22% water	49.8	DTA (2nd run)	

(Continued)

TABLE 2-3 *Continued*

Components	T_g (°C)	Method of T_g Determination	References
37% Corn starch/37% methyl cellulose/ 26% water	34.5	DTA (2nd run)	
35% Corn starch/35% methyl cellulose/ 30% water	17.1	DTA (2nd run)	
78% Corn starch/5% methyl cellulose/ 17% water	58.3	DTA (2nd run)	
58% Corn starch/25% methyl cellulose/ 17% water	65.2	DTA (2nd run)	
28% Corn starch/55% methyl cellulose/ 17% water	77.0	DTA (2nd run)	
47% Corn starch/47% methyl cellulose/ 6% water	109.4	DMTA (E')	
45% Corn starch/45% methyl cellulose/ 10% water	91.5	DMTA (E')	
42% Corn starch/42% methyl cellulose/ 16% water	68.4	DMTA (E')	
39% Corn starch/39% methyl cellulose/ 22% water	48.5	DMTA (E')	
37% Corn starch/37% methyl cellulose/ 26% water	36.7	DMTA (E')	
35% Corn starch/35% methyl cellulose/ 30% water	18.5	DMTA (E')	
78% Corn starch/5% methyl cellulose/ 17% water	59.5	DMTA (E')	
58% Corn starch/25% methyl cellulose/ 17% water	67.1	DMTA (E')	
28% Corn starch/55% methyl cellulose/ 17% water	76.1	DMTA (E')	

Composition	Value	Method
47% Corn starch/47% methyl cellulose/6% water	123.6	DMTA (tan δ)
45% Corn starch/45% methyl cellulose/10% water	110.2	DMTA (tan δ)
42% Corn starch/42% methyl cellulose/16% water	79.6	DMTA (tan δ)
39% Corn starch/39% methyl cellulose/22% water	63.5	DMTA (tan δ)
37% Corn starch/37% methyl cellulose/26% water	45.6	DMTA (tan δ)
35% Corn starch/35% methyl cellulose/30% water	23.0	DMTA (tan δ)
78% Corn starch/5% methyl cellulose/17% water	62.8	DMTA (tan δ)
58% Corn starch/25% methyl cellulose/17% water	76.0	DMTA (tan δ)
28% Corn starch/55% methyl cellulose/17% water	89.3	DMTA (tan δ)
47.5% Corn starch/47.5% MCC/0% glycerol/5% water	118.5	DTA
42.5% Corn starch/42.5% MCC/10% glycerol/5% water	59.4	DTA
38% Corn starch/38% MCC/19% glycerol/5% water	−2.8	DTA
45% Corn starch/45% MCC/5% sorbitol/5% water	91.8	DTA
42.5% Corn starch/42.5% MCC/10% sorbitol/5% water	57.0	DTA
39.5% Corn starch/39.5% MCC/16% sorbitol/5% water	23.8	DTA
44.5% Corn starch/44.5% MCC/6% sucrose/5% water	87.4	DTA

(Continued)

33

TABLE 2-3 *Continued*

Components	T_g (°C)	Method of T_g Determination	References
40% Corn starch/40% MCC/15% sucrose/ 5% water	19.1	DTA	
37.5% Corn starch/37.5% MCC/ 20% sucrose/5% water	−10.0	DTA	
45% Corn starch/45% MCC/5% xylose/ 5% water	84.2	DTA	
39.5% Corn starch/39.5% MCC/ 16% xylose/5% water	11.3	DTA	
37.5% Corn starch/37.5% MCC/ 20% xylose/5% water	−8.5	DTA	
47.5% Corn starch/47.5% MCC/ 0% glycerol/5% water	113.4	DMTA (E')	
42.5% Corn starch/42.5% MCC/ 10% glycerol/5% water	52.3	DMTA (E')	
38% Corn starch/38% MCC/ 19% glycerol/5% water	−5.1	DMTA (E')	
45% Corn starch/45% MCC/5% sorbitol/ 5% water	84.1	DMTA (E')	
42.5% Corn starch/42.5% MCC/ 10% sorbitol/5% water	48.0	DMTA (E')	
39.5% Corn starch/39.5% MCC/ 16% sorbitol/5% water	17.8	DMTA (E')	
44.5% Corn starch/44.5% MCC/ 6% sucrose/5% water	80.1	DMTA (E')	
40% Corn starch/40% MCC/15% sucrose/ 5% water	11.8	DMTA (E')	
37.5% Corn starch/37.5% MCC/ 20% sucrose/5% water	−22.5	DMTA (E')	

Material	Method	Value	Reference
45% Corn starch/45% MCC/5% xylose/5% water	DMTA (E')	77.1	
39.5% Corn starch/39.5% MCC/16% xylose/5% water	DMTA (E')	7.4	
37.5% Corn starch/37.5% MCC/20% xylose/5% water	DMTA (E')	−14.5	
47.5% Corn starch/47.5% MCC/0% glycerol/5% water	DMTA (tan δ)	130.0	
42.5% Corn starch/42.5% MCC/10% glycerol/5% water	DMTA (tan δ)	70.3	
38% Corn starch/38% MCC/19% glycerol/5% water	DMTA (tan δ)	6.8	
45% Corn starch/45% MCC/5% sorbitol/5% water	DMTA (tan δ)	97.5	
42.5% Corn starch/42.5% MCC/10% sorbitol/5% water	DMTA (tan δ)	66.9	
39.5% Corn starch/39.5% MCC/16% sorbitol/5% water	DMTA (tan δ)	30.3	
44.5% Corn starch/44.5% MCC/6% sucrose/5% water	DMTA (tan δ)	93.5	
40% Corn starch/40% MCC/15% sucrose/5% water	DMTA (tan δ)	58.3	
37.5% Corn starch/37.5% MCC/20% sucrose/5% water	DMTA (tan δ)	−4.5	
45% Corn starch/45% MCC/5% xylose/5% water	DMTA (tan δ)	88.0	
39.5% Corn starch/39.5% MCC/16% xylose/5% water	DMTA (tan δ)	17.3	
37.5% Corn starch/37.5% MCC/20% xylose/5% water	DMTA (tan δ)	−1.2	
Cellulose diacetate	–	114.3	Lee et al. (2006)

(Continued)

35

TABLE 2-3 *Continued*

Components	T_g (°C)	Method of T_g Determination	References
10% Starch/80% plasticized cellulose diacetate	111.9	—	Glasser et al. (2000)
20% Starch/80% plasticized cellulose diacetate	101.7	—	
30% Starch/70% plasticized cellulose diacetate	98.8	—	
40% Starch/60% plasticized cellulose diacetate	95.2	—	
50% Starch/50% plasticized cellulose diacetate	91.5	—	
CP (clear–pure) cellulose	130	DSC	
Cellulose propionate difluoroethoxy acetate	67	DSC	
Cellulose octafluoropentoxy acetate	113	DSC	
Cellulose propionate octafluoropentoxy acetate	53	DSC	
Medium–high DS hydroxypropyl cellulose	105	Oscillatory rheometry	Gómez-Carracedo et al. (2003)
Hydroxypropylmethyl cellulose	170–198	Oscillatory rheometry	
Methylcellulose	184–197	Oscillatory rheometry	
Low DS hydroxypropyl cellulose	220	Oscillatory rheometry	
Plasticized wheat starch	31	DMTA (tanδ)	Avérous et al. (2001b)
PWS/cellulose fibers	59–64	DMTA (tan δ)	

Composition	Value	Method	Reference
74% Starch/10% glycerol/26% water	43	DSC	Avérous and Fringant (2001)
70% Starch/18% glycerol/12% water	8	DSC	
67% Starch/24% glycerol/9% water	−7	DSC	
65% Starch/35% glycerol/0% water	−20	DSC	
45% Corn starch/45% methyl cellulose/0% glycerol/10% water	84.9	DTA	Arvanitoyannis and Biliaderis (1999)
42.5% Corn starch/42.5% methyl cellulose/5% glycerol/10% water	56.8	DTA	
37.5% Corn starch/37.5% methyl cellulose/15% glycerol/10% water	22.9	DTA	
30% Corn starch/30% methyl cellulose/30% glycerol/10% water	–	DTA	
42.5% Corn starch/42.5% methyl cellulose/5% sorbitol/10% water	43.9	DTA	
37.5% Corn starch/37.5% methyl cellulose/15% sorbitol/10% water	17.6	DTA	
30% Corn starch/30% methyl cellulose/30% sorbitol/10% water	–	DTA	
42.5% Corn starch/42.5% methyl cellulose/5% xylose/10% water	48.0	DTA	
37.5% Corn starch/37.5% methyl cellulose/15% xylose/10% water	21.3	DTA	
100% Plasticized starch/0% tunicin whiskers	13.2 (0% RH)	DSC	Mathew and Dufresne (2002)
95% Plasticized starch/5% tunicin whiskers	16.3 (0% RH)	DSC	
90% Plasticized starch/10% tunicin whiskers	18.7 (0% RH)	DSC	
85% Plasticized starch/15% tunicin whiskers	17.9 (0% RH)	DSC	

(*Continued*)

TABLE 2-3 *Continued*

Components	T_g (°C)	Method of T_g Determination	References
75% Plasticized starch/25% tunicin whiskers	10.5 (0% RH)	DSC	
100% Plasticized starch/0% tunicin whiskers	−30.3 (58% RH)	DSC	
95% Plasticized starch/5% tunicin whiskers	−31.5 (58% RH)	DSC	
90% Plasticized starch/10% tunicin whiskers	−32.0 (58% RH)	DSC	
85% Plasticized starch/15% tunicin whiskers	−25.1 (58% RH)	DSC	
75% Plasticized starch/25% tunicin whiskers	−34.5 (58% RH)	DSC	
100% Plasticized starch/0% tunicin whiskers	−64.7 (98% RH)	DSC	
95% Plasticized starch/5% tunicin whiskers	−61.6 (98% RH)	DSC	
90% Plasticized starch/10% tunicin whiskers	−56.5 (98% RH)	DSC	
85% Plasticized starch/15% tunicin whiskers	−49.2 (98% RH)	DSC	
75% Plasticized starch/25% tunicin whiskers	−54.5 (98% RH)	DSC	

is the preferred additive for increasing the adhesion of starch sizes for cotton. However, CMC must be classed as being very difficult to degrade. Only 20% of the initial amount is eliminated after 7 days. However, it is also reported that CMC can be reduced after long periods of adaptation (>4 weeks) and favorable conditions (especially higher temperatures) (URL3).

2.3.1 Mechanical Properties

Pure cellulose diacetate (CDA) has a tensile strength of 62 MPa and an elongation of 4%. The glass transition temperature of pure CDA was higher than that of plasticized CDA (Lee et al., 2006). Wibowo et al. (2006) examined the mechanical properties of biodegradable nanocomposites from cellulose acetate. Two methods of composite processing were used: Process A (extrusion at 200 rpm, followed by compression molding) and Process B (extrusion at 200 rpm, followed by injection molding). In both methods 0% and 5% (wt) clay was added to fabricated cellulose acetate plastic matrix. An increase of 60% in tensile strength and 80% in tensile modulus was reported for Process B samples compared to Process A samples. Furthermore, an increase of 38% in tensile strength and 33% in tensile modulus was observed for nanocomposites with 5% (wt) clay reinforcement compared to that of matrix in Process B.

Samples of commercial cleanex (fibrous delignified native cellulose) were pressed and annealed together with unpressed ones for 6 h at 120–265°C. An increase in tensile strength and tensile modulus of the pressed cleanex annealed at 180–230°C was observed, while a decrease of these parameters was reported when the annealing temperature was 265°C. Maximum values of tensile strength of 3.9 MPa and tensile modulus of 182 MPa were obtained after annealing under pressure at 230°C. The values of tensile strength and tensile modulus decreased continuously in the case of unpressed cleanex annealed at 120–265°C and they were lower than those of the pressed samples annealed at temperatures above 130°C (Avramova and Fakirov, 1990) (see Table 2-3).

2.3.2 Thermal Properties

The glass transitions of novel cellulose derivatives ranged between 53°C and 113°C, and are distinctly below that of cellulose propionate. The influence of the larger octa-fluoropentoxy group compared with the difluoroethoxy group is revealed by the significant T_g reduction of 14°C (from 67°C to 53°C) (Glasser et al., 2000). Values of T_g for various nonionic cellulose ethers were analyzed by DSC, modulated temperature differential scanning calorimetry (TMDSCO), and oscillatory rheometry. The best resolution was obtained using oscillatory rheometry since these cellulose ethers undergo considerable changes in their storage and loss moduli when reaching the T_g. Values of T_g were in the order: medium–high DS (degree of substitution) hydroxypropyl cellulose 105°C < hydroxypropylmethyl cellulose 170–198°C < methyl cellulose 184–197°C < low DS hydroxypropyl cellulose 220°C. For hydroxypropylmethyl celluloses, the T_g increases as the methoxyl/hydroxypropoxyl content ratio decreases (Gómez-Carracedo et al., 2003) (see Table 2-3).

2.4 STARCH–CELLULOSE BLENDS

In order to overcome the inherent hydrophilicity of starch, blends with conventional hydrophobic synthetic polymers have been considered for the production of plastic bags and commodity products (Corti et al., 1992). In order to reduce the sensitivity of starch to water, it is blended with more hydrophobic, thermoplastic materials such as polycaprolactone, polyhydroxybutyrate/valerate and cellulose acetate. Such blends are completely biodegradable under composting and other biologically active environments (Fang and Fowler, 2003). Both cellulose acetate and starch acetate have been known since the nineteenth century. However, a blend of these two common materials has not before been available because of processing problems (URL6).

2.4.1 Mechanical Properties

Natural cellulose fibers with different initial average lengths [60 (SF), 300 (MF) and 900 (LF) μm] and paper pulp fibers (PPF) were added to different plasticized starch matrixes (TPS$_1$ and TPS$_2$) (Avérous and Boquillon, 2004). When 10% of natural cellulose fibers with initial average lengths 60 (SF) and 300 (MF) was added to TPS$_1$ and TPS$_2$, the tensile modulus was greater and the elongation ratios were lower for TPS$_1$ than for TPS$_2$. On the other hand, the tensile modulus and the elongation ratios were greater when 20% of natural cellulose fibers [with initial average lengths 900 (LF) μm] and PPF was added to the TPS$_2$ than for the blend of 20% of natural cellulose fibers/PPF and TPS$_2$.

The tensile strength and Young's modulus of the plasticized starch (PS)/cellulose nanocrystallites and PS/ramie cellulose nanocrystallites (RN) increased significantly from 2.5 to 7.8 MPa and from about 36 to 301 MPa with increasing filler (RN) content from 0% to 30% (wt), whereas the elongation at break decreased from 105% to 46% (Lu et al., 2005, 2006).

The mechanical behavior of cellulose microfibril–plasticized starch composites was investigated by Dufresne et al. (2000). It was shown that cellulose microfibrils appreciably reinforce the starch matrix, regardless of the glycerol content. The evolution of the tensile modulus as a function of the cellulose content was nearly linear. In addition, the higher modulus was observed for the less plasticized material.

According to Psomiadou et al. (1996), with regard to mechanical properties the corn starch/methyl cellulose (MC)/water or polyol (glycerol, D-glucose, sucrose, xylose) blends exhibited similar behavior to the corn starch/MCC/water blends. Sucrose exerts an antiplasticizing action, resulting in considerably lower tensile strength and percentage elongation, whereas both sorbitol and glucose impart lower tensile strength but higher percentage elongation (plasticizing action). The incorporation of MC in the starch matrix was found to have a rather beneficial effect resulting in a gradual increase of tensile strength and percentage elongation. When MCC was employed, the tensile strength increased but the percentage elongation increased only slightly.

Acetylated starch and 15% cellulose acetate were pretreated with acetone solution, and the pretreated mixture was extruded as a composite. The tensile strength was effectively improved due to addition of cellulose acetate compared with the acetylated starch plastic (Chen et al., 2003). A series of water-resistant cellulose films was prepared by coating castor oil-based polyurethane/benzyl starch (BS) semi-interpenetrating polymer networks. The tensile strength and elongation at break of cellulose-coated films increased simultaneously with the increasing BS concentration on the whole and reached maximum values of 102 MPa and 14%, respectively (Cao et al., 2006).

Lee et al. (2006) examined the mechanical properties of plasticized cellulose diacetate (CDA) and starch, which were prepared using melt processing methods with triacetine (TA) as the primary plasticizer and epoxidized soybean oil (ESO) as the secondary plasticizer. At 7% (wt) starch in a blend, the tensile strength decreased from 62 MPa to 49 MPa for pure CDA, and the elongation increased from 4% to 6% (wt). The tensile strength was 16 MPa with 18% elongation at a 35% (wt) starch content. In general, the tensile strength and Young's modulus decreased and the level of elongation increased with increasing starch content in the CDA matrix.

Arvanitoyannis and Biliaderis (1999) estimated the tensile strength and percentage elongation of polyol-plasticized edible blends (polyols: glycerol, sorbitol, xylose) made from soluble starch–MC at different water contents. Percentage elongation exhibited a considerable increase, up to three times the initial values, while the tensile strength dropped to lower than 50% of the original values.

The mechanical properties of plasticized wheat starch (PWS)/cellulose fibers were examined by Avérous et al. (2001b). The characteristics of the cellulose fibers were: B600 [fiber length (L) 60 µm, diameter (d) 20 µm, L/d 3]; BC200 (L 300 µm, d 20 µm, L/d 15); and B400 (L 900 µm, d 20 µm, L/d 45). Young's modulus (E modulus) was increased by a factor of 5–9 (order of magnitude). For the lowest fiber content (15%), the modulus increases with the fiber length. The best results are obtained with B400 fibers that have the greatest length. Strength improvement depends both on the fiber length and content. Elongation at break decreased with the addition of fiber.

Avérous et al. (2001a) investigated the potential effect of the addition of cellulose fibers in a TPS matrix and reported the subsequent properties. Modulus or maximum strength results showed a strong improvement on addition of fibers to the TPS matrix.

Anglés and Dufresne (2001) showed that no significant reinforcing effect was observed on the tensile modulus of glycerol-plasticized starch/tunicin whisker composite films regardless of the moisture content up to 16.2% (wt) tunicin whiskers. When the cellulose content reached 25% (wt) an increase in the modulus was observed. No significant evolution for either the tensile strength or the elongation at break was reported, regardless of the whisker loading. On the contrary, the strength of the composite conditioned at 35% RH increased as the cellulose content increased.

The mechanical behavior of plasticized wheat starches referred to as PWS 1 to 3 with increasing glycerol/starch ratios, i.e., 0.14, 0.26 and 0.54, respectively, at 50% and 70% RH was also investigated by Avérous et al. (2001b). At 50% RH, almost the same modulus is found for PWS 1 as for the system PWS 3 fibers. From 50% to 75% RH, the modulus of PWS 1 is decreased by a factor of 10, whereas that of the composites PWS 3 fibers is only reduced by a factor of 6. PWS 2 and 3 have similar tensile strength and modulus but very different impact strength resistance.

The mechanical properties of various starch–cellulose blends are shown in Table 2-2.

2.4.2   Thermal Properties

It is sometimes difficult to determine the glass transition of plasticized starch by DSC analysis because the heat capacity change is quite low at the glass transition. The T_g determined by dynamic mechanical analysis (DMA) (tan δ) was different from the T_g determined by DSC. Dynamic mechanical thermal analysis (DMTA) determination is used to obtain this transition, which is clearly demonstrated by a broad tan δ peak. Tan δ drawing shows two relaxations (α and β). The main relaxation (termed α), associated with a large tan δ peak and an important decrease of the storage modulus, can be attributed to the TPS glass relaxation (Avérous and Boquillon, 2004). After the addition of cellulose fibers to PWS, a shift of the glass transition temperature (DSC analysis) of the matrix toward higher temperatures was observed for reinforced materials (maximum of tan δ at 59–64°C instead of 31°C for PWS alone) (Avérous et al., 2001b).

Plasticized starch film exhibited the T_g transition at about 22°C. The T_g transition of the glycerol-rich phase of PS film/cellulose nanocrystallites ranged from -80°C to -50°C, but could not be determined because of the limitation of refrigerated cooling systems in DSC. By adding cellulose crystallites into starch matrix, the T_g transition of the starch-rich phase is shifted to higher temperatures of 23–48°C (Lu et al., 2005). According to Lu et al. (2006), the PS film exhibited a T_g transition at about 26.8°C. By incorporating RN fillers of 0–40% (wt) into PS, the T_g transition for starch rich-phase is shifted to a higher temperature of 26.8–55.7°C.

The T_g of cellulose microfibril–plasticized starch composites was decreased to temperatures lower than room temperature, and the effect of adding filler became more significant when the glycerol content was increased (Dufresne et al., 2000). An increase of the main T_g of about 30°C was reported for the addition of cellulose fibers to the TPS matrix (Avérous et al., 2001a). Moreover, the presence of TA and ESO, as plasticizers, in CDA/starch decreased the T_g of the composites (Lee et al., 2006).

Psomiadou et al. (1996) investigated the thermal properties of various aqueous blends of microcrystalline cellulose (MCC) or methyl cellulose (MC) and corn starch with or without polyols (glycerol, D-glucose, sucrose, xylose). The results showed that the higher the water content, the greater the depression of the T_g value (DTA, DMTA, and permeability measurements).

TABLE 2-4   Melting Temperatures (T_m) of Starch–Cellulose Blends

Components	T_m (°C)	Conditions	References
100% Plasticized starch/0% tunicin whiskers	133.4	43% RH	Anglés and Dufresne (2001)
96.8% Plasticized starch/3.2% tunicin whiskers	132.9		
93.8% Plasticized starch/6.2% tunicin whiskers	131.9		
83.3% Plasticized starch/ 16.7% tunicin whiskers	132.9		
75% Plasticized starch/25% tunicin whiskers	135.4		
100% Plasticized starch/0% tunicin whiskers	157.9	58% RH	
96.8% Plasticized starch/3.2% tunicin whiskers	159.4		
93.8% Plasticized starch/6.2% tunicin whiskers	165.9		
83.3% Plasticized starch/ 16.7% tunicin whiskers	160.9		
75% Plasticized starch/25% tunicin whiskers	170.3		
100% Plasticized starch/0% tunicin whiskers	156.9	75% RH	
96.8% Plasticized starch/3.2% tunicin whiskers	156.9		
93.8% Plasticized starch/6.2% tunicin whiskers	156.9		
83.3% Plasticized starch/ 16.7% tunicin whiskers	157.4		
75% Plasticized starch/25% tunicin whiskers	159.4		
100% Plasticized starch/0% tunicin whiskers	146.5	43% RH	Mathew and Dufresne (2002)
95% Plasticized starch/5% tunicin whiskers	142.6		
90% Plasticized starch/10% tunicin whiskers	149.2		
85% Plasticized starch/15% tunicin whiskers	139		
75% Plasticized starch/25% tunicin whiskers	143.2		
100% Plasticized starch/0% tunicin whiskers	142.8	58% RH	
95% Plasticized starch/5% tunicin whiskers	140.0		

(Continued)

TABLE 2-4   *Continued*

Components	T_m (°C)	Conditions	References
90% Plasticized starch/10% tunicin whiskers	141.4		
85% Plasticized starch/15% tunicin whiskers	137.6		
75% Plasticized starch/25% tunicin whiskers	138.1		
100% Plasticized starch/0% tunicin whiskers	123.1	98% RH	
95% Plasticized starch/5% tunicin whiskers	124.8		
90% Plasticized starch/10% tunicin whiskers	135.6		
85% Plasticized starch/15% tunicin whiskers	126.3		
75% Plasticized starch/25% tunicin whiskers	130.1		
Cellulose propionate difluoroethoxy acetate	218	–	Glasser et al. (2000)

Mathew and Dufresne (2002) examined the thermal properties of sorbitol-plasticized starch conditioned at 0% up to 98% RH. It was observed that T_g was decreased with moisture content. Furthermore, an increase of T_m was observed when the RH increased from 43% to 58%. For higher moisture content (up to 75% RH), T_m tended to stabilize. For the more moist sample (98% RH), an effective plasticization was reported because T_m decreased (see Table 2-4).

According to Arvanitoyannis and Biliaderis (1999), at total plasticizer content (water, glycerol, and sugars) higher than 15%, a substantial depression of the T_g was observed. Glycerol had a greater depressive effect on T_g than sorbitol.

For glycerol-plasticized starch/tunicin whiskers composite films, the higher moisture content, the higher the melting temperature (at 75% RH the T_m was about 215°C) (Table 2-4) (Anglés and Dufresne, 2001). The T_g values of starch–cellulose blends are presented in Table 2-3.

2.4.3   Water Vapor Transmission Rate and Gas Permeability

The gas permeability (GP) measurements of plasticized starch/MC blends showed increases in GP proportional to the plasticizer content due to weakening of the intermolecular forces between adjacent polymeric chains, which facilitates chain mobility and redistribution of the originally existing voids (Arvanitoyannis, 1999). Furthermore, the GP of the soluble starch/MC blends increased proportionally to the total plasticizer content (water, glycerol, and sugars) (Arvanitoyannis and Biliaderis, 1999).

TABLE 2-5 Water Vapor Transmission Rate (WVTR) and Gas Permeability (GP) of Various Starch–Cellulose Blends

Components	WVTR (g m^{-1} s^{-1} Pa^{-1}) × 10^{-11}	GP$_{O_2}$ (cm^2 s^{-1} Pa^{-1})	GP$_{N_2}$ (cm^2 s^{-1} Pa^{-1})	GP$_{CO_2}$ (cm^2 s^{-1} Pa^{-1})	References
45% Corn starch/45% methyl cellulose/5% glycerol/5% water	3.2	9.8×10^{-15}	6.0×10^{-16}	8.5×10^{-14}	Arvanitoyannis and Biliaderis (1999)
40% Corn starch/40% methyl cellulose/15% glycerol/5% water	6.5	6.5×10^{-14}	5.3×10^{-15}	3.6×10^{-13}	
32.5% Corn starch/32.5% methyl cellulose/30% glycerol/5% water	20.7	4.7×10^{-11}	2.0×10^{-12}	4.2×10^{-10}	
45% Corn starch/45% methyl cellulose/5% sorbitol/5% water	4.5	3.4×10^{-14}	6.9×10^{-15}	7.6×10^{-13}	
40% Corn starch/40% methyl cellulose/15% sorbitol/5% water	7.9	8.8×10^{-14}	3.6×10^{-14}	5.4×10^{-12}	
32.5% Corn starch/32.5% methyl cellulose/30% sorbitol/5% water	26.5	5.7×10^{-11}	4.3×10^{-12}	9.0×10^{-10}	
45% Corn starch/45% methyl cellulose/5% xylose/5% water	4.9	1.5×10^{-14}	5.7×10^{-15}	2.1×10^{-14}	
40% Corn starch/40% methyl cellulose/15% xylose/5% water	8.6	9.9×10^{-14}	9.1×10^{-15}	8.5×10^{-13}	
32.5% Corn starch/32.5% methyl cellulose/30% xylose/5% water	28.7	6.5×10^{-11}	4.9×10^{-12}	7.0×10^{-10}	
45% Corn starch/45% methyl cellulose/0% glycerol/10% water	0.65	5.1×10^{-14}	3.9×10^{-15}	2.1×10^{-13}	
42.5% Corn starch/42.5% methyl cellulose/5% glycerol/10% water	7.2	8.0×10^{-14}	9.5×10^{-14}	8.4×10^{-13}	
37.5% Corn starch/37.5% methyl cellulose/15% glycerol/10% water	9.6	4.7×10^{-13}	2.5×10^{-13}	3.8×10^{-12}	

(Continued)

TABLE 2-5 *Continued*

Components	WVTR (g m⁻¹ s⁻¹ Pa⁻¹) × 10⁻¹¹	GP_{O_2} (cm² s⁻¹ Pa⁻¹)	GP_{N_2} (cm² s⁻¹ Pa⁻¹)	GP_{CO_2} (cm² s⁻¹ Pa⁻¹)	References
30% Corn starch/30% methyl cellulose/30% glycerol/10% water	25.9	3.0×10^{-10}	4.3×10^{-13}	3.2×10^{-9}	Psomiadou et al. (1996)
42.5% Corn starch/42.5% methyl cellulose/5% sorbitol/10% water	9.5	2.4×10^{-13}	8.6×10^{-14}	3.8×10^{-12}	
37.5% Corn starch/37.5% methyl cellulose/15% sorbitol/10% water	14.9	8.7×10^{-13}	4.5×10^{-13}	2.5×10^{-11}	
30% Corn starch/30% methyl cellulose/30% sorbitol/10% water	33.5	6.0×10^{-10}	6.3×10^{-11}	4.3×10^{-9}	
42.5% Corn starch/42.5% methyl cellulose/5% xylose/10% water	8.3	3.8×10^{-13}	7.0×10^{-14}	8.5×10^{-13}	
37.5% Corn starch/37.5% methyl cellulose/15% xylose/10% water	13.8	8.3×10^{-13}	4.9×10^{-13}	4.9×10^{-12}	
30% Corn starch/30% methyl cellulose/30% xylose/10% water	29.6	7.5×10^{-10}	4.0×10^{-11}	4.0×10^{-10}	
47.5% Corn starch/47.5% MCC/0% glycerol/5% water	0.2	1.8×10^{-17}	3.2×10^{-18}	1.2×10^{-16}	
42.5% Corn starch/42.5% MCC/10% glycerol/5% water	3.4	3.6×10^{-15}	6.5×10^{-16}	3.1×10^{-14}	
34% Corn starch/34% MCC/27% glycerol/5% water	11.4	3.0×10^{-13}	3.7×10^{-14}	0.7×10^{-12}	
45% Corn starch/45% MCC/5% sorbitol/5% water	2.5	4.0×10^{-16}	4.3×10^{-17}	2.5×10^{-15}	
39.5% Corn starch/39.5% MCC/16% sorbitol/5% water	7.0	3.7×10^{-15}	2.8×10^{-16}	1.3×10^{-14}	
33.5% Corn starch/33.5% MCC/28% sorbitol/5% water	14.2	4.5×10^{-12}	0.7×10^{-13}	1.5×10^{-11}	

Composition				
44.5% Corn starch/44.5% MCC/6% sucrose/5% water	3.8	1.5×10^{-16}	0.9×10^{-17}	1.4×10^{-15}
40% Corn starch/40% MCC/15% sucrose/5% water	8.4	1.1×10^{-15}	0.8×10^{-16}	1.1×10^{-14}
34.5% Corn starch/34.5% MCC/26% sucrose/5% water	16.3	3.0×10^{-12}	0.7×10^{-13}	1.9×10^{-11}
45% Corn starch/45% MCC/5% xylose/5% water	2.8	1.6×10^{-16}	1.1×10^{-17}	1.0×10^{-15}
39.5% Corn starch/39.5% MCC/16% xylose/5% water	7.4	1.5×10^{-15}	0.9×10^{-16}	1.2×10^{-14}
35% Corn starch/35% MCC/25% xylose/5% water	13.6	3.0×10^{-13}	5.5×10^{-14}	2.9×10^{-12}
47.5% Corn starch/47.5% methyl cellulose/0% glycerol/5% water	–	2.2×10^{-17}	1.8×10^{-17}	2.0×10^{-16}
42.5% Corn starch/42.5% methyl cellulose/10% glycerol/5% water	–	4.5×10^{-15}	1.8×10^{-17}	4.7×10^{-14}
34% Corn starch/34% MCC/27% glycerol/5% water	–	4.9×10^{-13}	1.8×10^{-17}	1.5×10^{-12}
45% Corn starch/45% methyl cellulose/5% sorbitol/5% water	–	6.1×10^{-16}	1.8×10^{-17}	3.8×10^{-15}
39.5% Corn starch/39.5% methyl cellulose/16% sorbitol/5% water	–	6.5×10^{-15}	1.8×10^{-17}	2.6×10^{-14}
33.5% Corn starch/33.5% methyl cellulose/28% sorbitol/5% water	–	6.4×10^{-12}	1.8×10^{-17}	2.8×10^{-11}
44.5% Corn starch/44.5% methyl cellulose/6% sucrose/5% water	–	2.8×10^{-16}	1.8×10^{-17}	1.7×10^{-15}
40% Corn starch/40% methyl cellulose/15% sucrose/5% water	–	2.0×10^{-15}	1.8×10^{-17}	2.2×10^{-14}
34.5% Corn starch/34.5% methyl cellulose/26% sucrose/5% water	–	5.1×10^{-13}	1.8×10^{-17}	4.5×10^{-12}

Psomiadou et al. (1996) revealed that the higher the starch content in edible fibers made from MCC, MC and corn starch and polyols, the higher the water vapor transmission rate (WVTR). In addition, the high MCC/starch or MC/starch contents in the blend decreased the GP of the blend, whereas the incorporation of high water or polyol contents resulted in substantial enhancement of GP values.

WVTR and GP values of various starch–cellulose blends are presented in Table 2-5.

2.5 APPLICATIONS

Depending on the degree of substitution (DS) of the free hydroxyl groups, starch acetate may be used in a number of commercially important ways. Low-DS acetates are important in food applications while highly derivatized starches with a DS of 2–3 are useful because of their solubility in organic solvents and ability to form films and fibers (URL6).

Chemically modified starches have a wide spectrum of applications, but the most important areas are in the paper, food and textile industries. Hydroxyethyl starch is used extensively in paper manufacturing, in the pharmaceutical industry as a plasma volume extender, and as adhesives. Hydroxypropyl-modified starches are of great importance in food and food-related products, primarily as thickeners, and as excipients in foods and pharmaceuticals. Cationic starches are used in large quantities in paper manufacturing. Cationic ethers with tertiary aminoalkyl groups or with quaternary ammonium groups are used in hair care products and as bactericides (Richardson and Gorton, 2003; URL2).

Cellulose acetate of certain degrees of acetylation is commonly used to manufacture cigarette tow, textile fibers, films, plastics, and other materials (URL6). Cellulose acetate is also used for wrapping baked goods and fresh produce. Although cellulose acetate requires the addition of plasticizers for production of films, the resulting product demonstrates good gloss and clarity, good printability, rigidity, and dimensional stability. Although these films can tear easily, they are tough and resistant to punctures (Liu, 2006).

Carboxymethyl cellulose (CMC) is used in coatings, detergents, food, toothpaste, adhesives, and cosmetics applications. Hydroxyethyl cellulose and its derivatives are used as thickeners in coatings and drilling fluids. Methyl cellulose is used in foods, adhesives, and cosmetics. Cellulose acetate is employed in packaging, fabrics, and pressure-sensitive tapes (URL1).

The major markets for CMC are in detergents, in textiles and in paper manufacturing. Another cellulose ether much used in industry is hydroxyethyl cellulose (HEC). This product acts, for example, as a thickener in drilling fluids, as a moisture-retaining agent and retarder in cements, and as a thickener in the oil industry and a suspension aid in paints. Ethyl(hydroxyethyl) cellulose (EHEC) is a mixed ether that is widely used in the paint and building industries; in the former as protective colloids, thickeners, and pigment suspension aids, and in the latter as dispersion agents in cement formulations (Richardson and Gorton, 2003). Cellulose ethers, such as methylcellulose, are widely used as gelling and dispersing agents in foods, cosmetics,

TABLE 2-6   Applications of Cellulose and Starch Derivatives and Starch–Cellulose Blends

Components	Applications	References
Starch acetate	Food applications, films, fibers	URL6
2-Hydroxyethyl starch	Paper industry, textiles, films, plasma extenders	Richardson and Gorton (2003); URL2
2-Hydroxypropyl starch	Food products, films, excipients in foods and pharmaceuticals	
Cationic starches	Paper industry, textiles, hair care products, bactericides	
Succinate starch	Films, emulsions	Richardson and Gorton (2003)
Acetyl starch	Food products, textiles, paper industry	Richardson and Gorton (2003)
Crosslinked starch	Food products, emulsions	Richardson and Gorton (2003)
Anionic starches	Paper, adhesives	Richardson and Gorton (2003)
Cellulose acetate	Cigarette tow, textile fibers, films, plastics, wrapping goods, packaging, fabrics, pressure-sensitive tapes	URL6; Liu (2006)
Ethyl cellulose	Paints, lacquers	Richardson and Gorton (2003)
2-Hydroxyethyl cellulose	Paints, emulsions, drilling mud	Richardson and Gorton (2003)
2-Hydroxypropyl cellulose	Building materials, paints, tablets	Richardson and Gorton (2003)
Sodium carboxymethyl cellulose	Detergents, textiles, food products	Richardson and Gorton (2003)
Methyl cellulose	Foods, adhesives, cosmetics, building materials, paint removers, gelling and dispersing agents in foods, cosmetics, pharmaceuticals and lacs	URL2
Ethyl cellulose	Carriers for the controlled release of nitrogen fertilizers	URL2
Hydroxypropyl cellulose	Coating of paper and tablets	URL2
Hydroxypropyl methyl cellulose	Agent for lowering blood cholesterol levels	URL2
Cellulose acetate propionate or butyrate	Toys, sport goods, ophthalmic and medical applications	URL2
Cellulose acetate phthalate	Tablet production	URL2
Carboxymethyl cellulose	Coatings, detergents, food, toothpaste, adhesives, cosmetics applications	URL1
Hydroxyethyl cellulose and derivatives	Thickeners in coatings and drilling fluids	URL1

(Continued)

TABLE 2-6   *Continued*

Components	Applications	References
50–99% Cellulose acetate/ 1–50% starch acetate blend	Fibers, filaments, yarns, fabrics, plastic materials	URL6
Oxidized starch and cellulose imine derivatives	Antimalarial activity	Taba Ohara et al. (1995)
70% Rice starch/30% microcrystalline cellulose	Compressible excipient	Limwong et al. (2004)

pharmaceuticals, and lacs. Ethyl cellulose has been evaluated for carriers for the controlled release of nitrogen fertilizers. Hydroxypropyl cellulose (HPC) has excellent surface and film-forming properties and is thus used for coating of paper and tablets. Hydroxypropylmethyl cellulose (HPMC) is a promising agent for lowering blood cholesterol levels. Cellulose acetate propionate (CAP) or butyrate (CAB) are used to make toys, sport goods, and in ophthalmic and medical applications. Cellulose acetate phthalate is used for tablet production (URL2).

A blend of 50–99% cellulose acetate with 1–50% starch acetate offers the possibility of combining a cheap raw material, starch, with the production of fibers, filaments, yarns, fabrics, plastic materials, and other uses (URL6).

Oxidized starch imine derivatives of sulfonamides or pyrimidine—derivatives of sulfisoxazole (ML8), sulfameter (ML11) and trimethoprim (ML13)—and oxidized cellulose imine derivatives of dapsone (ML14), sulfadiazine (ML17), sulfamethoxazole (ML18), sulfisoxazole (ML19), sulfamethoxypyridazine (ML20) and sulfameter (ML22) were submitted to in-vivo biological assays with mice infected with *Plasmodium berghei*. Only ML11 was 100% curative, while ML17 showed the same effect as its prototype (Taba Ohara et al., 1995).

Rice starch (RS) and microcrystalline cellulose in the ratio 7 : 3 was proposed as a suitable combination with respect to its properties for use as a new co-processed directly compressible excipient. The tablets made from these co-processed composite particles exhibited high compressibility, good flowability, and self-disintegration (Limwong et al., 2004).

Various applications of cellulose and starch derivatives, and starch–cellulose blends are presented in Table 2-6.

REFERENCES

Ahmad FB, Williams PA, Doublier JL, Durand S, Buleon A. 1999. Physico-chemical characterisation of sago starch. *Carbohydr Polym* 38:361–370.

Anglés MN, Dufresne A. 2001. Plasticized/tunicin whiskers nanocomposites materials. 2. *Mech Behav Macromol* 34:2921–2931.

Arvanitoyannis IS. 1999. Totally and partially biodegradable polymer blends based on natural and synthetic macromolecules: preparation, physical properties, and potential as food packaging materials. *JMS Rev Macromol Chem Phys* C39(2):205–271.

Arvanitoyannis I, Biliaderis CG. 1999. Physical properties of polyol-plasticized edible blends made of methyl cellulose and soluble starch. *Carbohydr Polym* 38:47–58.

Avérous L, Boquillon N. 2004. Biocomposites based on plasticized starch: thermal and mechanical behaviours. *Carbohydr Polym* 56:111–122.

Avérous L, Fringant C. 2001. Association between plasticized starch and polyesters: processing and performances of injected biodegradable systems. *Polym Eng Sci* 41:727–734.

Avérous L, Fringant C, Moro L. 2001a. Plasticized starch-cellulose interactions in polysaccharide composites. *Polymer* 42(15):6571–6578.

Avérous L, Fringant C, Moro L. 2001b. Starch-based biodegradable materials suitable for thermoforming packaging. *Starch/Stärke* 53:368–371.

Avramova N, Fakirov S. 1990. Study of the healing process in native cellulose. *Makromol Chem Rapid Commun* 11:7–10.

Bastioli C. 2000. Global status of the production of biobased packaging materials. In: Conference Proceedings. The Food Biopack Conference; 2000 August 27–29; Copenhagen. p. 2–7.

Battista OA. 1975. *Microcrystal Polymer Science*. New York: McGraw-Hill.

Briassoulis D. 2004. An overview on the mechanical behaviour of biodegradable agricultural films. *J Polym Environ* 12(2):65–81.

Cao X, Deng R, Zhang L. 2006. Structure and properties of cellulose films coated with polyurethane/benzyl starch semi-ipn coating. *Ind Eng Chem Res* 45:4193–4199.

Chen Y, Ishikawa Y, Zhang ZY, Maekawa T. 2003. Properties of extruded acetylated starch plastic filled with cellulose acetate. *Trans Am Soc Agric Eng* 46:1167–1174.

Choi EJ, Kim, CH, Park JK. 1999. Structure-property relationship in PCL/starch blend compatibilized with starch-g-PCL copolymer. *J Polym Sci Part B: Polym Phys* 37:2430–2438.

Corti A, Vallini G, Pera A, Cioni F, Solaro R, Chiellini E. 1992. Composting microbial ecosystem for testing the biodegradability of starch-filled polyethylene films. Proceedings of the Second International Scientific Workshop on Biodegradable Polymers and Plastics, Montpellier, France. p. 245–248.

Curvelo AAS, De Carvalho AJF, Agnelli JAM. 2001. Thermoplastic starch–cellulosic fibers composites: Preliminary results. *Carbohydr Polym* 45:183–188.

Dufresne A, Dupeyre D, Vignon MR. 2000. Cellulose microfibrils from potato tuber cells: processing and characterization of starch–cellulose microfibril composites. *J Appl Polym Sci* 76:2080–2092.

Ebeling T, Paillet M, Borsali R, et al. 1999. Shear-induced orientation phenomena in suspensions of cellulose microcrystals, revealed by small angle X-ray scattering. *Langmuir* 15:6123–6126.

Engelhardt J. 1995. Sources, industrial derivatives and commercial application of cellulose. *Carbohydr Eur* 12:5–14.

Fang J, Fowler P. 2003. The use of starch and its derivatives as biopolymer sources of packaging materials. *Food Agric Environ* 1(3&4):82–84.

Ganjyal GM, Reddy N, Yang YQ, Hanna MA. 2004. Biodegradable packaging foams of starch acetate blended with corn stalk fibers. *J Appl Polym Sci* 93:2627–2633.

Glasser WG, Becker U, Todd JG. 2000. Novel cellulose derivatives. Part VI. Preparation and thermal analysis of two novel cellulose esters with fluorine-containing substituents. *Carbohydr Polym* 42:393–400.

Gómez-Carracedo A, Alvarez-Lorenzo C, Gómez-Amoza JL, Concheiro A. 2003. Chemical structure and glass transition temperature of non-ionic cellulose ethers. DSC, TMDSC®, Oscillatory rheometry study. *J Therm Anal Calorim* 73:587–596.

Hon DNS. 2006. Cellulose: chemistry and technology. *Encyclopedia of Materials: Science and Technology*. Elsevier, pp. 1039–1045.

Kennedy JF, Phillips GO, Wedlock DJ, Williams PA. 1985. *Cellulose and Its Derivatives*, 1st ed. Ellis Horwood, Chichester. p. 3.

Kerr RW. 1950. Occurrence and varieties of starch. In: *Chemistry and Industry of Starch*, 2nd ed. New York: Academic Press. Chapter I. p. 3–25.

Lee SY, Cho MS, Nam JD, Lee Y. 2006. Melting processing of biodegradable cellulose diacetate/starch composites. *Macromol Symp* 242:126–130.

Limwong V, Sutanthavibul N, Kulvanich P. 2004. Spherical composite particles of rice starch and microcrystalline cellulose: A new coprocessed excipient for direct compression. *AAPS PharmSciTech* 5(2):article 30.

Liu L. 2006. Bioplastics in food packaging: Innovative technologies for biodegradable packaging. San Jose State University, February, 2006. [Online: www.iopp.orgpagesindex. cfm?pageid 1160].

Lourdin N, Della Valle G, Collona P, Poussin P. 1999. Polyméres biodégradables: mise en oeuvre et propértiés de l'amidon. *Caoutchoucs et Plastiques* 760:39–42.

Lu Y, Weng L, Cao X. 2005. Biocomposites of plasticized starch reinforced with cellulose crystallites from cottonseed linter. *Macromol Biosci* 5:1101–1107.

Lu Y, Weng L, Cao X. 2006. Morphological, thermal and mechanical properties of ramie crystallites—reinforced plasticized starch biocomposites. *Carbohydr Polym* 63:198–204.

Mathew AP, Dufresne A. 2002. Morphological investigations of nanocomposites from sorbitol elasticised starch and tunicin whiskers. *Biomacromolecules* 3(3):609–617.

Mohanty AK, Misra M, Hinrichsen G. 2000. Biofibres, biodegradable polymer and composites: An overview. *Macromol Mater Eng* 276/277:1–24.

Orford PD, Ring SG, Carroll V, Miles MJ, Morris VJ. 1987. The effect of concentration and botanical source on the gelation and retrogradation of starch. *J Sci Food Agric* 9:169–177.

Parker R, Ring SG. 1996. Starch structure and properties. *Carbohydr Eur* 15:6–10.

Poutanen K, Forsell P. 1994. Modification of starch properties with plasticizers. *Trends Polym Sci* 4:128–132.

Psomiadou E, Arvanitoyannis I, Yamamoto N. 1996. Edible films made from natural resources; microcrystalline cellulose (MCC), methylcellulose (MC) and corn starch and polyols—Part 2. *Carbohydr Polym* 31:193–204.

Radley JA. 1953. Some physical properties of starch. In: Gray AG, editor. *Starch and Its Derivatives*, 3rd ed. London: Chapman & Hall. p. 58–80.

Ramkumar D, Vaidya UR, Bhattacharya M, Hakkarainen M, Albertsson AC, Karlsson S. 1996. Properties of injection moulded starch/synthetic polymer blends—I. Effect of processing parameters on physical properties. *Eur Polym J* 32:999–1010.

Rexen F, Petersen B, Munck L. 1988. Exploitation of cellulose and starch polymers from annual crops. *Trends Biotechnol* 6:204–205.

Richardson S, Gorton L. 2003. Characterisation of the substituent distribution in starch and cellulose derivatives. *Anal Chim Acta* 497:27–65.

Sperling LH. 2006. Introduction to physical polymer science. Chapter 1. *Introduction to Polymer Science*, 4th ed. New York: Wiley.

Taba Ohara M, Sakuda T, Cruz M, Ferreira E, Korolkovas A. 1995. Antimalarial activity of oxidized starch and cellulose imine derivatives. *Boll Chim Farm* 134:522–527.

Vilpoux O, Avérous L. 2004. Starch-based plastics in technology, use and potentialities of Latin American starchy tubers. In: Cereda MP, Vilpoux O, editors. *Collection Latin American Starchy Tubers.* São Paolo, Brazil: NGO Raízes and Cargill Foundation. Book no. 3. Chapter 18. p. 521–553.

Wibowo AC, Misra M, Park HM, Drzal LT, Schalek R, Mohanty AK. 2006. Biodegradable nanocomposites from cellulose acetate: Mechanical, morphological, and thermal properties. *Composites: Part A*, 37:1428–1433.

Yu L, Petinakis S, Dean K, Bilyk A, Wu D. 2007. Green polymeric blends and composites from renewable resources. *Macromol Symp* 249–250:535–539.

Electronic Sources

URL 1. http://ms-biomass.org/conference/2004/rebirth%20of%20bio-based%20polymer%20development.ppt (accessed 28 July 2007)

URL 2. http://www.degradable.net/downloads/ics_info_pack_2001.pdf (accessed 25 July 2007)

URL 3. http://aida.ineris.fr/bref/bref_text/breftext/anglais/bref/BREF_tex_gb40.html (accessed 30 July 2007)

URL 4. http://ift.confex.com/direct/ift/2000/techprogram/paper_4379.htm (accessed 30 July 2007)

URL 5. http://www.indianplasticportal.com/cp.html (accessed 28 July 2007)

URL 6. http://www.patentstorm.us/patents/5446140-fulltext.html (accessed 25 July 2007)

CHAPTER 3

Starch–Sodium Caseinate Blends

IOANNIS S. ARVANITOYANNIS and PERSEFONI TSERKEZOU

Department of Agriculture, Icthyology and Aquatic Environment, School of Agricultural Sciences, University of Thessaly, Fytokou Str., 38446 Nea Ionia Magnesias, Volos, Greece

3.1 INTRODUCTION

The continuously increasing interest of consumers in quality, convenience, and food safety has encouraged further research into edible films and coatings (Krochta et al., 1994; Herald et al., 1995). The reinvention of "edible films" was due mainly to their numerous applications such as coatings for sausages, fruits and vegetables, chocolate coatings for nuts, and occasionally wax coatings (Camirand et al., 1992; Park et al., 1994). Although the use of edible films has multiple objectives, among the most important may be considered restriction of moisture loss, control of gas permeability,

Biodegradable Polymer Blends and Composites from Renewable Resources. Edited by Long Yu
Copyright © 2009 John Wiley & Sons, Inc.

control of microbial activity (e.g., chitin has an antimicrobial action), preservation of the structural integrity of the product, and gradual release of enrobed flavors and anti-oxidants in food (El Ghaouth et al., 1991; Wong et al., 1992). Edible films are primarily composed of polysaccharides, proteins and lipids, alone or in combination. Such films should possess suitable gas and aroma barrier and mechanical properties, to protect foodstuffs from deterioration. Combining polysaccharides and proteins in the form of blends or layers, with varying ratios of polymers, offers the possibility of creating different films with improved characteristics (Kristo et al., 2007).

Starch constitutes more than 60% of cereal kernels and is relatively easy to separate from the other chemical components. The other components, such as fibers, protein and fat, have a market value of their own as food or feed (Rexen et al., 1988). Starch (starch : water 30% w/w) was rendered amorphous by gelatinizing and roller drying (in one process) at 140°C, pressure 32 psi (221 kPa), and width 150 mm (Arvanitoyannis et al., 1996). Apart from some high-amylose and modified starches (crosslinked, methylated), starch in general does not form tough, pliable, and unsupported films. Nevertheless, it has found extensive use in the formulation of adhesives and sizings to improve the mechanical properties of items such as paper and textiles, in which the film-forming ability is of great importance (Young, 1984). Starch is the polysaccharide most commonly used to produce the protein–polysaccharide gels used in the food industry. In the dairy industry, the use of starch in the formulation of some dairy products has improved their texture and viscosity and promoted significant cost savings. Interactions between starch and milk proteins are anticipated to greatly affect the rheological properties of food systems (Lelievre and Husbands, 2002), markedly changing the gel network structure and the rheological profile (Goel et al., 1999).

Casein and casein derivatives have found many applications in the food and pharmaceutical industries (Southward, 1989). The unique characteristics of milk proteins make them excellent candidates for incorporation into edible films and coatings to control mass transfer in food systems (Avena-Bustillos and Krochta, 1993). Sodium caseinate was selected for these experiments because it is commercially available, has satisfactory thermal stability, and can easily form films from aqueous solutions due to its random coil nature and its ability to form extensive intermolecular hydrogen bonds, electrostatic bonds, and hydrophobic bonds (Arvanitoyannis and Biliaderis, 1998). The physical properties of edible films based on blends of sodium caseinate with starches of different origin (corn and wheat) plasticized with water, glycerol, or sugars were also reported by Arvanitoyannis et al. (1996). Casein and casein derivatives vary in molecular weight (19,000–23,900) and have been used extensively in the food industry (dairy, meat, and confectionery) and in medical and pharmaceutical applications (URL1). Sodium caseinate readily forms films, owing to its high water solubility, its random-coil structure, and its capacity to form chain aggregates via electrostatic forces, van der Waal's forces, and hydrophobic interactions (McHugh and Krochta, 1994). Despite the fact that starch and casein are involved in many food formulations, very few studies have focused on their interactions. The understanding of caseinate–starch interactions is an important tool that could be used to manipulate the rheological properties of food systems and to provide alternatives in the development of new products (Bertolini et al., 2005).

3.2 STARCH AND STARCH DERIVATIVES

Starch is mostly water soluble, difficult to process, and brittle when used without the addition of a plasticizer. Furthermore, its mechanical properties are very sensitive to moisture content, which is difficult to control. In principle, some properties of starch can be significantly improved by blending it with synthetic polymers (Dufresne et al., 2000). Starch alone cannot form films with satisfactory mechanical properties (high percentage elongation and tensile and flexural strength) unless it is plasticized, blended with other materials, chemically modified, or modified with a combination of these treatments (Liu, 2006).

Starch derivatives include dextrins, starch esters, and starch ethers. The starch esters most commonly used are esters with phosphoric acid (phosphate starches) and acetic acid (acetyl starches). The three most important starch ethers are the hydroxyethyl, hydroxypropyl, and carboxymethyl starches. Starch derivatives based on esters can be biologically reduced, while starches based on ethers are more difficult to biodegrade. However, the great variety of possible chemical modifications makes it possible to produce good slashing agents, which can be virtually completely biodegraded (URL2).

Starch is a highly hydrophilic material that contains anhydroglucose units linked by α-D-1,4-glycosidic bonds. The water absorbed by starch granules is mainly bound in the amorphous phase. At certain temperatures, shear stress, and pressure in an extruder, the gelatinization processing disrupts the crystalline and ordered structure in starch granules to produce an amorphous phase (Yu et al., 2007). Starch granules vary in size from about 2 to 150 mm (Kerr, 1950). Pure starch, as distinct from commercial starch, is a white, odorless, tasteless, neutral powder, insoluble in cold water or organic solvents (Radley, 1953). Most starch granules are composed of a mixture of two polymers (Ahmad et al., 1999), namely, amylose and amylopectin. The amylose content can vary over a broad range, from 0% to about 75%, but typically is 20–25% (w/w) (Parker and Ring, 1996). In addition, starch is easily hydrolyzed to fermentable substrates (Rexen et al., 1988).

3.2.1 Mechanical Properties

Dufresne et al. (2000) examined the mechanical behavior of an unfilled starch–glycerol matrix at room temperature as a function of glycerol content and relative humidity (RH). With increasing glycerol content, a decrease in the modulus was observed regardless of RH. This decrease was linear up to $\sim$15% glycerol. At a higher glycerol content, the drop in modulus drop increased strongly.

The tensile modulus of unfilled starch–glycerol matrix films ranged between 0.24 and 20 MPa, depending greatly on the water content. The 35% RH conditioned sample displayed a typical rubberlike behavior with a high strain at break

($\sim 140\%$). As the water content increased, starch tended to crystallize, and the elongation at break decreased strongly (up to 18% at 43% RH). No significant difference was reported for more moist specimens. The strength of the material was almost moisture independent and ranged between 0.25 and 1 MPa (Anglés and Dufresne, 2001).

Avérous and Boquillon (2004) investigated the thermal and mechanical behavior of two different plasticized starch matrixes [TPS$_1$: dried wheat starch (70 wt%), glycerol (18 wt%) and water (12 wt%); TPS$_2$: dried wheat starch (65 wt%) and glycerol (35 wt%)]. The elongation at break was 124% for TPS$_1$ and 60% for TPS$_2$. The maximum tensile stress was 3.6 MPa and 1.4 MPa for TPS$_1$ and TPS$_2$, respectively, while the tensile modulus was 87 MPa for TPS$_1$ and 12 MPa for TPS$_2$. It can be concluded that the less plasticized matrix shows higher mechanical behavior. Avérous et al. (2001b) reported that the plasticized wheat starch (PWS) tensile modulus and strength was 52 and 3 MPa, respectively, while the elongation at break was 126%. According to Avérous et al. (2001a), the tensile strength and modulus of TPS was 3.6 MPa and 87 MPa, respectively, and the elongation at break was 124%.

Starch acetate was blended with corn stalk fibers at different concentrations (0%, 2%, 6%, 10%, and 14% w/w) for the production of biodegradable extruded foams. The force required for shearing the extrudates increased with an increase in fiber content from 1% to 10%. There was a drastic increase in the force (0.06–0.075 N/mm^2) required to shear the extrudates with 14% fiber content (Ganjyal et al., 2004).

3.2.2 Thermal Properties

Native starch (NS) was extracted from natural corn grain or from alkaline temperature-treated corn (95°C/40 min in a 10% lime solution) (ATS). The NS and the ATS starches were stored (30°C) at different relative humidities (%RH) (11–84%) (URL3). In the experiment conducted by Avérous and Boquillon (2004), the glass transition temperature (T_g) (by differential scanning calorimetry [DSC] analysis) was 87°C for TPS$_1$ and 12°C for TPS$_2$. On the other hand, Avérous and Fringant (2001) found that the T_g values of various TPS formulations were below ambient temperature except for the 74% starch/10% glycerol/16% water blend.

The T_g and melting temperature (T_m) (by DSC) of the starches were determined at different storage times. The T_g of the starches in 80% water dispersion was between 60°C and 65°C. However, after the thermal treatment, T_g was increased (-14°C to 0°C) (URL3).

3.3 SODIUM CASEINATE DERIVATIVES

The effect of sodium caseinate (NaCas) on the rheological properties of starch pastes has also been studied (Kelly et al., 1995). Although NaCas was found to decrease the viscosity of potato starch, its addition to maize starch resulted in an increase in viscosity. These studies showed that, in distilled deionized water, NaCas increased the swelling properties of the starch granules. More recently, Bertolini et al. (2005)

studied the rheological properties of NaCas–starch gels. They reported that, when added to cassava, amylomais corn, waxy corn, wheat, or rice starch, NaCas increased the elastic modulus and the viscosity of the gel compared with the gel made using starch alone. However, when added to potato starch, NaCas decreased the viscosity of the NaCas/potato starch mixture. In the case of potato starch, although no clear explanation for the decrease in viscosity was proposed, the authors suggested that the role of minerals, such as phosphorus, calcium, and sodium, might be important and thus should be investigated (Noisuwan et al., 2008).

Caseinate films exhibit resistance to thermal denaturation and/or coagulation, which means that the protein film remains stable over a wide range of pH, temperature, and salt concentration (Kinsella, 1984). The structural and shear characteristics of mixed monolayers formed by an adsorbed NaCas film and a spread monoglyceride (monopalmitin or monoolein) on a previously adsorbed protein film have been analyzed. Measurements of the surface pressure (π)–area (A) isotherm and surface shear viscosity were obtained at 20°C and at pH 7 in a modified Wilhelmy-type film balance. The structural and shear characteristics depend on the surface pressure and on the composition of the mixed film. At surface pressures lower than the equilibrium surface pressure of NaCas, both NaCas and monoglyceride coexist at the interface, with structural polymorphism or a liquid expanded structure due to the presence of monopalmitin or monoolein in the mixture, respectively. At higher surface pressures, collapsed NaCas residues may be displaced from the interface by monoglyceride molecules. For a NaCas–monopalmitin mixed film, the η_s value varies greatly with the surface pressure (or surface density) of the mixed monolayer at the interface. In general, the greater the surface pressure, the greater are the values of η_s. However, the values of η_s for a NaCas–monoolein mixed monolayer are very low and practically independent of the surface pressure (Rodriguez Patino et al., 2007) (see Table 3-1).

3.3.1 Mechanical Properties

The maximum load (ML) and tensile strength (TS) of NaCas films decreased linearly with increasing glycerol. Films with no glycerol were found to be the strongest with a mean TS value of 37 MPa. Elongation (%E) increased with increasing glycerol content. The load–displacement curve for films manufactured without plasticizer showed an inverted "V" shape, i.e., a sharp increase in load followed by a sharp decrease after reaching the ML value. These films were only slightly extensible and shattered into small pieces at the end of the test. In contrast, films containing glycerol were better able to withstand the gradual increase in extension during the elongation test. The standard deviation for the determination of %E increased with increasing glycerol content. Increasing the glycerol content did not have a significant effect on the water vapor permeability (WVP) of the films, with the exception of the films with Gly : Pro ratio of 0.32. The latter had a significantly higher WVP than films with Gly : Pro ratio of 0.16 or less (Shou et al., 2005).

Tensile strength (TS) and percentage elongation at break (%E) were measured at $20 \pm 1°C$ with an Instron universal testing instrument according to AFNOR standard

TABLE 3-1   Sodium Caseinate Analysis

Analyses		References
Typical Analysis		
Protein (dry basis): Nx6.38	91.0%	URL4
Moisture	6.0%	
Fat	2.5%	
Ash	5.0%	
Carbohydrate	1.0%	
pH	6.6–7.2	
Typical Microbiological Analysis		
Standard plate count	<20,000/g	URL4
Coliform	<10/g	
Escherichia coli	Negative	
Yeast/mold	<100/g	
Salmonella	Negative	
S. aureus	Negative	
Typical Amino Acid Content (g/100 g of protein)		
Alanine	2.6	URL5
Arginine	3.6	
Aspartic acid	6.3	
Cystine	0.3	
Glutamic acid	20.0	
Glycine	2.4	
Histidine	2.7	
Isoleucine	5.4	
Leucine	8.2	
Lysine	7.3	
Methionine	2.5	
Phenylalanine	4.4	
Proline	10.1	
Serine	5.6	
Threonine	4.3	
Tryptophan	1.1	
Tyrosine	5.6	
Valine	6.4	

method NF T 54-102 (AFNOR, 1971). Five specimens were cut from each precondi-tioned film (50% RH, 20°C) and uniaxially stretched at a constant rate of 100 mm/ min until breaking. Tests were carried out over a period of about 10 min to minimize exposure of the samples to the ambient environment.

The general trend of the stress–strain curve of triethanolamine (TEA)-plasticized samples crosslinked by 1% formaldehyde is similar to the shape of that for the noncrosslinked samples. For TEA-plasticized films crosslinked with 2–10% w/w formaldehyde (HCHO/ε-NH$_2$ = 1.35–6.75; ε-NH$_2$ = free amino groups), the stresses at break, about 8–9 MPa, were more than twice the values of TEA/NaCas films without formaldehyde and ultimate strain values were up to 110–125%. Compared with only plasticized samples, small amounts (1% w/w; HCHO/ε-NH$_2$ = 0.67) of HCHO did not initiate major changes in elastic modulus (about 46 MPa in both cases). For HCHO/ε-NH$_2$ ratios ranging from 1.35–6.75, the Young's modulus increased to about 100–110 MPa, which is more than twice the value of TEA/NaCas films without HCHO. The mechanical properties of crosslinked plasticized NaCas films increased with the concentration of crosslinking agent. An HCHO/ε-NH$_2$ ratio of 1.35 seemed to be optimal for enhancing mechanical properties of plasticized caseinate films. Although the mechanical properties dependent on the ratio of crosslinker to protein in a limited range, further crosslinking does not seem to affect the film performance from a mechanical point of view (Audic and Chaufer, 2005).

3.3.2 Thermal Properties

Differential scanning calorimetry (DSC) measurements of the starch samples were performed with a Pyris-1 Thermal Analytical System (Perkin-Elmer, Norwalk, CT, USA). Samples were weighed in steel pans and distilled water or NaCas solution was added (10 mg of starch, dry basis, to 20 μL of water or NaCas solution) before sealing. The sealed pans were equilibrated overnight at 4°C and then heated from 25 to 180°C at a heating rate of 3°C/min. Indium was used for calibration, and an empty pan was used as a reference. Enthalpy changes (ΔH), the gelatinization onset temperature (T_{onset}), the gelatinization range ($T_2 - T_1$), and the peak temperature (T_{peak}) were calculated automatically. Statistical analyses of RVA (rapid visco analyzer) and DSC data were assessed considering starch, NaCas, and their interactions. The RVA data were analyzed for peak viscosity, final viscosity, and peak temperature. The DSC data were analyzed for enthalpy, temperature of gelatinization, peak temperature, peak start temperature, and peak end temperature (Bertolini et al., 2005).

Pullulan, NaCas, and their mixtures in different ratios were dissolved in distilled water under continuous stirring, to obtain casting solutions of 4% (w/w) concentration. Sorbitol at a concentration of 25% dry basis (db), was added as a plasticizer to the polymer solutions and this concentration was kept constant throughout the study. Such a concentration of sorbitol was necessary because films containing NaCas were very difficult to handle without breaking when lower plasticizer levels were used. The solutions were subsequently filtered to remove any undissolved material and vacuum degassed to remove air bubbles, and 20 g samples of the solution was cast on plastic frames. The frames were then stored in an oven at 35°C, allowing them to dry slowly (Kristo et al., 2007).

The glass transition determination was also carried out with DSC; the T_g was determined as the middle temperature of the observed change in heat capacity from the second DSC scan to eliminate the effects of sample history (Kalichevsky and Blanshard, 1993). However, in some cases, it was difficult to determine the T_g accurately from the DSC scans, particularly for samples containing NaCas, because with this material the T_g value was often smeared out over a broad temperature range. Also, when low temperatures were involved, in some cases it was impossible to determine the T_g values (e.g., because of ice interference). These difficulties, mainly related to proteins and foodstuffs, are often reported in the literature (Le Meste et al., 1996).

3.4 STARCH–SODIUM CASEINATE BLENDS

Protein-based films were cast from film-forming solutions of wheat gluten, corn zein, egg albumin, and NaCas (Rhim et al., 1999). Edible films based on soluble starch/NaCas blends, plasticized with water and/or other polyols, were studied with regard to their thermal, mechanical, and gas and water vapor barrier properties. Macroscopically examined, the films appeared to be clear, suggesting a homogeneous distribution of the two or three components in the blend. At a total plasticizer content lower than 17%, both thermal and mechanical properties of the blends remained within acceptable limits for packaging or coating applications. A marked change was observed in the mechanical properties at higher than 17% total plasticizer content, possibly due to a disruption, at least in part, of the soluble starch/NaCas amorphous matrix (Arvanitoyannis and Biliaderis, 1998).

The shear characteristics of pure and caseinate–monoglyceride mixed films were sensitive to the composition of the mixed film and to the surface pressure. This is due to lateral interactions between the proteins due to hydrogen bonding, hydrophobic and covalent bonding, and/or electrostatic interactions (Rodriguez Patino and Carrera, 2004). These interactions between adsorbed protein molecules may vary strongly. Moreover, for a fully packed adsorbed layer (at higher surface pressures), the deformability (i.e., the mechanical properties) of the protein molecules may be an important factor (Gunning et al., 2004). Thus, differences between η_s values for various proteins (caseinate, β-casein, κ-casein, etc.) are quite large and reflect differences in, among other things, the protein structure and the potential for the formation of disulfide crosslinks and the formation of interfacial aggregates of significant sizes. The values of η_s were higher for proteins that form interfacial gels by crosslinking of disulfide residues and the formation of interfacial aggregates of significant size (κ-casein and caseinate, in that order) compared with β-casein, which can only form physical gels stabilized by intermolecular hydrogen bonds (Rodriguez Patino and Carrera Sanchez, 2004). The differences in shear rheology observed between monoglyceride and caseinate monolayers justify the use of this technique to analyze the interfacial characteristics (structure, interactions, miscibility, squeezing out phenomena, etc.) of caseinate–monoglyceride mixed films at the air–water interface and thus allow a global consistency check on the same monolayer.

Caseinate exhibits thermoplastic and film-forming properties due to its random coil nature and its ability to form weak intermolecular interactions—hydrogen bonds, electrostatic bonds, and hydrophobic bonds (Creighton, 1984). These properties make sodium caseinate an interesting raw material for several applications as substitutes for traditional synthetic polymers. For their transparency, biodegradability, and good technical properties (high barrier for gases such as oxygen) (Chick and

Ustunol, 1998), caseinate-based films can find applications in packaging (Krochta et al., 1990), in edible or protective films and coatings (Chen, 1995), or in mulching films (De Graaf and Kolster, 1998). Such films are easily obtained by casting aqueous solutions of sodium caseinate. The objective has been to achieve caseinate-based films with improved properties as close as possible to those of available packaging films based on synthetic polymers such as polyethylene (Chen, 1995) or plasticized PVC (Audic et al., 2001). Such films show good elongation at break (from 150% to about 400%) and a rather low tensile strength ranging from 20 to 30 MPa. However, compared with synthetic films, protein-based films have two major drawbacks (Mauer et al., 2000; Grevellec et al., 2001): poor mechanical properties (lower tensile strength and elongation at break); and high water sensitivity (i.e., high water solubility and water vapor permeability).

Edible films were cast from solutions of sodium caseinate (NaCas) and from emulsions of this protein with anhydrous milk fat (AMF). The moisture sorption isotherms, mechanical properties (tensile strength [TS] and elongation) and water vapor and oxygen permeabilities were determined for films based on NaCas as a function of AMF concentration. AMF concentration significantly affected TS ($P < 0.001$) and elongation ($P < 0.05$). The increase of lipid content led to a loss of mechanical efficiency but had little influence on water vapor barrier properties. There was no significant difference ($P > 0.05$) between oxygen permeability of films at each lipid concentration (Khwaldia et al., 2004a).

Bilayer films were prepared by successive casting of NaCas solutions onto preformed pullulan films. For bilayer films, different polymer ratios resulted in differences in polymer layers. A mixture–process variable experimental design was employed to investigate the effect of film composition on the tensile properties of blend and bilayer films comprising pullulan and NaCas and plasticized with 25% db sorbitol. Mixture experiments are a special class of response surface experiments in which the factors are the proportions of the components and the measured property response depends on such component ratios (Kristo et al., 2007).

Plasticized protein films were prepared by the casting method from water solution of NaCas and plasticizers with the aim of obtaining environmentally friendly materials for packaging applications. Mechanical properties (tensile strength, elongation, and Young's modulus) of caseinate-based films were determined versus ratio of protein to plasticizer, plasticizer type, and relative humidity conditions. Among the different polyol-type plasticizers tested, glycerol (Gly) and triethanolamine (TEA) were the most efficient for the improvement of mechanical properties (high strains for low stresses). Further, chemical crosslinking between formaldehyde and free amino groups of sodium caseinate was done to increase water resistance of TEA-plasticized films (Audic and Chaufer, 2005).

3.4.1 Mechanical Properties

Edible films, from wheat and corn starch with NaCas plasticized by water or several sugars or glycerol, were tested with regard to their compatibility, mechanical strength, and gas and water vapor permeabilities. Most films, with water contents $< 15\%$ w/w, had sufficient strength and acceptable gas barrier properties and showed no phase separation. When the water/sugar content exceeded the threshold of 15% total plasticizer content, both thermal and mechanical properties showed a pronounced change because of the disruption of the starch/NaCas matrix (Arvanitoyannis et al., 1996).

It is noteworthy that similar tensile strengths were obtained for films cast from corn starch and from a solution containing 75%, by weight, amylopectin and 25%, by weight, amylose (Young, 1984). However, the mechanical properties measured by other researchers (Warburton et al., 1993) for maize grit extrudates were considerably higher (10–20 times). Such variations were explained by differences in density as well as the size and number of cavities present (Lourdin et al., 1997). Further work has to be conducted with microscopy or image analysis in order to clarify the nature of these differences.

In plasticized starches, the efficiency of plasticization with the same amount of plasticizer is dependent on the amylose/amylopectin ratio (Arvanitoyannis et al., 1996). Sorbitol has almost the same effect as glycerol on the mechanical properties of corn starch. It should be noted, however, that the increase in elongation was slightly greater than that caused by incorporation of glycerol. A possible explanation might be the higher hydroxyl number of sorbitol.

Another factor that should be borne in mind is the effect of developed crystallinity during storage of the films. It is well known from synthetic polymers that high crystallinity, in conjunction with chain orientation, has a positive influence upon the tensile strength, whereas the gas diffusion and permeation are substantially reduced (Arvanitoyannis and Blanshard, 1993; Arvanitoyannis et al., 1994). The starch-based films appeared to have adequate strength, especially when the amylopectin content was low since the branched molecules showed very limited orientation and packing ability compared to amylose (Young, 1984). Novamont tried to modify starch (within the frame of an innovatory R&D program) into a mechanically superior form devoid of crystallinity (not detectable with X-rays and dynamic thermal analysis [DTA]) and marketed several products possessing good mechanical properties (Bastioli, 1995). The upward curvature of the isotherms also suggests the formation of water clusters in the polymer matrix at high water activities. Indeed, the concomitant plasticization and swelling of the polymer matrix as the moisture content of the film increased provided more binding sites for water sorption (Lim et al., 1999). Several empirical and semiempirical equations have been proposed for the correlation of the equilibrium moisture content with the water activity of food products. Among them the Guggenheim–Anderson–de Boer (GAB) equation has been applied successfully to various foods (Van Den Berg, 1985). It is a relatively simple model with a small number of parameters that have physical meaning, and can be applied for water activities from 0 to 0.9 (Weisser, 1985).

The thermomechanical behavior of pullulan and NaCas films, either alone or with sorbitol (plasticized at 15% and 25% db level), was studied by dynamic mechanical thermal analysis (DMTA). Representative DMTA traces of plain pullulan and NaCas thick-film specimens containing or not containing sorbitol and conditioned at RH 64%. Similar behavior was observed for the samples equilibrated at other water activities. Usually the position of the tan δ peak or of the onset of E_0 drop is used as definition of glass transition temperature. The rationale behind such behavior is the substantially different structures of the two polymers (Kristo and Biliaderis, 2006).

Pullulan solids mainly exist in an amorphous state due to the presence of linkages on the polysaccharide chain, which allow molecular motions to take place around the three inter-residue bonds and provide extra distance between residues (due to the third glycosidic bond); the latter lead to an apparent lack of direct inter-residue hydrogen bonding (Biliaderis et al., 1999). For the protein component (NaCas), a chain rigidification mechanism may be provided by the electrostatic interactions involving the sodium cations (Gidley et al., 1993). The addition of sorbitol drastically diminished the T_g of pullulan, whereas it brought about a slight increase in T_g of NaCas. Furthermore, the storage modulus of the rubbery state decreases with increasing sorbitol content in both pullulan and NaCas films.

The Gordon–Taylor (G-T) equation was applied in a purely empirical sense to describe the water content dependence of the T_g of all the systems studied (Lazaridou and Biliaderis, 2002). It is obvious that the pullulan and sodium caseinate samples free of sorbitol exhibited increased sensitivity to plasticization at moisture contents below 8–10% (Gontard and Ring, 1996). Moreover, lower slopes of the G-T curves were observed when sorbitol was added, which is in agreement with the findings of Lourdin et al. (1997), who assumed a reduction of the amount of water available for efficient plasticization due to the presence of a second plasticizer.

The plasticizing effect of water was apparent in all cases, but pullulan samples exhibited greater T_g depression than the NaCas films at similar sorbitol and moisture content levels. The significantly lower plasticizing effect of water on NaCas than on pullulan could probably be explained by the greater hydrophobicity of the protein (Kalichevsky et al., 1993). Even though the water uptake of both polymers was very similar (as was shown from moisture sorption isotherms), part of the water could be preferably associated with sodium ions (in NaCas), allowing an uneven water distribution within the protein structure. The addition of sorbitol brought about a notable decrease of T_g for both pullulan and NaCas at low moisture content ($<10\%$ w/w). However, a different effect of sorbitol on these polymers was observed at higher water contents. Thus, the T_g of the sorbitol-plasticized pullulan samples continued to decline and approached the T_g of the polyol-free sample at equivalent water content, which concurs with the findings of Diab et al. (2001).

On the other hand, addition of sorbitol enhanced the T_g of sodium caseinate at water contents $>10\%$ (w/w). Also, Kalichevsky et al. (1992) observed an increase in T_g of gluten at higher water contents in the presence of fructose, glucose, and sucrose and attributed this behavior to an unequal distribution of water between polymer and sugar or to an increase of hydrophobic interactions in the protein structure. The last effect could arise from the ability of various sugars and polyols to stabilize proteins by increasing the structure of water, leading to intensification of hydrophobic interactions and stabilization of the protein conformation (Back et al., 1979). Kilburn et al. (2005) proposed two distinct mechanisms of polysaccharide plasticization by water and low molecular weight sugars. They assumed that water molecules form hydrogen bonds with the hydroxyl residues on the polymer chains, interrupting the polymer–polymer interactions and allowing the polymer chains to come apart. In contrast, the plasticizing action of the low molecular weight sugars is attributed to a strong reduction in molecular entanglements upon a shift of the "apparent" molecular weight distribution to lower values.

The low-temperature transition for both plain polymers at very low water content was close to but somewhat higher than the T_g of sorbitol alone, indicating probably the existence of a sorbitol-rich phase whose T_g was increased by the polymer domains (Moates et al., 2001) due to an increase in the average molecular weight (Cherian et al., 1995). Similarly, Kalichevsky et al. (1992) for gluten–sugar mixtures, Kalichevsky et al. (1993) for sodium caseinate–fructose systems, and Kalichevsky and Blanshard (1993) for amylopectin–fructose samples assigned the low-temperature transition to fructose and supported the conclusion given by Kolarik (1982) that such a type of low-temperature transition is frequently observed in compatible polymer–diluent mixtures, not necessarily implying phase separation but being due to the onset of short-range motions.

The mechanical properties of various starch–NaCas blends are shown in Table 3-2.

3.4.2 Thermal Properties

Considering that the properties of film-forming solutions must be reflected in the behavior of the solid film, the glass transition temperature, which can be interpreted as the range of temperatures at which segment motion of macromolecules becomes thermally activated, was analyzed. The T_g of proteins increases with the chain rigidity and the intensity of both inter- and intramolecular interactions, including hindrance to internal rotation along the macromolecular chain. An effective plasticizer has to shield the intermolecular and intramolecular interactions, facilitating the molecular mobility and decreasing the internal friction in the biopolymer material (Kalichevsky et al., 1993). The presence of sorbitol had a significant plasticizing effect on the NaCas film and the observation of a single glass transition reflects compatibility between sorbitol and sodium caseinate.

The interaction and, in consequence, the plasticizing effect, could be attributed to the low molecular weight of sorbitol and the presence of hydroxyl groups, leading to the formation of NaCas–sorbitol interactions, despite the polymer–polymer interactions, increasing the intermolecular spacing as a result. The six hydroxyl groups of sorbitol might interact with lateral residues of NaCas amino acids through hydrogen bonds. These interactions would decrease the partial specific volume of the protein, allowing a greater backbone chain segmental mobility of the NaCas and leading to lower values of the glass transition temperatures (Barreto et al., 2003a).

The thermal degradation of edible films based on pure NaCas, whey, and gelatin, and on these proteins in the presence of sorbitol as a plasticizer, was studied by thermogravimetry and infrared spectroscopy in nitrogen atmosphere. The presence of sorbitol significantly reduced the activation energy of the degradation of the edible protein films. This behavior was in agreement with the decrease of the initial and maximum temperatures of degradation observed by thermogravimetry. The decrease in the thermal stability is apparently associated with the effect of sorbitol on the inter- and intramolecular hydrogen bonds of the proteins. The Fourier transform infrared

TABLE 3-2 Mechanical Properties of Starch–Sodium Caseinate Blends

Components	% Elongation	Tensile Strength (MPa)	Young's Modulus (MPa)	References
85% Corn starch / 0% sodium caseinate / 15% water	6.2 ± 0.4	38.3 ± 2.1	23.8 ± 2.4	Arvanitoyannis et al. (1996)
80% Corn starch / 5% sodium caseinate / 15% water	6.9 ± 0.5	33.4 ± 2.2	24.7 ± 2.1	
75% Corn starch / 10% sodium caseinate / 15% water	9.0 ± 0.7	30.7 ± 1.9	25.6 ± 1.8	
70% Corn starch / 15% sodium caseinate / 15% water	11.8 ± 1.1	26.9 ± 1.6	26.4 ± 1.9	
60% Corn starch / 25% sodium caseinate / 15% water	14.9 ± 1.2	20.2 ± 2.2	28.0 ± 2.3	
55% Corn starch / 30% sodium caseinate / 15% water	19.2 ± 1.7	17.1 ± 1.3	28.8 ± 2.5	
45% Corn starch / 40% sodium caseinate / 15% water	28.0 ± 2.5	13.3 ± 1.2	29.5 ± 2.2	
47.5% Corn starch / 47.5% sodium caseinate / 5% water	30.2 ± 2.6	22.0 ± 0.4	30.1 ± 2.7	
45% Corn starch / 45% sodium caseinate / 10% water	32.6 ± 2.5	21.4 ± 0.2	29.0 ± 2.5	
42.5% Corn starch / 42.5% sodium caseinate / 15% water	34.0 ± 1.8	20.7 ± 0.5	27.7 ± 1.9	
40% Corn starch / 40% sodium caseinate / 20% water	38.0 ± 3.2	17.8 ± 0.3	25.9 ± 2.3	
37.5% Corn starch / 37.5% sodium caseinate / 25% water	37.1 ± 1.9	13.0 ± 0.3	24.0 ± 2.2	

(Continued)

TABLE 3-2 *Continued*

Components	% Elongation	Tensile Strength (MPa)	Young's Modulus (MPa)	References
35% Corn starch / 35% sodium caseinate / 30% water	39.5 ± 0.6	11.1 ± 1.0	21.7 ± 2.4	Arvanitoyannis and Biliaderis (1998)
90% Soluble starch / 0% sodium caseinate / 10% water	2.5 ± 0.3	38.5 ± 3.4	–	
80% Soluble starch / 10% sodium caseinate / 10% water	4.4 ± 0.3	33.6 ± 4.3	–	
70% Soluble starch / 20% sodium caseinate / 10% water	7.2 ± 0.6	28.2 ± 3.1	–	
60% Soluble starch / 30% sodium caseinate / 10% water	10.0 ± 1.2	23.1 ± 2.2	–	
50% Soluble starch / 40% sodium caseinate / 10% water	13.6 ± 1.5	19.3 ± 2.1	–	
47.5% Soluble starch / 47.5% sodium caseinate / 5% water	2.0 ± 0.1	22.1 ± 1.2	–	
45% Soluble starch / 45% sodium caseinate / 10% water	6.3 ± 0.5	19.5 ± 2.2	–	
40% Soluble starch / 40% sodium caseinate / 20% water	13.0 ± 1.1	17.3 ± 2.1	–	
35% Soluble starch / 35% sodium caseinate / 30% water	19.0 ± 2.0	12.2 ± 1.1	–	
50% Whey powder / 50% sodium caseinate	64.5 ± 26.8	14.5	–	Cho et al. (2002)

60% Whey powder / 40% sodium caseinate	87.8 ± 13.5	7.9 ± 1.4	–
70% Whey powder / 30% sodium caseinate	128.0 ± 18.8	4.7 ± 1.3	–
50% Whey powder / 50% sodium caseinate	2.6 ± 0.7	26.8 ± 3.8	–
60% Whey powder / 40% sodium caseinate	4.5 ± 0.6	17.0 ± 0.9	–
70% Whey powder / 30% sodium caseinate	3.2 ± 0.7	14.0 ± 1.0	–
4% Sodium caseinate	62.95 ± 10.61	3.10 ± 0.22	–
8% Sodium caseinate	69.19 ± 10.04	3.27 ± 0.25	–
8% Whey protein isolate	22.46 ± 6.61	5.26 ± 0.54	–
12% Whey protein isolate	25.14 ± 6.76	5.42 ± 0.29	–
4% Gelatin	45.26 ± 3.6	5.69 ± 0.02	–
8% Gelatin	89.69 ± 3.21	6.60 ± 0.253	–

(FT-IR) spectra showed that the effective degradation began at $\sim300°C$ with the formation of gaseous products such as CO_2 and NH_3, suggesting that the reaction mechanism included at the same time the scission of the $C-N$, $C(O)-NH$, $C(O)-NH_2$, $-NH_2$ and $C(O)-OH$ bonds of the proteins. The suggested mechanism of reaction was supported by the high values of the activation energy ($E > 100$ kJ/mol), which are probably associated with a process that occurred by random scission of the chain (Barreto et al., 2003a).

Although binary systems, such as starch/water and starch/sugars, have been quite extensively studied (Trommsdorff and Tomka, 1995), a three-component system tends to become even more complicated with regard to the interactions between the different components. It was previously shown that the gelatinized starch–water interactions are strong and particularly localized at the early stages of sorption whereas, at high water contents ($>20\%$), a certain proportion of water exhibited liquidlike properties. Sodium caseinate was preferred over casein for use in the preparation of blends with starch because it has a higher water uptake than casein (3–7% higher when conditioned at the same RH). Therefore, water will be more homogeneously distributed throughout the entire blend mass, which may favorably influence the mechanical properties of the blend. The presence of water and sugars had a significant plasticizing effect on starch and on NaCas. According to the extent of the induced plasticization, the sugars could be classified in the order glycerol < sorbitol < xylose < sucrose.

A general observation for all DMTA and DTA traces is that the incorporation of sugars, in conjunction with the presence of water within the starch and/or the protein matrix, resulted in substantial broadening of the tan peak or of the step transition. The limited plasticizing effect of sugars such as sucrose and xylose on starch/ water blends, reported elsewhere (Kalichevsky et al., 1993) and attributed to the preferential hydration of sugar molecules, was not confirmed in our case. On the contrary, sucrose and xylose had an even more pronounced plasticizing effect than glycerol and sorbitol on starch/water blends provided that their content did not exceed 15%. The plasticization of the composite food matrix with polyols or glycerol could possibly be due to changes in the polymer network, mainly related to the creation of highly mobile regions, which allow even more pronounced moisture uptake (Cherian et al., 1995).

No double peaks were recorded below 15% sugar content, implying that the starch/NaCas blends do not phase separate. When the polyol content exceeded the threshold of 15%, double peaks were recorded as previously found for natural and synthetic polymers (Bazuin and Eisenberg, 1986). This behavior is indicative of component incompatibility in the blend (i.e., between starch and polyol). The previously observed high- and low-temperature peaks (recorded by dynamic mechanical analysis [DMA]) in the case of binary (gluten–glycerol or gluten–sucrose) or ternary (gluten–glycerol–sucrose/sorbitol) systems (Cherian et al., 1995) was not confirmed by our experiments (DTA and DMTA) involving starch–NaCas–plasticizer systems.

The interactions of molecules in the wheat starch–water or corn starch–water–sugar systems may be attributed to the interactions between hydroxyl groups of starch chains, starch–water, and starch–sugar molecules, but also to sugar–sugar or water–sugar interactions (Tolstoguzow, 1994). The possibility of hydrogen bonding of starch within the blends increases considerably with the introduction of comparatively small molecules such as water and sugars. Previous studies on protein–protein (casein–ovalbumin, casein–soybean globulin), water, and protein–polysaccharide systems have shown that the compatibility of the blend components is greatly affected by thermal treatment or their previous thermal history (Tolstoguzow et al., 1985). It was also found that the protein–polysaccharide systems are characterized by limited compatibility between their components, occasionally resulting in phase separation.

The influence of NaCas on the thermal and rheological properties of starch gels at different concentrations and from different botanical sources was evaluated. In NaCas–starch gels, for all starches with the exception of potato starch, the NaCas promoted an increase in the storage modulus and in the viscosity of the composite gel when compared with starch gels. The addition of NaCas resulted in an increase in the onset temperature, the gelatinization temperature, and the end temperature, and there was a significant interaction between starch and NaCas for the onset temperature, the peak temperature, and the end temperature. Microscopy results suggested that NaCas promoted an increase in the homogeneity of the matrix of cereal starch gels (Bertolini et al., 2005).

In DSC analysis, the samples containing NaCas showed a decrease in the enthalpy of gelatinization. Statistical analysis of DSC data showed that there is no significant interaction between starch and NaCas addition, suggesting that NaCas addition had the same effect in decreasing the enthalpy for all starches. In samples with NaCas addition, a shift in the gelatinization peak was observed, with increases in the onset temperature, the peak gelatinization temperature, and the end temperature, in agreement with the results obtained by Erdogdu et al. (1995) for casein–wheat starch samples.

Changes in the gelatinization onset and peak temperatures appeared to be greater in cereal starches, even if the changes did not always follow this trend. In contrast to the wheat and rice starch samples, the endothermic peak at 100°C (relative to the amylase–lipid complex in the cereal starch samples) was not observed in the NaCas–starch samples. However, it is not clear whether these results were, at least in part, caused by a dilution effect. It has been reported that the changes in the thermal behavior of starch caused by the addition of nonstarch polysaccharide solution to the starch is markedly high at starch/solvent ratios of 1 : 10 (or above) and that the end temperature increases as the nonstarch polysaccharide concentration increases in the system (Tester and Sommerville, 2003).

There has long been debate regarding the homogeneous distribution of water in binary and ternary systems, particularly at a microstructural level. Although several new methods such as CP-MAS and nuclear magnetic resonance (NMR) imaging have been employed, no conclusive evidence has yet been published regarding this phenomenon. Therefore, in our study we assume that our binary or ternary blends are characterized by a homogeneous distribution of water. In all blends examined, a single glass transition was detected for the polymeric constituents (starch–caseinate blends), presumably due to the close proximity of the T_s (Glass Transition

Temperature) values of the individual components and a similar plasticization response in the presence of water and other polyols. The T_g of our binary system based on soluble starch/NaCas decreased as water content increased due to plasticization of the polymer blend. In general, sorbitol, despite its greater molecular weight, exhibited a more pronounced plasticizing effect (greater temperature depression) than glycerol on the polymer matrix. This is in agreement with other findings on starch-based polymer blends (Arvanitoyannis et al., 1996; Psomiadou et al., 1996).

Pullulan, NaCas, and their mixture in a ratio of 1 : 1 (w/w) were dissolved in distilled water under continuous stirring to obtain casting solutions of either 4% (w/w) concentration for determination of the moisture sorption isotherms or 7.5% (w/w) for preparing specimens for DSC and DMTA. The polyol-containing samples were prepared by adding sorbitol (15% or 25% db) to the polymer solutions. The solutions were then filtered through a coarse glass filter to remove any unsolubilized material (usually less than 0.5% of the dry initial matter), vacuum degassed to remove air bubbles, and cast on plastic frames. They were subsequently stored in an oven at 35°C and allowed to dry slowly. Bilayer films were prepared by successive casting of NaCas solutions onto preformed pullulan films.

In addition to the plain systems of pullulan and NaCas, the study of the composite films produced by blending or laminating both polymers could be of great interest, because this approach exploits the advantages of the distinct functional characteristics of the film-former components. Thus, the behavior of blends and bilayers was also studied by means of DMTA and DSC in the context of comparison with the plain polymer systems. As expected, bilayer films containing 25% sorbitol showed two distinct peaks corresponding to pullulan and NaCas alone. Peaks were sharper than in blend samples and their locations were near those of pure polymers.

Surprisingly, only one tan δ peak located near to the peak of pullulan was observed in bilayer systems free of sorbitol or containing 15% polyol. This is probably due to the difficulty of assigning a peak after the first one, since the DMTA signal after the first peak was very noisy. The transition temperatures corresponding to the two distinct a-relaxations observed in blend and bilayers with 25% sorbitol were determined and fitted separately to the G–T equation (Kristo and Biliaderis, 2006).

The T_g values of starch–NaCas blends are presented in Table 3-3.

3.4.3   Irradiation and Gas Permeability

Gamma irradiation was found to be an effective method for the improvement of both barrier and mechanical properties of edible films and coatings based on calcium and sodium caseinates alone or combined with some globular proteins. Our current studies concern the influence of gamma irradiation on the physical properties of calcium caseinate–whey protein isolate–glycerol (1 : 1 : 1) solutions and gels used for film preparation (Ciesla et al., 2006).

Irradiation of solutions was carried out with cobalt-60 gamma rays at doses of 0 and 32 kGy. The increase in the viscosity of solutions was found after irradiation was connected to induce crosslinking. Lower viscosity values were detected, however, after heating of the solutions irradiated with a 32 kGy dose than after

TABLE 3-3   Glass Transition Temperature (T_g) of Starch–Sodium Caseinate Blends[a]

Components	T_g (°C)		
	DTA	DMTA	MTA
85% Corn starch / 0% sodium caseinate / 15% water	61.2 ± 1.3	62.0 ± 1.4	64.9 ± 1.4
80% Corn starch / 5% sodium caseinate / 15% water	62.9 ± 0.9	63.5 ± 0.8	67.2 ± 1.6
75% Corn starch / 10% sodium caseinate / 15% water	64.5 ± 1.2	65.2 ± 1.6	70.0 ± 1.3
70% Corn starch / 15% sodium caseinate / 15% water	66.0 ± 1.4	67.0 ± 1.3	73.8 ± 1.2
60% Corn starch / 25% sodium caseinate / 15% water	68.0 ± 0.9	68.6 ± 0.7	75.2 ± 1.4
55% Corn starch / 30% sodium caseinate / 15% water	69.1 ± 1.1	70.0 ± 1.2	78.3 ± 1.5
45% Corn starch / 40% sodium caseinate / 15% water	70.9 ± 1.5	71.5 ± 1.4	81.2 ± 1.6
47.5% Corn starch / 47.5% sodium caseinate / 5% water	122.0 ± 2.1	122.8 ± 2.2	134.5 ± 1.0
45% Corn starch / 45% sodium caseinate / 10% water	93.5 ± 1.8	94.7 ± 1.5	106.2 ± 1.7
42.5% Corn starch / 42.5% sodium caseinate / 15% water	71.3 ± 1.9	70.6 ± 1.0	86.4 ± 1.6
40% Corn starch / 40% sodium caseinate / 20% water	54.5 ± 1.5	55.2 ± 1.4	67.8 ± 1.5
37.5% Corn starch / 37.5% sodium caseinate / 25% water	40.4 ± 2.0	39.6 ± 1.6	54.5 ± 1.8
35% Corn starch / 35% sodium caseinate / 30% water	19 ± 1.5	20.4 ± 1.2	35.0 ± 1.7
78% Wheat starch / 2% sodium caseinate / 17% water	56.4 ± 0.8	54.9 ± 1.5	60.3 ± 1.7
71% Wheat starch / 12% sodium caseinate / 17% water	57.6 ± 0.6	58.2 ± 0.9	63.5 ± 1.8
65% Wheat starch / 18% sodium caseinate / 17% water	59.2 ± 1.1	59.9 ± 1.3	66.7 ± 1.3
58% Wheat starch / 25% sodium caseinate / 17% water	63.1 ± 1.3	62.0 ± 1.2	71.0 ± 1.5
51% Wheat starch / 32% sodium caseinate / 17% water	65.9 ± 0.9	67.2 ± 1.4	74.2 ± 1.6
43% Wheat starch / 40% sodium caseinate / 17% water	69.3 ± 1.2	70.4 ± 1.1	77.5 ± 0.9
28% Wheat starch / 55% sodium caseinate / 17% water	74.6 ± 1.1	76.2 ± 0.8	81.9 ± 1.4
48% Wheat starch / 48% sodium caseinate / 4% water	114.6 ± 2.1	112.5 ± 1.8	119.5 ± 1.8

(Continued)

TABLE 3-3 *Continued*

	T_g (°C)		
Components	DTA	DMTA	MTA
46% Wheat starch / 46% sodium caseinate / 8% water	95.6 ± 1.8	98.0 ± 1.5	107.5 ± 1.2
42.3% Wheat starch / 42.5% sodium caseinate / 15% water	66.8 ± 1.7	64.3 ± 0.9	75.2 ± 1.4
40% Wheat starch / 40% sodium caseinate / 20% water	49.7 ± 1.8	51.2 ± 1.7	59.4 ± 1.5
37% Wheat starch / 37% sodium caseinate / 26% water	33.2 ± 1.5	31.5 ± 1.3	41.6 ± 1.8
35% Wheat starch / 35% sodium caseinate / 30% water	14.4 ± 0.7	21.7 ± 1.4	21.7 ± 1.4

[a]Arvanitoyannis et al. (1996); Kristo and Biliaderis (2006).

heating of the nonirradiated ones. Differences in the structure of gels resulted in different temperature–viscosity curves recorded for the irradiated and the nonirradiated samples during heating and cooling. Creation of less stiff but better ordered gels after irradiation probably arises from reorganization of the aperiodic helical phase and β-sheets, in particular from increase of β-strands, as detected by FT-IR.

Films obtained from these gels are characterized by improved barrier properties and mechanical resistance and are more rigid than those prepared from the non-irradiated gels. The route of gel creation was investigated for control and irradiated samples during heating and the subsequent cooling (Ciesla et al., 2004). The irradiation of protein solutions also improved puncture strength of films prepared with potato starch, soluble potato starch, or sodium alginate additions (at a level of 50 g/kg of total proteins). Addition of potato starch did not influence the mechanical properties of films but significantly improved barrier properties. Addition of sodium alginate improved both puncture strength and barrier properties. Addition of sodium alginate to irradiated protein solution resulted in films showing the greatest improvement in properties. Gels prepared using irradiated protein solutions mixed with calcium salt were stronger than gels prepared using nonirradiated solutions due to the preferential binding of calcium ions to the crosslinked protein network (Ciesla et al., 2006).

Stronger films with improved barrier properties were prepared from irradiated solutions than from nonirradiated ones, as shown by the larger puncture strength and smaller water vapor permeability. Simultaneously, smaller deformation values and larger viscoelasticities indicate higher rigidity of the irradiated films compared to the nonirradiated ones. Statistical analysis showed that all the functional properties measured for all the irradiated films differ significantly from those determined for the control films. Puncture tests were carried out using a Stevens LFRA Texture Analyzer Model TA/100 (Texture Technology Corporation, Scarsdale, NY, USA) according to the method described by Gontard et al. (1992) and Brault et al. (1997).

A cylindrical probe (2 mm in diameter) was moved perpendicularly to the film surface at a constant speed (1 mm/s) until it passed through the film. Strength and deformation values at the puncture point were used to determine the hardness and deformation capacity of the film. Force–deformation curves were recorded. The puncture strength was related to the film thickness to avoid any variations associated with inhomogeneity in thickness. For this purpose, the measured value was divided by the film thickness (measured directly at the probe path). Viscoelastic properties were evaluated using the force relaxation curves. The same procedure was applied, but the probe was stopped and maintained at 3 mm deformation (Ciesla et al., 2006).

Control and ultraviolet (UV)-irradiated (253.7 nm, $51.8 \, J/m^2$) films were evaluated for tensile strength, total soluble matter, water vapor permeability, and Hunter L, a, and b color values. UV treatment increased the tensile strength of gluten, zein, and albumin films, suggesting the occurrence of UV radiation-induced cross-linking within film structures. For caseinate films, UV-curing did not affect tensile strength but substantially reduced total soluble matter. Small but significant decreases in total soluble matter were also noted for UV-treated zein and albumin films. UV irradiation reduced water vapor permeability of albumin films but did not affect water vapor permeability of the other types of films. Gluten, albumin, and caseinate films had increased yellowness as a result of UV treatment. In contrast, UV treatment decreased the yellowness of zein films, possibly due to destruction of zein pigments by UV radiation (Rhim et al., 1999).

The gas permeability (GP) measurements of plasticized starch/NaCas blends showed increases in GP proportional to the plasticizer content due to weakening of the intermolecular forces between adjacent polymeric chains, which facilitates chain mobility and redistribution of the originally existing voids (Arvanitoyannis, 1999). The GP of soluble starch/NaCas blends also increased proportionally to the total plasticizer content (water, glycerol, and sugars) (Arvanitoyannis and Biliaderis, 1999).

3.5 APPLICATIONS

Films were cut into dumbbell-shaped strips and stored at an appropriate RH for 10 days, to obtain films with different moisture contents. Film thickness was measured at three different points with a hand-held micrometer and an average value was calculated. Samples were analyzed with a TA-XT2i instrument (Stable Microsystems, Godalming, Surrey, UK) in the tensile mode, operated at ambient temperature and a crosshead speed of 60 mm/min. Young's modulus (E), tensile strength (r_{max}) and percent elongation at break (%EB) were calculated from load–deformation curves of tensile measurements. The measurements represented an average of at least eight samples. The moisture content of samples, after storage, was determined by drying at $110°C$ for 2 h (Kristo et al., 2007).

Optimal mechanical properties (elastic modulus of 105 MPa, tensile strength of 8–9 MPa for elongation at break about 110–125%) were obtained for HCHO/ε-NH$_2$ ratios higher than 1.35. Protein specific water solubility was determined from the 280 nm absorbance. For convenient crosslinker (HCHO) content, NaCas solubility can be lowered to less than 5% wt after 24 hours immersion in water (Audic and Chaufer, 2005).

The physical properties of thin films (25–30 mm) made from mixtures of NaCas and whey protein isolate (WPI) were investigated. Films were formed by mixing solutions of NaCas (2.5% w/w protein), plasticized with glycerol (NaCas–Gly) at a glycerol : protein ratio of 0.32, with WPI solutions (2.3% w/w protein), plasticized (WPI-Gly) at a glycerol : protein ratio 0.37. Tensile and water barrier properties of films formed from mixtures of NaCas–Gly and WPI-Gly were similar to those of films containing NaCas–Gly only. Films containing only WPI-Gly had higher maximum load and elastic modulus values than the mixed films. Increasing the NaCas–Gly content of the films from 25% to 100% greatly increased solubility. This increased film solubility may increase the number of food applications for protein-based films (Longares et al., 2005).

All mechanical tests were performed at 50% $\pm$ 5% RH and a temperature of 23 $\pm$ 2°C using an Instron Universal Testing Instrument Model 4301 (Canton, MA, USA) fitted with a 100 N static load cell. The films were cut into strips 25.4 mm wide and 130 mm long using a scalpel and mounted between cardboard grips (26 $\times$ 70 mm) using double-sided adhesive tape so that the final area exposed was 100 $\times$ 25.4 mm. A minimum of eight strips were prepared from each film. The tensile properties of the films were measured according to the standard testing method 882–95a (ASTM, 1985). The mounted film samples were clamped into the metal grips of the tensile geometry of the Instron and stretched at an overhead crosshead speed of 10 mm/min.

Dynamic mechanical thermal analysis of films was conducted with a Perkin Elmer apparatus DMA-7 (Norwalk, CT, USA) equipped with a cryogenic system fed with liquid nitrogen (Air Liquide, Aix-en-Provence, France). A variable-amplitude, sinusoidal mechanical stress was applied to the sample (frequency 1 Hz) to produce a sinusoidal strain of selected amplitude. The compression mode of deformation was chosen for use with the sample geometry. Temperature scans from −30 to 260°C were performed at a heating rate of 3°C/min. The furnace temperature was calibrated with indium (Perkin Elmer Standard) and was flushed with dry nitrogen gas during analysis to avoid hydration of the films. The samples were dehydrated before testing to avoid the plasticizing effect of the water content. Three samples were tested for each material. During analysis, the stored values were the storage modulus (E') and the loss tangent (tan δ). The glass transition temperature (T_g) was determined from the maximum of the tan δ peak (Khwaldia et al., 2004a).

The surface pressure (π) at equilibrium (surface pressure isotherm) and surface dynamic properties (dynamic surface pressure and surface dilatational characteristics) of diglycerol ester (diglycerol monocaprinate, diglycerol monolaurate, diglycerol monostearate, and diglycerol monooleate) and protein (NaCas) as emulsifiers, at different concentrations in the aqueous phase, were measured using tensiometry and a dynamic drop tensiometer, respectively. Alvarez Gomez and Rodriguez Patino (2007) have observed that at equilibrium the value of critical micelle concentration (CMC) decreases and the maximum surface excess (Γ_{max}) increases as the hydrocarbon chain length increases because the hydrophobic character of the lipid also increases. The presence of a double bond in the hydrocarbon chain also increases the value of CMC and decreases that of Γ_{max}. Caseinate presents higher adsorption efficiency but the surface activity is between the values for lipids. The surface pressure isotherm of mixed systems is dependent on the emulsifier concentration and the protein/lipid ratio in the mixture. The adsorption of pure emulsifiers at the air–water interface increases with emulsifier concentration in the aqueous phase via diffusion and penetration of the emulsifier at the interface. For mixed films, the rate of adsorption depends on the concentration and composition of the mixture. Competitive or cooperative phenomena were observed during the adsorption of both emulsifiers at the interface. The surface dilatational characteristics of mixed films are viscoelastic. The surface dilatational modulus reflects the amount of emulsifier adsorbed at the interface and confirms the idea that the protein–lipid interactions at the air–water interface are somewhat weak, there even being the possibility of phase separation. The sorption curves of NaCas-based films have a sigmoid shape characteristic of water vapor-sensitive polymers such as cellulose (De Leiris, 1985) and wheat gluten (Gontard et al., 1993).

The curves showed a relatively slight slope at low water activity a_w. The exponential increase of equilibrium moisture contents with increasing a_w indicated that water sorption in the polymer did not follow Henry's law (Rogers, 1985); that is, the solubility of water in the polymer varies with the partial pressure of water vapor (Brown, 1992).

In general, the DSC transition temperatures (when observed) fell between the onset of the E_0 drop and the peak $\tan \delta$ (at 3 Hz) for an unplasticized pullulan sample containing 16.4% w/w water, which is in agreement with the observations of Kalichevsky and Blanshard (1993) on amylopectin/fructose and of Biliaderis et al. (1999) on pullulan/starch blends.

The viscosity increase is in fact associated with the protein–water interactions, leading to a higher solvent immobilization than the protein–protein interaction. Similar results were obtained by Konstance and Strange (1991) for solutions of casein and caseinates in the presence of various salt types and pH.

The effects of four milk protein ingredients, namely, skim milk powder (SMP), milk protein concentrate (MPC), sodium caseinate (NaCas) and whey protein isolate (WPI), on the pasting behavior of a 10% normal rice starch or waxy rice starch solution were assessed using rheological and DSC measurements and confocal scanning laser microscopy (CSLM). It was found that all these milk protein ingredients affected the pasting behavior of the two rice starches markedly and differently. For instance, for normal rice starch, SMP and NaCas shifted the temperature of peak viscosity to higher temperatures, whereas MPC and WPI reduced the temperature of peak viscosity. For waxy rice starch, SMP, NaCas and WPI increased the temperature of peak viscosity, but MPC did not affect it. The value of the peak viscosity was also either increased or decreased, depending on the type of milk protein and its concentration. The onset temperature measured by DSC correlated very well with that measured by rheology. However, there was no correlation between the peak temperature measured by DSC and that measured by rheology (Noisuwan et al., 2008).

Starches are widely used in dairy-based food products, such as yogurt, because of their thickening and gelling properties. They are also added to dairy-based products to achieve cost savings by reducing the amount of milk protein. A systematic study of mixtures of starches and different dairy ingredients, such as SMP or whey proteins, would greatly advance our understanding of how starches and milk proteins interact in typical dairy product formulations. Despite starches being extensively studied, little is found in the literature about their pasting behavior when mixed with milk proteins. Most studies involving starches and milk proteins are related either to the heat stability of milk or to the effect of starch on the gelation of milk proteins. Tziboula and Muir (1993) studied the effect of the addition of starches of various botanical origins to skim milk and skim milk concentrates. They reported that the heat coagulation time was decreased when the concentration of added starch was increased and that the heat stability of milk and milk concentrate containing starch could be improved by modifying the molecular properties of the starches by acid hydrolysis or by cross-linking. Matser and Steeneken (1997) investigated the rheological properties of highly crosslinked waxy maize starch in aqueous suspensions of skim milk components. It was found that, although the salts present in milk did not increase the final gel strength of the starch gels, lactose did affect the gelatinization temperature of the starch.

The differences in response of 1% potato and 4% maize starch pastes to inclusion of NaCas were investigated. Pasting of the starches was performed at 95°C for 1 hour over a range of concentrations of NaCas. Caseinate levels as low as 0.01% dramatically reduced the swelling volume of potato starch and hence the viscosity of the system. Since addition of sodium chloride shows similar effects, it appears that caseinate acts through a nonspecific ionic strength effect. The influence of caseinate on maize starch was less clear since it depended on the solvent medium. In distilled, deionized water, there was an increase in viscosity with increasing caseinate concentration, which may simply be explained by a contribution of the caseinate to the viscosity of the continuous phase. However, in 0.1 M, pH 7.0 buffer the results suggest that caseinate may inhibit retrogradation as the viscosity of the system after aging is reduced by its inclusion. It is suggested that phase separation between starch and caseinate is encouraged at high salt concentrations. As a consequence, both starch granule swelling and subsequent retrogradation are discouraged by caseinate in the buffer system, but not when pasting is carried out in distilled, deionized water (Kelly et al., 1995).

It may be concluded that potato starch is unique in its response to low levels of electrolyte and that the very large NaCas effect is related to this rather than a specific protein–starch interaction. Potato starch is known to contain a high level of phosphate groups compared with other starches (Swinkels, 1985; Muhrbeck and Tellier, 1991). It will, therefore, have a significant polyelectrolyte character. It seems probable that the dramatic ionic strength dependence is due to Donnan effects and that the gelatinized potato starch granule is behaving as a "super-swelling" polyelectrolyte gel maintained by a relatively low density of entanglements between amylopectin molecules. The effect of caseinate on maize starch is less clear.

The large viscosity reduction reported with potato starch was not repeatable and the solvent medium seemed to play an important part in determining the effects observed. There was a higher viscosity in buffer than in water in the absence of caseinate, suggesting that the extent of granule swelling was greater in buffer. This could reflect the lower pH (5.2) of starch alone pasted in water. In addition, in the pH 7 buffer sodium caseinate addition had no apparent effect on fresh paste viscosity. In distilled, deionized water, however, the viscosity of the fresh system increased steadily with increasing caseinate concentration. The lack of change in viscosity with increasing NaCas in buffer may reflect a slight decrease in starch swelling volume. This decrease in swelling volume would compensate for the increase in viscosity of the continuous phase with increasing caseinate concentration. On aging, maize starch will retrograde far more readily than potato starch (Kelly et al., 1995).

3.6 COMPARISION BETWEEN SODIUM CASEINATE AND OTHER EDIBLE FILMS

The plasticizer size and the effectiveness of the interaction between plasticizer and protein are important factors in determining the physical properties of plasticized NaCas films. Similar ideas have been invoked in the literature to explain the differences in film strength and permeability of other plasticized biopolymer systems, including gamma-irradiated caseinate (propylene glycol and triethylene glycol) (Mezgheni, 1998), whey/protein (glycerol vs. sorbitol) (McHugh and Krochta, 1994), and grain/protein (glycerol vs. PEG) (Park et al., 1994) films. The ability of the plasticizer to change the physical and water permeability properties of the film depends largely on the compatibility between the plasticizing material and the protein. The actual effect of two plasticizers on a protein is difficult to deduce in most of the published literature because comparisons are usually carried out on a weight protein/weight plasticizer basis, which does not necessarily take into account the properties and number of active sites on each plasticizer molecule. In addition, some properties (such as elongation) are affected by variations in film thickness (Sinew et al., 1999).

For caseinate adsorbed films, at the air–water interface the values of the surface dilatational modulus and its elastic component are similar but not identical to those for diglycerol esters. Thus, the values of the viscous component (E_{va}) are not zero, and the film presents viscoelastic behavior with values of tan δ higher than zero. The increase in surface dilatational elasticity, or the decrease in the phase angle tangent, with time should be associated with adsorption of proteins at the air–water interface (Liu et al., 1999) as was observed for proteins in general at the air–water (Boss and van Viet, 2001) and oil–water (Benjamin's, 2000) interfaces. In fact, the results of time-dependent surface dilatational property measurements are consistent with the existence of protein–protein interactions which are thought to be due to protein adsorption at the interface via diffusion, penetration, and rearrangement (looping of the amino acid residues) as both the adsorption time and the protein concentration in the bulk phase increase (Alvarez and Rodriguez Patino, 2006). Films and variable experimental designs with level of factors are given in Table 3-4.

TABLE 3-4 Films and Variable Experimental Design with Level of Factors (°C, %RH)

Films	Process Variables		References
	°C	% RH	
WPI / glycerol (5.7 : 1)	23	50	Siew et al. (1999)
WPI / glycerol (2.3 : 1)	23	50	
WPI / sorbitol (2.3 : 1)	23	50	
WPI / sorbitol (1 : 1)	23	50	
WG / PEG (2.48 : 1)	23	50	
WG / glycerol (2.2 : 1)	23	50	
sodium caseinate / glycerol (4 : 1)	23	50	
sodium caseinate / glycerol (2 : 1)	23	50	
sodium caseinate / PEG (4.54 : 1)	23	50	
sodium caseinate / PEG (1.9 : 1)	23	50	
LDPE	23	50	
HDPE	23	50	
0.25 Pullulan / 0.75 sodium caseinate	–	53	Kristo et al. (2007)
0.625 Pullulan / 0.375 sodium caseinate	–	53	
0.75 Pullulan / 0.25 sodium caseinate	–	53	
0.375 Pullulan / 0.625 sodium caseinate	–	53	
0.5 Pullulan / 0.5 sodium caseinate	–	53	
0.75 Pullulan / 0.25 sodium caseinate	–	75	
0.625 Pullulan / 0.375 sodium caseinate	–	75	
0.5 Pullulan / 0.5 sodium caseinate	–	75	
0.375 Pullulan / 0.625 sodium caseinate	–	75	
0.25 Pullulan / 0.75 sodium caseinate	–	75	

REFERENCES

AFNOR. 1971. Determination des caracteristiques en traction pour matieres plastiques, NFT 54–102. Paris: Association Francaise de Normalisation.

Ahmad FB, Williama PA, Doublier JL, Durand S, Buleon A. 1999. Physico-chemical characterisation of sago starch. *Carbohydr Polym* 38:361–370.

Alvarez JM, Rodrıguez Patino JM. 2006. Formulation engineering of food model foams containing diglycerol esters and α-lactoglobulin. *Ind Eng Chem Res* 45:7510–7519.

Alvarez Gomez JM, Rodriguez Patino JM. 2007. Interfacial properties of diglycerol esters and caseinate mixed films at the air–water interface. *J Phys Chem* 111:4790–4799.

Anglés MN, Dufresne A. 2001. Plasticized/tunicin whiskers nanocomposites materials. 2. Mechanical behaviour. *Macromolecules* 34:2921–2931.

Arvanitoyannis I, Biliaderis CG. 1998. Physical properties of polyol-plasticized edible films made from sodium caseinate and soluble starch blends. *Food Chem* 62(3):333–342.

Arvanitoyannis I, Blanshard JMV. 1993. Anionic copolymer of octanelactam with laurolactam (nylon 8/12): VII. Study of diffusion and permeation of gases in undrawn and uniaxially drawn (conditioned at different relative humidities) polyamide films. *J Appl Polym Sci* 47(1):1933–1959.

Arvanitoyannis I, Kalichevsky MT, Blanshard JMV, Psomiadou E. 1994. Study of diffusion and permeation of gases in undrawn and uniaxially drawn films made from potato and rice starch conditioned at different relative humidities. *Carbohydr Polym* 24:1–15.

Arvanitoyannis IS, Psomiadou E, Nakayama A. 1996. Edible films made of sodium caseinate, starches, sugars or glycerol. Part 1. *Carbohydr Polym* 31:179–192.

Arvanitoyannis IS. 1999. Totally and partially biodegradable polymer blends based on natural and synthetic macromolecules: preparation, physical properties, and potential as food packaging materials. *JMS Rev Macromol Chem Phys* C39(2):205–271.

Arvanitoyannis I, Biliaderis CG. 1999. Physical properties of polyol-plasticized edible blends made of methyl cellulose and soluble starch. *Carbohydr Polym* 38:47–58.

ASTM. 1985. Standard test method for tensile properties of thin plastic sheeting, Standards Designation: D882. In: *Annual Book of American Standard Testing Methods.* Philadelphia: ASTM. p. 182–188.

Avena-Bustillos RJ, Krochta JM. 1993. Water vapour permeability of caseinate-based edible films as affected by pH, calcium crosslinking ligand and lipid content. *J Food Sci* 58(4): 904–907.

Avérous L, Boquillon N. 2004. Biocomposites based on plasticized starch: thermal and mechanical behaviours. *Carbohydr Polym* 56:111–122.

Avérous L, Fringant C. 2001. Association between plasticized starch and polyesters: processing and performances of injected biodegradable systems. *Polym Eng Sci* 41:727–734.

Avérous L, Fringant C, Moro L. 2001a. Plasticized starch-cellulose interactions in polysaccharide composites. *Polymer* 42(15):6571–6578.

Avérous L, Fringant C, Moro L. 2001b. Starch-based biodegradable materials suitable for thermoforming packaging. *Starch/Stärke* 53:368–371.

Audic JL, Chaufer B. 2005. Influence of plasticisers and crosslinking on the properties of biodegradable films made from sodium caseinate. *Eur Polym J* 41:1934–1942.

Audic JL, Poncin-Epaillard F, Reyx D, Brosse JC. 2001. Cold plasma surface modification of conventionally and nonconventionally plasticized PVC-based flexible films. Global and specific migration of additives into isooctane. *J Appl Polym Sci* 79:1384–1393.

Back JF, Oakenfull D, Smith MB. 1979. Increased thermal stability of proteins in the presence of sugars and polyols. *Biochemistry* 18:5191–5196.

Barreto PLM, Roeder J, Crespo JS, et al. 2003a. Effect of concentration, temperature and plasticiser content on rheological properties of sodium caseinate/sorbitol solutions and glass transition of their films. *Food Chem* 82:425–431.

Barreto PLM, Pires ATN, Soldi V. 2003b. Thermal degradation of edible films based on milk proteins and gelatine in inert atmosphere. *Polym Degrad Stabil* 79:147–152.

Bastioli C. 1995. Starch–polymer composites. In: Scott G, Gilead D (eds.), *Degradable Polymers: Principles and Applications.* Chapman and Hall, London. p. 112–137.

Bazuin CG, Eisenberg A. 1986. Dynamic mechanical properties of plasticized polystyrene based ionomers. I. Glassy to rubbery zones. *J Polym Sci, Part B Polym Phys* 24:1137–1153.

Benjamins J. 2000. Static and dynamic properties of proteins adsorbed at liquid interfaces. Ph.D. thesis, University of Wageningen, Wageningen The Netherlands.

Bertolini AC, Creamer LK, Eppink M, Boland M. 2005. Some rheological properties of sodium caseinate-starch gels. *J Agric Food Chem* 53(6):2248–2254.

Biliaderis CG, Lazaridou A, Arvanitoyannis I. 1999. Glass transition and physical properties of polyol–plasticized pullulan–starch blends at low moisture. *Carbohydr Polym* 40:29–47.

Bos MA, van Vliet T. 2001. Interfacial rheological properties of adsorbed protein layers and surfactants: A review. *Adv Colloid Interface Sci* 91:437–471.

Brault D, D'Aprano G, Lacroix M. 1997. Formation of free standing sterilised edible films from irradiated caseinates. *J Agric Food Chem* 45:2964–2969.

Briassoulis D. 2004. An overview on the mechanical behaviour of biodegradable agricultural films. *J Polym Environ* 12(2):65–81.

Brown WE. 1992. *Plastics in Food Packaging: Properties, Design and Fabrication.* New York: Marcel Dekker. p. 298–299, 323–324.

Camirand W, Krochta JM, Pavlath AE, Wong D, Cole ME. 1992. Properties of some edible carbohydrate polymer coatings for potential use in osmotic dehydration. *Carbohydr Polym* 17:39–49.

Chen H. 1995. Functional properties and applications of edible films made of milk proteins. *J Dairy Sci* 78:2563–2583.

Cherian G, Gennadios A, Weller C, Chinachoti P. 1995. Thermomechanical behaviour of wheat gluten films: Effect of sucrose, glycerin and sorbitol. *Cereal Chem* 72:1–6.

Chick J, Ustunol Z. 1998. Mechanical and barrier properties of lactic acid and rennet precipitated casein-based edible films. *J Food Sci* 63(6):1024–1027.

Cho SY, Park JW, Rhee C. 2002. Properties of laminated films from whey powder and sodium caseinate mixtures and zein layers. *Lebensm-Wiss u Technol* 35:135–139.

Ciesla K, Salmieri S, Lacroix M, Le Tien C. 2004. Gamma irradiation influence on physical properties of milk proteins. *Radiat Phys Chem* 71:93–97.

Ciesla K, Salmieri S, Lacroix M. 2006. Modification of the properties of milk protein films by gamma radiation and polysaccharide addition. *J Sci Food Agric* 86:908–914.

Creighton TE. 1984. In: Freeman WH, editor. *Proteins: Structure and Molecular Properties.* New York.

Curvelo AAS, De Carvalho AJF, Agnelli JAM. 2001. Thermoplastic starch–cellulosic fibers composites: Preliminary results. *Carbohydr Polym* 45:183–188.

De Graaf LA, Kolster P. 1998. Industrial proteins as a green alternative for "petro" polymers: potentials and limitations. *Macromol Symp* 127:51–58.

De Leiris JP. 1985. Water activity and permeability. In: Mathlouthi M, editor. *Food Packaging and Preservation.* New York: Elsevier Applied Science. p. 213–233.

Diab T, Biliaderis CG, Gerasopoulos D, Sfakiotakis E. 2001. Physico-chemical properties and application of pullulan edible films and coatings in fruit preservation. *J Sci Food Agric* 81:988–1000.

Dufresne A, Dupeyre D, Vignon MR. 2000. Cellulose microfibrils from potato tuber cells: processing and characterization of starch–cellulose microfibril composites. *J Appl Polym Sci* 76:2080–2092.

El Ghaouth A, Arm J, Ponnamplalan R, Boulet M. 1991. Chitosan coating effect on storability and quality of fresh strawberries. *J Food Sci* 56:1618–1624.

Erdogdu N, Czuchajowska Z, Pomeranz Y. 1995. Wheat flour and defatted milk fractions characterised by differential scanning calorimetry. II. DSC of interaction products. *Cereal Chem* 72:76–79.

Ganjyal GM, Reddy N, Yang YQ, Hanna MA. 2004. Biodegradable packaging foams of starch acetate blended with corn stalk fibers. *J Appl Polym Sci* 93:2627–2633.

Gidley MJ, Cooke D, Ward-Smith S. 1993. Low moisture polysaccharide systems: Thermal and spectroscopic aspects. In: Blanshard JMV, Lillford PJ, editors *The Glassy State in Foods*. Nottingham, UK: University Press. p. 303–316.

Goel PK, Singhal RS, Kulkarni PR. 1999. Studies on interactions of corn starch with casein and casein hydrolysates. *Food Chem* 64:383–389.

Gontard N, Guilbert S, Cuq JL. 1992. Edible wheat gluten films: influence of the main process variables on film properties using surface response methodology. *J Food Sci* 57:190–195.

Gontard N, Guilbert S, Cuq JL. 1993. Water and glycerol as plasticisers affect mechanical and water vapour barrier properties of an edible wheat film. *J Food Sci* 58:206–211.

Gontard N, Ring S. 1996. Edible wheat gluten film: influence of water content on glass transition temperature. *J Agric Food Chem* 44:3474–3478.

Grevellec J, Marquie C, Ferry L, Crespy A, Vialettes V. 2001. Processability of cottonseed proteins into biodegradable materials. *Biomacromolecules* 2:1104–1109.

Gunning PA, Mackie AR, Gunning AP, Woodward NC, Wilde PJ, Morris VJ. Effect of surfactant type on surfactant-protein interactions at the air–water interface. 2004. *Biomacromolecules* 5(3):984–991.

Herald TJ, Gnanasambandam R, Mcguire BH, Hachmeister KA. (1995). Degradable wheat gluten films: preparation, properties and applications. *J Food Sci* 60(5):1147–1150.

Kalichevsky MT, Blanshard JMV. 1993. The effect of fructose and water on the glass transition of amylopectin. *Carbohydr Polym* 20:107–113.

Kalichevsky MT, Blanshard JMV, Tokarczuk PF. 1993. Effect of water-content and sugars on the glass transition of casein and sodium caseinate. *Int J Food Sci Technol* 28:139–151.

Kalichevsky MT, Jaroskiewicz EM, Blanshard JMV. 1992. Glass transition of gluten. 1. Gluten and gluten-sugar mixtures. *Int J Biol Macromol* 14:257–266.

Kelly RJ, van Wangenberg M, Latham J, Mitchell JR. 1995. A rheological comparison between the effects of sodium caseinate on potato and maize starch. *Carbohydr Polym* 28:347–350.

Kerr RW. 1950. Occurrence and varieties of starch. In: *Chemistry and Industry of Starch*, 2nd ed. New York: Academic Press. p. 3–25.

Khwaldia K, Banon S, Perez C, Desobry S. 2004a. Properties of sodium caseinate films – forming dispersions and films. *J Dairy Sci* 87:2011–2016.

Khwaldia K, Banon S, Desobry S, Hardy J. 2004b. Mechanical and barrier properties of sodium caseinate-anhydrous milk fat edible films. *Int J Food Sci Technol* 39:403–411.

Kilburn D, Claude J, Schweizer T, Alam A, Ubbink J. 2005. Carbohydrate polymers in amorphous states: An integrated thermodynamic and nanostructural investigation. *Biomacromolecules* 6:864–879.

Kinsella JE. 1984. Milk proteins: Physicochemical and functional properties. *Crit Rev Food Sci Nutr* 21:197–262.

Kolarik J. 1982. Secondary relaxations in glassy polymers: Hydrophilic polymethacrylates and polyacrylates. *Adv Polym Sci* 46:119–161.

Konstance RP, Strange ED. 1991. Solubility and viscous properties of casein and caseinates. *J Food Sci* 56:556–559.

Kristo E, Biliaderis CG. 2006. Water sorption and thermo-mechanical properties of water/ sorbitol-plasticized composite biopolymer films: caseinate-pullulan bilayers and blends. *Food Hydrocolloids* 20:1057–1071.

Kristo E, Biliaderis CG, Zampraka A. 2007. Water vapour barrier and tensile properties of composite caseinate-pullulan films: biopolymer composition effects and impact of beeswax lamination. *Food Chem* 101:753–764.

Krochta JM, Pavlath AE, Goodman N. 1990. Edible films from casein-lipid emulsions for lightly-processed fruits and vegetables. In: Spiess WEL, Schubert H, editors. *Engineering and Food*. Vol. 2, *Preservation Processed Fruits and Vegetables*. New York: Elsevier Applied Science. p. 329–340.

Krochta JM, Baldwin EA, Nisperos-Carriedo MO. 1994. *Edible Coatings and Films to Improve Food Quality*. Lancaster, PA: Technomic.

Lazaridou A, Biliaderis CG. 2002. Thermophysical properties of chitosan, chitosan-starch and chitosan-pullulan films near the glass transition. *Carbohydrate Polymers* 48(2):170–190.

Lelievre J, Husbands J. 1989. Effects of sodium caseinate on the rheological properties of starch pastes. *Starch/Staerke* 41:236–238.

Le Meste M, Roudaut G, Davidou S. 1996. Thermomechanical properties of glassy cereal foods. *J Thermal Anal* 47:1361–1375.

Lim LT, Mine Y, Tung MA. 1999. Barrier and tensile properties of transglutaminase cross-linked gelatine films as affected by relative humidity, temperature and glycerol content. *J Food Sci* 64:616–622.

Liu L. 2006. Bioplastics in food packaging: Innovative technologies for biodegradable packaging. San Jose State University, February, 2006. [On line: www.iopp.org/pages/index. cfm?pageid = 1160].

Liu M, Lee DS, Damodaran S. 1999. Emulsifying properties of acidic subunits of soy [11]S globulin. *J Agric Food Chem* 47:4970–4975.

Longares A, Monahan FJ, O'Riordan ED, O'Sillivan M. 2005. Physical properties of edible films made from mixtures of sodium caseinate and WPI. *Int Dairy J* 15:1255–1260.

Lourdin D, Coignard L, Bizot H, Colonna P. 1997. Influence of equilibrium relative humidity and plasticizer concentration on the water content and glass transition of starch materials. *Polymer* 38(21):5401–5406.

Lourdin N, Della Valle G, Collona P, Poussin P. 1999. Polyméres biodégradables: mise en oeuvre et propértiés de l'amidon. *Caoutchoucs et Plastiques* 760:39–42.

Matser AM, Steeneken PAM. 1997. Rheological properties of highly cross-linked waxy maize starch in aqueous suspensions of skim milk components. Effects of the concentration of starch and skim milk components. *Carbohydr Polym* 32(3–4):297–305.

Mauer LJ, Smith DE, Labuza TP. 2000. Water vapor permeability, mechanical, and structural properties of edible beta-casein films. *Int Dairy J* 10:353–358.

McHugh TH, Krochta JM. 1994. Permeability properties of edible films. In: Krochta JM, Baldwin EA, Nisperos-Carriedo MO, editors. *Edible Coatings and Films to Improve Food Quality*. Lancaster, PA: Technomic.

Mezgheni E, D'Aprano G, Lacroix M. 1998. Formation of sterilized edible films based on caseinates: Effects of calcium and plasticisers. *J Agric Food Chem* 46:318–324.

Moates GK, Noel TR, Parker R, Ring SG. 2001. Dynamic mechanical and dielectric characterization of amylose–glycerol film. *Carbohydr Polym* 44:247–253.

Muhrbeck P, Tellier C. 1991. Determination of starch from native potato varieties by ^{31}P NMR. *Starch/Staerke* 43:25–27.

Noisuwan A, Bronlund J, Wilkinson B, Hemar Y. 2008. Effect of milk protein products on the rheological and thermal (DSC) properties of normal rice starch and waxy rice starch. *Food Hydrocolloids* 22:174–183.

Park HJ, Chiman MS, Shewfelt RL. 1994. Edible coating effects on storage life and quality of tomatoes. *J Food Sci* 59(3):568–570.

Parker R, Ring SG. 1996. Starch structure and properties. *Carbohydr Eur* 15:6–10.

Psomiadou E, Arvanitoyannis I, Yamamoto N. 1996. Edible films made from natural resources; microcrystalline cellulose (MCC), methyl cellulose (MC) and corn starch and polyols—Part 2. *Carbohydr Polym* 31:193–204.

Radley JA. 1953. Some physical properties of starch. In: *Starch and Its Derivatives*, 3rd ed. London: Chapman & Hall. p. 58–80.

Rexen F, Petersen B, Munck L. 1988. Exploitation of cellulose and starch polymers from annual crops. *Trends Biotechnol* 6:204–205.

Rhim JW, Gennadios A, Fu D, Weller CL, Hanna MA. 1999. Properties of ultraviolet irradiated protein films. *Lebensm-Wiss u Technol* 32:129–133.

Rodriguez Patino JM, Carrera C. 2004. *Langmuir* 20:4530.

Rodriguez Patino JM, Carrera Sanchez C. 2004. Shear characteristics, miscibility and topography of sodium caseinate-monoglyceride mixed films at the air-water interface. *Biomacromolecules* 5:2065–2072.

Rodriguez Patino JM, Cejudo Fernandez M, Carrera Sanchez C, Rodriguez Nino MR. 2007. Structural and shear characteristics of adsorbed sodium caseinate and monoglyceride mixed monolayers at the air-water interface. *J Colloid Interface Sci* 313:141–151.

Rogers CE. 1985. Permeation of gases and vapors in polymers. In: Comyn J, editor. *Polymer Permeability*. New York: Elsevier Applied Science. p. 11–73.

Shou M, Longares A, Montesinos-Herrerro C, Monahan FJ, O'Riordan D, O'Sullivan M. 2005. Properties of edible sodium caseinate films and their application as food wrapping. *Lebensm-Wiss u Technol* 38:605–610.

Siew DCW, Heilmann C, Easteal AJ, Cooney RP. 1999. Solution and film properties of sodium caseinate/glycerol and sodium caseinate/polyethylene glycol edible coating systems. *J Agric Food Chem* 47:3432–3440.

Southward CR. 1989. Uses of casein and caseinates. In: Fox PF, editor. *Developments in Dairy Chemistry—4*. London: Elsevier Applied Science. p. 173–244.

Swinkels JM. 1985. Composition and properties of commercial native starches. *Starch/Staerke* 37:1–15.

Tester RF, Sommerville MD. 2003. The effects of non-starch polysaccharides on the extent of gelatinization, swelling and R-amylase hydrolysis of maize and wheat starches. *Food Hydrocolloids* 17:41–54.

Tolstoguzow VB. 1994. Some physicochemical aspects of protein processing in foods. In: Philips GO, Williams PA, Wedlock DJ, editors. *Gums and Stabilizers for the Food Industry*. Vol. 7. Oxford: IRL Press. p. 115–154.

Tolstoguzow VB, Grinberg YV, Gurov AN. 1985. Some physico-chemical approaches to the problem of protein texturization. *J Agric Food Chem* 33(2):151–159.

Trommsdorff U, Tomka I. 1995. Structure of amorphous starch. Vol. 2. Molecular interactions with water. *Macromolecules* 28:6138–6150.

Tziboula A, Muir DD. 1993. Effect of starches on the heat stability of milk. *Int J Food Sci Technol* 28(1):13–24.

Van Den Berg C. 1985. Development of B.E.T. like models for sorption of water on foods; theory and relevance. In: Simatos D, Multon JL, editors. *Properties of Water in Foods.* Dordrecht: Martinus Nijhoff. p. 119–131.

Warburton SC, Donald AM, Smith AC. 1993. *Carbohydr Polym* 21:17–21.

Weisser H. 1985. Influence of temperature on sorption equilibria. In: Simatos D, Multon JL, editors. *Properties of Water in Foods.* Dordrecht: Martinus Nijhoff. p. 95–118.

Wong DWS, Gastineau FA, Gregorski KS, Tillin ST, Pavlath AE. 1992. Chitosan-lipid films: microstructure and surface energy. *J Agric Food Chem* 40:540–544.

Young AH. 1984. Fractionation of starch. In: Whistler RL, Bemiller IN, Paschall EF, editors. *Starch: Chemistry and Technology.* 2nd ed. New York: Academic Press p. 285.

Yu L, Petinakis S, Dean K, Bilyk A, Wu D. 2007. Green polymeric blends and composites from renewable resources. *Macromol Symp* 249–250:535–539.

Electronic Sources

URL1. http://www.americancasein.com/docs/Sodium%20Caseinate.doc   (accessed   10 September 2007)

URL2. http://aida.ineris.fr/bref/bref_text/breftext/anglais/bref/BREF_tex_gb40.html (accessed 10 September 2007)

URL3. http://ift.confex.com/direct/ift/2000/techprogram/paper_4379.htm   (accessed   10 September 2007)

URL4. http://www.americancasein.com/docs/Sodium%20Caseinate.doc   (accessed   12 September 2007)

URL5. http://www.idb.ie/products/SODCASE.HTM (accessed 12 September 2007)

URL6. http://www.degradable.net/downloads/ics_info_pack_2001.pdf   (accessed   14 September 2007)

URL7. http://ms-biomass.org/conference/2004/rebirth%20of%20bio-based%20polymer% 20development.ppt (accessed 14 September 2007)

CHAPTER 4

Novel Plastics and Foams from Starch and Polyurethanes

YONGSHANG LU

Department of Chemistry, Iowa State University, Ames, Iowa, USA

LAN TIGHZERT

École Supérieure d'Ingénieurs en Emballage et Conditionnement, Université de Reims Champagne-Ardenne, Reims Cedex 2, France

4.1 INTRODUCTION

Worldwide potential demand for replacing petroleum-derived raw materials by renewable resources in the production of valuable biodegradable polymeric materials is significant from both social and environmental viewpoints (Mohanty et al., 2000; Lu et al., 2004; Yu et al., 2006). Among the several candidates including aliphatic polyesters, natural polymers, and their derivatives, starch—a polysaccharide obtained in granular form from corn, cereal grain, rice, and potatoes—is one of the most promising materials for biodegradable polymeric materials, due to its easy availability and low price (Wang et al., 2003; Halley, 2005). The starch granule is essentially composed of two main polysaccharides of amylose and amylopectin (see Scheme 4-1). Amylose is a linear molecule with an extended helical twist and generally has a molecular weight 1.0–1.5 million. Amylopectin is a branched molecule with a

Biodegradable Polymer Blends and Composites from Renewable Resources. Edited by Long Yu
Copyright © 2009 John Wiley & Sons, Inc.

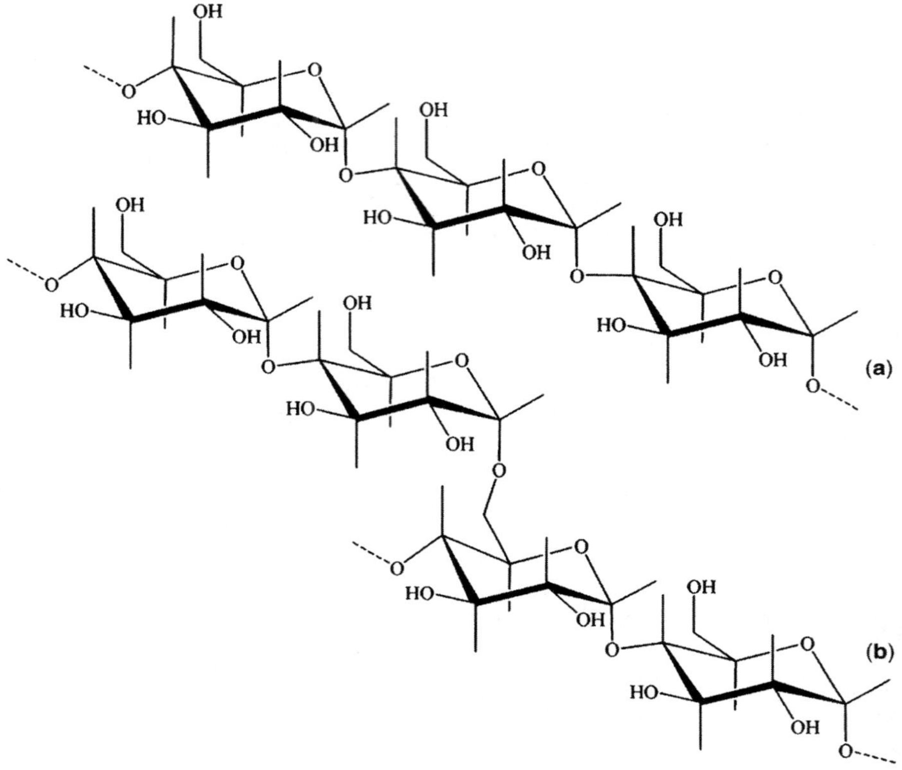

Scheme 4-1 Chemical structure of amylose (a) and amylopectin (b) (Wang et al., 2003).

much higher molecular weight in the range 50–500 million. Starch has been widely used in several nonfood sectors, most notably in the sizing and coating of papers, as an adhesive, as a thickener, and as a "green strength" additive to simple composite materials (Petersen, 1999). With the increased interest of biobased polymeric materials as an alternative to those produced from nonrenewable resources, research and development on starch-based materials used for plastics and foams has attracted much attention (Guan and Hanna, 2004; Lu et al., 2005a, 2006; Xu and Hanna, 2005).

Polyurethane plastics were first synthesized by Otta Bayer and co-workers in 1937 and were introduced into the market in the late 1940s (Oertel, 1994). Polyurethane, consisting of soft segments (hydroxyl-terminated oligomers) and hard segments that are the result of the introduction of short diols or diamine chain extenders with diisocyanates forming urethane or urea linkages, is a unique material that offers the elasticity of rubber combined with the toughness and durability of metals. The availability of polyurethane in a very broad hardness range allows it to replace rubber, plastic, and metal with favorable abrasion resistance and physical properties. Due to their high performance, wide variety of raw materials, and adaptable synthetic techniques, the polyurethanes have been widely used in many applications, such as foam, cast resin, coatings, adhesives, and sealants (Król, 2007). In recent years, combinations of natural polymers with synthetic polymers have been used in many scientific disciplines because these possess better physical properties and biocompatibility than do the individual components themselves. The renewable starch-based systems are traditionally of low cost but suffer from poor processability and final product properties. Whereas the synthetic biodegradable polymers are traditionally easier to process and have excellent properties, their cost is high. Thus synergizing the advantages of starch-based and synthetic biodegradable polymers represents one key strategy for creating more applications and larger markets for biodegradable polymeric materials with high performance (Halley, 2005). The purpose of this chapter is to summarize some issues of the manufacturing processes, properties, and potential applications of biodegradable starch/polyurethane-based materials.

4.2 STARCH-FILLED POLYURETHANE ELASTOMERS AND PLASTICS

The presence of many hydroxyl groups in starch granules makes it a good filler for reacting with isocyanate groups of polyurethane, resulting an increase in the interfacial adhesion between the starch granules and polyurethane matrix. Chen and Li (1997) synthesized starch-filled (10–30 wt%) polyurethane sheets with good elasticity recovery by reacting a mixture of starch and poly(tetramethylene oxide) (PTMEG, $M_w = 2000$) with toluene diisocyanate (TDI). After one month of embedment under soil, the strength of the moldy sheets containing starch decreased by 20–40%, suggesting better biodegradability of the starch-modified polyurethanes than in those without starch. In addition to better biodegradation, the phosphate starch-filled polyurethane sheets also showed improved flame resistance when compared with pure polyurethane (Li et al., 1999). For polyurethane elastomers prepared using poly(propylene glycol) (PPG, $M_w = 2000$) as the soft segment and starch as a multifunctional crosslinker, the materials exhibited high mechanical properties when compared with polyurethanes using 1,1,1-trimethylolpropane (TMP) as the crosslinker. The tensile strength and elongation at break increased with increasing starch content. The filled polyurethanes exhibited two glass transitions, whereas only one glass transition temperature (T_g) was observed for the TMP-based polyurethane, indicating phase separation between starch–TDI and PPG–TDI segments. Greater biodegradability was observed in the starch-based polyurethanes than in the TMP-based ones (Desai et al., 2000).

The polyester-based polyurethanes are much more susceptible to biodegradation than those derived from polyether diols (Huang, 1989). Degradation studies indicated that when the polyols used were poly(hexamethylene adipate) or polycaprolactone triol (PCL) diols, the resulting polyurethanes exhibit maximum biodegradation rates under composting conditions. Surface hydrophobicity, which is related to adhesion of bacteria on the polymer surface, is considered to be a factor in biodegradation rate under composting conditions (Huang et al., 1981; Kim and Kim, 1998).

For PCL/4,4-diphenylmethane diisocyanate (MDI)-based polyurethane filled with various amounts of starch granules (Ha and Broecker, 2002, 2003), the starch granules were well dispersed as a grafted state in the continuous polyurethane phase to form a three-dimensional network as shown in Scheme 4-2. The grafted proportion of polyurethane to starch granules increased to a maximum when the starch content was about 20 wt%, resulting in increased tensile strength and elongation for the resulting materials. However, these properties decreased rapidly with a further increasing starch content, caused by the phase separation between starch granules and polyurethane phase, the lower grafting percentage, and the fracturing of the starch granules.

Renewable resource (castor oil)-based polyurethane materials filled with starch of 5–25 wt% were synthesized and investigated by Swamy and Siddaramaiah (2003). The sheets obtained were tough, rigid, and opaque. It was observed that incorporating starch into castor oil-based polyurethane improved the physicomechanical properties of the resulting materials because of the presence of multifunctional starch in the polyurethane and the formation of extensive hydrogen bonds

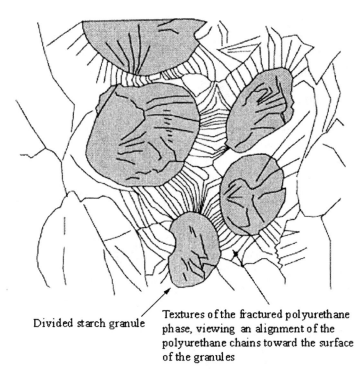

Divided starch granule

Textures of the fractured polyurethane phase, viewing an alignment of the polyurethane chains toward the surface of the granules

Scheme 4-2 Sketch of the surface textures for the starch granule-filled polyurethane (Ha and Broecker, 2003).

between polyurethane and starch. According to the authors, 5 wt% starch filler was the optimized composition to obtain high-performance material. Addition of starch into the polyurethane did not alter the crystalline region or the extent of crystallinity, but chain conformation in the amorphous region might change.

4.3 STARCH-FILLED POLYURETHANE FOAMS

Polyurethane (PU) foams can be considered as composite structures resulting from the controlled entrapment of the gases that are generated during the polymerization reaction between polyfunctional alcohols and polyisocyanate to form urethane linkages (Alfani et al., 1998). PU foams, compared with other materials, exhibit physical and mechanical properties related to their chemical composition and density, and are widely used in many fields as structural, cushioning, insulation, electrical, flotation, and packaging materials. High volumes of conventional polyurethane foams are produced every year, resulting in serious environmental problems. Government regulation regarding waste disposal in Europe and the United States will result in a reduction of the use of landfill and an increase of the recycling and incineration of these materials. However, recycling is difficult because of the thermosetting characteristics of polyurethanes.

Corn flour was once considered a possible source of starch for the reaction of an organic polyisocyanate with degraded starch polyoxalkylene ethers to make polyurethane foams (Otey and Mehltretter, 1965). Cunningham et al. (1991) evaluated the use of corn starch flour with different contents (5, 10, 20, 30, and 40 wt%, based on weight of polyethyl polyol) as filler-extenders in polyurethane foams, and found that 10 wt% corn flour added to the formulation served not only as filler but as a promising contributor to maintaining specific properties of polyurethane foams, especially compressive strength ($192 \pm 17\,kN/m^2$) and insulating properties ($0.0232\,W/mK$) as well as dimensional stability when subjected to thermal and humid conditions. They also studied the influence of unmodified starch and modified starch (20 wt% based on the polyols) on the compressive strength, density, dimensional stability, open-cell content and thermal conductivity of the resulting starch-filled polyurethane foams (Cunningham et al., 1992). The modified cornstarches used were waxy, acid-modified waxy, maltodextrin, and canary dextrin, respectively. The dextrin-filled foams exhibited higher compressive strength than that filled with starch ($141-158$ vs. $86-137\,kN/m^2$, $196\,kN/m^2$ for control) and responded to compressive stress like the control foam with a yield point before 10% deformation. The unmodified cornstarch-filled foams exhibited higher compressive strength than those containing either of the waxy starches (137 vs. $86-97\,kN/m^2$). The open cell content of foams filled with dextrins was slight lower than that of the control (12%), foams filled with unmodified corn starch (13%), and those with modified waxy and acid-modified waxy starches (15%). According to the authors, the canary dextrin as high as 40 wt% could be incorporated into rigid polyurethane foam formulations to provide foams with similar or improved thermal stability, open-cell content, and thermal conductivity compared with the control.

Fantest is a starch–oil composite prepared by a jet cooking that uses the high temperature and turbulence within the cooker to uniformly disperse the oil component within the starch–water matrix as small droplets (Knutson et al., 1996). These droplets will not separate or coalesce, even after prolonged standing and after the product is dried. Fantest of 5–40 wt% (based on polyester polyol) blended with a polyester polyol (Lexorez 1102-50A) has been reacted with isocyanate to produce polyurethane foams, where the starch reacted with the isocyanate to form urethane bonds, resulting in a highly branched and interconnected molecular structure (Cunningham et al., 1997). Uniform mixing of the ingredients was difficult due to high viscosity of this system, resulting in the cell structure of the blend foams deteriorating with increasing Fantest content. Therefore, low-viscosity glycols, such as ethylene glycol, poly(ethylene glycol) (PEG), and propylene glycol, were used to replace the polyester polyol for preparation of polyurethane foams, in which Fantest content was maintained at 20 wt% (Cunningham et al., 1998). These three liquid glycols permitted the Fantest to blend well with the other ingredients in the foam formulations, resulting in foams with uniform, continuous cell networks with essentially transparent wall and connections as shown in Fig. 4-1. Foams containing polyethylene glycol (PEG) showed a greater number of larger cells and hence had lower densities than foams containing either ethylene or propylene glycol. NMR results indicated that oil droplets in Fantest were crosslinked with isocyanate. Intact flakes of the Fantest were observed in the foam, and their presence would make the foams more susceptible to biodegradation.

Using a mixture of starch and PCL, Alfani et al. (1998) synthesized a series of polyurethane foams with various starch contents of 0–40 wt%. The starch participated in the chemical reaction, resulting in an increase in T_g, an increase in material modulus and, at the same time, a reduction in density of open-cell foams obtained. The properties of the resulting open-cell foams could be modulated by adding a certain amount of PEG which, introducing flexible chains into the macromolecular structure, reduced the T_g value of the materials and generated foams with higher flexibility. In contrast to the open-cell foams, Kwon and co-workers obtained foams with closed-cell structure using polyols of fully gelatinized starch (30–40 wt% based on polyols) mixed with PEG, glycerol, and 1,4-butanediol. The cell size and compressive moduli increased with NCO/OH molar ratio. The resulting foams with NCO/OH molar ratio of 0.8 exhibited the maximum absorbency for organic solvents (Kwon et al., 2007).

Many efforts have also been made to liquefy biomass in the presence of some organic reagents, such as polyhydric alcohols and phenols. It was found that polyhydric alcohols with appropriate molecular weights could be used in the liquefaction of biomass (starch, wood, etc.) with sulfuric acid as catalyst. The resulting liquefied mixtures could be used directly as polyols to prepare polyurethane foams without any additional reaction or treatment, offering a simple method to incorporate substantial amounts of biomass into the polyurethane foams (Yao et al., 1996; Lin et al., 1997; Lee et al., 2000).

Poly(ethylene glycol) shows great advantage due to its large capacity for liquefying biomass and appropriate price (Lin et al., 1997; Lee et al., 2000). Using polyols

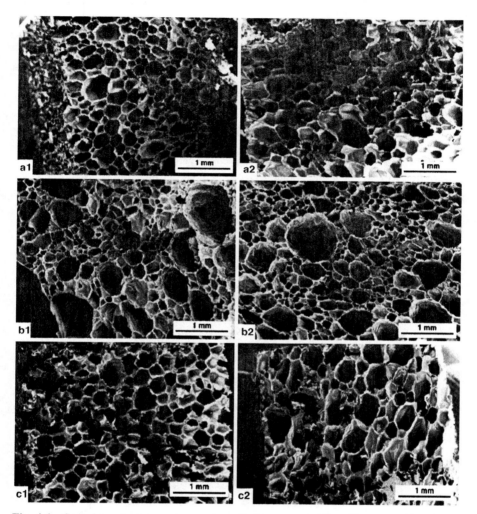

Fig. 4-1 Scanning electron micrographs of foams containing Fantest with glycols (a) ethylene, (b) polyethylene, and (c) propylene: (1) transverse or cross sections; (2) longitudinal sections (Cunningham et al., 1998).

prepared from liquefaction of starch in the presence of PEG ($M_w = 400$)-dominant reaction reagents, Yao et al. (1996) demonstrated the synthesis of water-absorbing polyurethane foams. It was found that about 1 wt% sulfuric acid was enough to obtain complete liquefaction even at 50 wt% starch content. The hydroxyl numbers of the liquefied starch polyols were in the range 270–369 mg KOH/g, suitable for preparation of polyurethane foams. The resulting liquefied starch polyol-based polyurethane foams exhibited continuous cell structures when a cell-opening type surfactant and a small amount of high molecular triol (poly(ether polyol), $M_w = 3000$, functionality of 3) were employed in the formulations. The resulting foams could absorb water up to 2000 wt% within several minutes and showed good water-retention properties and substantial mechanical properties.

Starch-containing polyurethane foams were also studied by Ge and co-workers using polyols from liquefaction of the bark of *Acacia mearnsii* (BK) and corn starch (CS) in PPG, glycerol, and sulfuric acid (Ge et al., 2000). When BK was replaced partially with CS, the density and compressive strength of the foams decreased with an increase in CS content, whereas the resilience value reached its maximum when the weight ratio of CS to BK was 1 : 1, leading to possible application of the resulting materials in car-seat cushions. In addition, incorporating CS contributed to a better biodegradation of the foam materials. Using these foams as coating materials for controlled release of fertilizer, the release ratio could be controlled more effectively when BK was partially replaced with CS. Moreover, the remaining fertilizer in the foams seemed to be released completely because the foams were degradable, to some extent, by soil microorganisms (Ge et al., 2002).

4.4   STARCH GRAFTED WITH POLYURETHANES

The nature of brittleness caused by the relatively high glass transition temperature (T_g) and lack of a sub-T_g main-chain relaxation (β-transition) limits the use of starch in plastics. Moreover, the brittleness increases with time due to free volume relaxation and retrogradation (Wang et al., 2003). To overcome the inherent problems with starch, Kweon and co-workers synthesized starch-g-polycaprolactone copolymers by a reaction of a diisocyanate-terminated polycaprolactone-based prepolymer (NCO-PCL) with corn starch at a weight ratio of starch to NCO-PCL of 2 : 1 as shown in Scheme 4-3 (Kweon et al., 2000). On grafting of NCO-PCL (35–38 wt%) prepared with TDI or 4,4-diphenylmethane diisocyanate (MDI) onto starch, the T_g values of both copolymers were 238°C. However, when TDI was replaced by hexamethylene diisocyanate (HDI), the T_g of the material was found to be around 195°C. On grafting NCO-PCL (PCL-1250), the degradation temperature (T_d) changed depending on the type of isocyanate used. The starch-g-PCL copolymers prepared with the MDI intermediate were most stable, to the point of thermal degradation when compared with those from TDI and HDI. Barikani and Mohammadi (2007) found that the T_g values of the starch-g-PCL copolymers decreased with increasing percentage of the urethane prepolymer and depended on the crosslinking effect of prepolymer between two chains of starch and strong hydrogen bonding between molecules, which affected chain mobility of starch-modified urethane in different ways. Hydrophobicity of the starch grafted with a PCL-based urethane prepolymer increased with increasing amount of urethane prepolymer. According to the authors, this modified starch could be used as filler in biodegradable starch-based polyethylene due to better dispersion and compatibility.

Wilpiszewska and Spychaj synthesized starch–urethane polymers via chemical modification of potato starch with urethane and urea derivatives of HDI, as shown in Scheme 4-4, with efficiency of substitution usually above 70% (Wilpiszewska

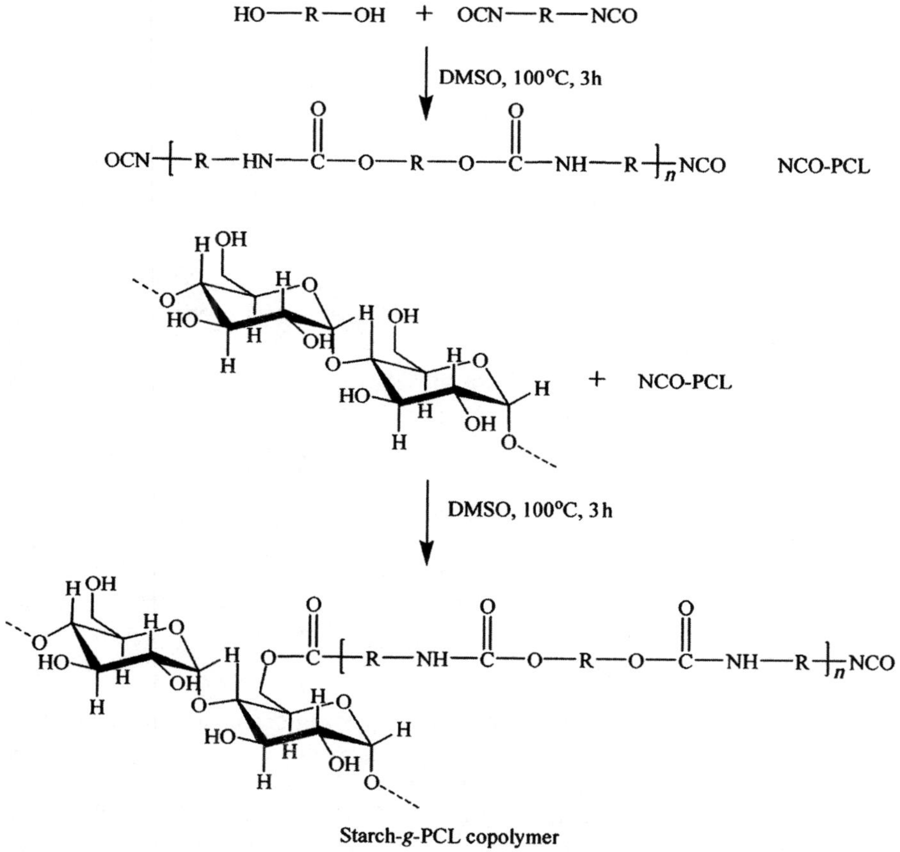

HO—R—OH + OCN—R—NCO

DMSO, 100°C, 3h

$$\text{OCN} \left[\text{R—HN—}\overset{\overset{\textstyle O}{\|}}{\text{C}}\text{—O—R—O—}\overset{\overset{\textstyle O}{\|}}{\text{C}}\text{—NH—R} \right]_n \text{NCO} \qquad \text{NCO-PCL}$$

+ NCO-PCL

DMSO, 100°C, 3h

Starch-*g*-PCL copolymer

Scheme 4-3 Synthesis of starch-*g*-PCL copolymers (Kweon et al., 2000).

and Spychaj, 2007). The presence of additional urethane or urea bonds in short modifying branches can alter the properties of urethane–starch derivatives compared with those found for polymers obtained with the usually applied monoalkyl isocyanate modifiers. From the viewpoint of application, starch polymers with the degree of substitution (DS) in the range of 1.6–1.8 (for theoretical DS = 2) exhibited acceptable bulk hydrophobic properties (swellability in water in the range 2.0–2.5 cm^3/g) and melt flow features in hot press test, making them good candidates for manufacturing through reactive extrusion processes.

4.5 THERMOPLASTIC STARCH/POLYURETHANE BLENDS

Beside the copolymers of starch grafted with urethane prepolymer, many plasticizers, such as glycerol, water, sorbitol, urea, amide, sugars, sodium lactate, and oligomeric

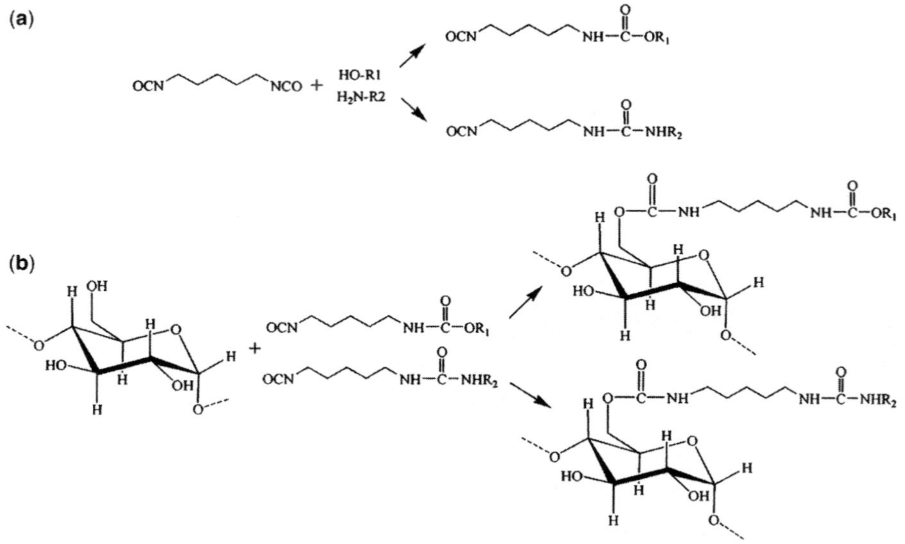

Scheme 4-4 Reaction between HMDI and a monoalcohol or primary amine (a), and reaction between starch polysaccharide unit and HMDI/monoalcohol or HMDI/monoamine modifiers (b) (Wilpiszewska and Spychaj, 2007).

polyglycol, have also been used to lower the T_g and melting temperature of the starch (van Soest et al., 1996; Lourdin et al., 1997). With the aid of plasticizer, the starch is suitable for processing through mechanical shear at elevated temperature to become an essentially homogeneous material termed thermoplastic starch (TPS) or plasticized starch (Suvorova et al., 2000). Research on thermoplastic, destructed, or melted starch has been reviewed by Roper and Koch (1990), Swanson et al. (1993), Lai and Kokini (1991), Wang et al. (2003), and Chiou et al. (2005a, b, c). Thermoplastic starch-based products are moisture sensitive and become brittle, especially after aging, when compared to typical synthetic plastics. To obtain commercially acceptable products, starch is usually blended with other materials to improve its physical properties and to minimize its water sensitivity (Chiou et al., 2005a, b, c). Examples include starch blended with other biopolymers or totally biodegradable synthetic polymers, such as starch/pectin blends (Coffin and Fishman, 1994; Fishman et al., 1996), starch/cellulose crystalline biocomposites (Lu et al., 2005, 2006), starch/poly(vinyl alcohol) blends (Mao et al., 2000), starch/polyester blends (Bastioli et al., 1995; Fang and Hanna, 2001; Ha and Cho, 2002; Matzinos et al., 2002), and starch/poly(D,L-lactic acid) blends (Kim et al., 1998; Park et al., 1999).

Because of the excellent properties of polyurethanes, blending starch with polyurethanes for preparation of high-performance biodegradable materials has also received much attention. In order to improve water resistance of the starch and to enhance compatibility between hydrophilic starch and hydrophobic polyurethane,

Santayanon and Wootthikanokkhan produced compression-molded blend plastics from acrylated starch (20–60 wt%) and thermoplastic polyurethane comprising of 1,4-butanediol in the hard segment and a polyester-polyol in the soft segment (Santayanon and Wootthikanokkhan et al., 2003). The blends exhibited improved water-resistance and better interfacial adhesion compared with those based on normal starch and polyurethane. However, little difference in the mechanical properties, such as tensile strength, elongation, and toughness, was observed for the blends modified with acrylated starch and with normal starch. The blends with modified starch experienced slower biodegradation than those containing normal starch.

Blend materials based on thermoplastic poly(ester-urethanes) and starch can be tailored to be biodegradable (Seidenstücker and Fritz, 1999). Seidenstücker and Fritz described a compounding procedure for production of blends consisting of thermoplastic poly(ester-urethane) (TPU) and TPS using co-rotating twin-screw extruders, and showed that the TPU/ST blends could be easily processed to give excellent flat and tubular films. Table 4-1 compared the properties of the new TPU/TPS blends with polyolefins. It can be clearly seen that many properties of the TPU/TPS blends are in the range of polyolefin properties. Further, incorporation of starch into thermoplastic poly (ester-urethane) can significantly increase the rate of biodegradation of the resulting materials. The outstanding properties of such blends, according to the authors, are summarized below:

1. They consist of high amounts of renewable resources (up to 80–85 wt%) in cases where the polyol is based upon renewable resources as well.
2. Compared with most biodegradable materials such blends have good to excellent mechanical properties that can be varied within wide limits.
3. Such blends have a favorable price range compared with other biodegradable polymers.
4. Films as well as molded parts can easily be produced.
5. Compared with polyolefin films and sheets, films made of such new blends show significantly higher steam permeability. Their oxygen permeability lies in the same range.
6. They possess a fixed morphology whereby no coalescence occurs in downstream molding processes.
7. These TPU/TPS blends can be designed to be compostable.
8. The blends show good printability and colorability.
9. They are easily weldable and sealable.
10. The blends possess high resistance to abrasion.

Conventional polyurethane products, such as coatings and adhesives, contain significant amount of organic solvents and some also contain free isocyanate monomers (Lee et al., 2006). To meet the increasing concerns about health, safety, and the environment, the conventional polyurethanes have gradually been replaced by the

TABLE 4-1 Properties of the New TPU/TPS Blends Compared with Polyolefins and TPU/Starch Compounds Seidenstücker and Fritz (1999)

Properties	TPU/Native Starch[a]	TPU/TPS[a]	Polyethylene	Polypropylene
Tensile strength (MPa)	16–35	18–25	8–29	21–37
Elongation at break (%)	4–180	4–800	40–500	20–800
Elasticity modulus (MPa)	100–1750	20–500	200–1400	1100–1300
Hardness				
Shore A	–	>70	>70	–
Shore D	≈55	–	–	–
Max. temperature (°C)	<80	<80	60–80	80–120
Density (g/cm^3)	1.31–1.39	1.30–1.37	0.916–0.96	0.9–0.907
Impact resistance (kJ/m^3)	No break at room temperature			

[a]50 wt% TPU.

waterborne polyurethanes. Waterborne polyurethanes present many advantages in relation to conventional solvent-borne polyurethanes, including low viscosity at high molecular weight and good applicability (Modesti and Lorenzetti, 2001). Due to their versatility and environmental friendliness, waterborne polyurethanes are now one of the most rapidly developing and active branches of polyurethane chemistry and technology.

Wu and Zhang prepared a series of compression-molded sheets based on TPS and waterborne polyurethanes from TDI, poly(1,4-butylene glycol adipate) (PBA) and dimethylol propionic acid (DMPA) (Wu and Zhang, 2001a). When the waterborne polyurethane content was in the range of 5–30 wt%, the tensile strengths of the molded blend sheets were higher than those of starch and polyurethane, while the elongations at break were between those of the pure components and much higher than that of TPS, indicating that waterborne polyurethane could effectively improve the mechanical properties of TPS. The water resistance of the blend sheets also significantly increased with an increase in the content of waterborne polyurethane. For TPS, the R value (R = wet strength/dry strength) is 0.01. However, this value increased significantly to 0.24, 24 times higher than that of TPS, when 30 wt% waterborne polyurethane was incorporated. These improvements could be explained by strong interactions in the molded sheets. In this system, the waterborne polyurethane played an important role in formation of new morphology and in the performance enhancement of the blends, providing a possible application in the field of biodegradable materials. Similar results have also been observed in the casting blend systems consist of starch and polyurethane from TDI, poly(oxypropylene glycol) and DMPA (Wu and Zhang, 2001b).

In casting blends of the aqueous mixture of plasticized starch and polyester-based waterborne polyurethanes with different NCO/OH molar ratios, Cao et al. (2003) found that the thermal and mechanical properties of the starch/PU blends depend not only on the starch content but also on the microstructure of the waterborne polyurethanes. The resulting blends containing 20 wt% starch exhibit higher T_g than the corresponding waterborne polyurethane and other blends with starch content higher than 50 wt%, implying that the starch in the soft segment matrix of polyurethane restricted the mobility of soft segment due to the relative strong hydrogen-bonding interaction. The polyurethane (WPU2) with higher NCO/OH molar ratio exhibited higher tensile strength than WPU1 with lower NCO/OH molar ratio, because of the formation of hard segment ordered structure. However, when starch content was lower than 50 wt%, the resulting starch blends from WPU1 exhibited higher tensile strength and elongation at break when compared with that from WPU2, because the incorporation of the appropriate content of starch not only filled the soft-segment matrix to reinforce the materials but also hindered the formation of hard-segment ordered structure, resulting in the increase of tensile strength of the starch/WPU1 blends. The blend with an optimized composition (WPU1/starch = 80/20, weight ratio) exhibited simultaneously good tensile strength (27 MPa) and elongation at break (949%).

Most polyols used for production of waterborne polyurethanes are based on petroleum-based resources. Recently, Lu et al., synthesized novel waterborne polyurethane dispersions from rapeseed oil polyols (Lu et al., 2005b) and castor oil (Lu et al., 2005c). The use of vegetable oils to synthesize environmentally friendly waterborne polyurethanes has superb environmental credentials, such as the products being inherently biodegradable, having low ecotoxicity and low toxicity towards humans, being derived from renewable resources, and contributing no volatile organic chemicals (Erhan, 2005). Transparent films cast from waterborne polyurethane dispersions (molar ratio of NCO to OH of vegetable oil-based polyol is 2.0) exhibited high tensile strength (8–15 MPa) and elongation at break (280–520%). The extruded blend sheets show good interfacial adhesion between polyurethane and starch, as shown in Fig. 4-2, due to the hydrogen bonding interactions between the urethane groups of the polyurethane and the hydroxyl groups on starch. These interactions tend to lower the interfacial tension between polyurethane and starch, making them more compatible.

For starch/castor oil-based waterborne polyurethane blends, the tensile strength first increased with increasing polyurethane content and reached its maximum (5.1 MPa) at 15 wt% content of polyurethane, then decreased to 2.6 MPa at 30 wt% polyurethane. Simultaneously, the elongation at break increased from 120% to 176% with increasing polyurethane content from 0 to 10 wt%, and then decreased slightly to 140% with further increasing polyurethane (Lu et al., 2005c). This indicates that

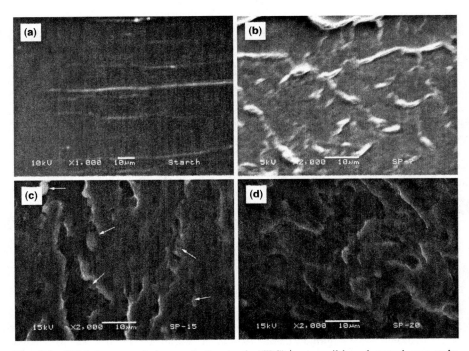

Fig. 4-2 SEM images of thermoplastic starch (TPS)/castor oil-based waterborne polyurethane blends. (a) (TPS), (b) (blend with 7 wt% polyurethane), (c) (blend with 15 wt% polyurethane), and (d) (blend with 20 wt% polyurethane) (Lu et al., 2005c).

incorporating an appropriate content of polyurethane into starch matrix can improve the mechanical properties of the blends in both tensile strength and elongation at break. However, the mechanical behavior of these blends was different from that of the blends of both TPS/PCL (Averous et al., 2000) and TPS/PLA (Martin and Averous, 2001), whose elongations at break decreased significantly when PCL or PLA was added. After aging of 30 weeks, an increase in strength was observed for the aged blends compared with that of non-aged ones, due to the retrogradation of starch during the storage. Although there was a decrease in elongation at break after aging, the blends with 4–15 wt% PU still exhibited relatively high flexibility (elongation at break 92–121%).

The water resistance of starch can also be significantly improved by blending castor oil- or rapeseed oil-based waterborne polyurethane. The plasticized starch absorbed about 60% water at equilibrium. This value decreased with an increase of content of vegetable oil-based polyurethanes in the blends. For example, only about 25% of water uptake was observed for blends containing 50 wt% rapeseed oil-based waterborne polyurethane (Lu et al., 2005b). This can be ascribed to the presence of strong hydrogen-bonding interactions between starch and polyurethanes, which tend to stabilize and prevent the swelling of the starch in high moisture environments. The resulting blend materials show high potential for packaging applications.

4.6 CONCLUDING REMARKS

Research conducted in recent decades has demonstrated that incorporating starch into polyurethane imparts good physical properties and higher biodegradability on the materials. Their mechanical properties, close to those of traditional plastics such as polyethylene and polypropylene, and their improved water resistance make these materials particularly suitable for the production of films, molded items, and foams, leading to novel materials with high potential for practical application. However, the starch–polyurethane materials are still at an early stage of development with many opportunities still to be exploited—for example, the development of advanced high starch loading foams and the compatibilization between starch and polyurethane. Moreover, effective environmental assessments of the end-life materials have not been undertaken. Thus, the future development of advanced starch–polyurethane materials will rely on continued research, especially in the mechanisms and kinetics of reaction and compatibilization, advanced processing techniques, and environmental assessments.

REFERENCES

Alfani R, Iannace S, Nicolais L. 1998. Synthesis and characterization of starch-based polyurethane foams. *J Appl Polym Sci* 68:739–745.

Averous L, Moro L, Dole P, Fringant C. 2000. Properties of thermoplastic blends: starch-polycaprolactone. *Polymer* 41:4157–4167.

Barikani M, Mohammadi M. 2007. Synthesis and characterization of starch-modified polyurethane. *Carbohydr Polym* 68:773–780.

Bastioli C, Cerutti A, Guanella I, Romano GC, Tosin M. 1995. Physical state and biodegradation behavior of starch-polycaprolactone systems. *J Environ Polym Degrad* 3:81–95.

Cao XD, Zhang LN, Huang J, Yang G, Wang YX. 2003. Structure-properties relationship of starch/waterborne polyurethane composites. *J Appl Polym Sci* 90:3325–3332.

Chen D, Li Y. 1997. Biodegradable polyurethane modified by starch. *China Synthetic Rubber Industry* 20:244–244.

Chiou B-S, Glenn GM, Imam SH, Inglesby MK, Wood DF, Orts WJ. 2005. Starch polymers: chemistry, engineering, and novel products. In: Mohanty AK, Misra M, Drzal LT, editors. *Natural Fibers, Biopolymers, and Biocomposites.* Boca Raton, FL: CRC Press. p. 639–669.

Coffin DR, Fishman ML. 1994. Physical and mechanical properties of highly plasticized pectin/starch films. *J Appl Polym Sci* 54:1311–1320.

Cunningham RL, Carr ME, Bagley EB. 1991. Polyurethane foams extended with corn flour. *Cereal Chem* 68:258–261.

Cunningham RL, Carr ME, Bagley EB. 1992. Preparation and properties of rigid polyurethane foams containing modified corn starch. *J Appl Polym Sci* 44:1477–1482.

Cunningham RL, Gordon SH, Felker FC, Eskins K. 1997. Jet-cooked starch–oil composite in polyurethane foams. *J Appl Polym Sci* 64:1355–1361.

Cunningham RL, Gordon SH, Felker FC, Eskins K. 1998. Glycols in polyurethane foam formulations with a starch-oil composite. *J Appl Polym Sci* 69:957–964.

Desai S, Thakore IM, Sarawade BD, Surekha D. 2000. Structure–property relationship in polyurethane elastomers containing starch as a crosslinker. *Polym Eng Sci* 40:1200–1210.

Erhan SZ. 2005. *Industrial Uses of Vegetable Oils.* Champaign, IL: AOCS Press.

Fang Q, Hanna MA. 2001. Preparation and characterization of biodegradable copolyester-starch based foams. *Biores Tech* 78:115–122.

Fishman ML, Coffin DR, Unruh JJ, Ly T. 1996. Pectin/starch/glycerol films: blends or composites. *J Macromol Sci Pure Appl Chem* 33:639–654.

Ge JJ, Zhong W, Guo ZR, Li WJ, Sakai K. 2000. Biodegradable polyurethane materials from bark and starch. I. Highly resilient foams. *J Appl Polym Sci* 77:2575–2580.

Ge JJ, Wu R, Shi XH, Yu H, Wang M, Li WJ. 2002. Biodegradable polyurethane materials from bark and starch. II. Coating material for controlled-release fertilizer. *J Appl Polym Sci* 86:2948–2952.

Guan J, Hanna MA. 2004. Functional properties of extruded foam composites of starch acetate and corn cob fiber. *Ind Crops Prod* 19:255–269.

Ha CS, Cho WJ. 2002. Miscibility, properties, and biodegradability of microbial polyester containing blends. *Prog Polym Sci* 27:759–809.

Ha S-K, Broecker HC. 2002. Characteristics of polyurethanes incorporating starch granules. *Polymer* 43:5227–5234.

Ha S-K, Broecker HC. 2003. The cross-linking of polyurethane incorporated with starch granules and their rheological properties: Influences of starch content and reaction conditions. *Macromol Mater Eng* 288:569–577.

Halley PJ. 2005. Themoplastic starch biodegradable polymers. In: Smith R, editor. *Biodegradable Polymers for Industrial Applications*. Cambridge, UK: Woodhead. p. 140–162.

Huang SJ. 1989. Biodegradation. In: Allen G, Bevington JC, editors. *Comprehensive Polymer Science: The Synthesis, Characterization, Reactions and Applications of Polymers*. Oxford, UK: Pergamon Press. p. 597–606.

Huang SJ, Macri C, Roby M, Benedict C, Cameron JA. 1981. Biodegradation of polyurethanes derived from polycaprolactonediols. In: Edwards KN, Editor. *Urethane Chemistry and Applications*. Washington, DC: American Chemical Society. p. 471–487.

Kim SH, Chin IJ, Yoon JS, Kim SH, Huang JS. 1998. Mechanical properties of biodegradable blends of poly(D,L-lactic acid) with starch. *Korea Polym J* 6:422–427.

Kim YD, Kim SC. 1998. Effect of chemical structure on the biodegradation of polyurethanes under composting conditions. *Polym Degrad Stab* 62:343–352.

Knutson A, Eskins K, Fanta GF. 1996. Composition and oil-retaining capacity of jet-cooked starch-oil composites. *Cereal Chem* 73:185–188.

Król P. 2007. Synthesis methods, chemical structures and phase structures of linear polyurethanes. Properties and applications of linear polyurethanes in polyurethane elastomers, copolymers and ionomers. *Prog Mater Sci* 52:915–1015.

Kweon DK, Cha DS, Park HJ, Lim ST. 2000. Starch-g-polycaprolactone copolymerization using diisocyanate intermediates and thermal characteristics of the copolymers. *J Appl Polym Sci* 78:986–993.

Kwon O-J, Yang S-R, Kim D-H, Park J-S. 2007. Characterization of polyurethane foam prepared by using starch as polyols. *J Appl Polym Sci* 103:1544–1553.

Lai LS, Kokini JL. 1991. Physicochemical changes and rheological properties of starch during extrusion. *Biotechnol Prog* 7:251–266.

Lee S-H, Yoshioka M, Shiraishi N. 2000. Liquefaction of corn bran (CB) in the presence of alcohols and preparation of polyurethane foam from its liquefied polyol. *J Appl Polym Sci* 78:319–325.

Lee H-T, Wu S-Y, Jeng R-J. 2006. Effects of sulfonated polyol on the properties of the resultant aqueous polyurethane dispersions. *Colloid Surf A: Physicochemical and Engineering Aspects* 276:176–185.

Li Y, Chen D, Li Y. 1999. Flame-resistant and biodegradable polyurethanes. *Polyurethane Ind* 14(3):12–14.

Lin L, Yao Y, Yoshioka M, Shiraishi N. 1997. Molecular weights and molecular weight distributions of liquefied wood obtained by acid-catalyzed phenolysis. *J Appl Polym Sci* 64:351–357.

Lourdin D, Coignard L, Bizot H, Colonna P. 1997. Influence of equilibrium relative humidity and plasticizer concentration on the water content and glass transition of starch materials. *Polymer* 38:5401–5406.

Lu Y, Weng L, Zhang L. 2004. Morphology and properties of soy protein isolate thermoplastics reinforced with chitin whiskers. *Biomacromolecules* 5:1046–1051.

Lu Y, Weng L, Cao X. 2005a. Biocomposites of plasticized starch reinforced with cellulose crystallites from cottonseed linter. *Macromol Biosci* 5:1101–1107.

Lu Y, Tighzert L, Berzin F, Rondot S. 2005b. Innovative plasticized starch films modified with waterborne polyurethane from renewable resources. *Carbohydr Polym* 61:174–182.

Lu Y, Tighzert L, Dole P, Erre D. 2005c. Preparation and properties of starch thermoplastics modified with waterborne polyurethane from renewable resources. *Polymer* 46:9863–9870.

Lu Y, Weng L, Cao X. 2006. Morphological, thermal and mechanical properties of ramie crystallites-reinforced plasticized starch biocomposites. *Carbohydr Polym* 63:198–204.

Mao LJ, Imam S, Gordon S, Cinelli P, Chiellini E. 2000. Extruded corn starch-glycerol-polyvinyl alcohol blends: mechanical properties, morphology, and biodegradability. *J Polym Environ* 4:205–211.

Martin O, Averous L. 2001. Poly(lactic acid): plasticization and properties of biodegradable multiphase systems. *Polymer* 42:6209–6219.

Matzinos P, Tserki V, Kontoyiannis A, Panayiotou C. 2002. Processing and characterization of starch/polycaprolactone products. *Polym Degrad Stab* 77:17–24.

Modesti M, Lorenzetti A. 2001. An experimental method for evaluating isocyanate conversion and trimer formation in polyisocyanate-polyurethane foams. *Eur Polym J* 37:949–954.

Mohanty AK, Misra M, Hinrichsen G. 2000. Biofibres, biodegradable polymers and biocomposites: An overview. *Macromol Mater Eng* 276/277:1–26.

Oertel G. 1994. *Polyurethane Handbook*. Munich: Hanser Publishers.

Otey FH, Mehltretter CL. 1965. Degraded starch polyoxyalkylene ether compositions and process for producing the same. U.S. Patent 3,165,508.

Park JW, Lee DJ, Yoo ES, Im SS, Kim SH, Kim YH. 1999. Biodegradable polymer blends of poly(lactic acid) and starch. *Korea Polym J* 7:93–101.

Petersen K, Nielsen PV, Bertelsen G, Lawther M, Olsen MB, Nilsson NH, Mortensen G. 1999. Potential of biobased materials for food packaging. *Trends Food Sci Technol* 10:52–68.

Roper H, Koch H. 1990. The role of starch in biodegradable thermoplastic materials. *Starch/Staerke* 42:123–130.

Santayanon R, Wootthikanokkhan J. 2003. Modification of cassava starch by using propionic anhydride and properties of the starch-blended polyester polyurethane. *Carbohydrate Polymers* 51:17–24.

Seidenstücker T, Fritz H-G. 1999. Compounding procedure, processing behaviour and property profiles of polymeric blends based on thermoplastic poly(ester-urethanes) and destructurized starch. *Starch/Stärke* 51:93–102.

Swamy BKK, Siddaramaiah. 2003. Structure-property relationship of starch-filled chain-extended polyurethanes. *J Appl Polym Sci* 90:2945–2954.

Swanson CL, Shogren RL, Fanta GF, Iman SH. 1993. Starch-plastic materials-preparation, physical properties, and biodegradability (a review of recent USDA research). *J Environ Polym Degrad* 1:155–166.

Suvorova AI, Tyukova IS, Trufanova EI. 2000. Biodegradable starch-based polymeric materials. *Russ Chem Rev* 69:451–459.

van Soest JJG, Benes K, de Wit D, Vliegenthart JFG. 1996. The influence of starch molecular mass on the properties of extruded thermoplastic starch. *Polymer* 37:3543–3552.

Wang X-L, Yang K-K, Wang Y-Z. 2003. Properties of starch blends with biodegradable polymers. *J Macromol Sci Part C: Polym Rev* 43:385–409.

Wilpiszewska K, Spychaj T. 2007. Chemical modification of starch with hexamethylene diisocyanate derivatives. *Carbohydr Polym* 70:334–340.

Wu QX, Zhang LN. 2001a. Preparation and characterization of thermoplastic starch mixed with waterborne polyurethane. *Ind Eng Chem Res* 40:558–564.

Wu QX, Zhang LN. 2001b. Structure and properties of casting films blended with starch and waterborne polyurethane. *J Appl Polym Sci* 79:2006–2013.

Xu Y, Hanna MA. 2005. Preparation and properties of biodegradable foams from starch acetate and poly(tetramethylene adipate-co-terephthalate). *Carbohydr Polym* 59:521–529.

Yao YG, Yoshioka M, Shiraishi N. 1996. Water-absorbing polyurethane foams from liquefied starch. *J Appl Polym Sci* 60:1939–1949.

Yu L, Dean K, Li L. 2006. Polymer blends and composites from renewable resources. *Prog Polym Sci* 31:576–602.

Cationic

▬ CHAPTER 5

Chitosan—Properties and Application

NILDA DE FÁTIMA FERREIRA SOARES

Universidade Federal de Viçosa, Food Technology Department, Viçosa-MG, Brazil

5.1 SOURCES

Chitosan is a polymer derived from chitin when the degree of deacetylation (DD) of chitin reaches values of about 50% and then becomes soluble in aqueous acidic media. The solubilization occurs by protonation of the $-NH_2$ function on the C-2 position of the D-glucosamine repeat unit, whereby the polysaccharide is converted to a polyelectrolyte in acidic media.

Chitosan is inexpensive, biodegradable, and nontoxic for mammals. This makes it suitable for use as an additive in the food industry (Koide, 1998; Shahidi et al., 1999), as a hydrating agent in cosmetics, and more recently as a pharmaceutical agent in biomedicine (Dodane and Vilivalam, 1998; Illum, 2003; Khor and Lim, 2003). This biopolymer is synthesized by an enormous number of living organisms; and, considering the amount of chitin produced annually in the world, it is the most abundant polymer after cellulose. Chitin occurs in nature as ordered crystalline microfibrils forming structural components in the exoskeletons of arthropods or in the cell walls of fungi such as *Aspergillus* and *Mucor* (Qin et al., 2006) and yeast. It is also

Biodegradable Polymer Blends and Composites from Renewable Resources. Edited by Long Yu
Copyright © 2009 John Wiley & Sons, Inc.

produced by a number of other living organisms in the lower plant and animal king-doms, serving in many functions where reinforcement and strength are required.

In industrial processing, chitin is extracted from crustaceans by acid treatment to dissolve calcium carbonate followed by alkaline extraction to solubilize proteins. In addition a decolorization step is often added to remove leftover pigments and obtain a colorless product. These treatments must be adapted to each chitin source, owing to differences in the ultrastructure of the initial materials (the extraction and pretreatment of chitin are not described here). The resulting chitin needs to be graded in terms of purity and color because residual protein and pigment can cause problems in further utilization, especially for biomedical products. By partial deacetylation under alkaline conditions, one obtains chitosan, which is the most important chitin derivative in terms of applications.

5.2 STRUCTURE

Chitin, poly(β-(1−4)-*N*-acetyl-D-glucosamine), is a natural polysaccharide of major importance, first identified in 1884 (see Fig. 5-1).

Depending on its source, chitin occurs as two allomorphs, namely, the α and β forms (Allen, 1963; Alves et al., 1999), which can be differentiated by infrared and solid-state NMR spectroscopy together with X-ray diffraction. A third allomorph γ-chitin has also been described (Allen, 1993; Anker, 2002) but, from a detailed

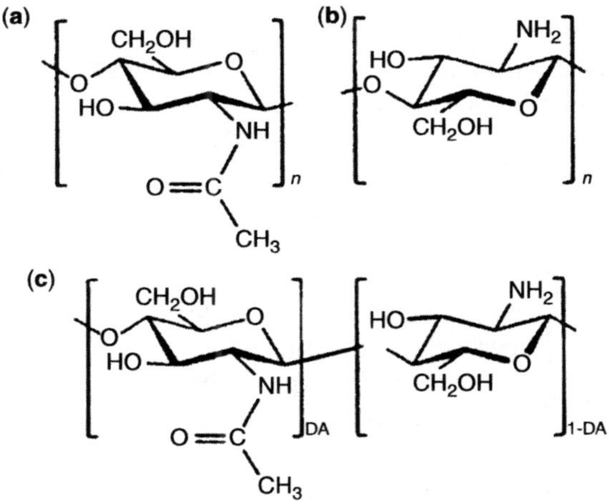

Fig. 5-1 Chemical structure of (a) chitin, poly(*N*-acetyl-β-D-glucosamine) and (b) chitosan, poly(D-glucosamine) repeat units. (c) Structure of partially acetylated chitosan, a copolymer characterized by its average degree of acetylation.

analysis, it seems that it is just a variant of the α family. α-Chitin is by far the most abundant; it occurs in fungal and yeast cell walls, in krill, in lobster and crab tendons and shells, and in shrimp shells, as well as in insect cuticle. It is also found in or produced by various marine living organisms. The rarer β-chitin is found in association with proteins in squid pens (Allen, 1963; Anker, 2002) and in the tubes synthesized by pogonophoran and vestimetiferan worms (Butler, 1996; Cha and Cooksey, 2001). It also occurs in *Aphrodite chaetae* (Chen, 1995) as well as in the lorica built by some seaweeds or protozoa (Debeaufort et al., 1993; Coma et al., 2002). A particularly pure form of β-chitin is found in the monocrystalline spines excreted by the diatom *Thalassiosira fluviatilis* (Debeaufort et al., 1993, 1998, 2000). To date, it has not been possible to obtain β-chitin either from solution or by *in vitro* biosynthesis.

The most important derivative of chitin is chitosan, obtained by (partial) deacetylation of chitin in the solid state under alkaline conditions (concentrated NaOH) or by enzymatic hydrolysis in the presence of a chitin deacetylase. Because of the semicrystalline morphology of chitin, chitosans obtained by solid-state reaction have a heterogeneous distribution of acetyl groups along the chains. In addition, it has been demonstrated that β-chitin exhibits much higher reactivity in deacetylation than α-chitin (Lerdthanangkul and Krochta, 1996). The influence of this distribution was examined by Aiba (1991), who showed that the distribution, random or blockwise, is very important in controlling solution properties. Reacetylation, up to 51%, of a highly deacetylated chitin in the presence of acetic anhydride gives a water-soluble derivative, whereas a heterogeneous product obtained by partial deacetylation of chitin is soluble only under acidic conditions, or is even insoluble. It was demonstrated from NMR measurements that the distribution of acetyl groups must be random to achieve the higher water solubility around 50% acetylation. Homogeneously deacetylated samples were obtained recently by alkaline treatment of chitin under dissolved conditions (Petersen et al., 1999). Toffey and Glasser, 1999 transformed chitosan films cast from aqueous acetic acid into chitin by heat treatment. Chitosan must be dissolved in an acid solution in order to activate its antimicrobial properties, and acetic acid is the best acid for this purpose (Romanazzi et al., 2005).

Chitosan is the only pseudonatural cationic polymer and thus it finds many applications that follows from this unique character (flocculants for protein recovery, depollution) (Rinaudo, 2006). An important parameter for the use of chitosan is its solubility, which depends not only on the average degree of deacetylation but also on the distribution of the acetyl groups along the main chain, in addition to molecular weight (Aiba, 1991; Kubota and Eguchi, 1997; Rinaudo and Domard, 1999). Chitosan solubility is usually tested in acetic acid by dissolving it in 1% or 0.1 M acetic acid. It has also been demonstrated that the amount of acid needed depends on the quantity of chitosan to be dissolved. The concentration of protons needed is at least equal to the concentration of $-NH_2$ units involved (Rinaudo et al., 1999). Chitosan also has good complexing ability due to the $-NH_2$ groups present in the polymer. Better chelation is related to higher degree of deacetylation, i.e., to the content of $-NH_2$ groups, as well as to their distribution.

The cationic nature of chitosan is very important in its chelation mechanism; the affinity of chitosan follows the order:

$$Cu^{2+} \gg Hg^{2+} > Zn^{2+} > Cd^{2+} > Ni^{2+} > Co^2 \sim Ca^{2+}$$

In the solid state, chitosan is a semicrystalline polymer. Its morphology has been investigated and many polymorphs are described in the literature. Single crystals of chitosan were obtained using fully deacetylated chitin of low molecular weight.

Characteristics of particular interest are broad antimicrobial activity (Sugarcane et al., 1992), antioxidant properties (Kamala et al., 2002), use as a curing agent for color and flavor development (Park et al., 1997), and excellent film-forming ability (Kittery et al., 1998). Chitosan is able to extend storage life and to control decay of strawberries, apples, peaches, pears, kiwifruit, cucumbers, litchis, sweet cherries, and citrus fruit. The biopolymer has a dual mechanism of action: it inhibits the growth of decay-causing fungi (Allan and Headwater, 1979) and it induces defense responses in host tissues (Shibuya and Minami, 2001).

In particular, the use of chitosan as films and edible coatings to extend shelf-life and preserve quality of fruits and vegetables has received considerable attention in recent years (Shahidi et al., 1999). For use as an effective coating, the films should exhibit adequate mechanical properties and create a gas barrier. However, the aqueous acid media necessary to solubilize the chitosan present a serious drawback, particularly to the mechanical properties of the cured films, since mechanical features of chitosan films are strongly solvent-dependent: different aqueous acid media employed for chitosan dissolution generate films with variable degrees of brittleness and toughness. Several researchers have developed methods to improve the properties of chitosan using chemical and enzymatic modifications.

One strategy to improve the mechanical properties and alter the hydrophilic character of chitosan is to synthesize alkyl derivatives (Descriers et al., 1996). The alkyl moiety may reduce the inter- and intracanial hydrogen bonds, introducing plastic characteristics to the derivative, as well as improving its barrier properties against water loss (Lazaridou and Biliaderis, 2002). Besides this, there is evidence that the alkyl moiety also plays important role in increasing the antimicrobial activity of chitosan derivative (Jail et al., 2001).

Films of water-soluble quaternary salts of chitosan with different alkyl moieties were successfully obtained using diethyl sulfate as a ethylating agent. For food use, the quaternary salt represents an advance as it obviates the need for an acid medium (Britton and Assist, 2007).

Another strategy to improve chitosan properties is to form the chitosan–glucose complex (CGC), which is a better preservative than chitosan alone. Research on this complex showed superior antioxidant activity compared with chitosan/glucose alone. The antimicrobial activity of the CGC was identical to that of chitosan against the common food spoilers and pathogens such as *E. coli*, *Pseudomonas* sop., *S. aurous*, and *B. cereus*. Thus CGC seems to be a novel natural preservative endowed with both antibacterial and antioxidant activity and may find applications in the food industry.

5.3 APPLICATION IN THE FOOD INDUSTRY

In native as well as modified forms both chitin and chitosan are used in a wide range of applications, such as in food, biotechnology, materials science, drugs and pharmaceuticals, and recently in gene therapy (Kop et al., 2002; Agulló et al., 2003). The biocompatibility and the biodegradability of chitosans make them a very interesting polymer for all these applications. The net catholicity and the presence of reactive functional groups (1 amino and 2 hydroxyl groups per glucose-N residue) in the molecule make chitosan an attractive bimolecular (Peasant and Tharanathan, 2007). The free amino group present in each monomeric unit affords an ammonium group, due to protonation, in aqueous acidic media. This offers scope for manipulation for preparing a broad spectrum of derivatives for specific end use applications in diversified areas such as agriculture and medicine.

In the food industry, chitosan has been studied as an antioxidant, as an antimicrobial (Prashanth and Tharanathan, 2007), for recovering soluble proteins from surimi waste (Wibowo et al., 2005), for edible coatings (Ribeiro et al., 2007), for films (Britto and Assis, 2007), for clarification of fruit juices, for control of enzymatic browning in fruits, and for purification of water (Shahidi et al., 1999). No and co-workers gave an extensive review of food applications of chitosan (No et al., 2007). Chitosan has also been used as wall material for encapsulation of some sensitive core ingredients such as lipophilic drugs (Ribeiro et al., 1999), vitamin D_2 (Shi and Tan, 2002), astaxanthin (Higuera-Ciapara et al., 2004), and olive oil extract (Kosaraju et al., 2006).

Chitin is widely used to immobilize enzymes and whole cells; enzyme immobilization has applications in the food industry, such as clarification of fruit juices and processing of milk when α- and β-amylases or invertase are grafted onto chitin. On account of its biodegradability, lack of toxicity, physiological inertness, antibacterial properties, hydrophilicity, gel-forming properties, and affinity for proteins, chitin has found applications in many areas other than food, such as in biosensors (Krajewska, 2004). Chitin-based materials are also used for the treatment of industrial waste (Rinaudo, 2006). Chitin can be processed in the form of films and fibers: Fibers were first developed by Austin and Brine (1977) and then by Hirano and Midorikawa (1998). The chitin fibers, obtained by wet-spinning of chitin dissolved in a 14% NaOH solution, can also after post chemical modifications give rise to a series of biofibers: chitin cellulose, chitin-silk, chitin-glycosaminoglyccus (Hirano and Midorikawa, 1998; Hirano et al., 1999). They are nonallergic, deodorizing, antibacterial, and moisture controlling (Tatsumi et al., 1991). Regenerated chitin derivative fibers are used as binders in the paper making process; the addition of 10% *n*-isobutylchitin fiber improves the breaking strength of paper (Kobayashi et al., 1992).

5.4 ANTIMICROBIAL PROPERTIES

The antimicrobial property of chitosan and its derivatives has received considerable attention in recent years as a potential food preservative of natural origin due to its antimicrobial activity against a wide range of foodborne filamentous fungi, yeast, and bacteria (Sagoo et al., 2002), as well as to emerging problems associated with synthetic chemical agents. The mechanism of the antimicrobial activity of chitosan has not yet been fully elucidated, but several hypotheses have been proposed. The most feasible is a change in cell permeability due to interactions between the positively charged chitosan molecules and the negatively charged components (lipopolysaccharides and proteins) of the microbial cell membranes. The cationic charge of the chitosan molecule gives rise to aggressive binding to the microbial cell surface, leading to gradual shrinkage of the cell membrane and finally death of the cell. This interaction leads to the leakage of proteinaceous and other intracellular constituents (Leuba and Stössel, 1986; Papineau et al., 1991; Sudarshan et al., 1992; Young et al., 1982; Fang et al., 1994). Other mechanisms are the interaction of diffused hydrolysis products with microbial DNA, which leads to the inhibition of mRNA and protein synthesis (Hadwiger and Loschke, 1981; Hadwiger et al., 1986; Sudarshan et al., 1992) and the chelation of metals, and essential nutrients (Cuero et al., 1991).

Chitosan shows a broad-spectrum antimicrobial activity against both gram-positive and gram-negative bacteria and fungi (Kumar et al., 2005). According to Muzzarelli et al. (1990), chitosan's antimicrobial activity against bacteria could be due to the polycationic nature of its molecule, which allows interaction and formation of polyelectrolye complexes with acid polymers produced at the bacteria cell surface (lipopolysaccharides, teichoic and teichunoric acids, or capsular polysaccharides). Chitosan-based films and coatings tested on *Listeria monocytogenes* were found to inhibit the growth of this microorganism (Coma et al., 2002). El Ghaouth et al. (1992b) showed that coatings based on 1% and 2% chitosan reduced the incidence of tomato deterioration, mainly that caused by *Botrytis cinerea*. Chitosan films were produced from a solution of starch yam (4%) and glycerol (2%) gelatinized in a viscoamylograph and with addition of chitosan at concentrations of 1%, 3%, and 5%. Films incorporating chitosan at different concentrations showed similar antimicrobial efficiency on *Salmonella enteritidis* suspension. The chitosan-containing films caused a reduction of 1 to 2 log cycles in the number of microorganisms, whereas the pure chitosan solution gave a reduction of 4 to 6 log cycles log compared to the control and starch films (Durango et al., 2005).

In a study on the mode of antimicrobial action of chitosan (250 ppm at pH 5.3) by monitoring the uptake of the hydrophobic probe 1-*N*-phenylnaphthylamine, *Escherichia coli*, *Pseudomonas aeruginosa*, and *Salmonella typhimurium* showed significant uptake, which was reduced (*E. coli*, salmonellae) or abolished (*P. aeruginosa*) by $MgCl_2$. Chitosan also sensitized *P. aeruginosa* and salmonellae to the lytic effect of sodium dodecyl sulfate. Electrophoretic and chemical analyses of the cell-free supernatants revealed no release of lipopolysaccharides or other membrane lipids. Electron-microscopic observations showed that chitosan caused extensive cell surface alterations and covered the outer membrane with vesicular structures, resulting in the loss of barrier functions (Helander et al., 2001).

However, many other food compounds such as carbohydrate, protein, fat, minerals, vitamins, salts, and others may interact with chitosan and lead to loss or enhancement of antibacterial activity.

Devlieghere et al. (2004) studied extensively the influence of different food components (starch, protein, oils, and NaCl) on the antimicrobial effect of chitosan. For this, the media were inoculated with *Candida lambica* (2 log CFU/ml) and incubated at 7°C with varying chitosan concentrations (43 kDa, DD = 94%; 0%, 0.005%, and 0.01%) and with the separate addition of the following food components: starch (0%, 1%, and 30% water-soluble starch), proteins (0%, 1%, and 10% whey protein isolate), oil (0%, 1%, and 10% sunflower oil) and NaCl (0%, 0.5%, and 2%). Starch, whey protein, and NaCl had a negative effect on the antimicrobial activity of chitosan. The oil had no influence. The same authors tested the antimicrobial activity against several microorganisms at different chitosan concentrations and pH values at 7°C for 60 days. The results are summarized in Table 5-1.

The major postharvest losses of fruits are due to fungal infection, physiological disorders, and physical injuries (El Ghaouth et al., 1992a, b). Edible coatings might extend the storability of perishable commodities by acting as a protective barrier to reduce respiration and transpiration rates through fruit surfaces, retard microbial growth and color changes, and improve texture quality of fruits (Kester and Fennema, 1986). Studies have shown that chitosan-based coatings have the potential to increase the shelf-life of fresh fruits and vegetables, inhibiting the growth of microorganisms, reducing ethylene production, increasing internal carbonic gas, and decreasing oxygen levels (Lazaridou and Biliaderis, 2002).

Soares et al. (2007) evaluated guava subjected to four treatments (starch coating with 1.0% and 1.5% of chitosan acetic acid, starch coating, and acetic acid/starch coating) during 12 days of storage at room temperature. The skin color of the fruit remained more greenish when treated with chitosan (see Fig. 5-2) and the pulp was more consistent (see Fig. 5-3).

Geraldine et al. (2007) found that peeled garlic coated with chitosan-agar solution had a significant reduction in respiration rate compared with peeled cloves (see Fig. 5-4).

Durango et al. (2006) developed an edible antimicrobial coating based on a starch–chitosan matrix and evaluated its effect on minimally processed carrot by means of microbiological analysis. Carrot slices were immersed into four coatings (1, no coating; 2, yam starch + glycerol; 3, yam starch + glycerol + 0.5% chitosan; 4, yam starch + glycerol + 1.5% chitosan) and stored in trays wrapped with poly(vinylchloride) films at 10°C for 15 days. The presence of 1.5% chitosan in the coating inhibited the growth of lactic acid bacteria and total coliforms throughout the storage period (see Fig. 5-5).

Fruits and vegetables, in general, are wrapped in their own natural packaging. This barrier regulates the transport of oxygen, carbon dioxide, and moisture

TABLE 5-1 Time for Turbidity (Days) at 7°C for Different Microorganisms in Traditional Growth Media with Different Chitosan Concentrations

| | | Turbidity Time (days) | | | | | | | | | | |
| | | Chitosan (mg/ml) | | | | | | | | | | |
Microorganism	pH	0	40	60	80	100	200	250	300	400	500	750
Candida lambica	5.5	5	5	5	5	6	14	ND[a]	40	–[b]	–	ND
Cryptococcus humiculus	4	12	12	12	–	–	–	ND	–	–	–	ND
Photobacterium phosphoreum	5.5	14	14	21	–	–	–	ND	–	–	–	ND
Pseudomonas fluorensces	5.5	8	18	28	–	–	–	ND	–	–	–	ND
Enterobacter aeromonas	5.5	39	47	–	–	–	–	ND	–	–	–	ND
Bacillus cereus	5.5	11	11	–	–	–	–	ND	–	–	–	ND
Brochothrix thermosphacta	5.5	5	5	5	–	–	–	ND	–	–	–	ND
Listeria monocytogenes	5.5	8	10	18	32	33	–	ND	–	–	–	ND
Lactobacillus sakei	5.5	4	4	4	4	6	12	ND	13	26	30	ND
Lactobacillus plantarum 1	5.0	24	ND	ND	ND	24	ND	49	ND	ND	–	–
Lactobacillus plantarum 2	5.0	23	ND	ND	ND	25	ND	25	ND	ND	25	–
Lactobacillus curvatus	5.0	22	ND	ND	ND	22	ND	22	ND	ND	ND	–
Pediococcus acidilactiei	5.0	40	ND	ND	ND	40	ND	40	ND	ND	–	–

[a]ND, not determined.
[b]–, ■■■■■.

114

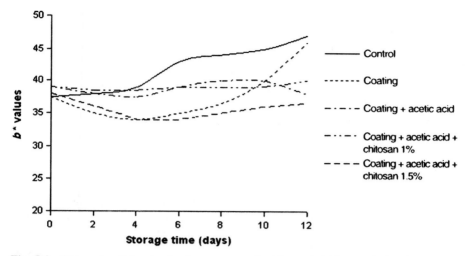

Fig. 5-2 Skin color (b^* values) of guava coated with starch/chitosan during 12 days of storage at room temperature.

and also reduces the loss of flavor and aroma. Fresh-cut vegetables and fruits lose their barrier (skin and peel), so that maintenance of the appropriate atmosphere for the product must be established by other means to maintain product shelf-life. Oxygen and CO_2 concentrations and also moisture loss must be taken into consideration. Interest in the quality and microbiological safety of minimally processed foods promotes the use of chitosan/starch coatings to preserve the products. Geraldine et al. (2007) showed significant inhibition of

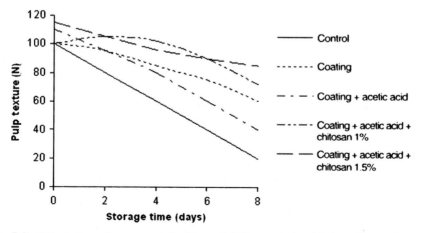

Fig. 5-3 Pulp texture of guava coated with starch/chitosan during 12 days of storage at room temperature.

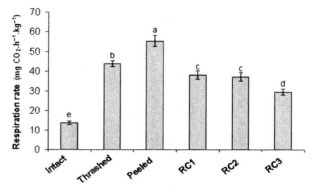

Fig. 5-4 Garlic respiration rates at 25°C for different treatments: bulb (intact); cloves thrashed; cloves peeled; cloves peeled and coated with agar-agar (RC1), agar-agar and acetic acid (RC2), agar-agar, and chitosan (RC3). Bars represent the mean standard error. Means followed by the same letter were not significantly different ($P \geq 0.05$).

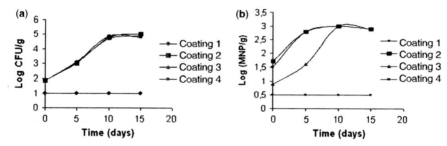

Fig. 5-5 Effect of coatings on lactic acid bacteria (a) and total coliforms (b) in minimally processed carrot stored at 10°C for 15 days. Coatings are as defined in the text.

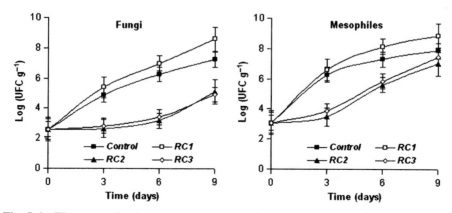

Fig. 5-6 Filamentous fungi and yeasts and mesophilic aerobic counts of minimally processed garlic without coating (Control) and coated with agar-agar (RC1), agar-agar and acetic acid (RC2), and agar-agar and chitosan (RC3), during storage at 25°C. Bars represent mean and standard errors.

Fig. 5-7 Left to right: Garlic cloves without coatings (control) and coated with agar-agar, agar-agar and acetic acid, and agar-agar and chitosan, after 9 days at 25°C.

growth of filamentous fungi and yeast and mesophiles on coated garlic cloves (see Fig. 5-6). Also, chitosan coatings affect root primordium formation and inhibit browning in garlic cloves (see Fig. 5-7).

5.5 OTHER PROPERTIES

Another important action attributed to chitosan is its antioxidant activity, which is dependent on molecular weight and concentration of chitosan. Kamil et al. (2002) investigated chitosan solutions of different viscosities (360, 57, and 14 cP; corresponding molecular weights of 1800, 960, and 660 kDa) with cooked, comminuted flesh of herring (*Clupea harengus*). The oxidative stability of fish flesh with added chitosans (50, 100, and 200 ppm) was compared with that with added conventional antioxidants, butylated hydroxyanisole + butylated hydroxytoluene (BHA + BHT, 200 ppm) and *tert*-butylhydroquinone (TBHQ, 200 ppm), during storage at 4°C. Among the three chitosans, 14 cP chitosan solution was most effective in preventing lipid oxidation evaluated by the formation of TBARS. At 200 ppm, 14 cP chitosan exerted an antioxidant effect similar to that of commercial antioxidants in reducing TBARS values in comminuted herring flesh. Similarly, Kim and Thomas (2007) also observed that the antioxidative effects of chitosan in salmon depended on its molecular weight (30, 90, and 120 kDa) and concentration (0.2%, 0.5%, and 1.0%). The 30 kDa chitosan showed higher scavenging activity than 90 and 120 kDa chitosan. The scavenging activities of chitosans increased with increasing concentration, except for the 120 kDa chitosan, which showed no significant difference with concentration.

Xue et al. (1998) suggested that the antioxidant mechanism of chitosan could be exerted by chelating action on metal ions and/or combination with lipids. Lipid oxidation was evaluated on skinless pink salmon (*Oncorhynchus gorbuscha*) with an edible chitosan coating (Sathivel, 2005). After 3 months of frozen storage the chitosan coating was effective in reducing moisture loss of fillets by about 50% compared with the control uncoated fillets, and in delaying lipid oxidation. After the period of storage, the edible coating showed no effect on whiteness values for cooked pink salmon fillets. In another study with cod patties the chitosan coating was effective as a protective film, retarding lipid oxidation and microbial spoilage, and acted as a barrier against oxygen (Jeon et al., 2002).

In many countries sodium nitrite is used as a curing agent for color and flavor development as well as preservative effect in various meat products. However, nitrite reacts with amines in meat and may produce nitrosoamine, a strong toxicant detrimental to human health. Some workers (Park et al., 1999; Youn et al., 1999, 2001) have investigated the possible role of chitosan in lieu of sodium nitrite as curing agent in sausages, and found that addition of chitosan could reduce or replace the use of nitrite without affecting preservative effect and color development. A mixture of 0.2% chitosan (120 kDa, DD = 85%, dissolved in 0.3% lactic acid) and 0.005% sodium nitrite or 0.5% chitosan alone exhibited the same preservative effect as did 0.01% sodium nitrite alone when added in meat sausage. The addition of 0.2% chitosan (30 kDa, DD = 92%) dissolved in 0.3% lactic acid reduced sodium nitrite by half of the standard amount (150 ppm) without affecting the quality and storage stability of sausage. In addition to the preservative effect, chitosan also notably reduced the concentration of residual nitrite in sausage. The concentration of residual nitrite in sausage decreased with increasing molecular weight and concentration of chitosan.

Geraldine (2004) evaluated the thermal treatment (121°C for 15 min) of chitosan solution (1% chitosan–1% acetic acid), for the preparation of a coating solution (0.24% chitosan) and reported a low effect on antifungal activity (*Aspergillus niger*). In the same study the author also investigated the effect of chitosan and acetic acid concentration on colony diameter and germination of *A. flavus* (see Table 5-2). The fungus sporulated but showed no growth in the culture medium.

5.6 CHITOSAN DERIVATIVES

After chitosan, the most studied derivative of chitin is carboxymethylchitin (CM-chitin), a water-soluble anionic polymer. The carboxymethylation of chitin is done similarly to that of cellulose; chitin is treated with monochloroacetic acid in the presence of concentrated sodium hydroxide (Rindlav-Westling et al., 1998). The method for cellulose derivatization is also used to prepare hydroxypropylchitin, a water-soluble derivative used for artificial lachrymal drops (Rico-Peña and Torres, 1990; Rindlav-Westling et al., 1998). Other derivatives such as fluorinated chitin (Robertson, 1993), *O*-sulfated chitin (Redl et al., 1996; Sotero, 2000; Sebti and Coma, 2002), (diethylamino)ethylchitin (Sothornvit and Krochta, 2001), phosphoryl-chitin (Stuchell and Krochta, 1995), mercaptochitin (Tatsumi et al., 1991), and chitin carbamates (Tharanathan and Farooqahmed, 2003) have been described in the

TABLE 5-2 Effect on the Growth of A. *niger* of the Increase of Acetic Acid in 1% Agar-Agar Solution Incorporating 0.36% Chitosan

Chitosan Content	Parameter	Concentration of Acetic Acid (%)				
		0.12	0.13	0.14	0.15	0.16
Without chitosan	pH	3.63	3.60	3.57	3.55	3.53
	Diameter (mm)[a,b]	13.97 ± 1.42^{a}	11.91 ± 1.12^{ab}	10.29 ± 0.84^{b}	7.82 ± 0.45^{c}	0^{d}
	Germination	+	+	+	+	—
With chitosan (0.36%)	pH	5.00	4.93	4.87	4.82	4.78
	Diameter (mm)	0	0	0	0	0
	Germination	+	+	+	+	+

[a]Values are mean ± standard error.

[b]Means followed by the same superscript letter within the row are not significantly different by the Tukey test ($P \geq 0.05$).

literature. Modification of chitin is also often effected via water-soluble derivatives of chitin (mainly CM-chitin).

The same type of chemical modifications (etherification and esterification) as for cellulose can be performed on the available C-6 and C-3 hydroxyl groups of chitin (Prashanth and Tharanathan, 2007). Chitin can be used in blends with natural or synthetic polymers; it can be crosslinked by the agents used for cellulose (epichlorhydrin, glutaraldehyde, etc.) or grafted in the presence of ceric salt (Prashanth and Tharanathan, 2007) or after selective modification (Williams et al., 1978). Chitin is partially degraded by acid to obtain series of oligochitins (Lerdthanangkul and Krochta, 1996; Wong et al., 1994). These oligomers, as well as those derived from chitosan, are recognized for their bioactivity, including antitumor, bactericidal, and fungicidal activity, eliciting chitinase and regulating plant growth. They are used in testing for lysozyme activity and are also used as active starting blocks to be grafted onto protein and lipids to obtain analogues of glycoproteins and glycolipids.

To enhance the antibacterial potency of chitosan, thiourea chitosan was prepared by reacting chitosan with ammonium thiocyanate followed by its complexation with silver (Chen et al., 2005). It has been reported that quaternary ammonium salts of chitosan exhibit good antibacterial activities; for example, diethylmethyl chitosan chloride showed higher antibacterial activity than chitosan.

With the same aim of improving antibacterial activity, Xie et al. (2007) reported novel synthetic procedures for the preparation of chitosan derivatives with a quaternary ammonium salt, expecting to improve its water solubility and anti-bacterial activity. The quaternary derivative (ethylamine hydroxyethyl chitosan [EHC]) was prepared by treating hydroxyethyl chitosan (HEC) with chloro-ethylamine hydrochloride in sodium hydroxide solution. This work suggested that EHC decreased the intermolecular interactions, such as van der Waals forces, and increased water solubility. Derivates showed good inhibitory effects against *Escherichia coli*, and $NH_2\%$ was the factor that most strongly affected the anti-bacterial activity.

Novel *N,O*-acylchitosan derivatives were more active than chitosan itself against the gray mould fungus *Botrytis cinerea* and the rice leaf blast fungus *Pyricularia oryzae*, and the effect was concentration dependent (Badawy et al., 2004). Hydroxypropyl chitosan grafted with maleic acid sodium killed over 99% of *Staphylococcus aureus* and *E. coli* within 30 min of contact at a concentration of 100 ng/ml. It was a potent inhibitor of *Azotobacter mali*, *Clostridium diplodiella*, *Fusarium oxysporum*, and *Pyricularia piricola*. The degree of hydroxypropyl group substitution also influenced the antifungal activity. With regard to their antifungal mechanisms, it was reported that these chitosan derivatives directly interfered with fungal growth and activated several defense processes, such as accumulation of chitinases, synthesis of proteinase inhibitors and induction of callous synthesis. It was also noted that the antibacterial activity of chitosan derivatives increased with increasing chain length of the alkyl substituent, and this was attributed to the increased hydrophobicity (Prashanth and Tharanathan, 2007).

Some applications for food coating or film require better controlled release of the incorporated components, such as additives, antioxidants, and aromas from the film/coating into the packed food. To enhance this property, nanocomponents has been added to chitosan-based films and coatings. Chitosan/organic rectorite (chitosan/OREC) nanocomposite films with different mass ratios of chitosan to organic rectorite and corresponding drug-loaded films were obtained by a casting/solvent evaporation method. Rectorite is a kind of layered silicate, with structure and characteristics much like those of montmorillonite. The addition of OREC to pure chitosan film influenced many of the properties. The enhancement of properties was related to the amount and the interlayer distance of the layered silicate in the chitosan/OREC nanocomposite films. In-vitro studies of controlled drug release showed a slower and more continuous release from the nanocomposite films than from pure chitosan film. Chitosan/OREC nanocomposite films provide promise as applications as antimicrobial agents, water-barrier compounds, antiultraviolet compounds, and controlled drug release carriers in antimicrobial food packaging and drug-delivery system (Wang et al., 2007).

Copolymer hydrogels have similarly been studied for control of compound delivery. These thermosensitive hydrogels, which are triggered by changes in environmental temperature to give in-situ hydrogel formation, have recently attracted the attention of many investigators for biomedical applications. Tang et al. (2007) aimed to develop an injectable and thermosensitive system based upon a blend of chitosan and poly(vinyl alcohol) (PVA) that can serve as a therapeutic drug-delivery system promoting tissue repair and regeneration through controlled release of loaded drugs. The thermosensitive hydrogel was prepared by mixing chitosan, PVA, and sodium hydrogencarbonate. The mixture was a liquid aqueous solution at low temperature (about 4°C) but a gel under physiological conditions, at which bioactive species can be safely and uniformly incorporated. These copolymer hydrogels are potentially suited for a wide range of in-vivo biomedical applications, such as drug release and tissue repair and regeneration. The system has been tested for drug delivery but it can also be applied for release of food additives.

The presence of *N*-sulfofurfuryl and quaternary ammonium groups has a remarkable impact on the response of the chitosan film to charged proteins in terms of adsorbed quantity and selectivity.

Taking advantage of the availability of functional groups for chemical reactions at the chitosan surface and the diverse bioactivity of its charged derivatives, Hoven et al. (2007) aimed to tailor protein adsorption at the chitosan surface by chemically introducing charged functionalities specifically to amino groups under heterogeneous conditions. The charged-modified chitosan surfaces are exposed to proteins having different molecular weights and isoelectric points. Besides hydrophobic interactions and hydrogen bonding, it is hypothesized that the extent of protein adsorption should depend also on electrostatic interaction between protein molecules and the modified groups on the chitosan surface. This surface modification should expand the applicability of chitosan in biomedical-related fields.

Positive and negative charges were introduced to chitosan surfaces via methylation using methyl iodide (MeI) and reductive alkylation using 5-formyl-2-furan sulfonic acid (FFSA). The chitosan films bearing negative charges of N-sulfofurfuryl groups on their surface (SFC films) exhibited selective protein adsorption against both negatively charged proteins (albumin and fibrinogen) and positively charged proteins (ribonuclease, lysozyme). The adsorption can be explained in terms of electrostatic attraction and repulsion. In contrast, the adsorption behavior of chitosan films bearing positive charges of quaternary ammonium groups on their surface (QAC films) was anomalous. The quantity of the adsorbed protein tended to increase as a function of the swelling ratio of the QAC films regardless of the charge characteristics of the protein. The ability to sustain their charges in a broader pH range should make these surface-charged chitosan films more versatile in applications than native chitosan films for which the charge is altered as a function of environmental pH (Jia et al., 2001).

REFERENCES

Agulló E, Rodríguez MS, Ramos V, Albertengo L. 2003. Present and future role of chitin and chitosan in food. *Macromol Biosci* 3:521–530.

Aiba S. 1991. Studies on chitosan: 3. Evidence for the presence of random and block copolymer structures in partially N-acetylated chitosans. *Int J Biol Macromol* 13:40–44.

Allan CR, Hadwiger LA. 1979. The fungicidal effect of chitosan on fungi of varying cell wall composition. *Exp Mycol* 3:285–287.

Allen L, Nelson AI, Steinber GMP, McGill JN. 1963. Edible carbohydrates food coating II: Evaluation on fresh meat products. *Food Technol* 17:1442–1446.

Alves RML, Grossmann MVE, Silva RSSF. 1999. Gelling properties of extruded yam (*Dioscorea alata*) starch. *Food Chem* 67:123–127.

Anker M, Berntsen J, Hermansson AM, Stading M. 2002. Improved water vapor barrier of whey protein films by addition of an acetylated monoglyceride. *Innov Food Sci Emer Technol* 3:81–92.

Austin PR, Brine J. 1977 Chitin films and fibers. US Patent 4,029,727.

Badawy ME, Rabea EI, Rogge TM, et al. 2004. Synthesis and fungicidal activity of new N,O-acyl chitosan derivatives. *Biomacromolecules* 5:589–595.

Britto D, Assis BG. 2007. A novel method for obtaining a quaternary salt of chitosan. *Carbohydr Polym* 69:305–310.

Butler BL, Vergano PJ, Testin RF, Bunn JM, Wiles JL. 1996. Mechanical and barrier properties of edible chitosan films as affected by composition and storage. *J Food Sci* 61:953–956.

Cha DH, Cooksey K. 2001. Preparation and diffusion rate of a nisin incorporated antimicrobial film. In: Book of abstracts, IFT Annual Meeting Technical Program. New Orleans. Chicago, IL: IFT.

Chen H. 1995. Functional properties and applications of edible films made of milk proteins. *J Dairy Sci* 78:2563–2583.

Chen S, Wu G, Zeng H. 2005. Preparation of high antimicrobial activity thiourea chitosan–Ag$^+$ complex *Carbohydr Polym* 60:33–38.

Coma V, Martial-Giros A, Garreau S, Copinet A, Salin F, Deschamps A. 2002. Edible antimicrobial films based on chitosan matrix. *J Food Sci* 67:1162–1168.

Cuero RG, Osuji G, Washington ANC. 1991. *N*-Carboxymethylchitosan inhibition of aflatoxin production: role of zinc. *Biotechnol Lett* 13:441–444.

Debeaufort F, Martin-Polo MO, Voilley A. 1993. Polarity, homogeneity and structure affect water vapor permeability of model edible films. *J Food Sci* 58:426–429.

Debeaufort F, Quezada-Gallo JA, Voilley A. 1998. Edible films and coatings: tomorrow packaging: a review. *Crit Rev Food Sci* 38:299–313.

Debeaufort F, Quezada-Gallo JA, Delporte B, Voilley A. 2000. Lipid hydrophobicity and physical state effects on the properties of bilary edible films. *J Membr Sci* 180:47–55.

Desbrières J, Martinez C, Rinaudo M. 1996. Hydrophobic derivatives of chitosan: Characterization and rheological behaviour. *Int J Biol Macromol* 19:21–28.

Devlieghere F, Vermeulen A, Debevere J. 2004. Chitosan: antimicrobial activity, interactions with food components and applicability as a coating on fruit and vegetables. *Food Microbiol* 21:703–714.

Dodane V, Vililvalam VD. 1998. Pharmaceutical applications of chitosan. *Pharm Sci Technol Today* 1:246–253.

Durango AM, Soares NFF, Benevides S, Teixeira J, Carvalho MGX, Wobeto C, Andrade NJ. 2005. Development and evaluation of an edible antimicrobial film based on yam starch and chitosan. *Packag Technol Sci* 19:55–59.

Durango AM, Soares NFF, Andrade NJ. 2006. Microbiological evaluation of an edible antimicrobial coating on minimally processed carrots. *Food Control* 17:336–41.

El Ghaouth A, Arul J, Grenier J, Asselin A. 1992a. Antifungal activity of chitosan on two postharvest pathogens of strawberry fruits. *Phytopathology* 82:398–402.

El Ghaouth A, Ponnampalam R, Castaigne F, Arul J. 1992b. Chitosan coating to extend the storage life tomatoes. *Hort Sci* 27:1016–1018.

Fang SW, Li CF, Shih DYC. 1994. Antifungal activity of chitosan and its preservative effect on low-sugar candied kumquat. *J Food Protect* 57:136–140.

Geraldine RM. 2004. Embalagem ativa na conservação de alho minimamente processado. PhD Thesis. Universidade Federal de Viçosa.

Geraldine RM, Soares NFF, Botrel DA, Gonçalves LA. 2007. Characterization and effect of edible coatings on minimally processed garlic quality. *Carbohydr Polym*, DOI: 10.1016/j.carbpol.2007.09.012.

Hadwiger LA, Loschke DC. 1981. Molecular communication in host–parasite interactions: hexosamine polymers (chitosan) as regulator compounds in race-specific and other interactions. *Phytopathology* 71:756–62.

Hadwiger LA, Kendra DF, Fristensky BW, Wagoner W. 1986. Chitosan both activates genes in plants and inhibits RNA synthesis in fungi. In: Muzzarelli R, Jeuniaux C, Gooday GW, editors. *Chitin in Nature and Technology*. New York: Plenum Press. 209–214.

Helander IM, Nurmiaho-Lassila EL, Ahvenainen R, Rhoades J, Roller S. 2001. Chitosan disrupts the barrier properties of the outer membrane of gram-negative bacteria. *Int J Food Microbiol* 71:235–244.

Higuera-Ciapara I, Felix-Valenzuela L, Goycoolea FM, Arguelles-Monal W. 2004. Microencapsulation of astaxanthin in a chitosan matrix. *Carbohydr Polym* 56:41–45.

Hirano S, Midorikawa T. 1998. Novel method for the preparation of *N*-acylchitosan fiber and *N*-acylchitosan-cellulose fiber. *Biomaterials* 19:293–297.

Hirano S, Nakahira T, Nakagaa M, Kim SK. 1999. The preparation and application of functional fibers from crac shell chitin. *J Biotechnol* 70:373–377.

Hoven VP, Tangpasuthadol V, Angkitpaiboon Y, Vallapa N, Kiatkamjornwong S. 2007. Surface-charged chitosan: Preparation and protein adsorption. *Carbohydr Polym* 68:44–53.

Illum L. 2003. Nasal drug delivery-possibilities, problems and solutions. *J Control Release* 87:187–198.

Jeon YJ, Kamil JYVA, Shahidi F. 2002. Chitosan as an edible invisible film for quality preservation of herring and Atlantic cod. *J Agric Food Chem* 50:5167–5178.

Jia Z, Shen D, Xu W. 2001. Synthesis and antibacterial activities of quaternary ammonium salt of chitosan. *Carbohydr Res* 333:1–6.

Kamil JYVA, Jeon YJ, Shahidi F. 2002. Antioxidative activity of chitosans of different viscosity in cooked comminuted flesh of herring (*Clupea harengus*). *Food Chem* 79:69–77.

Kester JJ, Fennema OR. 1986. Edible films and coatings: a review. *Food Technol* 60:47–59.

Khor E, Lim Y. 2003. Implantable applications of chitin and chitosan. *Biomaterials* 24:2339–2349.

Kim KW, Thomas RL. 2007. Antioxidative activity of chitosans with varying molecular weights. *Food Chem* 101:308–313.

Kittur FS, Kumar KR, Tharanathan RN. 1998. Functional packaging properties of chitosan films. *Z Lebensm Unters Forsch* 206:44–47.

Ko JA, Park HJ, Hwang SJ, Park JB, Lee JS. 2002. Preparation and characterization of chitosan microparticles intended for controlled drug delivery. *Int J Pharm* 249:165–174.

Kobayashi Y, Nishiyama M, Matsuo R, Tokura S, Nishi N. 1992. Application of chitin and its derivatives to paper industry. In: Hirano S, Tokura S, editors. *Chitin Chitosan*. Proceeding of the 2nd International Conference, Jpn Soc. Chitin Chitosan, Tattori, Japan. p. 244–247.

Koide SS. 1998. Chitin-chitosan, properties, benefits and risks. *Nutr Res* 18:1091–1101.

Kosaraju SL, D'ath L, Lawrence A. 2006. Preparation and characterization of chitosan microspheres for antioxidant delivery. *Carbohydr Polym* 64:163–167.

Krajewska B. 2004 Application of chitin and chitosan-based materials for enzyme immobilization: a review. *Enzyme Microbiol Technol* 35:126–39.

Kubota N, Eguchi Y. 1997. Facile preparation of water-soluble *N*-acetylated chitosan and molecular weight dependence of its water-solubility. *Polym J* 29:123–127.

Kumar ABV, Varadaraj MC, Gowda LR, Tharanathan RN. 2005. Characterization of chito-oligosaccharides prepared by chitosanolysis with the aid of papain and pronase, and their bactericidal action. *Biochem J* 391:167–175.

Lazaridou A, Biliaderis CG. 2002. Thermophysical properties of chitosan, chitosan-starch and chitosan-pullulan films near the glass transition. *Carbohydr Polym* 48:179–190.

Lerdthanangkul S, Krochta JM. 1996. Edible coating effects on postharvest quality of green bell peppers. *J Food Sci* 61:176–179.

Leuba JL, Stössel P. 1986. Chitosan and other polyamines: antifungal activity and interaction with biological membranes. In: Muzzarelli R, Jeuniaux C, Gooday GW, editors. *Chitin in Nature and Technology.* New York: Plenum Press. p. 215–222.

Muzzarelli R, Tarsi R, Filippini O, Giovanetti F, Biagini G, Varaldo PR. 1990. Antimicrobial properties of *N*-carboxybutyl chitosan. *Antimicrob Agents Chemother* 34:2019–2023.

No HK, Meyers SP, Prinyawiwatkul W, Xu Z. 2007. Applications of chitosan for improvement of quality and shelf life of foods: a review. *J Food Sci* 72:R87–R100.

Papineau AM, Hoover DG, Knorr D. 1991. Antimicrobial effect of water-soluble chitosans with high hydrostatic pressure. *Food Biotechnol* 5:59–66.

Park SM, Youn SK, Kim HJ, Ahn DH. 1999. Studies on the improvement of storage property in meat sausage using chitosan-I. *J Korean Soc Food Sci* 28:167–71.

Petersen K, Nielsen PV, Bertelsen G, et al. 1999. Potential of biobased materials for food packaging. *Food Sci Technol* 10:52–68.

Prashanth KVH, Tharanathan RN. 2007. Chitin/chitosan: modifications and their unlimited application potential—an overview. *Trends Food Sci Technol* 18:117–131.

Qin C, Li H, Xiao Q, Liu Y, Zhu J, Du Y. 2006. Water-solubility of chitosan and its antimicrobial activity. *Carbohydr Polym* 63:367–374.

Redl A, Gontard N, Guilbert S. 1996. Determination of sórbico acid diffusivity in edible wheat gluten and lipid films. *J Food Sci* 61:116–120.

Ribeiro AJ, Neufeld RJ, Arnaud P, Chaumei JC. 1999. Microencapsulation of lipophilic drugs in chitosan-coated alginate microspheres. *Int J Pharm* 187:115–123.

Ribeiro C, Vicente AA, Teixeira JA, Miranda C. 2007. Optimization of edible coating composition to retard strawberry fruit senescence. *PostHarvest Biol Technol* 44:63–70.

Rico-Peña DC, Torres JA. 1990. Edible methylcellulose-base films a moisture impermeable barrier in ice cream cones. *J Food Sci* 55:1468–1469.

Rinaudo M. 2006. Chitin and chitosan: properties and applications. *Prog Polym Sci* 31:603–632.

Rinaudo M, Domard A. 1989. Solution properties of chitosan. In: Skjak-Braek G, Anthosen T, Sandford P, editors. *Chitin and Chitosan: Sources, Chemistry, Biochemistry, Physical Properties and Applications.* London and New York: Elsevier, p. 71–76.

Rinaudo M, Pavlov G, Desbrières J. 1999. Influence of acetic acid concentration on the solubilization of chitosan. *Polymer* 40:7029–7032.

Rindlav-Westling A, Stading M, Hermansson AM, Gatenholm P. 1998. Structure, mechanical and barrier properties of amylase and amylopectin films. *Carbohydr Polym* 36:217–224.

Robertson GL. 1993. *Food Packaging: Principles and Practice.* New York: Marcel Dekker.

Romanazzi G, Mlikota-Gabler F, Smilanick JL. 2005. Chitosan treatment to control postharvest gray mold of table grapes. *Phytopathology* 95:S90.

Sagoo S, Board R, Roller S. 2002. Chitosan inhibits growth of spoilage micro-organisms in chilled pork products. *Food Microbiol* 19:175–182.

Sathivel S. 2005. Chitosan and protein coatings affect yield, moisture loss, and lipid oxidation of pink salmon (*Oncorhynchus gorbuscha*) fillets during frozen storage. *J Food Sci* 70:455–459.

Sebti I, Coma V. 2002. Active edible polysaccharide coating and interactions between solution coating compounds. *Carbohydr Polym* 49:139–144.

Shahidi F, Arachchi JKV, Jeon YJ. 1999. Food applications of chitin and chitosan. *Trends Food Sci Technol* 10:37–51.

Shi X.-Y, Tan T.-W. 2002. Preparation of chitosan/ethylcellulose complex microcapsule and its application in controlled release of vitamin D2. *Biomaterials* 23:4469–4473.

Shibuya N, Minami E. 2001. Oligosaccharide signalling for defence responses in plant. *Physiol Mol Plant Pathol* 59:223–233.

Soares NFF, Camilloto GP, Pinheiro NMS, Oliveira CP, Silva DFP. 2007. Desenvolvimento e avaliação de revestimento comestível antimicrobiano na conservação de goiaba. Proceedings of Anais do 24° Congresso Brasileiro de Microbiologia, CD-ROM.

Sotero AP. 2000 (August). Plásticos biodegradáveis trazem melhoria ambiental. *Jornal de plásticos*. Unicamp. http://www.jorplast.com.br/jpago00/ago006.html.

Sothornvit R, Krochta JM. 2001. Plasticizer effect on mechanical properties of β-lactoglobulin films. *J Food Eng* 50:149–155.

Stuchell YM, Krochta JM. 1995. Edible coating on frozen king salmon: effect of whey protein isolate and acetylated monoglycerides on moisture loss and lipid oxidation. *J Food Sci* 60:28–31.

Sudarshan NR, Hoover DG, Knorr D. 1992. Antibacterial action of chitosan. *Food Biotechnol* 6:257–272.

Tang YF, Du YM, Hu XW, Shi X.-W, Kennedy JF. 2007. Rheological characterisation of a novel thermosensitive chitosan/poly(vinyl alcohol) blend hydrogel. *Carbohydr Polym* 67:491–499.

Tatsumi Y, Watada AE, Wergin WP. 1991. Scanning electron microscopy of carrots stick surface to determine cause of white translucent appearance. *J Food Sci* 56:1357–1359.

Tharanathan RN, Farooqahmed SK. 2003. Chitin—the undisputed biomolecule of great potential. *Crit Rev Food Sci* 43:61–87.

Toffey A, Glasser WG. 1999. Chitin derivatives. II. Time–temperature-transformation cure diagrams of the chitosan amidization process. *J Appl Polym Sci* 73:79–89.

Wang X, Du Y, Luo J, Lin B, Kennedy JF. 2007. Chitosan/organic rectorite nanocomposite films: Structure, characteristic and drug delivery behaviour. *Carbohydr Polym* 69:41–49.

Wibowo S, Velazquez G, Savant V, Torres JA. 2005. Surimi wash water treatment for protein recovery: Effect of chitosan-alginate complex concentration and treatment time on protein adsorption. *Bioresource Technol* 96:665–671.

Williams SK, Oblinger JI, West RI. 1978. Evaluation of calcium alginate films for use on beef cuts. *J Food Sci* 43:292.

Wong DWS, Tillin SJ, Hudson JS, Pavlath AE. 1994. Gas exchange in cut apples with bilayer coatings. *J Agric Food Chem* 42:2278–2285.

Xie Y, Liu X, Chen Q. 2007. Synthesis and characterization of water-soluble chitosan derivate and its antibacterial activity. *Carbohydr Polym* 69:142–147.

Xue C, Yu G, Hirata T, Terão J, Lin H. 1998. Antioxidative activities of several marine polysaccharides evaluated in a phosphatidylcholine-liposomal suspension and organic solvent. *Biosci Biotechnol Biochem* 62:206–209.

Youn SK, Park SM, Kim YJ, Ahn DH. 1999. Effect on storage property and quality in meat sausage by added chitosan. *Chitin Chitosan Res* 4:189–195.

Youn SK, Park SM, Kim YJ, Ahn DH. 2001. Studies on substitution effect of chitosan against sodium nitrite in pork sausage. *Korean J Food Sci Technol* 33:551–559.

Young DH, Kohle H, Kauss H. 1982. Effect of chitosan on membrane permeability of suspension cultured *Glycine max* and *Phaseolus vulgaris* cells. *Plant Physiol* 70:1449–1454.

■■■■ **CHAPTER 6**

Blends and Composites Based on Cellulose and Natural Polymers

YIXIANG WANG and LINA ZHANG

Department of Chemistry, Wuhan University, Wuhan, China

6.1 INTRODUCTION

In the twenty-first century, science and technology have moved toward renewable raw materials and more environmentally friendly and sustainable resources and processes (Schurz, 1999). The Technology Road Map sponsored by the US Department of Energy (DOE) has targeted the achievement of 10% of basic chemical building blocks arising from plant-derived renewable sources by 2020 (Mohanty et al., 2002). Natural polymers from plants, such as cellulose, starch, and protein, have been reevaluated not only as sustainable resources but also as fascinating safer

Biodegradable Polymer Blends and Composites from Renewable Resources. Edited by Long Yu
Copyright © 2009 John Wiley & Sons, Inc.

chemicals with various uses as materials. The research and development of natural polymers is an ancient science, and it now attracts much attention for preparation of materials with new functions by the use of powerful modern tools. Renewable resource (RR) is a modern term for an old subject; it arose between 1973 and 1979 when the price of crude oil was lifted. RRs are products originating from plants and animals and used for industrial purposes; these also include food products that are not used for nutrition, as well as wastes and co-products of food processing (Zoebelein, 2001). Table 6-1 lists some natural polymers (Yu et al., 2006). However, one of the main disadvantages of biodegradable polymers obtained from RRs limits their applications; this involves their dominant hydrophilic character, fast degradation rate, and, in some cases, unsatisfactory mechanical properties, particularly under wet conditions (Yu et al., 2006). Thus, the research into and applications of natural polymers are focused on chemical and physical modifications to improve their mechanical properties, water resistance, and flexibility. Novel biodegradable materials can be obtained by blending cellulose and its derivatives with other natural polymers, and these products have promising applications in a number of fields.

Cellulose is one of the oldest and most abundant natural polymers. It is renewable, biodegradable, biocompatible, and derivatizable, and it possesses several further advantages such as low density, high modulus and high strength, high stiffness, little damage during processing, few requirements on processing equipment, and relatively low price (Zadorecki and Michell, 1989; Joly et al., 1996). As a chemical raw material, cellulose has been used for more than 150 years, and is considered an almost inexhaustible source of raw material for the increasing demand for environmentally friendly and biocompatible products (Klemm et al., 2002). However, the full potential of cellulose has not yet been exploited because of the lack of environmentally friendly methods and the limited number of common solvents that readily dissolve cellulose (Zhu et al., 2006). The traditional cuprammonium and xanthate processes are often cumbersome or expensive and require the use of unusual solvents, typically with high ionic strength, and use relatively harsh conditions (Heinze and Liebert, 2001; Zhu et al., 2006). It is notable that the traditional viscose route for producing

TABLE 6-1   List of Natural Polymers (Kalplan DL, 1998)

Polysaccharides	
Plant/algal	Starch, cellulose, pectin, konjac, alginate, carageenan, gums
Animal	Hyluronic acid
Fungal	Pulluan, elsinan, scleroglucan
Bacterial	Chitin, chitosan, levan, xanthan, polygalactosamine, curdlan, gellan, dextran
Proteins	Soy, zein, wheat gluten, casein, serum, albumin, collagen/gelatin, silks, polylysine, polyamino acids, poly(γ-glutamic acid), elastin, polyarginyl-polyaspartic acid
Lipids/surfactants	Acetoglycerides, waxes, surfactants, emulsan
Speciality polymers	Lignin, shellac, natural rubber

regenerated cellulose fibers, films, and nonwoven fabrics still dominates the current processing route (Klemm et al., 2005). Carbon disulfide (CS_2), a toxic gas, has been used to prepare cellulose products in the viscose process, but this can enter the human body through the breath, skin, and digestive tract, leading to the destabilization of protein. Moreover, waste containing CS_2 and hydrogen sulfide (H_2S) will rot the roots of plants and destroy the equilibrium of the soil. There is therefore a growing urgency to develop novel nonpolluting processes for fiber production in the cellulose industries to meet environmental concerns. Recently, the new and powerful organic solvent N-methylmorpholine-N-oxide (NMMO), has been developed and used for preparing regenerated cellulose films and fibers (Schurz, 1999; Fink et al., 2001). Ionic liquids are a group of new organic salts that are termed "green" solvents of cellulose (Swatloski et al., 2002; Turner et al., 2004; Zhang H et al., 2005; Abbott et al., 2006; Zhu et al., 2006). Rogers RD was a winner of the 2005 US Presidential Green Chemistry Challenge Awards (United Environmental Protection Agency, EPA744-K-05-001) for his great contribution to the dissolution of cellulose in ionic liquids and its regeneration (Zhu et al., 2006). We have recently developed new types of solvent mixtures for cellulose/NaOH/urea, NaOH/thiourea, and LiOH/urea aqueous solutions, precooled to low temperatures (between -5 and $-12°C$)—in which cellulose can be dissolved rapidly to obtain a transparent cellulose solution (Cai and Zhang, 2006; Cai et al., 2006; Ruan et al., 2006). Moreover, we have successfully produced multifilament fibers from cellulose solutions in the NaOH/urea and NaOH/thiourea aqueous systems with pilot machinery (Ruan et al., 2006; Cai et al., 2007a). Unsurprisingly, regenerated cellulose films and fibers prepared from "green" process have attracted much attention because of their great potential for sustainable development.

It is well known that polymer/polymer blending is an effective method for improving the original physical properties of one or both of the components, or for preparing new polymeric materials that exhibit widely variable properties without parallel in homopolymers (Nishio, 2006). Usually, two basic means are used to make a blend: mixing the components in the softened or molten state, and blending of components from their solutions. A great deal of attention has been paid to composites based on cellulose and other natural polymers, which are totally environmentally friendly and can be considered substitutes for the polymers obtained from petroleum. However, cellulose has poor solubility in organic solvents and has a low thermal decomposition temperature, which lies below its melting point. These properties make it difficult to disperse cellulose in melted polymers and a suitable solvent system is urgently required. Usually, chemical modification is used to prepare cellulose derivatives, and these are then blended with other natural polymers to produce new materials. Meanwhile, with the great efforts put into solvent systems for cellulose, a series of cellulose/natural polymer blends have been successfully obtained. In this chapter, an attempt is made to review the state of the art of composites of cellulose blended with chitin, chitosan, alginate, protein, and konjac glucomannan over past decade. Blends containing cellulose derivatives (cellulose esters and ethers) and natural polymers are also discussed.

6.2   CELLULOSE: STRUCTURE AND SOLVENTS

6.2.1   Structure of Cellulose

In 1838 the French chemist Anselme Payen (1795–1871) produced a resistant fibrous compound by treating various plant tissues alternately with nitric acid and sodium hydroxide (NaOH) solutions; this plant constituent was named "cellulose" in 1839 (Heuser, 1944). Cellulose is a polydisperse, linear-chain carbohydrate homopolymer, that is generated from repeating cellobiose units covalently linked through acetal functions between the OH group of the C4 and the C1 carbon atoms (β-1, 4-glucan). There is a large number of hydroxyl groups in the cellulose, and they are placed at positions C2 and C3 (secondary, equatorial) as well as C6 (primary) in a β-1,4-D-glucan unit, in general accessible to the typical conversions of primary and secondary alcoholic OH groups. These three different OH groups are considered to be an important factor influencing the physical properties of cellulose; they react easily with various reagents, allowing the synthesis of derivatives (Cazacu and Popa, 2005). The cellulose chain consists at one end of a D-glucose unit with an original C4-OH group (the nonreducing end); the other end is terminated with an original C1-OH group, which is in equilibrium with the aldehyde structure (the reducing end) (Klemm et al., 2005). The molecular structure of cellulose is shown in Fig. 6-1 (Heinze and Liebert, 2001). Its molecular formula is determined to be $C_6H_{10}O_5$, and the anhydroglucose unit (AGU) presents a $^4C_1$ conformation (Nehls et al., 1994).

6.2.2   Solvents of Cellulose

Cellulose is difficult to process in solution or as a melt because of the large numbers of intra- and intermolecular hydrogen bonds in cellulose, which form a highly organized network system surrounding the single polyglucan chain and interrupt the dissolution of cellulose solid into solution. Thus the key to the application of cellulose is to search for a solvent that can effectively destroy the intra- and interchain hydrogen bonding in cellulose. A nontoxic and easy dissolution system for cellulose has been developed, and it includes indirect and direct solvent systems. In indirect solvent systems, such as dimethylformamide/pyridine, dimethylformamide/N_2O_4, and dimethyl sulfoxide/N_2O_4, cellulose forms derivatives during dissolution. Direct solvent systems such as trifluoroacetic acid/dichloromethane, liquid ammonia/NH_4SCN, dimethylacetamide/LiCl, and NMMO/H_2O may form complexes with

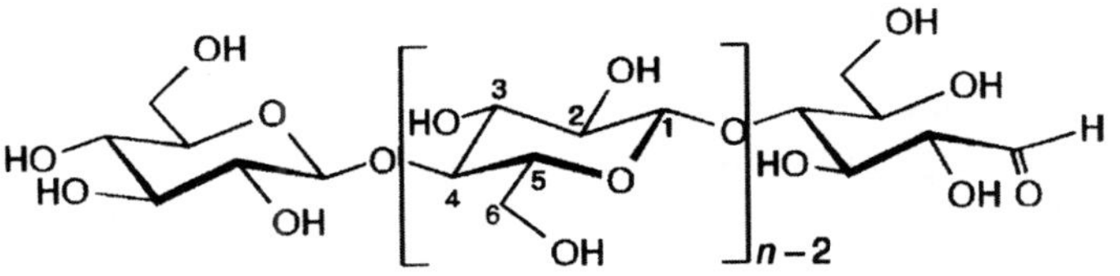

Fig. 6-1   Molecular structure of cellulose including numbering of carbon atoms (n = DP, degree of polymerization) (Heinze and Liebert, 2001).

cellulose, but the molecular structure of cellulose is not altered (Kim et al., 1999). A suitable classification of cellulose solvents identifies five types of systems as follows.

NMMO Solvent System The most advanced development took place in the 1980s with a process based on the *N*-methylmorpholine-*N*-oxide (NMMO) monohydrate solvent system (Firgo et al., 1995). Owing to its strong N—O dipole, NMMO in combination with water can dissolve cellulose typically as the monohydrate (about 13% water) at about 100°C (Michael et al., 2000) without prior activation or derivatization. Furthermore, solutions with high cellulose content of up to 23% can be produced by dispersing conventional cellulose into NMMO with high water content (such as 50%) and then subsequently removing water with an applied vacuum until the dissolution of the cellulose. This novel, environmentally friendly, direct solvent system leads to a new class of man-made cellulose fibers with the generic name of Lyocell (Woodings, 2001). Lyocell fibers show better performance qualities, but the Lyocell process suffers from uncontrolled thermal stability of the NMMO/cellulose/H_2O system (a runaway reaction), high evaporation costs (energy costs), and high tendency to fibrillation of the Lyocell fiber (Fink et al., 2001), which limit the enlargement of production. Meanwhile, the solvent systems NMMO/H_2O/DMSO (Chanzy et al., 1979) and NMMO/H_2O/DETA (Drechsler et al., 2000) have proved to be thermodynamically good solvents for cellulose and suitable for samples of various origins. A solution of 32.6 wt% NMMO, 10.0 wt% H_2O, and 57.4 wt% DETA can dissolve cellulose at room temperature, and a slightly elevated temperature (40°C) at the beginning of the dissolution process will lead to considerable shorter dissolution time (Heinze and Liebert, 2001).

LiCl/DMAc Solvent System Around 1980 it was discovered that *N,N*-dimethylacetamide (DMAc) containing lithium chloride (~8–9 wt%) can dissolve cellulose (McCormick, 1981; Turbak et al., 1981). This system shows enormous potential for cellulose in organic syntheses (Dawsey and McCormick, 1990) as well as for analytical purposes (Burchard et al., 1994) because the solvent is colorless and dissolution succeeds without or at least with negligible degradation even in the case of high molecular weight polysaccharides such as cotton linters or bacterial cellulose (Heinze and Liebert, 2001). The content of cellulose in solution can reach to 15 wt%, while that of LiCl is 5–9 wt% after dissolution for 6 h at 100°C (Conio et al., 1984). Cellulose of high molecular weight can be dissolved and the dissolution time can be shortened if the start temperature of the dissolution process is 150°C and the system is cooled slowly. Empirically determined solvatochromic polarity parameters for the cellulose/LiCl/DMAc system indicate that the ability to keep cellulose in solution is due to the very strong chloride–cellulose interaction. The chloride–cellulose interaction contributes about 80% to the dipolar–dipolar interactions between DMAc and cellulose, whereas the specific $Li^+(DMAc)_n$–cellulose interaction contributes about 10% (Spange et al., 1998).

Aqueous Metal-Based Solvent System Aqueous solutions of a number of metal complexes have been found to dissolve cellulose. The best known solvent of this group is cupric hydroxide in aqueous ammonia, which is often called cuoxam. Cellulose can be dissolved to a molecular level in cuoxam, and most effective is coordinative binding of the metal complex to the deprotonated hydroxyl groups in the C2 and C3 positions of the AGU in the chain (Fuchs et al., 1993; Burchard et al., 1994). However, cuoxam has some disadvantages, which consist of easy degradation of the cellulose chain, a deep blue color, and a restricted dissolution power that is limited to degrees of polymerization of $DP_w < 5000$. Metal ions such as Cu^{2+}, Ni^{2+}, Cd^{2+}, Fe^{2+}, and Co^{2+} have been used to form complexes with ethylenediamine (en) and other polydentate ligands and all these reagents gave clear solutions, indicating full solubility of the cellulose. A number of modern aqueous metal complex solvents, such as aqueous solutions of Ni-tren and Cd-tren (tren = tris(2-aminoethyl)amine), have been produced, and the dissolution of a large number of samples, cotton linters, various pulp celluloses, and bacterial celluloses has been studied. Both these solvents exhibited good solution properties, but only Cd-tren was capable of dissolving cotton linters and bacterial cellulose of the highest degrees of polymerization ($DP_w = 9700$) (Saalwächter et al., 2000).

Ionic Liquid Solvent System Room-temperature ionic liquids (ILs) have recently received significant attention due to their favorable of properties including low melting points, wide liquid ranges, and lack of vapor pressure, which have encouraged researchers to explore known chemical reactions and processes in place of volatile organic solvents (Turner et al., 2004; Heinze et al., 2005). The IL 1-butyl-3-methylimidazolium chloride (BMIMCl) can be used as a nonderivatizing solvent for cellulose. It has been shown that ILs incorporating anions that are strong hydrogen bond acceptors are most effective, especially when combined with microwave heating, whereas ILs containing "noncoordinating" anions, including $(BF_4)^-$ and $(PF_6)^-$ are nonsolvents. Chloride-containing ILs appear to be the most effective solvents, presumably solubilizing cellulose through hydrogen-bonding from hydroxyl functions to the anions of the solvent (Swatloski et al., 2002). More recently, a novel IL, 1-allyl-3-methylimidazolium chloride (AMIMCl) has been used to carry out homogeneous esterification of cellulose (Wu J et al., 2004; Zhang et al., 2005).

NaOH/Urea Aqueous Solvent System A novel solvent system that has been developed for cellulose is a NaOH/urea aqueous solution precooled to $-12°C$. Dissolution of cellulose could be achieved rapidly (about 5 min) at ambient temperatures (below 20°C), and the resulting solution is colorless and transparent (Zhang L et al., 2001, 2005). Interestingly, cellulose with a relatively high molecular weight could not be dissolved in the solvent without being precooled to $-12°C$ or without urea being added. The result from ^{13}C NMR indicated that this system is a good direct solvent of cellulose with nonderivatizing processes (Cai and Zhang, 2005). The addition of urea and the low temperature play important roles in the improvement of cellulose dissolution because low temperature creates a large and stable inclusion complex associated with cellulose, NaOH, urea, and H_2O clusters through hydrogen-bonding, which destroys effectively the interchain hydrogen bonding in cellulose and bring cellulose into aqueous solution. Multifilament fibers have been produced successfully from the cellulose dope using a pilot machine (Cai et al., 2007a). The cellulose dope could remain in a liquid state for a long period (more than a week) at about 0–5°C (Cai and Zhang, 2006). This solvent system has been shown to be an economical and environmentally friendly cellulose-fiber fabrication process on an industrial scale. In addition, NaOH/ thiourea (Ruan et al., 2004, 2006) and LiOH/urea (Cai et al., 2007a, b) aqueous systems have been used to rapidly dissolve cellulose, and these exhibit greater solubility for cellulose than NaOH/urea.

6.3 CELLULOSE/NATURAL POLYMER BLENDS

6.3.1 Cellulose/Chitin and Cellulose/Chitosan Blends

Chitin, a polymer composed of *N*-acetyl-D-glucosamine residues that is extracted from crab and shrimp shells, is the second most abundant resource (next to cellulose) in nature (Parisher and Lombardi, 1989); chitosan is obtained by *N*-deacetylation of chitin. Both chitin and chitosan have good biocompatibility (Shigemasa and Minami, 1995) and biodegradability (Shigemasa and Minami, 1995), along with various biofunctionalities including antithromobogenic, homeostatic, immunity enhancing and wound healing (Hirano, 1996; Muzzarelli, 1997), and they are recognized as excellent metal ligands, forming stable complexes with many metal ions (Chui et al., 1996). The similarity in chemical structure of cellulose, chitin, and chitosan fundamentally determines their compatibility and makes it possible to prepare homogeneous blends combining the unique properties of chitin and chitosan with the availability of cellulose (Rogovina et al., 2006). Due to the high adhesiveness between them (Isogai et al., 1992), addition of even relatively low amounts of chitin or chitosan to cellulose-based products allows significant increases in their mechanical strength, bactericidal properties, and other characteristics (Yunlin et al., 1998).

Reinforced Materials Cellulose and chitin are polymers that show miscibility in the solid state. Fibers made from cellulose and chitin in various proportions have been wet spun from their solutions in dimethylacetamide containing 5 wt% LiCl (Marsano et al., 2002). Polymer concentration was purposely maintained as low as possible (2.4 wt%) and the dope temperature was 60°C to have give spinability. The mechanical behaviors of blends highlighted a synergistic interaction between the components. Thus, the elastic modulus (E) of fibers increased when chitin was added to cellulose dope and vice versa, reaching the maximum value of 15 GPa at a cellulose/chitin ratio of about 1/1 w/w, while the values for neat cellulose and chitin fibers were 9 and 11 GPa, respectively. Application of a pull-off ratio during the coagulation allowed the E value to be increased further. These results, as well as electron microscopy of the fiber sections, concur with complete miscibility of two polymers in the solid state, and it is believed that the DMAc/LiCl system can induce a high degree of interaction between chitin and cellulose and a consequent physical network able to be stretched and to realize a high molecular orientation level inside the cellulose/chitin fibers.

A 6 wt% NaOH/4 wt% urea aqueous solution (Zhang L et al., 2000a; Zhou and Zhang, 2000) and a 1.5 M NaOH/0.65 M thiourea aqueous solution (Zhang L et al., 2000; Zhang et al., 2002) have been used to prepare regenerated cellulose/chitin blend films (RCCH) (Zhang L et al., 2002a; Zheng et al., 2002b). The cellulose/chitin blends exhibit a moderate miscibility, and achieved maximum tensile strength in dry and wet states of 89.1 and 43.7 MPa, respectively, when the chitin content was 10 wt%. The tensile strength (σ_b) and water resistivity of the RCCH film are higher than those of the regenerated cellulose (RC) film unblended with chitin. In addition, compared to the mechanical properties of chitin film, those of the RCCH films containing 10–50 wt% chitins are significantly improved (Zheng et al., 2002). Two coagulation systems have been used to prepare cellulose–chitin blend films; one is 5 wt% H_2SO_4 (system H), and the other is 5 wt% $CaCl_2$ followed by 5 wt% H_2SO_4 (system C) (Zhang et al., 2002a). The system H blend membranes have relatively dense structure, high crystallinity, and strong intermolecular hydrogen bonding between cellulose and chitin, leading to higher mechanical properties and thermal stability. When the chitin content is 7.5 wt%, the σ_b value of the blend membrane is 90.4 MPa and even higher than that of the system H unblended RC membrane, and the values of ε_b for the blend membranes H and C are all higher than unblended membranes H-RC and C-RC, respectively. This provides promising application in the pharmaceutical and biomedical fields for chitin in functional materials retaining its characteristics. The blend membranes C had a mesh structure and relatively large pores ($2r_e = 210 \, \mu m$) due to a loss of chitin in the process, resulting in lower σ_b and higher ε_b. The intermolecular hydrogen bonds between the cellulose and the chitin in the blends can be broken in 5 wt% $CaCl_2$ aqueous solution, in which a water-soluble calcium complex of chitin as pore former is removed from membranes.

Antibacterial Materials   A solution of a blend of cellulose xanthate with chitin xanthate has been successfully prepared, and fibers were subsequently wet spun from this solution (Pang et al., 2003). According to the bacteriostatic test GB15981–1995 I Pasteurization and Supervision Technical Standard of Pasteurization Product, both the blend fibers and the woven fabric made from the blend fibers after processing and several machine washings contained 3.54 wt% chitin and have an effective bacteriostatic effect $>26\%$ on *Staphylococcus aureus*, *Escherichia coli*, and *Corynebacterium michiganense*. Chitosan is known to be able to accelerate the healing of wound in humans (Cho et al., 1999; Ueno et al., 1999). Chitosan shows considerable antibacterial activity against a broad spectrum of bacteria (Muzzarelli et al., 1990). Polysaccharide-based membranes of chitosan and cellulose blends have been prepared using trifluoroacetic acid as a cosolvent (Wu YB et al., 2004). The intermolecular hydrogen bonding of cellulose is supposed to break down to form cellulose–chitosan hydrogen bonding; however, the intramolecular and intrastrand hydrogen bonds hold the network flat. The reduced water vapor transpiration rate through the chitosan/cellulose membranes indicates that the membranes used

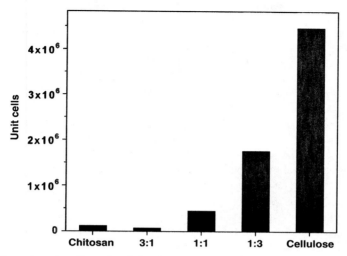

Fig. 6-2 Bacteriostatic effect of cellulose, chitosan, and chitosan/cellulose blends on the growth of *Escherichia coli* (Wu YB et al., 2004).

as wound dressings may prevent excessive dehydration of the wound. In addition, chitosan/cellulose blend membranes exhibited effective antimicrobial activity against *Escherichia coli* and *Staphylococcus aureus*, shown in Fig. 6-2. Thus, chitosan/cellulose blend membranes may be suitable for use as wound dressings with antibacterial properties.

The introduction of —CH_3COO— groups into the —OH of chitosan can give an improved antibacterial activity (Liu XF et al., 2001). Two types of *O*-carboxymethylated chitosan (*O*-CMCh)/cellulose polyblends have been prepared by mixing cellulose in LiCl/DMAc solution with *O*-CMCh aqueous solution or DMAc emulsion, and their corresponding films were regenerated in water (Li et al., 2002). FT-IR analyses showed that amino groups of *O*-CMCh were not affected during the film formation, and that there was some interaction between *O*-CMCh and cellulose in the blends, improving their compatibility. From the results of SEM observations, the *O*-CMCh/cellulose polyblend displays a heterogeneous microstructure. *O*-CMCh microdomains disperse in the cellulose matrix of the blend film. The addition of *O*-CMCh did not significantly influence the crystallinity and thermal properties. The antibacterial activity of the films against *Escherichia coli* was also measured by an optical density method. Both blend films exhibited satisfying antibacterial activity against *E. coli*, even at *O*-CMCh concentration of only 2 wt%. Due to the coagulation effect of water on the polyblend, *O*-CMCh water solution is suitable for the preparation of the low *O*-CMCh-content blend film, while *O*-CMCh/DMAc emulsion is used for high *O*-CMCh-concentration blends.

Sorption Active Materials   Cellulose/chitin beads prepared in 6 wt% NaOH−5 wt% thiourea aqueous solution by coagulating with 5 wt% H_2SO_4 have been found to be an effective biosorbent for the removal of heavy metals such as Pb^{2+}, Cd^{2+}, and Cu^{2+} (Zhou et al., 2004a). Interestingly, the sorption activity for cellulose–chitin on heavy metals is significantly higher than that of both pure cellulose and chitin beads. The beads exhibit microporous structure (shown in Fig. 6-3), large surface area, and high affinity for metals. The uptakes of Pb^{2+}, Cd^{2+}, and Cu^{2+} ions on cellulose/chitin beads are 0.33 mmol/g at pH 4, 0.32 mmol/g at pH 5, and 0.30 mmol/g at pH 4, respectively.

The adsorption equilibrium is well described by the Langmuir adsorption isotherm, which is valid for monolayer adsorption onto a surface with a finite number of identical sites. The temperature and ionic strength of the solution did not exhibit a significant effect on the amount of metal ions adsorbed over the tested temperature range of 10–40°C. This suggests that chemisorption mechanisms play the controlling part in the adsorption. The adsorption of these heavy metals is selective in the order $Pb^{2+} > Cd^{2+} > Cu^{2+}$ in solution of low ionic concentration. Adsorption equilibrium was established within 4–5 h for these ions on beads. However, the cellulose/chitin beads can easily be regenerated within 5–10 min using 1 mol/l HCl with up to 98% recovery. The adsorption mechanisms of heavy metals by cellulose/chitin beads mainly involve complexation between heavy metal ions and chitin at low pH as well as hydrolysis adsorption and surface microprecipitation at high pH. In addition, the network structure and hydrophilic skeleton of the cellulose/chitin beads could promote adsorption. Further research on the ability of these beads to adsorb Pb^{2+} in aqueous solution has been done with a fixed-bed column (Zhou et al., 2004b). The resulting breakthrough curves for the adsorption behavior indicated that the column performance is improved with decreasing initial lead concentration, ionic

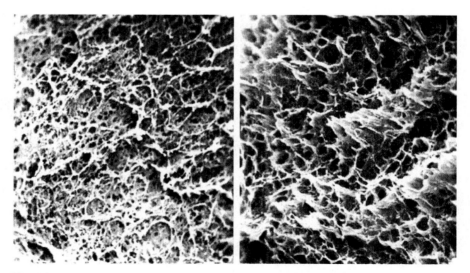

Fig. 6-3   SEM of surface (left) and cross-section (right) of cellulose/chitin beads (Zhou et al., 2004a).

strength, flow velocity, or bead size, as well as increasing pH and bed height. It is notable that the efficiency of the column for the removal of lead is not significantly reduced (by not more than 5%) after four adsorption–desorption cycles. Preparation of cellulose/chitin beads is simple and low cost; they have potential application in industrial wastewater treatment for removal and recovery of heavy metals; and the materials are biodegradable after use, giving favorable environmental credentials.

Cellulose/chitosan blends also display sorption activity. Novel natural polymer cellulose/chitosan blend beads have been obtained via homogeneous dissolution of chitosan and cellulose in NMMO (Two et al., 2003). Chitosan microspheres with particle size below 3 μm, prepared after a spray-drying process, dissolved readily in NMMO. The cellulose/chitosan blend beads show a rough and folded surface morphology and an interior pore structure. Tests of their deodorizing property against trim ethylamine and of their metal ion sorption properties for Cu^{2+}, Fe^{3+}, and Ni^{2+} ions indicate that the cellulose/chitosan blend beads have potential applications for odor treatment as well as metal ion adsorption.

An important phenomenon relevant to sorption applications is the solute transport in the membranes or beads; however, the transport behavior of these biomaterial products remains unknown. For large-scale application a deep understanding of the parameters governing solute diffusion within these products is needed. A series of biodegradable cellulose/chitin blend membranes have been successfully prepared from blend solution of cellulose and chitin in 9.5 wt% NaOH–4.5 wt% thiourea aqueous solution by coagulating with 5.0 wt% $(NH_4)_2SO_4$ (Liang et al., 2007). Using a double-cell method and a solution depletion method, the permeability and partition coefficients of three model drugs (ceftazidine, cefazolin sodium, and thiourea) were determined in phosphate buffer solution to clarify the diffusion mechanism governing transport of solutes in these membranes. The membranes exhibited high permeability to the model drugs. The diffusion coefficients of the model drugs increase with increasing temperature and decrease in the size of drug molecule. The permeability of the membranes might improved by increasing the chitin content because of its great contribution to the water-filled pore volume. The transport of drugs through the membranes is primarily by a pore mechanism. The partition mechanism and hindrance effects have also been shown to contribute. Thus, the diffusion of drugs through these membranes occurs by a dual transport mechanism (pore mechanism and partition mechanism) with some hindrance of molecular diffusion via polymer obstruction. The changes of diffusion coefficients and permeability coefficients of the model drugs in the membranes are attributed to the overall effects of these mechanisms.

6.3.2 Cellulose/Alginate Blends

Alginic acid is a heteropolysaccharide containing mannuronic acid and guluronic acid groups, and is commonly found in seaweeds (Muzzarelli, 1973). Alginate has been widely used in the fields of controlled release, ion exchange, and the vapor-permeation membrane-separation technique (Chandra and Rustgi, 1998). Alginate membranes show outstanding performance for the dehydration of ethanol–water mixtures (Yeom et al., 1996; Huang et al., 1999). Moreover, their performance exceeds that of poly(vinyl alcohol) (PVA) and some other polysaccharides with similar chemical structure such as chitosan and cellulose (Uragami and Takigawa, 1990; Shi et al., 1996). The excellent pervaporation performance of alginate membrane is attributed mainly its the extraordinary permselectivity to water in the sorption step (Yeom and Lee, 1998). However, hydrophilicity of the membrane material is not essential for the dehydration process as it can result in a low selectivity and poor strength in aqueous solution at the cost of high flux. Thus, polymer blending techniques should be useful for the preparation of new alginate membranes.

The miscibility and interactions of cellulose/alginate have been studied in cadoxen using the technique of dilute solution viscosity, and a modified treatment basing on Chee's method (Chee, 1990) has been used to determine the miscibility between a stiff polymer (cellulose) and another stiff polymer (alginate) (Zhang L et al., 1998). When the weight ratio of cellulose to alginate is 1 : 0.1, the blend is completely miscible and the interaction between molecules of cellulose and alginate is significantly enhanced; when the ratio of cellulose to alginate is lower than 1 : 0.75, the blend is immiscible. The strong interactions between the molecules of cellulose and alginate reduce the dominant contribution of the molecular extension of alginate to the viscosity. Blend membranes prepared from regenerated cellulose and sodium alginate in 6 wt% NaOH–4 wt% urea aqueous solution, pre-cooled to low temperature, by coagulating with 5 wt% $CaCl_2$ aqueous solution are miscible at all weight ratios of cellulose to alginate because of strong intermolecular hydrogen bonds between the two polymers and the formation of a Ca^{2+} bridge (Zhou and Zhang, 2001). The membrane surface exhibits a homogeneous mesh structure, and the mesh size increases from 200 to 2000 nm with an increase of alginate from 0 to 80 wt% in the blend membranes, as shown in Fig. 6-4. The mesh structure in the blend membranes is woven from both alginate and cellulose molecules. The crystalline state of the alginic acid membrane prepared from 6 wt% NaOH–4 wt% urea aqueous solution is broken completely, and the crystallinity of the blend membranes decrease with increase of alginic acid. The mechanical properties and thermal stability of the blend membranes are markedly improved compared with the alginate membrane. They thus show promis as separate and functional materials used in the wet state.

Many applications of the ion exchange equilibrium principle have been considered to solve two important environmental problems: the recovery and enrichment of valuable ions, and the removal of undesirable ions from waste water (Huang and Wang, 1993), especially to extract toxic metal ions (Kedem and Bromberg, 1993). Based on the good miscibility and mechanical properties, cellulose and alginate have been utilized as raw materials to make ion exchange membranes (Zhang L et al., 1997, 1999). Ion exchange membranes have been prepared by coagulating a blend of 6.4 wt% cellulose cuoxam (I) and 3 wt% aqueous alginate solution (II) with a weight ratio of I/II >1 : 1.4 (Zhang L et al., 1997). The tensile strength of both the dry and wet blend membranes is clearly higher than that of membranes made of pure alginate. The ion exchange capacity is 1.25 meq/g dry membrane for the blend membrane with I/II = 1 : 1.4 by weight, and the ion exchange membrane has excellent reproducibility. In addition, the blends exhibit a moderate miscibility, and the

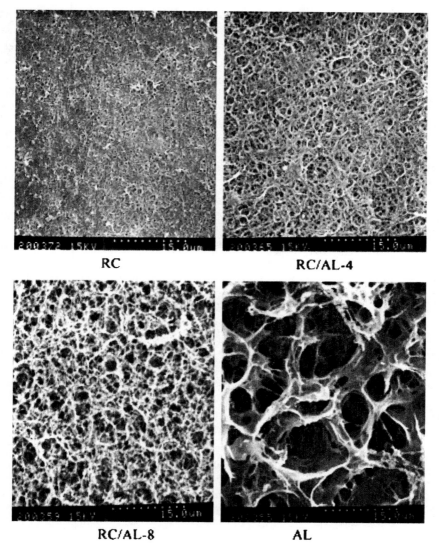

Fig. 6-4 SEM images of cellulose/alginate membrane surfaces with different sodium alginate contents: RC (0 wt%). RC/AL-4 (40 wt%), RC/AL-8 (80 wt%) and AL (100 wt%).
(Zhou and Zhang, 2001).

strong interaction between the cellulose and alginic acid is due to intermolecular
hydrogen bonding. The mechanical properties of alginic acid membranes in water
are significantly improved by blending with cellulose cuoxam; thus the blend
membranes can be used as ion exchange membranes in the water-swollen state.

The adsorption abilities of cellulose/alginic acid ion-exchange membranes have
been investigated (Zhang L et al., 1999). The adsorbed amounts (c_A) of Cd^{2+} and

Sr^{2+} are 0.79 meq/g at pH 5.8 and 0.75 meq/g at pH 6.0. However, the c_A value of Sr^{2+} for competitive adsorption on the membrane is larger than that of Cd^{2+}. The c_A values of Cd^{2+} or Sr^{2+} for single adsorption on the membrane decrease with increasing ionic strength. When EDTA or oxalic acid was added to solutions containing Cd^{2+}, Sr^{2+}, Cu^{2+}, and Fe^{3+}, the c_A values of Cd^{2+} and Sr^{2+} for competitive adsorption on the membrane were much larger than those of Cu^{2+} and Fe^{3+}, which were captured by the complexing agents to form stable complexes. EDTA enhances the c_A values of Cd^{2+} and Sr^{2+} on the membrane than the others, and the selectivity of metal ions on the membrane is in the order $Sr^{2+} > Cd^{2+} > Fe^{3+} > Cu^{2+}$ in this case. The ion exchange equilibrium was established within 30 min for Cd^{2+} and 90 min for Sr^{2+}. The ions adsorbed on the RC-AC membrane are released in 2 mol/l HCl aqueous solution within 10 min.

Pervaporation is a process for the separation of organic mixtures with an azeotropic point, those with similar physical and chemical properties, and aqueous–organic mixtures. It is therefore especially useful for the dehydration of aqueous alcohol mixtures because of the energy-saving efficiency of pervaporation compared with conventional azeotropic distillation (Huang, 1991). Cellulose and cellophane prepared from xanthate cellulose (viscose method) have higher permeation flux in pervaporation of ethanol–water mixtures, but wide application is limited by lack of a good separation factor. Thus, 8 wt% cellulose cuoxam and 8 wt% aqueous sodium alginate solution have been mixed and crosslinked using a Ca^{2+} bridge in 5 wt% $CaCl_2$ aqueous solution to prepare new membranes. The pervaporation separation parameter of ethanol–water mixtures through these blend membranes can be measured with a pervaporation apparatus. For the Ca^{2+} crosslinked blend membranes, which have relatively higher density and crystallinity, the permeation flux hardly changed with decrease of the pore size measured by the flow rate method. Moreover, cellulose/alginate (8 : 2, by weight) blend membranes crosslinked with Ca^{2+} showed higher pervaporation separation factor for 90% alcohol feed mixture at 60°C and higher permeation flux at lower temperature than did noncrosslinked membrane, owing to the Ca^{2+} bridge. The mechanical properties of alginate membranes are significantly improved by introducing cellulose and a Ca^{2+} bridge, and their tensile strength is 12 times that of the pure alginate membrane.

The recent recognition of cellulose as a smart material (Kim et al., 2006a, b) has opened the way for cellulose to be used in biomimetic sensor/actuator devices and microelectromechanical systems. This smart cellulose is termed Electro-Active Paper (EAPap). Its performance is sensitive to humidity, but an EAPap has been made with cellulose and sodium alginate in an aqueous solution of NaOH/urea, which produces its maximum displacement at a lower humidity level (Kim et al., 2007). The d.c. voltage activation shows that sodium ions in the actuator moved to the negative electrode, which defined the ion migration effect in the actuator. When the EAPap actuators were activated with a.c. voltage, the tip displacement output increased with voltage, sodium alginate content, and humidity level. In terms of durability, the displacement of the EAPap actuators decreased with time. The blocked force of the EAPap actuator increased with sodium alginate content. Since this output performance is achieved at room temperature and at moderate relative humidity levels, an EAPap actuator that is less sensitive to humidity level might be possible. The actuation principle is a combination of two mechanisms: ion migration and the piezoelectric effect associated with dipolar orientation. This combination has the potential for enhancing the properties of this EAPap as a new smart material with a low actuation voltage and fast response along with the many advantages of cellulose material.

6.3.3 Cellulose/Protein Blends

Cellulose/Silk Fibroin Blends   Silk fibroin (SF), a fibrous protein consisting of glycine, alanine, and serine as the main amino acid residues, is one of the most extensively studied materials because it has good compatibility with the living body (Tanaka et al., 1998) and exhibits oxygen permeability in the wet state (Freddi et al., 1995). SF fibers and membranes have also been used in a glucose biosensor (Liu et al., 1996) and in contact lenses (Minoura et al., 1990), and other applications. However, SF membranes are very brittle (Yamaura et al., 1985) and their changeability and solubility in alkali solution limit wide application. Blending SF with cellulose is a handy and important way to improve their mechanical properties or provide new function.

The effects of coagulants (Yang et al., 2000) and posttreatment with alkali (Yang et al., 2002b) on the structure and micropore formation of cellulose/SF blend membranes have been investigated. Two kinds of blend membranes of regenerated cellulose and SF have been prepared by coagulating solutions of mixtures of cellulose and SF in cuoxam with acetone–acetic acid (4 : 1 by vol) and 10% NaOH aqueous solutions, coded as RCF1 and RCF2, respectively. The coagulant plays an important role in the structure and micropore formation of the blend membrane (Yang et al., 2000). On coagulation with acetone–acetic acid, RCF1 is composed of cellulose and SF in the ratio of the solution before coagulation. This was attributed to strong interaction caused by intermolecular hydrogen bonding between cellulose and SF, resulting in good miscibility of the blends. The crystallinities of the blend membranes obtained are decreased, and the SF in RCF1 exhibits mainly β-sheet conformation. However, the intermolecular hydrogen bonds between the cellulose and SF in the RCF2 blends are broken in 10% NaOH aqueous solution, which allows most of the SF content to be removed by extraction. Therefore, the cellulose/SF blends prepared by coagulating with 10% NaOH aqueous solution form microporous membranes that as a result of the selective removal of SF become open porous structures. The pore size and water permeability of the microporous RCF2 membranes are greater than those of the RCF1 blend membranes, and increase sharply with increasing of SF content. Since the pore size distribution can be controlled by the conditions of coagulation, the microporous products have promising potential in separation, chromatographic, and biocompatible materials applications.

Furthermore, the RCF1 blend membranes have been subjected to posttreatment with 10% NaOH aqueous solution (Yang et al., 2002b). Alkalies are good solvents for SF, but strong hydrogen bonding between cellulose and SF inhibits the removal of SF, making RCF1 types alkali resistant. In addition, the crystallinity

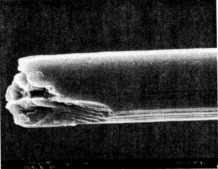

Fig. 6-5   SEM images of a cellulose/SF blend fiber (Marsano et al., 2007).

and the mean pore size of the blend membranes decrease slightly with increasing posttreatment time, indicating that the cellulose/silk blend membrane can be used under alkaline medium conditions.

Cellulose and silk fibroin are natural, biorenewable, and biodegradable polymers that are widely used in the textile industry. Fibers made of cellulose and SF have been wet spun from solutions in N,N-dimethylacetamide containing 7 wt% LiCl, by using different coagulation baths (water and ethanol) and spinning conditions (Marsano et al., 2007). By using water as the coagulant, a partial dissolution of SF can be achieved and there is negligible alteration of the mechanical properties of the cellulose–SF fibers with respect to cellulose fibers. However, fibers coagulated in ethanol are dimensionally homogeneous and have better properties. A modulus of about 13 GPa and elongation to break of 16% for the blend containing 30 wt% SF was obtained with a 20-mm air gap. In particular, X-ray results showed that cellulose–SF fibers are amorphous with a homogeneous dispersion of small SF domains (1.3 nm) in the cellulose matrix. The morphology of a cellulose/SF blend fiber is shown in Fig. 6-5. These results confirmed the good compatibility between the two natural polymers.

Cellulose/Soy Protein Isolate Blends   Interestingly, soy protein isolate (SPI) coated onto paper can enhance its grease resistance and mechanical properties (Park et al., 2000), suggesting that an interaction between SPI and cellulose. There is therefore a rationale to blend SPI with cellulose for preparation of novel materials with improved properties and biofunctionality. New blend membranes of cellulose/SPI have been prepared in aqueous solution of 6 wt% NaOH and 5 wt% thiourea by coagulating with 5 wt% H_2SO_4 solution (Chen and Zhang, 2004). SPI and cellulose are miscible when the SPI content is less than 40 wt%, and the blend membranes exhibit a mesh structure woven of cellulose and protein chains. The pore structure and properties of the blend membranes are significantly improved by incorporation of SPI into cellulose. With an increase in W_{SPI} from 10 to 50 wt%, the apparent size of the pore ($2r_e$) measured by SEM for the blend membranes increases from 115 nm to 2.43 μm, and the pore size ($2r_f$) measured by the flow rate method increases from 43 to 59 nm. The tensile strength (σ_b) and thermal stability of blend membranes with lower than 40 wt% W_{SPI} are higher than those of the pure cellulose membrane, as a result of the strong hydrogen bonding between hydroxyl groups of cellulose and amido groups of SPI molecules. The values of tensile strength and elongation at break for the blend membranes with 10 wt% W_{SPI} reach 136 MPa and 12%, respectively. It is notable that blend membranes containing protein can be used in water because their σ_b values in the wet state remain in the range 10–37 MPa. These blend membranes (CS1) can be hydrolyzed with 5 wt% NaOH aqueous solution to obtain membranes designated CS2 (Chen et al., 2004). The CS2 membranes possess microporous structure as a result of the removal of SPI during the NaOH-hydrolysis process, and their pore size increases with an increase of SPI content. The values of $2r_f$ (47.7–77.2 nm) and UFR (28–53.5 ml h^{-1} m^{-2} mmHg^{-1}) of the CS2 microporous membranes are higher than those of the corresponding CS1 membranes and pure cellulose membrane, and the microporous membranes also retained high tensile strength in both dry and wet states. In addition, the CS2 membranes containing a small amount of SPI are suitable for the culture of Vero cells. SEM images of Vero cells cultured on the free surfaces of the membranes are shown in Fig. 6-6. These membranes could be candidates for application in separation technology and biomedical fields.

Cellulose/Casein Blends Blend membranes of cellulose/casein were prepared from their cuprammonium hydroxide solution to produce membranes having good permeability and excellent mechanical strength (Zhang LN et al., 1995; Yang et al., 1997). The blends are miscible when the casein content is less than 15 wt%, and the mechanical properties of both the dry and wet blend membranes are superior to those of the pure regenerated cellulose membranes; their mean pore diameters and permeabilities are also improved. Extensive research has been done using circular dichroism (CD) measurements on the mixed solutions of cellulose/casein cuoxam and on their gel membranes obtained by alkali coagulation directly or after vaporization of ammonia to detect the phase-separation by optical anisotropy (Yang et al., 1997). The blend membranes with 10 wt% casein show the longest-wavelength CD peak for the charge transfer excitation of the cellulose/cuprammonium complex, while these gel membranes give the highest tensile strength and crystalline orientation. The CD measurements on the mixed solution and gel membranes reveal that the mixed solution with more than 30 wt% casein formed an independent casein/cuprammonium hydroxide complex, giving optically anisotropic phase-separated states; this upper limit shifts toward lower casein content during coagulation.

6.3.4 Cellulose/Konjac Glucomannan Blends

Konjac glucomannan (KGM), a natural polysaccharide is composed of β-1,4-pyranoside bond linked mannose and glucose (Shimahara et al., 1975; Maeda et al., 1980) is

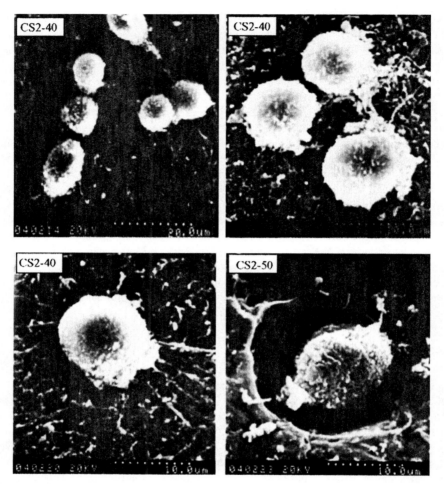

Fig. 6-6   SEM images of Vero cells cultured on the free surfaces of CS2-40 (40 wt% SPI) and CS2-50 (50 wt% SPI) membranes (Chen et al., 2004).

similar to cellulose. It is known that KGM can be gelled by salts and alkalies, and has potential uses as a thickener and food additive (Nishinari et al., 1992). The crystal structure of KGM has been found to depend strongly on glucose content and temperature. These physical-chemical characteristics and properties of KGM have prompted series of studies on the KGM/cellulose blend system.

Cellulose/KGM blend membranes were first prepared in cuprammonium solution with different coagulations (aqueous 10 wt% NaOH solution or water) at different temperatures (20 or 40 °C) (Yang et al., 1998). The blend membranes are miscible with 0–30 wt% content of KGM, and the tensile strength, pore size, and water permeability of cellulose/KGM blend membranes can be affected by the coagulating conditions. Similar research on preparing cellulose/KGM blend membranes has used precooled NaOH/thiourea aqueous solutions by coagulating with 5 wt% $CaCl_2$ aqueous solution (Yang et al., 2002a). In this solvent system, cellulose and KGM are miscible over the whole range of their weight ratios. The blend membranes are "alloys" of cellulose and KGM, and form mesh structures woven of two polymers. New crystalline planes form due to the similarity of structures and interaction between cellulose and KGM. With an increase of the KGM content, pore size and water permeability of the blend membranes rapidly increase, then reach a plateau at about 50 nm at $w_{KGM} \geq 30\%$. The tensile strength and breaking elongation of the blend membranes in the dry state decrease slightly and then decrease significantly at $w_{KGM} \geq 50\%$. The content of through-pores of the cellulose membranes blended with KGM increases, and the pore size distribution becomes wider, compared with pure cellulose membrane. KGM plays an important role in the formation of through-pores in the microporous membranes, leading to higher water permeability. This is thus a novel way to prepare membranes and porous gel particles with various pore sizes for application in the field of separation.

Size-exclusion chromatography (SEC) is an important technique in the analysis and quality control of polymers (Churms, 1996; Lundahl et al., 1999). The chromatographic packing materials for separation of biopolymer are mostly prepared from polysaccharides (Gustavsson and Larsson, 1996; Noel et al., 1996) or inorganic packings coated with polysaccharides (Petro and Berek, 1993, 1994; Castells and Carr, 2000; Chen et al., 2002). Cellulose has great potential for use as a chromatographic packing because of its characteristics of good solvent resistance, biocompatibility, biodegradability, and relatively low cost. Recently, novel microporous beads of particle size about 90 μm have been prepared, for the first time, from cellulose and konjac glucomannan (RC/KGM) in 1.5 M NaOH–0.65 M thiourea aqueous solution by an emulsification method (Xiong et al., 2005). The microporous beads were then modified with silane to avoid the adsorption of polymers containing hydroxyl groups; these are coded as RC/KGM-Si. The morphology of these beads with and without modification is shown in Fig. 6-7. A preparative SEC column (500 mm × 20 mm) is packed with RC/KGM-Si; its exclusion limit and fractionation range of the stationary phase are, respectively, weight-average molecular mass (M_w) of 4.8×10^5 g/mol and 5.3×10^3–4.8×10^5 g/mol for polystyrene in tetrahydrofuran. A preparative SEC column has been used to fractionate poly(ε-caprolactone) (PCL; $M_w = 8.31 \times 10^4$ g/mol, polydispersity index $d = 1.55$) in tetrahydrofuran and a polysaccharide (PC3-2; $M_w = 1.21 \times 10^5$ g/mol, $d = 1.70$) in 0.05 M NaOH aqueous solution. The M_w values of the fractions determined by analytical SEC combined with laser light scattering range from 1.2×10^4 to 1.84×10^5 for PCL and from 8.5×10^4 to 2.13×10^5 for PC3-2; d values are in the range 1.2–1.5. Preparative SEC has good fractionation efficiency in both organic solvents and alkaline aqueous solutions for the various polymers, leading to the promise of application in polymer separation.

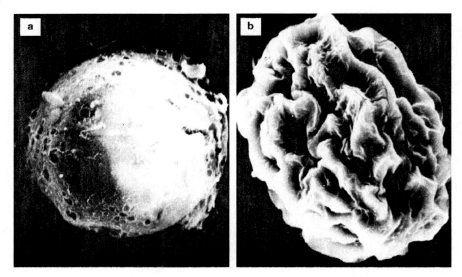

Fig. 6-7   SEM images of RC/KGM (a) and RC/KGM-Si (b) (Xiong et al., 2005).

6.4   CELLULOSE DERIVATIVE/NATURAL POLYMER BLENDS

Cellulose esters of inorganic and organic acids, as well as cellulose ethers, are pioneer compounds of cellulose chemistry and remain the most important technical derivatives of cellulose. In recent years, various cellulose derivatives (esters or ethers) mixed with natural polymers have been studied and developed in order to obtain new blend materials with multiple applications.

6.4.1   Blends Containing Cellulose Ethers

The synthesis of cellulose ethers is an important aspect of commercial cellulose derivatization. Cellulose ethers have outstanding properties, good solubility, and high chemical stability, and are toxicologically innocuous. The especially important parameter of water solubility can be controlled to a certain extent by the constitution and combination of the ether groups, the degree of substitution, and the distribution of substituents. Cellulose ethers, such as hydroxyethyl cellulose (HEC), methyl cellulose (MC), ethyl cellulose (EC), hydroxypropylmethyl cellulose (HPMC), and carboxymethyl cellulose (CMC), can be also blended with various natural polymers to obtain new composite materials with multiple applications.

The miscibility of blends of chitosan and a series of cellulose ethers—HEC (Wali et al., 2005), MC (Yin et al., 2006), and HPMC (Jayaraju et al., 2006; Yin et al., 2006)—in acetic acid solution has been studied. Blends of chitosan/HEC are completely miscible in all proportions. However, blends of chitosan/MC are not fully miscible in the dry state, and those of chitosan/HPMC are miscible when the chitosan content is >50%, although weak hydrogen bonding exists between the polymer functional groups.

HEC, a nonionic water-soluble and water-swellable cellulose ether, is compatible with a wide range of other water-soluble polymers (Adeyeye and Price, 1991; Dalal and Narurkar, 1991; Downs et al., 1992). It is a commercially useful polymer that finds applications as a thickener in latex paints and paper finishes. Pervaporation separation membranes of sodium alginate and HEC have been prepared by solution casting, crosslinked with glutaraldehyde and urea/formaldehyde/sulfuric acid mixtures (Vijaya et al., 2005a, b). Membranes were tested for pervaporation separation of feed mixtures ranging from 10–50 wt% water in water/1,4-dioxane and water/tetrahydrofuran mixtures at 30°C. For 10 wt% of the feed mixture, pervaporation experiments were also done at higher temperatures (40 and 50°C). On increasing the temperature, a slight increase in flux with a considerable decrease in selectivity is observed for all the membranes and for both the mixtures. The blend membranes exhibit different pervaporation performance for both the binary mixtures investigated. For water/1,4-dioxane mixture the pervaporation performance did not improve much after blending, whereas for water/tetrahydrofuran mixture the pervaporation performance was improved considerably over that of plain sodium alginate membrane. Similarly, blend membranes of sodium alginate and HEC have been synthesized and ionically crosslinked with phosphoric acid for the separation of t-butanol/water mixtures (Kalyani et al., 2006). The crosslinked polymers have good potential for breaking the azeotrope of 88 wt% t-butanol by giving high selectivity of 3237 and substantial water flux of 0.2 kg m^{-2} h^{-1}. The pervaporation performance is evaluated by varying experimental parameters such as feed composition, membrane thickness, and permeating pressure, and is promising for t-butanol dehydration.

A variety of polymers have been used for developing controlled-release formulations to enhance the release rates of drugs. Among various polymers employed, hydrophilic biopolymers are suitable in oral applications because of their inherent advantages over synthetic polymers. Controlled release of diclofenac sodium (DS) and ibuprofen (IB) drugs through sodium alginate–HEC blend polymeric beads was investigated (Krishna et al., 2006). Beads are prepared by precipitating the viscous solution of sodium alginate/HEC blend in alcohol followed by crosslinking with calcium chloride. The beads formed are smooth with nonporous surfaces, as seen in the SEM image shown in Fig. 6-8. The DS-loaded beads show better release performance than the IB-loaded beads. Diffusion parameters have been evaluated from Fick diffusion theory. Mathematical modeling studies and drug release characteristics through bead matrices involve solving Fick's diffusion equation. Carbohydrate polymeric blend microspheres with controlled release properties, consisting of sodium alginate and MC have been prepared by a water-in-oil (W/O) emulsion method (Ramesh et al., 2007). These microspheres are cross-linked with glutaraldehyde and loaded with nifedipine (NFD), an anti-inflammatory drug. The drug can be released in a controlled manner. Swelling studies of microspheres show that water uptake decreases with an increasing amount of MC in the microspheres. This effect is correlated with the release rates of the drug though the microspheres containing different amounts of MC. The microspheres have lower

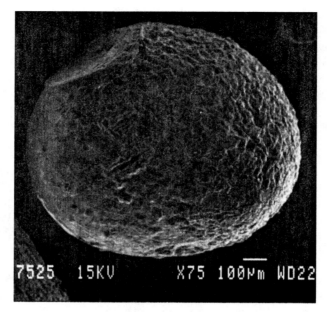

Fig. 6-8 SEM image of a sodium alginate/HEC bead (Krishna et al., 2006).

densities and hence should be retained in the gastric environment for more than 12 hours, which would help to improve the bioavailability of nifedipine.

Chitosan can also be blended with a hydrophobic coating material, in ratios appropriate to provide desirable drug release rates from core tables or pellets. Thus, EC can be chosen as the major coating component and chitosan as the minor component to form blended coating films determine the most promising proportion range for practical application (He et al., 2006). Dry films have been produced by a casting/solvent evaporation method, with different volume ratios of EC and chitosan solution and various plasticizers. Wet films can be prepared by immersing dry films in pH 6.8 phosphate-buffered saline (PBS) for 24 hours. Promising ratios of EC to chitosan are below 20 : 5 or 20 : 6 with various plasticizers, determined by comparing the viscosity of the blended solutions and the morphology of the blended films. All the plasticizers tested have good compatibility with EC or chitosan; dibutyl phthalate (DBP) has the greatest efficacy, inducing the lowest T_g (39.9°C) of the film. The release rates of tetramethylpyrazine phosphate (TMPP) from pellets coated with the blended films of EC/chitosan (20 : 6 v/v) with various plasticizers show that the more water-soluble the plasticizer is, the more quickly is TMPP dissolved from the coated pellets. This further indicates that the water-insoluble plasticizers (such as DBP) could be more suitable for producing the sustained-release or controlled-release property of the blended films in wet state.

CMC, a cellulose derivative formed by the carboxymethylation of the hydroxyl group in cellulose, can act as an ion-exchange material for carboxyl groups (Lali et al., 2000; Aruna and Lali, 2001). The adsorption of lysozyme has been investigated with novel macroporous chitosan/CMC blend membranes, which can be prepared by a simple solution-blending method with glutaraldehyde as a crosslinking agent for chitosan and with silica particles as porogens (Chen et al., 2005). The maximum adsorption capacity of the macroporous chitosan/CMC blend membranes is as high as 240 mg/g (170 mg/ml), and more than 95% of the adsorbed lysozyme is desorbed in a pH buffer at 11.8, indicating that the chitosan/CMC blend membranes could act as cation-exchange membranes. In addition, the lysozyme adsorption capacity of the blend membranes increased with an increase in the initial lysozyme concentration and the adsorption temperature. The blend membranes also have good reusability after several adsorption–desorption cycles.

At the same time, interpenetrating network polymeric beads of sodium alginate blended with sodium carboxymethyl cellulose had been prepared by crosslinking with a common crosslinking agent, glutaraldehyde, for the release of the insecticide carbaryl (Işiklan, 2006). It was observed that the carbaryl release decreased with increase in crosslinking of the network, while it increased with increase in carboxymethyl cellulose content in the blend beads and with temperature. Diffusion coefficients were calculated for the transport of insecticide through the polymeric beads, using the initial-time approximation method. These values are consistent with the carbaryl release data, which followed the Fick trend.

6.4.2 Blends Containing Cellulose Esters

Among industrially established cellulosic products, organic ester derivatives of cellulose form a valuable family, finding general acceptance in various fields of application such as packaging, coating, release-controllable excipients, molded plastics, optical films, fibers, membranes, and other separation media (Edgar et al., 2001). It is a further attraction that polymer blending may offer opportunities not only to improve the processability and modify the physical properties of cellulose esters, but also to alter the thermal instability and mechanical brittleness of the second component polymers.

It has long been known that cellulose acetate (CA) has good mechanical strength in fabricating hollow fibers. Thus, novel chitosan/CA blend hollow fibers have been prepared by the wet spinning method, in which CA acted as a matrix polymer and chitosan as a functional polymer to provide the product with coupling or reactive sites for affinity-based separations (Liu and Bai, 2005). Formic acid can be used as the cosolvent for both CA and chitosan to prepare the dope solution and NaOH solution is used as the external and internal coagulant in the wet spinning fabrication process. The hollow blend fibers display good tensile stress, though tensile stress reduces with the increase of the chitosan content in the blend. Furthermore, highly porous chitosan/CA adsorptive hollow fiber membranes have been prepared, which exhibit the ability to remove copper ions from aqueous solutions (Liu and Bai, 2006b).

The effects of polymer concentrations and coagulant compositions on the structures and morphologies of blend membranes (Liu and Bai, 2006a) have been investigated (see Fig. 6-9). For CA concentrations of 12–18 wt% and chitosan

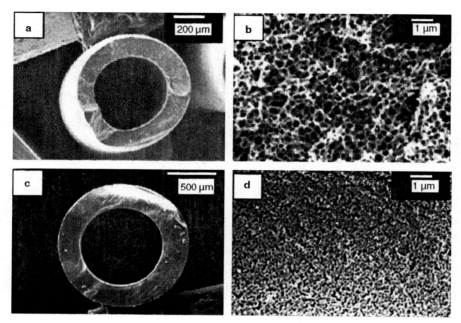

Fig. 6-9   SEM images showing the overall (a and c) and cross-sectional (b and d) structures of chitosan/cellulose acetate blend hollow fibers prepared from spinning solutions containing 12 wt% cellulose acetate and 2 wt% chitosan (a and b) and 18 wt% cellulose acetate and 2 wt% chitosan (c and d) (water was used as both the external and internal coagulant (Liu and Bai, 2006a).

concentration up to 4 wt% in the spinning dope solutions, hollow blend fibers have successfully been prepared with outer surface pore sizes, specific surface areas, and porosities in the range of 0.54–0.049 μm, 10.4–14.5 m²/g and 80.6–70.4%, respectively, depending on the amounts of CA and chitosan in the spinning dope solutions and the coagulants used. Water, a weaker coagulant, can be used as both the external and internal coagulant in the fabrication process. The resultant chitosan/CA blend hollow fibers show spongelike, macrovoid-free and, relatively uniform porous structures, which are desirable for adsorptive membranes. With increasing alkalinity of the coagulants, the coagulation rate of the hollow blend fibers increases and the fibers can be observed to form relatively denser surface layers and to have smaller surface pore sizes and slightly greater specific surface areas, due to the stronger coagulation effect. In particular, when NaOH solutions (1–3 wt%) were used as the internal coagulant, more and larger macrovoids were formed in the hollow fibers at the near-lumen side with increase of the NaOH concentration at low chitosan concentrations (<3 wt%). In addition, chitosan/CA blend hollow fiber membranes have good adsorption capacity (up to 35.3–48.2 mg/g), fast adsorption rates, and short adsorption equilibrium times (less than 20–70 min) for copper ions. Moreover, they can work effectively at low copper ion concentrations (<6.5 mg/l) to reduce the residual level to as low as 0.1–0.6 mg/l in the solution. The adsorption of copper ions on chitosan/CA blend hollow fiber membranes is attributed mainly to the formation of surface complexes with the nitrogen atoms of chitosan in the hollow fiber membranes; hence higher chitosan contents in the hollow blend fiber membranes render the membranes more adsorptive to copper ions. It has been found that the copper ions adsorbed on the hollow fiber membranes can be effectively desorbed in an EDTA solution (up to 99% desorption efficiency) and the membranes can be reused almost without loss of adsorption capacity for copper ions.

6.5 PROMISING APPLICATIONS OF CELLULOSE BLENDS

Cellulose is playing an increasingly important role in the development and application of natural polymer materials, and has attracted a great deal of attention. Because of the abound OH groups and good compatibilities, cellulose and its derivatives can be blended more easily with other natural polymers to obtain novel biodegradable materials with unique functions and good properties. These blends can be expected to substitute for a proportion of synthetic polymers. In the meantime, the development of solvent systems for cellulose, especially the invention of new

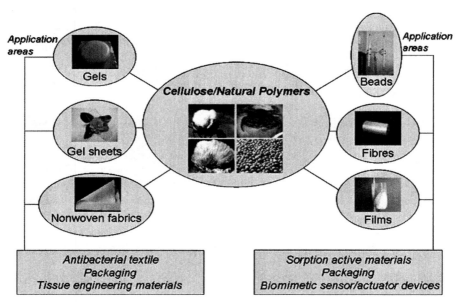

Scheme 6-1 Products and applications of cellulose and cellulose/natural polymer blends. The photographs show cellulosic beads, fibers, films, gels, and gel sheets prepared from cellulose solution in NaOH/urea or NaOH/thiourea solvent systems via "green" processes; the nonwoven fabrics are commercial products.

nonpolluting solvent systems, provides a "green" path for preparation of composites based on cellulose, and makes it possible to implement industrial production.

Scheme 6-1 shows potential applications of cellulose and cellulose/natural polymer blends. The cellulosic beads, fibers, films, gels, and gel sheets illustrated have been successfully prepared from our cellulose solution in NaOH/urea or NaOH/thiourea solvent systems by "green" processes; the nonwoven fabrics shown are commercial products. These cellulosic blends exhibit not only the native properties of the components but also a significant synergistic effect caused by the interactions between cellulose and other natural polymers. As mentioned above, cellulosic composites containing chitin/chitosan exhibit good bactericidal properties, and accordingly the corresponding fibers and cloth can be used to produce antibacterial textiles. Many of the cellulose/natural polymer blends form micro-porous structures and can be used for sorption active materials that are biosorbent for the removal of heavy metals, packings for size exclusion chromatography, and the controlled release of drugs. Cellulose/natural polymer blends also possess good biocompatibilities and thus can produce materials for packaging and tissue engineering. Recent discoveries provide a way for cellulose to be used in biomimetic sensor/actuator devices and microelectromechanical systems. This opens a new field for the application of cellulosic materials that can be expected to show further improvement in the future. Nowadays, the goal of the creation of a biobased economy is a challenge to agriculture, forestry, chemistry, and industry. Natural polymers from plants are renewable resources, and their products can be biodegraded in the soil or in lakes at the end of their use life. These new products blending cellulose and natural polymers meet present requirements concerning environmental protection and sustainable development.

REFERENCES

Abbott AP, Bell TJ, Handaa S, Stoddart B. 2006. Cationic functionalisation of cellulose using a choline based ionic liquid analogue. *Green Chem* 8:784–786.

Adeyeye CM, Price JC. 1991. Development and evaluation of sustained-release ibuprofen-wax microspheres. I. Effect of formulation variables on physical characteristics. *Pharm Res* 8:1377–1383.

Aruna N, Lali A. 2001. Purification of a plant peroxidase using reversibly soluble ion-exchange polymer. *Process Biochem* 37:431–437.

Burchard W, Habermann N, Klüfers P, Seger B, Wilhelm U. 1994. Cellulose in Schweizer's reagent: a stable, polymeric complex with high chain stiffness. *Angew Chem Int Ed Engl* 33:884–887.

Cai J, Zhang L. 2005. Rapid dissolution of cellulose in LiOH/urea and NaOH/urea aqueous solutions. *Macromol Biosci* 5:539–548.

Cai J, Zhang L. 2006. Unique gelation behavior of cellulose in NaOH/urea aqueous solution. *Biomacromolecules* 7:183–189.

Cai J, Liu Y, Zhang L. 2006. Dilute solution properties of cellulose in LiOH/urea aqueous system. *J Polym Sci Polym Phys* 44:3093–3101.

Cai J, Zhang L, Zhou J, et al. 2007a. Multifilament fibers based on dissolution of cellulose in NaOH/urea aqueous solution: structure and properties. *Adv Mater* 19:821–825.

Cai J, Zhang L, Chang C, Cheng G, Chen X, Chu B. 2007b. Hydrogen-bond-induced inclusion complex in aqueous cellulose/LiOH/urea solution at low temperature. *Chem Phys Chem* 8:1572–1579.

Castells CB, Carr PW. 2000. Fast enantioseparations of basic analytes by high-performance liquid chromatography using cellulose tris(3,5-dimethylphenylcarbamate)-coated zirconia stationary phases. *J Chromatogr A* 904:17–33.

Cazacu G, Popa VI. 2005. *Polysaccharides.* 2nd ed. 1151. New York: Marcel Dekker, Inc.

Chandra R, Rustgi R. 1998. Biodegradable polymers. *Prog Polym Sci* 23:1273–1335.

Chanzy H, Dubé M, Marchessault RH. 1979. Crystallization of cellulose with *N*-methyl-morpholine *N*-oxide: a new method of texturing cellulose. *J Polym Sci Polym Lett Ed* 17:219–226.

Chee KK. 1990. Determination of polymer–polymer miscibility by viscometry. *Eur Polym J* 26:423–426.

Chen X, Zhou H, Zhang Q, Ni J, Zhang Z. 2002. Synthesis of chemically bonded cellulose trisphenylcarbamate chiral stationary phases for enantiomeric separation. *J Chromatogr Sci* 40:315–320.

Chen X, Liu JH, Feng ZC, Shao ZZ. 2005. Macroporous chitosan/carboxymethylcellulose blend membranes and their application for lysozyme adsorption. *J Appl Polym Sci* 96:1267–1274.

Chen Y, Zhang L. 2004. Blend membranes prepared from cellulose and soy protein isolate in NaOH/thiourea aqueous solution. *J Appl Polym Sci* 94:748–757.

Chen Y, Zhang L, Gu J, Liu J. 2004. Physical properties of microporous membranes prepared by hydrolyzing cellulose/soy protein blends. *J Membr Sci* 241:393–402.

Cho YW, Cho YN, Chung SH, Ko W. 1999. Water-soluble chitin as a wound healing accelerator. *Biomaterials* 20:2139–2145.

Chui VWD, Mok KW, Ng CY, Luong BP, Ma KK. 1996. Removal and recovery of copper (II), chromium (III) and nickel (II) from solutions using crude shrimp chitin packed in small. columns. *Environ Int* 22:463–468.

Churms SC. 1996. Recent progress in carbohydrate separation by high-performance liquid chromatography based on size exclusion. *J Chromatogr A* 720:151–166.

Conio G, Corazzo P, Bianchi E, Tealdi A, Citerri A. 1984. Phase equilibria of cellulose in *N*, *N*-dimethylacetamide/lithium chloride solutions. *J Polym Sci Polym Lett Ed* 22:273–277.

Dalal PS, Narurkar MM. 1991. In vitro and in vivo evaluation of sustained-release suspensions of ibuprofen. *Int J Pharm* 73:157–162.

Dawsey TR, McCormick CL. 1990. The lithium chloride/dimethylacetamide solvent for cellulose: a literature review. *J Macromol Sci Rev Macromol Chem Phys Macromol Chem Phys* 30:405–440.

Downs EC, Robertson NE, Riss TL, Plunkett MIJ. 1992. Calcium alginate beads as a slow-release system for delivering angiogenic molecules in vivo and in vitro. *J Cell Physiol* 152:422–429.

Drechsler U, Radosta S, Vorwerg W. 2000. Characterization of cellulose in solvent mixtures with *N*-methylmorpholine-*N*-oxide by static light scattering. *Macromol Chem Phys* 201:2023–2031.

Edgar KJ, Buchanan CM, Debenham JS, et al. 2001. Advances in cellulose ester performance and application. *Prog Polym Sci* 26:1605–1688.

Fink HP, Weigel P, Purz HJ, Ganster J. 2001. Structure, formation of regenerated cellulose materials from NMMO-Solutions. *Prog Polym Sci* 26:1473–1524.

Firgo H, Eibl M, Eichinger D. 1995. Kritische Fragen zur Zukunft der NMMO-Technologie. *Lenzinger Ber* 75:47–50.

Freddi G, Romano M, Rosaria M, Tsukada M. 1995. Silk fibroin/cellulose blend films: Preparation, structure, and physical properties. *J Appl Polym Sci* 56:1537–1545.

Fuchs R, Habermann N, Klüfers P. 1993. Multinuclear sandwich-type complexes of deprotonated-cyclodextrin and copper(II) ions. *Angew Chem Int Ed Engl* 32:852–854.

Gustavsson PE, Larsson PO. 1996. Superporous agarose, a new material for chromatography. *J Chromatogr A* 734:231–240.

He W, Du YM, Fan LH. 2006. Study on volume ratio and plasticizer screening of free coating membranes composed of ethyl cellulose and chitosan. *J Appl Polym Sci* 100:1932–1939.

Heinze T, Liebert T. 2001. Unconventional methods in cellulose functionalization. *Prog Polym Sci* 26:1689–1762.

Heinze T, Schwikal K, Barthel S. 2005. Ionic liquids as reaction medium for cellulose functionalization. *Macromol Biosci* 5:520–525.

Heuser E. 1944. *The Chemistry of Cellulose*. New York: Wiley.

Hirano S. 1996. Chitin biotechnology applications. *Biotechnol Annu Rev* 2:237–258.

Huang RYM. 1991. *Pervaporation Membrane Separation Processes*. Amsterdam: Elsevier.

Huang RYM, Pal R, Moon GY. 1999. Characteristics of sodium alginate membranes for the pervaporation dehydration of ethanol–water and isopropanol–water mixtures. *J Membr Sci* 160:101–113.

Huang TC, Wang JK. 1993. Selective transport of metal ions through cation exchange membrane in the presence of a complexing agent. *Ind Eng Chem Res* 32:133–139.

Işiklan N. 2006. Controlled release of insecticide carbaryl from sodium alginate, sodium alginate/gelatin, and sodium alginate/sodium carboxymethyl cellulose blend beads crosslinked with glutaraldehyde. *J Appl Polym Sci* 99:1310–1319.

Isogai A, Hasegava M, Onabe F, Usuda M. 1992. Polysaccharide-based polymer blends: Methods of their production. *Gakkashi-Fiber* 48:655–659.

Jayaraju J, Raviprakash SD, Keshavayya J, Rai SK. 2006. Miscibility studies on chitosan/hydroxypropylmethyl cellulose blend in solution by viscosity, ultrasonic velocity, density, and refractive index methods. *J Appl Polym Sci* 102:2738–2742.

Joly C, Kofman M, Gauthier RJ. 1996. Polypropylene/cellulosic fiber composites: chemical treatment of the cellulose assuming compatibilization between the two materials. *J Macromol Sci* A33:1981–1996.

Kalyani S, Smitha B, Sridhar S, Krishnaiah A. 2006. Blend membranes of sodium alginate and hydroxyethylcellulose for pervaporation-based enrichment of *t*-butyl alcohol. *Carbohydr Polym* 64:425–432.

Kaplan DL. 1998. *Biopolymers from Renewable Resources*. New York: Springer.

Kedem O, Bromberg L. 1993. Ion-exchange membranes in extraction processes. *J Membr Sci* 78:255–264.

Kim J, Song CS, Yun SR. 2006a. Cellulose based electro-active papers: performance and environmental effects. *Smart Mater Struct* 15:719–723.

Kim J, Yun S, Ounaies Z. 2006b. Discovery of cellulose as a smart material. *Macromolecules* 39:4202–4206.

Kim J, Wang NG, Chen Y, Lee SK, Yun GY. 2007. Electroactive-paper actuator made with cellulose/NaOH/urea and sodium alginate. *Cellulose* 14:217–223.

Kim SO, Shin WJ, Cho H, Kim BC, Chung IJ. 1999. Rheological investigation on the aniso-tropic phase of cellulose-MMNO/H2O solution system. *Polymer* 40:6443–6450.

Klemm D, Schmauder HP, Heinze T. 2002. Cellulose. *Biopolymers* 6:275–319.

Klemm D, Heublein B, Fink HP, Bohn A. 2005. Cellulose: fascinating biopolymer and sustain-able raw material. *Angew Chem Int Ed* 44:3358–3393.

Krishna Rao KSV, Subha MCS, Vijaya Kumar Naidu B, Sairam M, Mallikarjuna NN, Aminabhavi TM. 2006. Controlled release of diclofenac sodium and ibuprofen through beads of sodium alginate and hydroxy ethyl cellulose blends. *J Appl Polym Sci* 102:5708–5718.

Lali A, Aruna N, John R, Thakrar D. 2000. Reversible precipitation of proteins on carboxy-methyl cellulose. *Process Biochem* 35:777–785.

Li Z, Zhuang XP, Liu XF, Guan YL, Yao KD. 2002. Study on antibacterial *O*-carboxymethy-lated chitosan/cellulose blend film from LiCl/N, *N*-dimethylacetamide solution. *Polymer* 43:1541–1547.

Liang S, Zhang L, Xu J. 2007. Morphology and permeability of cellulose/chitin blend mem-branes. *J Membr Sci* 287:19–28.

Liu CX, Bai RB. 2005. Preparation of chitosan/cellulose acetate blend hollow fibers for adsorptive performance. *J Membr Sci* 267:68–77.

Liu CX, Bai RB. 2006a. Preparing highly porous chitosan/cellulose acetate blend hollow fibers as adsorptive membranes: Effect of polymer concentrations and coagulant compo-sitions. *J Membr Sci* 279:336–346.

Liu CX, Bai RB. 2006b. Adsorptive removal of copper ions with highly porous chitosan/cellulose acetate blend hollow fiber membranes. *J Membr Sci* 284:313–322.

Liu H, Liu Y, Qian J, Yu T, Deng J. 1996. Fabrication and features of a Methylene Green-med-iating sensor for hydrogen peroxide based on regenerated silk fibroin as immobilization matrix for peroxidise. *Talanta* 43:111–118.

Liu XF, Guan YL, Yao KD. 2001. Antibacterial action of chitosan and carboxymethylated chit-osan. *J Appl Polym Sci* 79:1324–1335.

Lundahl P, Zeng CM, Hägglund CL, Schalk IG, Greijer E. 1999. Chromatographic approaches to liposomes, proteoliposomes and biomembrane vesicles. *J Chromatogr B* 720:103–119.

Maeda M, Shimahara H, Sugiyama N. 1980. Detail examination of the branched structure of Konjac glucomannan. *Agric Biol Chem* 44:245–252.

Marsano E, Conio G, Martino R, Turturro A, Bianchi E. 2002. Fibers based on cellulose-chitin blends. *J Appl Polym Sci* 83:1825–1831.

Marsano E, Canetti M, Conio G, Corsini P, Freddi G. 2007. Fibers based on cellulose–silk fibroin blend. *J Appl Polym Sci* 104:2187–2196.

McCormick CL. 1981. U.S. Patent 4,278,790.

Michael M, Ibbett RN, Howarth OW. 2000. Interaction of cellulose with amine oxide solvents. *Cellulose* 7:21–33.

Minoura N, Tsukada M, Nagura M. 1990. Physico-chemical properties of silk fibroin membrane as a biomaterial. *Biomaterials* 11:430–434.

Mohanty AK, Misra M, Drzal LT. 2002. Sustainable bio-composites from renewable resources: opportunities and challenges in the green materials world. *J Polym Environ* 10:19–26.

Muzzarelli RAA. 1973. *Natural Chelating Polymers: Alginic Acid, Chitin and Chitosan.* Oxford: Pergamon Press.

Muzzarelli RAA. 1997. Human enzymic activities related to the therapeutic administration of chitin derivatives. *Cell Mol Life Sci* 53:131–140.

Muzzarelli R, Tarsi R, Filippini O, Giovanetti E, Biagini G, Varaldo PE. 1990. Antimicrobial properties of *N*-carboxybutyl chitosan. *Antimicrob Agents Chemother* 34:2019–2023.

Nehls L, Wagenknecht W, Philipp B, Stscherbina D. 1994. Characterization of cellulose and cellulose derivatives in solution by high resolution ^{13}C-NMR spectroscopy. *Prog Polym Sci* 19:29–78.

Nishinari K, A Williams PA, Phillips GO. 1992. Review of the physicochemical characteristics and properties of konjac mannan. *Food Hydrocolloids* 6:199–222.

Nishio Y. 2006. Material functionalization of cellulose and related polysaccharides via diverse microcompositions. *Adv Polym Sci* 205:97–151.

Noel RJ, O'Hare WT, Street G. 1996. Thiophilic nature of divinylsulphone cross-linked agarose. *J Chromatogr A* 734:241–246.

Pang FJ, He CJ, Wang QR. 2003. Preparation and properties of cellulose/chitin blend fiber. *J Appl Polym Sci* 90:3430–3436.

Parisher E, Lombardi D. 1989. *Chitin Source Book.* New York: Wiley.

Park HJ, Kim SH, Lim ST, Shin DH, Choi SY, Hwang KT. 2000. Grease resistance and mechanical properties of isolated soy protein-coated paper. *J Am Oil Chem Soc* 77:269–273.

Petro M, Berek D. 1993. Polymers immobilized on silica gels as stationary phases for liquid chromatography. *Chromatographia* 37:549–561.

Petro M, Berek D, Novák I. 1994. Composite sorbents for liquid chromatography. A size exclusion study of dextran gel incorporated into porous solid particles. *Reactive Polym* 23:173–182.

President Green Chemistry Challenge, United Environmental Protection Agency (EPA744-K-05–001).

Ramesh Babu V, Sairam M, Hosamani KM, Aminabhavi TM. 2007. Preparation of sodium alginate–methylcellulose blend microspheres for controlled release of nifedipine. *Carbohydr Polym* 69:241–250.

Rogovina SZ, Vikhoreva GA. 2006. Polysaccharide-based polymer blends: methods of their production. *Glycoconj J* 23:611–618.

Ruan D, Zhang L, Zhou J, Jin H, Chen H. 2004. Structure and properties of novel fibers spun from cellulose in NaOH/thiourea aqueous solution. *Macromol Biosci* 4:1105–1112.

Ruan D, Zhang L, Lue A, et al. 2006. A rapid process for producing cellulose multi-filament fibers from a NaOH/thiourea solvent system. *Macromol Rapid Commun* 27:1495–1500.

Saalwächter K, Burchard W, Klüfers P, et al. 2000. Cellulose solutions in water containing metal complexes. *Macromolecules* 33:4094–4107.

Schurz J. 1999. Trends in polymer science. A bright future for cellulose. *Prog Polym Sci* 24:481–483.

Shi Y, Wang X, Chen G. 1996. Pervaporation characteristics and solution-diffusion behaviors through sodium alginate dense membrane. *J Appl Polym Sci* 61:1387–1394.

Shigemasa Y, Minami S. 1995. *Chitin/Chitosan: A Handbook of Chitin and Chitosan.* Tokyo: Gihodo Publishing.

Shimahara H, Suzuki H, Sugiyama N, Nisizawa K. 1975. Partial purification of β-mannanases from the konjac tubers and their substrate specificity in relation to the structure of konjac glucomannan. *Agric Biol Chem* 39:301–312.

Spange S, Reuter A, Vilsmeier E, Heinze T, Keutel D, Linert W. 1998. Synthesis and polymerization of functionalized dendritic macromonomers. *J Polym Sci Part A Polym Chem* 36:1945–1954.

Swatloski RP, Spear SK, Holbrey JD, Rogers RD. 2002. Dissolution of cellose with ionic liquids. *J Am Chem Soc* 124:4974–4976.

Tanaka T, Tanigami T, Yamaura K. 1998. Phase separation structure in poly(vinyl alcohol)/silk fibroin blend films. *Polym Int* 45:175–184.

Turbak A, El-Kafrawy A, Snyder FW, Auerbach AB. 1981. U.S. Patent 4,302,252.

Turner MB, Spear SK, Holbrey JD, Rogers RD. 2004. Production of bioactive cellulose films reconstituted from ionic liquids. *Biomacromolecules* 5:1379–1384.

Twu YK, Huang HI, Chang SY, Wang SL. 2003. Preparation and sorption activity of chitosan/cellulose blend beads. *Carbohydr Polym* 54:425–430.

Ueno H, Yamada H, Tanaka I, et al. 1999. Accelerating effects of chitosan for healing at early phase of experimental open wound in dogs. *Biomaterials* 20:1407–1414.

Uragami T, Takigawa K. 1990. Permeation and separation characteristics of ethanol-water mixtures through chitosan derivative membranes by pervaporation and evapomeation. *Polymer* 31:668–672.

Vijaya Kumar Naidu B, Krishna Rao KSV, Aminabhavi TM. 2005a. Pervaporation separation of water + 1,4-dioxane and water + tetrahydrofuran mixtures using sodium alginate and its blend membranes with hydroxyethylcellulose—A comparative study. *J Membr Sci* 260:131–141.

Vijaya Kumar Naidu B, Sairam M, Raju Kothapalli VSN, Aminabhavi TM. 2005b. Thermal, viscoelastic, solution and membrane properties of sodium alginate/hydroxyethylcellulose blends. *Carbohydr Polym* 61:52–60.

Wali AC, Vijaya Kumar Naidu B, Mallikarjuna NN, Sainkar SR, Halligudi SB, Aminabhavi TM. 2005. Miscibility of chitosan–hydroxyethylcellulose blends in aqueous acetic acid solutions at 35°C. *J Appl Polym Sci* 96:1996–1998.

Woodings C. 2001. *Regenerated Cellulose Fibers.* Cambridge, UK: Woodhead Publishing.

Wu J, Zhang J, Zhang H, He J, Ren Q, Guo M. 2004. Homogeneous acetylation of cellulose in a new ionic liquid. *Biomacromolecules* 5:266–268.

Wu YB, Yu SH, Mi FL, et al. 2004. Preparation and characterization on mechanical and anti-bacterial properties of chitsoan/cellulose blends. *Carbohydr Polym* 57:435–440.

Xiong X, Zhang L, Wang Y. 2005. Polymer fractionation using chromatographic column packed with novel regenerated cellulose beads modified with silane. *J Chromatogr A* 1063:71–77.

Yamaura K, Tanigami T, Matsuzawa S. 1985. A single large bubble consisting of a very thin film of native aqueous silk. *J Colloid Interface Sci* 106:565–566.

Yang G, Yamane C, Matsui T, Miyamoto I, Zhang L, Okajima K. 1997. Morphology and amorphous structure of blend membranes from cellulose and casein recovered from its cuprammonium solution. *Polym J* 29:316–332.

Yang G, Zhang L, Yamane C, Miyamoto I, Inamoto M, Okajima K. 1998. Blend membranes from cellulose/konjac glucomannan cuprammonium solution. *J Membr Sci* 139:47–56.

Yang G, Zhang L, Liu Y. 2000. Structure and microporous formation of cellulose/silk fibroin blend membranes I. Effect of coagulants. *J Membr Sci* 177:153–161.

Yang G, Xiong X, Zhang L. 2002a. Microporous formation of blend membranes from cellulose/konjac glucomannan in NaOH/thiourea aqueous solution. *J Membr Sci* 201:161–173.

Yang G, Zhang L, Cao X, Liu Y. 2002b. Structure and microporous formation of cellulose/silk fibroin blend membranes Part II. Effect of post-treatment by alkali. *J Membr Sci* 210:379–387.

Yeom CK, Lee KH. 1998. Characterization of sodium alginate membrane crosslinked with glutaraldehyde in pervaporation separation. *J Appl Polym Sci* 67:209–219.

Yeom CK, Jegal JG, Lee KH. 1996. Characterization of relaxation phenomena and permeation behaviors in sodium alginate membrane during pervaporation separation of ethanol-water mixture. *J Appl Polym Sci* 62:1561–1576.

Yin JB, Luo K, Chen XS, Khutoryanskiy VV. 2006. Miscibility studies of the blends of chitosan with some cellulose ethers. *Carbohydr Polym* 63:238–244.

Yu L, Dean K, Li L. 2006. Polymer blends and composites from renewable resources. *Prog Polym Sci* 31:576–602.

Yunlin G, Xiaofei L, Yingping Z, Kangde Y. 1998. Polysaccharide-based polymer blends: methods of their production. *J Appl Polym Sci* 67:1965–1972.

Zadorecki P, Michell AJ. 1989. Future prospects for wood cellulose as reinforcement in organic polymer composites. *Polym Compos* 10:69–77.

Zhang H, Wu J, Zhang J, He J. 2005. 1-Allyl-3-methylimidazolium chloride room temperature ionic liquid: A new and powerful nonderivatizing solvent for cellulose. *Macromolecules* 38:8272–8277.

Zhang L, Yang G, Xiao L. 1995. Blend membranes of cellulose cuoxam/casein. *J Membr Sci* 103:65–71.

Zhang L, Zhou D, Wang H, Cheng S. 1997. Ion exchange membranes blended by cellulose cuoxam with alginate. *J Membr Sci* 124:195–201.

Zhang L, Zhou D, Cheng S. 1998. Studies on cellulose/alginate miscibility in cadoxen by viscometry. *Eur Polym J* 34:381–385.

Zhang L, Zhou J, Zhou D, Tang YR. 1999. Adsorption of cadmium and strontium on cellulose/alginic acid ion-exchange membrane. *J Membr Sci* 162:103–109.

Zhang L, Ruan D, Gao S. 2000a. Pat. Appl. CN 00128162.3.

Zhang L, Zhou J, Ruan D. 2000b. Chin. Pat. Appl. CN 00114485.5.

Zhang L, Ruan D, Zhou J. 2001. Structure and properties of regenerated cellulose films prepared from cotton linters in NaOH/urea aqueous solution. *Ind Eng Chem Res* 40:5923–5928.

Zhang L, Guo J, Du Y. 2002a. Morphology and properties of cellulose/chitin blends membranes from NaOH/thiourea aqueous solution. *J Appl Polym Sci* 86:2025–2032.

Zhang L, Ruan D, Gao S. 2002b. Dissolution and regeneration of cellulose in NaOH/thiourea aqueous solution. *J Polym Sci Part B Polym Phys* 40:1521–1529.

Zhang L, Cai J, Zhou J. 2005. Chinese Patent ZL 03128386.1.

Zheng H, Zhou J, Du Y, Zhang L. 2002. Cellulose/chitin films blended in NaOH/urea aqueous solution. *J Appl Polym Sci* 86:1679–1683.

Zhou J, Zhang L. 2000. Solubility of cellulose in NaOH/urea aqueous solution. *Polym J* 32:866–870.

Zhou J, Zhang L. 2001. Structure and properties of blend membranes prepared from cellulose and alginate in NaOH/urea aqueous solution. *J Polym Sci Part B Polym Phys* 39:451–458.

Zhou D, Zhang L, Zhou J, Guo S. 2004a. Cellulose/chitin beads for adsorption of heavy metals in aqueous solution. *Water Res* 38:2643–2650.

Zhou D, Zhang L, Zhou J, Guo S. 2004b. Development of a fixed-bed column with cellulose/chitin beads to remove heavy-metal ions. *J Appl Polym Sci* 94:684–691.

Zhu S, Wu Y, Chen Q, et al. 2006. Dissolution of cellulose with ionic liquids and its application: a mini-review. *Green Chem* 8:325–327.

Zoebelein H. 2001. *Dictionary of Renewable Resources*. New York: Wiley.

PART II

ALIPHATIC POLYESTER BLENDS

■■■ CHAPTER 7

Stereocomplexation Between Enantiomeric Poly(lactide)s

HIDETO TSUJI

Department of Ecological Engineering, Faculty of Engineering, Toyohashi University of Technology, Aichi, Japan

YOSHITO IKADA

Department of Environmental Medicine, Faculty of Medicine, Nara Medical University, Nara, Japan

7.1 INTRODUCTION

Poly(lactide) or poly(lactic acid) (PLA) is producible from renewable resources such as starch, is biodegradable in the human body as well as in the environment, is compostable, and is not toxic to the human body and the environment (Coombes and Meikle, 1994; Doi and Fukuda, 1994; Kharas et al., 1994; Vert et al., 1995; Hartmann, 1998; Ikada and Tsuji, 2000; Garlotta, 2001; Albertsson, 2002; Scott,

Biodegradable Polymer Blends and Composites from Renewable Resources. Edited by Long Yu
Copyright © 2009 John Wiley & Sons, Inc.

2002; Södergård and Stolt, 2002; Tsuji, 2002a, 2007; Auras et al., 2004; Yu et al., 2006; Gupta et al., 2007). Because of these properties, PLA has been studied intensively for more than forty years in terms of scientific and industrial interest. The synthesis, recycling, and biodegradation of PLA are schematically represented in Fig. 7-1 (Tsuji, 2007). PLA has high mechanical performance similar to that of representative commercial polymers such as polyethylene and poly(ethylene terephthalate) (PET).

PLA-based materials have been used for biomedical, pharmaceutical, environmental, industrial, and commodity applications. Because of good biodegradability and very low toxicity, PLA and lactide (LA) copolymers have been used as biomedical material for fixation of fractured bone and as matrices for drug delivery systems. Poly(L-lactide) composite materials are also used in automobile parts and the casings for personal computers and mobile phones. The physical properties and biodegradation behavior of PLA can be modified depending on their applications. Manipulation of the physical characteristics and biodegradation kinetics will employ blending and composite formation by selection of the filler (second polymer) type and concentration, for reasons of simplicity and commercial advantage. From among the PLA-based polymer blends, this chapter focuses on blends of poly(L-lactide) (PLLA) with poly(D-lactide) (PDLA), as stereocomplex formation takes place upon blending of PLLA with PDLA. Since this stereocomplexation between enantiomeric PLAs (i.e., PLLA and PDLA) was first reported by us (Ikada et al., 1987), numerous studies have addressed the formation, structure, physical properties, degradation, and applications of the PLA stereocomplex (Slager and Domb et al., 2003a; Tsuji, 2005). Stereocomplexation enhances mechanical properties, thermal resistance, and hydrolytic degradation resistance of PLA-based materials. These improvements arise

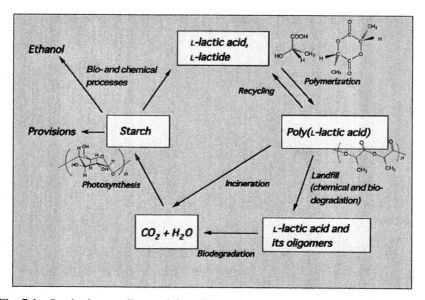

Fig. 7-1   Synthesis, recycling, and degradation of poly(L-lactide) (PLLA) (Tsuji, 2007).

from a strong interaction between L-lactyl and D-lactyl unit sequences, opening a new route for preparation of novel hydrogels and particles for biomedical applications. It has been shown that crucial parameters affecting the stereocomplexation include the mixing ratio and molecular weights of L-lactyl and D-lactyl unit sequences. The polymer pairs and stereoblock copolymers that may give rise to complexation are summarized in review articles (Slager and Domb, 2003a; Tsuji, 2005).

7.2 STEREOCOMPLEX FORMATION

The synthetic routes for LA or lactic acid homopolymers having a variety of tacticity and optical purity are summarized in Fig. 7-2 (Tsuji, 2005). The unit cell parameters of pure PLLA and PLA stereocomplex are summarized in Table 7-1 (2002a). Nonblended PLLA has been reported to crystallize in three forms: α-form (De Santis and Kovacs, 1968; Hoogsteen et al., 1990; Kobayashi et al., 1995), β-form (Hoogsteen et al., 1990; Puiggali et al., 2000), and γ-form (Cartier et al., 2000). The β-form of pure PLLA has a rather frustrated structure (Puiggali et al., 2000), but the stereocomplex crystal has a triclinic unit cell, in which PLLA and PDLA chains taking a 3_1 helical conformation are packed side-by-side in a parallel fashion. PLLA can crystallize in the temperature range $75-160°C$ and the radius growth rate of spherulite (G) shows the maximum at $130°C$ (Kalb and Pennings, 1981; Tsuji and Ikada, 1995; Abe et al., 2001; Di Lorenzo, 2005; Tsuji et al., 2005a, b). The crystallization mechanism changes from Regime III to Regime II at $120°C$ and Regime II to Regime I at $163°C$ (Kalb and Pennings, 1980; Abe et al., 2001; Di Lorenzo, 2005; Tsuji et al., 2005a, b). It has been suggested that the presence of two G maxima or the regime change from III to II is associated with the crystalline, structural change from α'-form to α-form at $110-130°C$ (Zhang et al., 2005a; Kawai et al., 2007; Pan et al., 2008).

On the other hand, as seen in Fig. 7-3, main peaks of PDLA ($X_D = 1$) film appear at 2θ values of $15°$, $17°$, and $19°$ (Ikada et al., 1987), which are comparable to those for the α-form of PLLA crystallized in a pseudo-orthorhombic unit cell with two 10_3 helices and the dimensions $a = 1.07$ nm, $b = 0.595$ nm, and $c = 2.78$ nm (Okihara et al., 1991). The most intense peaks of equimolar blended film ($X_D = 0.5$) are observed at 2θ values of $12°$, $21°$, and $24°$, where X_D is defined as

$$X_D = \frac{\text{weight of PDLA}}{\text{weight of PLLA and PDLA}} \qquad (7.1)$$

They correspond to those of the PLA stereocomplex crystallized in a triclinic unit cell of dimensions $a = 0.916$ nm, $b = 0.916$ nm, $c = 0.870$ nm, $\alpha = 109.2°$, $\beta = 109.2°$, and $\gamma = 109.8°$, in which L-lactide (LLA) and D-lactide (DLA) segments are packed parallel taking a 3_1 helical conformation (Okihara et al., 1991). The crystal structure of the PLA stereocomplex is demonstrated in Fig. 7-4. The lattice containing one PLLA or PDLA chain with a 3_1 helical conformation has the shape of an equilateral

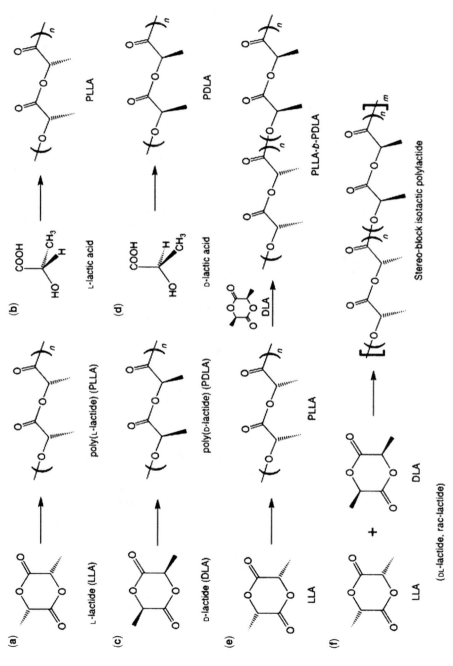

Fig. 7-2 Synthesis and molecular structures of PLLA (a and b), PDLA (c and d), and stereoblock isotactic PLA (e and f) (Tsuji, 2005).

TABLE 7-1 Unit Cell Parameters Reported for Non-Blended PLLA and Stereocomplex Crystals [Tsuji, 2002a]

	Space Group	Chain Orientation	Number of Helices Per Unit Cell	Helical Conformation	a (nm)	b (nm)	c (nm)	α (deg.)	β (deg.)	γ (deg.)
PLLA (α-form) (De Santis and Kovacs, 1968)	Pseudo-orthorhombic	–	2	10_3	1.07	0.645	2.78	90	90	90
PLLA (α-form) (Hoogsteen et al., 1990)	Pseudo-orthorhombic	–	2	10_3	1.06	0.61	2.88	90	90	90
PLLA (α-form) (Kobayashi et al., 1995)	Orthorhombic	–	2	10_3	1.05	0.61	2.88	90	90	90
PLLA (β-form) (Hoogsteen et al., 1990)	Orthorhombic	–	6	3_1	1.031	1.821	0.90	90	90	90
PLLA (β-form) (Puiggali et al., 2000)	Trigonal	Random up-down	3	3_1	1.052	1.052	0.88	90	90	120
PLLA (γ-form) (Cartier et al., 2000)	Orthorhombic	Antiparallel	2	3_1	0.995	0.625	0.88	90	90	90
Stereocomplex (Okihara et al., 1991)	Triclinic	Parallel	2	3_1	0.916	0.916	0.870	109.2	109.2	109.8

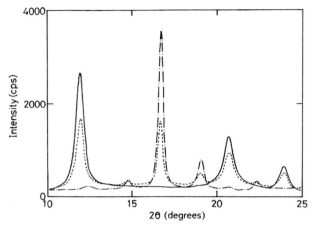

Fig. 7-3 WAXS profiles of PLLA/PDLA blends having different X_D values (Ikada et al., 1987). Solid line, dashed line, and dashed/single dotted line are for $X_D = 0.5$, 0.75, and 1, respectively.

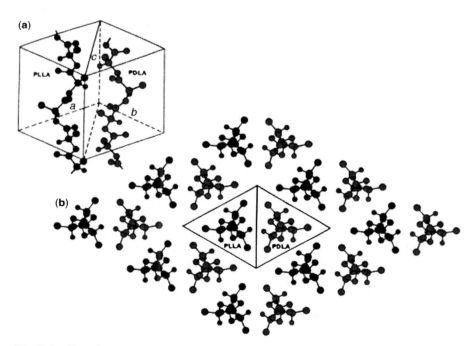

Fig. 7-4 Crystal structure of PLA stereocomplex (Okihara et al., 1991). The lines between PLLA and PDLA chains have been added to the original figure.

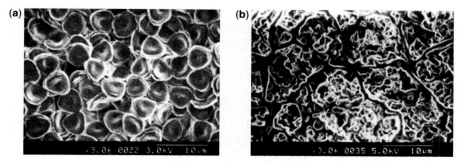

Fig. 7-5 PLA stereocomplex formed in dilute solution (single crystals) (a) (Tsuji et al., 1992) and concentrated solution (dried gel) (b) (Tsuji and Ikada, 1999).

triangle. This is expected to form equilateral-triangle-shaped single crystals of PLA stereocomplex. Equilateral-triangular single crystals [Okihara et al., 1991; Tsuji et al., 1992 (Fig. 7-5a); Brizzolara et al., 1996] and gels [Tsuji and Ikada, 1999 (Fig. 7-5b)] of PLA stereocomplex were formed when crystallized in a solution. Brizzolara et al. proposed the crystallization mechanism of PLA stereocomplex shown in Fig. 7-6 (Brizzolara et al., 1996).

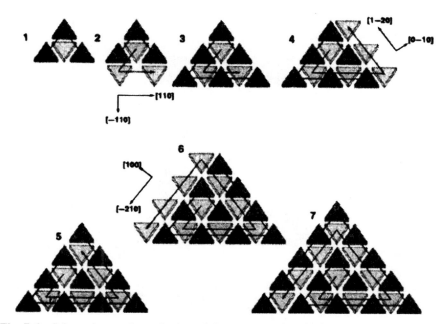

Fig. 7-6 Schematic growth mechanism of the stereocomplex single crystal (Brizzolara et al., 1996). Reprinted with permission from *Macromolecules*. Copyright (1996) American Chemical Society.

DSC and WAXS measurements of PLLA/PDLA blends having different X_D values first confirmed PLA stereocomplexation, as shown in Fig. 7-7 (Tsuji et al., 1991a) and Fig. 7-3 (Ikada et al., 1987), respectively. The specimens were prepared by precipitation of PLLA and PDLA in methylene chloride mixed solution with stirred methanol. The peak seen in Fig. 7-7 at 180°C for PLLA or PDLA ($X_D = 0$ or 1) is ascribed to the melting of PLLA or PDLA homocrystallites composed of either PLLA or PDLA chains alone. In contrast, a new melting peak appears at 230°C in the blend specimens, irrespective of X_D. The new peak is due to the melting of stereocomplex crystallites.

The difficulty of keeping PLLA/PDLA blends with molecular weights in the order of 10^3 g/mol in the amorphous state by melt-quenching strongly suggests that the overall crystallization rate of PLA stereocomplex is much higher than that of homocrystallites of PLLA or PDLA alone. This was evidenced by polarized optical microscopy (Tsuji and Tezuka, 2004), as represented in Fig. 7-8 for PLLA/PDLA blend and pure PLLA and PDLA films crystallized from their melt. Figure 7-9 shows G values of the spherulites of stereocomplex crystallites in PLLA/PDLA blends and homocrystallites in pure PLLA and PDLA films, and the

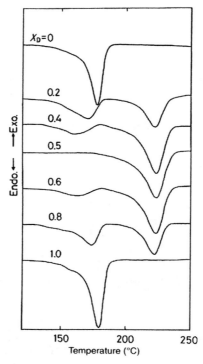

Fig. 7-7 DSC thermograms of PLLA/PDLA blends with different X_D values (Tsuji et al., 1991a). Viscosity-average molecular weight (M_v) of both PLLA and PLLA is 3.6 × 10^5 g/mol.

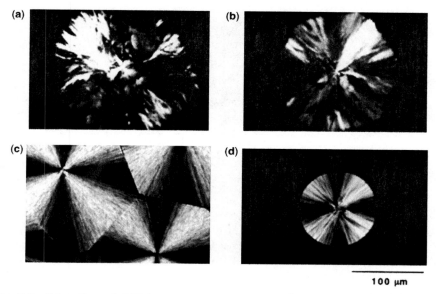

Fig. 7-8 Spherulites of PLLA ($X_D = 0$) ($M_w = 1.0 \times 10^4$ g/mol) (a), PDLA ($X_D = 1$) ($M_w = 2.2 \times 10^4$ g/mol) (b), and their blend ($X_D = 0.5$) (c, d) films crystallized at 140°C (a–c) and 190°C (d) from the melt at 250°C. Crystallization times were 11, 6, 0.5, and 12 min for (a), (b), (c), and (d), respectively (Tsuji and Tezuka, 2004).

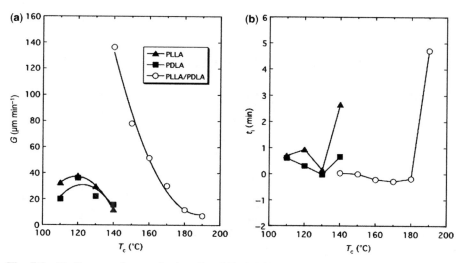

Fig. 7-9 Radius growth rate of spherulites (G) and induction period of spherulite formation (t_i) of PLLA ($X_D = 0$), PDLA ($X_D = 1$), and their blend ($X_D = 0.5$) films as functions of crystallization temperature (T_c) (Tsuji and Tezuka, 2004).

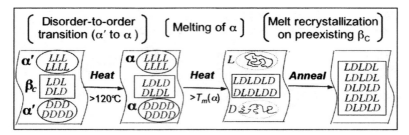

Fig. 7-10   Schematic illustration of the regularization from α'-to α-form, the melting of the α-form, and recrystallization/epitaxial growth on the oriented stereocomplex crystallites (β$_c$) form in the heating of a PLLA/PDLA blend sample (X$_D$ = 0.5) (Zhang et al., 2007).

induction period (t_i) for spherulite growth (Tsuji and Tezuka, 2004). These results indicate that the numbers of spherulites of stereocomplex crystallites per unit area are larger with higher G and lower t_i values than those of the spherulites of homocrystallites. This resulted in enhanced overall crystallization rate of stereocomplex crystallites compared with that of homocrystallites.

Zhang and co-workers investigated the crystalline phase change of PLLA/PDLA blends with high molecular weights using FT-IR spectroscopy by following three phase-transitions: the transition of α'-form to α-form at 120°C, melting of α-form, and recrystallization/epitaxial growth of stereocomplex crystallites above the melting temperature of homocrystallites (>170–180°C) (Fig. 7-10) (Zhang et al., 2007). Zhang and co-workers and Sarasua and co-workers revealed the hydrogen bonding between methyl hydrogen (or α-hydrogen) and carbonyl oxygen, which is expected to enhance rapid spherulite formation and growth of PLA stereocomplex crystallites (Zhang et al., 2005b; Sarasua et al., 2005).

7.3   METHODS FOR INDUCING STEREOCOMPLEXATION

Homocrystallization dominates stereocomplex crystallization, when the molecular weight of PLLA and PDLA is elevated. Figure 7-11 shows the melting enthalpies (ΔH_m) of stereocomplex crystallites and homocrystallites when a casting method was used for the specimen preparation. Clearly, ΔH_m of stereocomplex crystallites and homocrystallites decreased and increased, respectively, with increasing molecular weight, and ΔH_m of the stereocomplex became lower than that of homocrystallites when the weight-average molecular weight (M_w) increased above 3×10^5 g/mol (Tsuji et al., 1991a). The critical molecular weight, below which only stereocomplex crystallites are formed depends on the method and conditions of material preparation. Normally, the critical molecular weight is lower for melt processing than for solution processing (nonsolvent precipitation and solution casting), and higher for nonsolvent precipitation than for solution casting (Tsuji et al., 1991a; Tsuji and Ikada, 1993). Duan and co-workers reported the M_w dependence of stereocomplex formation for Langmuir monolayers on water (Duan et al., 2006). The result is similar to that for as-cast or melt-crystallized blends. PLLA and PDLA polymer pairs having high

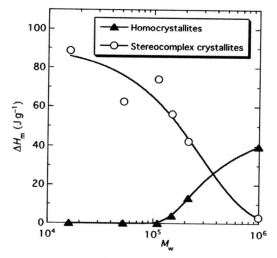

Fig. 7-11 Melting enthalpies (ΔH_m) of stereocomplex crystallites and homocrystallites for as-cast PLLA/PDLA blend films ($X_D = 0.5$) as a function of weight-average molecular weight (M_w) (Tsuji et al., 1991a).

molecular weights are required for high mechanical performance, whereas stereocomplex is readily formed when the molecular weight of either PLLA or PDLA is low. Therefore, the selection of processing method and conditions is crucial to resolve the conflict if high-molecular-weight PLLA/PDLA blend materials are required to contain only stereocomplex crystallites.

One method for enhancing stereocomplex formation is to increase the chain orientation of PLLA/PDLA (Tsuji et al., 1991a, 1994; Takasaki et al., 2003; Furuhashi et al., 2006, 2007; Sawai et al., 2006). This may be effected by the increased interaction between PLLA and PDLA chains due to shearing forces or thermal drawing. A novel method was developed to prepare well-stereocomplexed PLA materials by the use of electrospinning (i.e., by electrostatically induced high shearing force) of PLLA/PDLA concentrated solutions. The formation of homocrystallites was suppressed, while that of stereocomplex crystallites was enhanced (Tsuji et al., 2006a). This study further revealed that nano-order stereocomplex fibers could be formed by electrospinning, as shown in Fig. 7-12. On the other hand, stereocomplexation occurred upon compression of a monolayer of PLLA/PDLA blend films (Bourque et al., 2001; Pelletier and Pézolet, 2004). Study of the structural change of Langmuir films of PLLA/PDLA by the use of surface-pressure measurement and polarization modulation infrared reflecting-absorption spectroscopy (PM-IRRAS) indicated that a stereocomplex bilayer in equilibrium with the monolayer was formed at the air–water interface upon compression. Serizawa and co-workers reported that stepwise alternate immersion of a quartz crystal microbalance (QCM) substrate into respective acetonitrile solutions of PLLA and PDLA gave rise to stereocomplexation (Serizawa et al., 2003). They also indicated that the assembly of PLLA

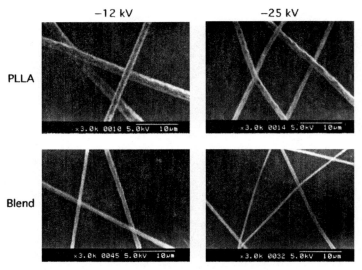

Fig. 7-12 SEM images of PLLA ($X_D = 0$) and PLLA/PDLA ($X_D = 0.5$) fibers electrospun at applied voltages of -12 and -25 kV (Tsuji et al., 2006a).

could grow epitaxially on the surface of stereocomplex crystallites, as shown also by Brochu et al. (1995).

A novel method was proposed by Spinu and co-workers for stereocomplexation between PLLA and PDLA. They conducted polymerization of LLA and DLA in the presence of PDLA and PLLA by mixing LLA and PDLA (or DLA and PLLA) (Spinu et al., 1996). This method successfully formed well-stereocomplexed PLA structures. In the strict sense, this method may not be identified as "template polymerization" (Challa and Tan, 1981), but it effectively utilizes the fact that polymerized chains have a strong interaction with the template chains.

Low-molecular-weight stereoblock PLA was incorporated in a 1 : 1 blend system of PLLA and PDLA having high molecular weights, acting as a compatibilizer to enhance stereocomplex formation (Fukushima et al., 2007).

7.4 PHYSICAL PROPERTIES

Physical properties of PLLA, PDLLA, and PLA stereocomplex are given in Table 7-2 (Tsuji, 2002a), together with those of poly((R)-3-hydroxybutyrate) (R-PHB), poly(ε-caprolactone) (PCL), and poly(glycolide) (PGA). The glass transition and melting of PLLA occur at 60°C and 180°C, respectively. As shown in Fig. 7-7, the T_m of the stereocomplex crystallites (220–230°C) was higher by about 50°C than that of homocrystallites of PLLA or PDLA (170–180°C). When stereocomplex formation takes place in concentrated PLA solutions, the stereocomplex microcrystallites formed serve as crosslinks accompanying a dramatic increase in solution viscosity and finally a three-dimensional gel is formed (Tsuji et al., 1991b; Tsuji and Ikada, 1999). Such a viscosity increase or gelation was also reported for blends made from various block and graft copolymers containing LLA or DLA chains and hydrophilic chains in aqueous media (Lim et al., 2000; De Jong et al., 2000, 2001a, b, 2002; Fujiwara et al., 2001; Watanabe et al., 2002a, b; Watanabe and Ishihara, 2003; Li and Vert, 2003; Li, 2003; Mukose et al., 2004; Hennink et al., 2004; van Nostrum, 2004; Hiemstra et al., 2006).

The tensile properties of PLA were reported to be enhanced by stereocomplex formation. For M_w exceeding 1×10^5 g/mol, tensile strength, Young's modulus, and elongation-at-break of PLLA/PDLA films were much higher than those of pure PLLA and PDLA films (Fig. 7-13) (Tsuji and Ikada, 1999). As noted above, the stereocomplex crystallites can act as crosslinks and the well-stereocomplexed materials reveal higher mechanical performance than the nonstereocomplexed materials (Fig. 7-14) (Tsuji and Ikada, 1999). Even in the melt state, PLLA and PDLA chains have a strong interaction with each other, resulting in high thermal stability, as traced by thermogravimetric isothermal measurements in the temperature range 230–250°C (Tsuji and Fukui, 2003). It was also found that PLLA/PDLA films had a lower water vapor transmission rate than pure PLLA and PDLA films (Tsuruno and Tsuji, 2007). The differences in the rates between blend and non-blended films were higher for amorphous than for crystallized films. These results reflect that the interaction is stronger between different polymer chains than between the same polymer chains.

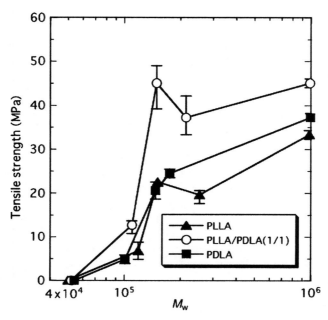

Fig. 7-13 Tensile strength of as-cast pure PLLA ($X_D = 0$), PDLA ($X_D = 1$), and PLLA/PDLA blend ($X_D = 0.5$) films as a function of M_w (Tsuji and Ikada, 1999).

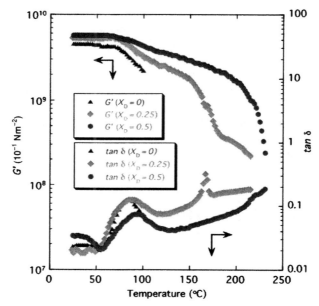

Fig. 7-14 Storage modulus (G') and loss tangent (*tan* δ) for PLLA ($X_D = 0$) and PLLA/PDLA blend ($X_D = 0.25$ and 0.5) films (Tsuji and Ikada, 1999).

7.5 BIODEGRADATION

Stereocomplexed PLA has a higher hydrolytic degradation resistance than non-blended PLLA or PDLA when they undergo hydrolytic degradation in phosphate-buffered solution at pH 7.4 and 37°C (Fig. 7-15) (Tsuji, 2000). It is surprising that although the decrease in tensile strength of nonblended specimens started at 4 months, stereocomplex specimens maintained their initial tensile strength for as long as 16 months. Similarly, De Jong and co-workers studied hydrolytic degradation of solution-cast LLA oligomers (degree of polymerization (DP) = 7 lactyl units) and the stereocomplex of LLA and DLA oligomers (DP = 7 lactyl units) at pH 7 and 37°C (De Jong et al., 2001a). They showed that the fraction of the LLA oligomer remaining approached zero within 4 h, whereas 50% of the stereocomplex remained even after 96 h of degradation. Although the LLA oligomer and stereocomplex are amorphous and crystalline, respectively, the observed results are in agreement with those for nonoligomeric PLA (Tsuji, 2000).

Stereocomplex crystallization must disturb the hydrolytic degradation of stereo-complexed specimens compared to that of nonblended PLA specimens, in spite of predominant hydrolytic degradation in the amorphous region between the stereocom-plex crystalline regions. This means that PLLA and PDLA chains should have a strong interaction between them even when they are in the amorphous state, as suggested by the aforementioned thermal degradation results (Tsuji and Fukui, 2003). To confirm this assumption, various types of equimolarly blended specimens

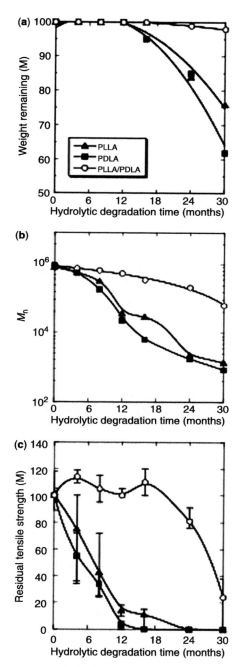

Fig. 7-15 Residual tensile strength (a), Young's modulus (b), and elongation-at-break (c) of PLLA ($X_D = 0$), PDLA ($X_D = 1$), and PLLA/PDLA blend ($X_D = 0.5$) films, as a function of hydrolytic degradation time (Tsuji, 2000).

were prepared from PLLA and PDLA—amorphous (Tsuji, 2002b) and homocrystallized (Tsuji, 2003)—and their hydrolytic degradation was carried out in phosphate-buffered solution at pH 7.4 and 37°C, together with nonblended PLLA and PDLA specimens. It was found that hydrolytic degradation of the blended specimens was retarded compared with that of the nonblended specimens, irrespective of their state (amorphous or homocrystallized), supporting the above hypothesis. Moreover, hydrolytic degradation of stereocomplexed fibers and films revealed that the morphology of the stereocomplexed materials had crucial effects on their hydrolytic degradation rate (Tsuji and Suzuki, 2001). Karst and Yang studied the molecular modeling of PLLA/PDLA blends, PLLA, and PDLA and found that hydrogen-bonding and dipole–dipole interactions were higher for PLLA/PDLA blends than for pure PLLA or PDLA, resulting in a higher hydrolytic degradation resistance of PLLA/PDLA blends (Karst and Yang, 2006). Possibly, hydrogen-bonding has a greater effect than the dipole–dipole interactions on the resistance to hydrolytic degradation.

Stereocomplexation between enantiomeric LLA and DLA unit sequences has been reported to retard hydrolytic degradation, for instance, of poly[2-hydroxyethyl methacrylate-*graft*-oligo(lactide)] (Lim et al., 2000) and A-B-A triblock copolymers of PLA (A) with poly(sebacic acid) (PSA) (B) (Slivniak and Domb, 2002). De Jong and co-workers indicated further that hydrolytic degradation of stereocomplex hydrogels from dextran (DS $= 3-12$)-*graft*-LLA and DLA oligomers (DP $= 6-12$ lactyl units) depended on the number of lactate grafts (DS), the length (DP) and polydispersity of the grafts, and the initial water content, with the hydrolytic degradation time varying from 1 to 7 days (De Jong et al., 2001b). van Nostrum and co-workers also showed that the hydrolytic degradation time of stereocomplex hydrogels from poly(2-hydroxypropyl methacrylamide) (pHPMAm)-*graft*-LLA and DLA oligomers could be readily tailored from 1 week to almost 3 weeks by changing the grafting density of the polymers and the structure of the terminal groups of side-chains (van Nostrum et al., 2004).

Another example was presented by proteinase K-catalyzed enzymatic degradation of PLLA/PDLA blends (Tsuji and Miyauchi, 2001a). Proteinase K is an endoprotease having broad specificity but with preference for the cleavage of the peptide bond C-terminal to aliphatic and aromatic amino acids, especially alanine (Sweeney and Walker, 1993). Similarity in chemical structures between lactic acid and alanine is expected to induce the proteinase K-catalyzed degradation of the C-terminal of PLLA or L-lactyl unit sequences. As is known, proteinase K can catalyze hydrolytic degradation of L-lactyl chains in the amorphous region (Tsuji and Miyauchi, 2001b, c). The proteinase K-catalyzed enzymatic degradation rate (R_{ED}) of PLLA/PDLA blends is plotted as a function of X_D in Fig. 7-16 (Tsuji and Miyauchi, 2001a). Proteinase K catalyzed the cleavage of L-lactyl unit sequences, when the average L-lactyl unit sequence length (l_L) was 4 units, and R_{ED} decreased with a decrease in l_L. Therefore, PLLA ($l_L = 57.1$) can be degraded in the presence of proteinase K, whereas degradation of PDLA ($l_L = 0$) is not catalyzed. In this study, the specimens were made amorphous to exclude the effects of highly ordered

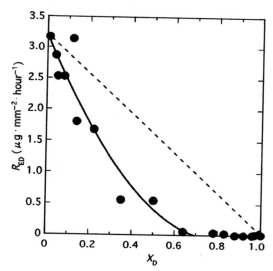

Fig. 7-16 Proteinase K-catalyzed enzymatic degradation rate (R_{ED}) of PLLA/PDLA blend films as a function of X_D (Tsuji and Miyauchi, 2001a).

structure such as crystallinity. As is seen in Fig. 7-16, R_{ED} of PLLA/PDLA blends is much lower than the expected, reflecting the mutual miscibility of PLLA and PDLA and the PDLA disturbance of enzymatic degradation of PLLA in the blends. If PDLA and PLLA were completely phase-separated, such disturbance would not be observed. Lee and co-workers also reported the disturbed proteinase K-catalyzed enzymatic degradation of PLLA/PDLA stereocomplex monolayer films compared with that of pure PLLA or PDLA monolayer film (Lee et al., 2005). However, in this case it seems difficult to define "stereocomplex" for the monolayer film.

7.6 APPLICATIONS

7.6.1 Biodegradable Films

Biodegradable PLA-based stereocomplex films can be prepared by a solution-casting method with organic solvents such as chloroform and methylene chloride (Murdoch and Loomis, 1988; Tsuji et al., 1991c; Tsuji and Ikada, 1999). In this method, the molecular weight of L-lactyl and D-lactyl unit sequences as well as the solvent evaporation rate are crucial. If the molecular weight and solvent evaporation rate are too high, stereocomplex formation is disturbed, resulting in formation of films containing a relatively large amount of homocrystallites or having low crystallinity (Tsuji et al., 1991c; Tsuji and Ikada, 1999). Although PLA stereocomplex films can also be prepared by a melt-molding method, it should be noted that the critical molecular weight of PLLA and PDLA, below which only stereocomplex crystallites

TABLE 7-2 Physical Properties of Some Biodegradable Aliphatic Polyesters (Tusji, 2002a)

Property	Polymer				
	PLLA	PLA Stereocomplex	PCL	R-PHB	PGA
T_m (°C)	170–190	220–230	60	180	225–230
T_m^0 (°C)	205–215	279	71, 79	188, 197	–
T_g (°C)	50–65	65–72	−60	5	40
ΔH_m^0 (J/g)	93, 135, 142, 203	142	142	146	180–207
Density (g/cm^3)	1.25–1.29	–	1.06–1.13	1.177–1.260	1.50–1.69
Solubility parameter (δ_p) (25°C) [(J/cm^3)$^{0.5}$]	19–20.5, 22.7	–	20.8	20.6	–
$[\alpha]_{589}^{25}$ in chloroform (25°C) (deg dm^{-1} g^{-1} cm^3)	−155 ± 1	–	0	+44[a]	–
WVTR[b] (g m^2 day^{-1})	82–172	–	177	13[c]	–
Tensile strength[d] (GPa)	0.12–2.3	0.92	0.1–0.8	0.18–0.20	0.08–1
Young's modulus[d] (GPa)	7–10	8.6	–	5–6	4–14
Elongation-at-break[d] (%)	12–26	30	20–120	50–70	30–40

[a] 300 nm, 23°C.
[b] Water vapor transmission rate at 25°C.
[c] P(HB-HV) (94/6).
[d] Oriented fiber.

are formed, decreases dramatically compared with that in the solution-casting method (Tsuji and Ikada, 1993). This indicates the difficulty of preparing well-stereocomplexed PLA with high molecular weights.

7.6.2   Biodegradable Fibers

Murdoch and Loomis 1988 prepared melt-spun PLA stereocomplex fibers from equimolar mixtures of PLLA and PDLA, while Tsuji and co-workers obtained wet- and dry-spun PLA stereocomplex fibers (Tsuji et al., 1994) from mixed chloroform solutions of equimolar PLLA and PDLA. The former study did not estimate the fraction of stereocomplex crystallites and homocrystallites (Murdoch and Loomis, 1988), whereas the latter estimated the fraction by DSC to reveal that hot-drawing of as-spun fibers increased the amount of stereocomplex crystallites but reduced that of homocrystallites, resulting in formation of fibers with stereocomplex as the main crystalline type (Tsuji et al., 1994). The PLA stereocomplex fibers used by Tsuji and Suzuki for hydrolytic degradation experiments were prepared by melt-spinning and subsequent two-stage hot-drawing, and contained only stereocomplex crystallites but no homocrystallites (Tsuji and Suzuki, 2001).

Takasaki and co-workers revealed that stereocomplexation was favored in melt-spinning from equimolar mixtures of PLLA and PDLA if the spinning was conducted under conditions of higher take-up velocity, lower throughput rate, and lower extrusion temperature (Takasaki et al., 2003). This finding suggests that these conditions can enhance the orientation-induced crystallization of the stereocomplex, as described above (Tsuji et al., 1994; Tsuji and Suzuki, 2001). They also confirmed that drawing at lower temperature and annealing between the T_m values of stereocomplex crystallites and homocrystallites enhanced stereocomplexation. This is in agreement with the above findings (Murdoch and Loomis, 1988; Tsuji and Ikada, 1993). The maximum tensile strength and Young's modulus were respectively 530 MPa and 7.4 GPa for melt-spun and drawn stereocomplex fibers (Murdoch and Loomis, 1988), 920 MPa and 8.6 GPa for solution-spun and drawn stereocomplex fibers (Tsuji et al., 1994), and 400 MPa and 4.7 GPa for as-spun stereocomplex fibers (Takasaki et al., 2003). These values are much lower than the maximum tensile strength and Young's modulus (2.1 GPa and 16 GPa) of melt-spun and drawn PLLA fibers (Leenslag and Pennings, 1987).

7.6.3   Biodegradable Microspheres for Drug Delivery Systems

Loomis and Murdoch (1990) prepared injectable stereocomplex microspheres containing a naltrexone base using an oil-in-water (O/W) solvent-evaporation method (O: methylene chloride; W: water with a surfactant). The release of naltrexone was delayed by a lag time of 200, 230, and 240 h in water, acid, and buffer solutions, respectively. After the lag, drug release occurred with zero-order kinetics up to 450 h with 46%, 48%, and 35% of the drug released in water, acid, and buffer solutions, respectively. In contrast, de Jong et al. (2001b) showed that stereocomplex hydrogels prepared from dextran (DS = 3−12)-*graft*-LLA and DLA oligomers

(DP $= 6-12$ lactyl units) released entrapped model proteins (IgG and lysozyme) during a period of 6 days.

Slager and Domb formulated "heterostereocomplex" drug delivery system (DDS) particles from L-configured peptides such as insulin with PDLA, PDLA-b-PEG, PDLA-b-PEG/PDLA, PLLA/PDLA, or PLLA-b-PEG/PDLA-b-PEG (Slager and Domb, 2002). Strong physical entrapment of peptides by the DLA unit sequence resulted in retarded release of the peptides. They also prepared heterostereocomplex DDS particles from an L-configured leuprolide and PDLA (Slager and Domb, 2003b, c; 2004). Various factors affected the release of leuprolide from the hetero-stereocomplex particles. The release rate of leuprolide from the particles increased with the decreasing molecular weight of PDLA and the increasing weight fraction of leuprolide. Continuous release of leuprolide over 100 days was observed at certain stereocomplex compositions (Slager and Domb, 2003b).

7.6.4   Biodegradable Hydrogels

Stereocomplexed PLA hydrogels can be prepared in aqueous media by blending block or graft copolymers possessing hydrophilic segments and L-lactyl or D-lactyl unit sequences. Such stereocomplex hydrogels were reported for enantiomeric A-B diblock and A-B-A triblock copolymers (Fujiwara et al., 2001; De Jong et al., 2002; Li and Vert, 2003; Li, 2003) and B-A-B triblock copolymers (Mukose et al., 2004) of PLA (A) with PEG (B); for enantiomeric graft copolymers of poly[2-hydroxyethyl methacrylate-$graft$-oligo(lactide)] (Lim et al., 2000); poly(2-hydroxy-propyl methacrylamide) (pHPMAm)-$graft$-LA oligomers (van Nostrum et al., 2004) and dextran-$graft$-LA oligomers (De Jong et al., 2000, 2001a,b, 2002; Hennink et al., 2004). Hiemstra and co-workers synthesized PEG-(PLLA)$_8$ and PEG-(PDLA)$_8$ star block copolymers with high storage moduli up to 14 kPa by mixing aqueous solutions with equimolar amounts of PEG-(PLLA)$_8$ and PEG-(PDLA)$_8$ (Hiemstra et al., 2006).

Watanabe, Ishihara and co-workers prepared porous stereocomplexed PLA films from graft-type copolymers containing LLA unit sequences or DLA sequences as side-chains (PMBLLA and PMBDLA, respectively), using an extraction method with water-soluble particles of NaCl (Watanabe et al., 2002a,b; Watanabe and Ishihara 2003). They showed that the cell adhesion and morphology of the porous scaffolds were correlated with the PLLA or PDLA content and the 2-methacryloyloxy-ethylphosphorylcholine (MPC) unit content, respectively (Watanabe et al., 2002b). Fibroblast cells adhered on the surface and intruded into the scaffolds through the connected pores after 24 h. The cell morphology became round from spreading when the PLLA or PDLA content in the scaffolds decreased.

7.6.5   Nucleation Agents

As reported in many articles (Brochu et al., 1995; Schmidt and Hillmyer, 2001; Yamane and Sasai, 2003; Anderson and Hillmyer 2006; and Tsuji et al. 2006b,c), stereocomplex crystallites formed by addition of small amounts of PDLA to

PLLA act as heterogeneous nucleation sites for PLLA crystallization, and PLLA homocrystallites are formed epitaxially on the stereocomplex crystallites. Such nucleation agents increase the number of PLLA spherulites per unit volume and the total crystallization rate, but do not alter the spherulite growth rate (Tsuji et al., 2006b, c).

REFERENCES

Abe H, Kikkawa Y, Inoue Y, Doi Y. 2001. Morphological and kinetic analyses of regime transition for poly((S)-lactide) crystal growth. *Biomacromolecules* 2:1007–1014.

Albertsson A-C, editor. 2002. *Degradable Aliphatic Polyesters* (Advances in Polymer Science, vol. 157). Berlin: Springer.

Anderson KS, Hillmyer MA. 2006. Melt preparation and nucleation efficiency of polylactide stereocomplex crystallites. *Polymer* 47:2030–2035.

Auras R, Harte B, Selke S. 2004. An overview of polylactides as packaging materials. *Macromol Biosci* 4:835–864.

Bourque H, Laurin I, Pézolet M. 2001. Investigation of the poly(L-lactide)/poly(D-lactide) stereocomplex at the air–water interface by polarization modulation infrared reflection absorption spectroscopy. *Langmuir* 17:5842–5849.

Brizzolara D, Cantow H-J, Diederichs K, Keller E, Domb AJ. 1996. Mechanism of stereocomplex formation between enantiomeric poly(lactide)s. *Macromolecules* 29:191–197.

Brochu S, Prud'homme RE, Barakat I, Jérôme R. 1995. Stereocomplexation and morphology of polylactides. *Macromolecules* 28:5230–5239.

Cartier L, Okihara T, Ikada Y, Tsuji H, Puiggali J, Lotz B. 2000. Epitaxial crystallization and crystalline polymorphism of polylactides. *Polymer* 41:8909–8919.

Challa G, Tan YY. 1981. Template polymerization. *Pure Appl Chem* 53:627–641.

Coombes AGA, Meikle MC. 1994. Resorbable synthetic polymers as replacements for bone graft. *Clin Mater* 17:35–67.

De Jong SJ, DeSmedt SC, Wahls MWC, Demeester J, Kettenes-van den Bosch JJ, Hennink WE. 2000. Novel self-assembled hydrogels by stereocomplex formation in aqueous solution of enantiomeric lactic acid oligomers grafted to dextran. *Macromolecules* 33:3680–3686.

De Jong SJ, van Eerdenbrugh B, van Nostrum CF, Kettenes-van den Bosch JJ, Hennink WE. 2001a. Biodegradable hydrogels based on stereocomplex formation between lactic acid oligomers grafted to dextran. *J Control Release* 72:47–56.

De Jong SJ, DeSmedt SC, Demeester J, van Nostrum CF, Kettenes-van den Bosch JJ, Hennink WE. 2001b. Physically crosslinked dextran hydrogels by stereocomplex formation of lactic acid oligomers: degradation and protein release behavior. *J Control Release* 71:261–275.

De Jong SJ, van Nostrum CF, Kroon-Batenburg LMJ, Kettenes-van den Bosch JJ, Hennink WE. 2002. Oligolactate-grafted dextran hydrogels: Detection of stereocomplex crosslinks by X-ray diffraction. *J Appl Polym Sci* 86:289–293.

De Santis P, Kovacs AJ. 1968. Molecular conformation of poly(S-lactic acid). *Biopolymer* 6:299–306.

Di Lorenzo ML. 2005. Crystallization behavior of poly(L-lactic acid). *Eur Polym J* 41:569–575.

Doi Y, Fukuda K, editors. 1994. *Biodegradable Plastics and Polymers*. Amsterdam: Elsevier.

Duan Y, Liu J, Sato H, Zhang J, Tsuji H, Ozaki Y, Yan S. 2006. Molecular weight dependence of the poly(L-lactide)/poly(D-lactide) stereocomplex at the air–water interface. *Biomacromolecules* 7:2728–2735.

Fujiwara T, Mukose T, Yamaoka T, Yamane H, Sakurai S, Kimura Y. 2001. Novel thermoresponsive formation of a hydrogel by stereo-complexation between PLLA-PEG-PLLA and PDLA-PEG-PDLA block copolymers. *Macromol Biosci* 1:204–208.

Fukushima K, Chang Y-H, Kimura Y. 2007. Enhanced stereocomplex formation of poly(L-lactic acid) and poly(D-lactic acid) in the presence of stereoblock poly(lactic acid). *Macromol Biosci* 7:829–835.

Furuhashi Y, Kimura Y, Yoshie N, Yamane H. 2006. Higher-order structures and mechanical properties of stereocomplex-type poly(lactic acid) melt spun fibers *Polymer* 47:5965–5972.

Furuhashi Y, Kimura Y, Yamane H. 2007. Higher order structural analysis of stereocomplex-type poly(lactic acid) melt-spun fibers. *J Polym Sci Part B: Polym Phys* 45:218–228.

Garlotta D. 2001. A literature review of poly(lactic acid). *J Polym Environ* 9:63–84.

Gupta B, Revagade N, Hilborn J. 2007. Poly(lactic acid) fiber: An overview. *Prog Polym Sci* 32:455–482.

Hartmann MH. 1998. In: Kaplan DL, editor. *Biopolymers from Renewable Resources*. Berlin: Springer; 1998. p. 367–411.

Hennink WE, De Jong SJ, Bos GW, Veldhuis TFJ, van Nostrum CF. 2004. Biodegradable dextran hydrogels crosslinked by stereocomplex formation for the controlled release of pharmaceutical proteins. *Int J Pharm* 277:99–104.

Hiemstra C, Zhong Z, Li L, Dijkstra PJ, Feijen J. 2006. In-situ formation of biodegradable hydrogels by stereocomplexation of PEG-(PLLA)$_8$ and PEG-(PDLA)$_8$ star block copolymers. *Biomacromolecules* 7:2790–2795.

Hoogsteen W, Postema AR, Pennings AJ, ten Brinke G, Zugenmaier P. 1990. Crystal structure, conformation, and morphology of solution-spun poly(L-lactide) fibers. *Macromolecules* 23:634–642.

Ikada Y, Tsuji H. 2000. Biodegradable polyesters for medical and ecological applications. *Macromol Rapid Commun* 21:117–132.

Ikada Y, Jamshidi K, Tsuji H, Hyon S-H. 1987. Stereocomplex formation between enantiomeric poly(lactides). *Macromolecules* 20:904–906.

Kalb B, Pennings AJ. 1980. General crystallization behaviour of poly(L-lactic acid). *Polymer* 21:607–612.

Kawai T, Rahman N, Matsuba G, Nishida K, Kanaya T, Nakano M, Okamoto H, Kawada J, Usuki A, Honma N, Nakajima K, Matsuda M. 2007. Crystallization and melting behavior of poly(L-lactic acid). *Macromolecules* 40:9463–9469.

Karst D, Yang Y. 2006. Molecular modeling study of the resistance of PLA to hydrolysis based on the blending of PLLA and PDLA. *Polymer* 47:4845–4850.

Kharas GB, Sanchez-Riera F, Severson DK. 1994. In Mobley DP, editors. Plastics from Microbes. New York: Hanser Publishers. p. 93–137.

Kobayashi J. Asahi T, Ichiki M, et al. 1995. Structural and optical properties of poly lactic acids. *J Appl Phys* 77:2957–2973.

Lee W-K, Iwata T, Gardella JA, Jr. 2005. Hydrolytic behavior of enantiomeric poly(lactide) mixed monolayer films at the air/water interface: stereocomplexation effects. *Langmuir* 21:11180–11184.

Leenslag JW, Pennings AJ. 1987. High-strength poly(L-lactide) fibres by a dry-spinning/hot-drawing process. *Polymer* 28:1695–1702.

Li S. 2003. Bioresorbable hydrogels prepared through stereocomplexation between poly(L-lactide) and poly(D-lactide) blocks attached to poly(ethylene glycol). *Macromol Biosci* 3:657–661.

Li S, Vert M. 2003. Synthesis, characterization, and stereocomplex-induced gelation of block copolymers prepared by ring-opening polymerization of L(D)-lactide in the presence of poly(ethylene glycol). *Macromolecules* 36:8008–8014.

Lim DW, Choi SH, Park TG. 2000. A new class of biodegradable hydrogels stereocomplexed by enantiomeric oligo(lactide) side chains of poly(HEMA-γ-OLA)s. *Macromol Rapid Commun* 21:464–471.

Loomis GL, Murdoch JR. 1990. Polylactide compositions. U.S. Patent, 4,902,515.

Mukose T, Fujiwara T, Nakano J, et al. 2004. Hydrogel formation between enantiomeric B-A-B-type block copolymers of polylactides (PLLA or PDLA: A) and Polyoxyethylene (PEG: B); PEG-PLLA-PEG and PEG-PDLA-PEG4. *Macromol Biosci* 361–367.

Murdoch JR, Loomis GL. 1988. Polylactide compositions. U.S. Patent, 4,719,246.

Okihara T. Tsuji M, Kawaguchi A, et al. 1991. Crystal structure of stereocomplex of poly(L-lactide) and poly(D-lactide). *Macomol Sci-Phys B* 30:119–140.

Pan P, Zhu B, Kai W, Dong T, Inoue T. 2008. Effect of crystallization temperature on crystal modifications and crystallization kinetics of poly(L-lactide). *J Appl Polym Sci* 107:54–62.

Pelletier I, Pézolet M. 2004. Compression-induced stereocomplexation of polylactides at the air/water interface. *Macromolecules* 37:4967–4973.

Puiggali J, Ikada Y, Tsuji H, Cartier L, Okihara T, Lotz B. 2000. The frustrated structure of poly(L-lactide). *Polymer* 41:8921–8930.

Sarasua J-R, López Rodrígues N, López Arraiza A, Meaurio E. 2005. Stereoselective crystallization and specific interactions in polylactides. *Macromolecules* 38:8362–8371.

Sawai D, Tamada M, Yokoyama T, Kanamoto T, Hyon S-H, Moon S. 2007. Stereocomplex crystal formation and development of morphology and mechanical property upon drawing of a blend of high molecular weight poly(L-lactic acid)/poly(D-lactic acid). *Sen'I Gakkaishi* 63:1–7.

Schmidt SC, Hillmyer MA. 2001. Polylactide stereocomplex crystallites as nucleating agents for isotactic polylactide. *J Polym Sci Part B: Polym Phys* 39:300–313.

Scott G, editor. 2002. *Biodegradable Polymers: Principles and Applications*, 2nd ed. Dordrecht: Kluwer Academic Publishers.

Serizawa T, Arikawa Y, Hamada K, et al. 2003. Alkaline hydrolysis of enantiomeric poly(lactide)s stereocomplex deposited on solid substrates, *Macromolecules* 36:1762–1765.

Slager J, Domb AJ. 2002. Stereocomplexes based on poly(lactic acid) and insulin: Formulation and release studies. *Biomaterials* 23:4389–4396.

Slager J, Domb AJ. 2003a. Biopolymer stereocomplexes. *Adv Drug Delivery Rev* 55:549–583.

Slager J, Domb AJ. 2003b. Heterostereocomplexes prepared from D-poly(lactide) and leuprolide. I. Characterization. *Biomacromolecules* 4:1308–1315.

Slager J, Domb AJ. 2003c. Heterostereocomplexes prepared from D-PLA and L-PLA and leuprolide. II. Release of leuprolide. *Biomacromolecules* 4:1316–1320.

Slager J, Domb AJ. 2004. Hetero-stereocomplexes of D-poly(lactic acid) and the LHRH analogue leuprolide. Application in controlled release. *Eur J Pharm Biopharm* 58:461–469.

Slivniak R, Domb AJ. 2002. Stereocomplexes of enantiomeric lactic acid and sebacic acid ester-anhydride triblock copolymers. *Biomacromolecules* 3:754–760.

Södergård A, Stolt M. 2002. Properties of lactic acid based polymers and their correlation with composition. *Prog Polym Sci* 27:1123–1163.

Spinu M, Jackson C, Keating MY, Gardner KH. 1996. Material design in poly(lactic acid) system: Block copolymers, star homo- and copolymers, and sereocomplexes. *J Macromol Sci-Pure Appl Chem* A33:1497–1530.

Sweeney PJ, Walker JM. 1993. Proteinase K. In: Burrell MM, editor. *Enzymes of Molecular Biology* (Methods in Molecular Biology, vol. 16). Totowa, NJ: Humana Press. p. 305–311.

Takasaki M, Ito H, Kikutani T. 2003. Development of stereocomplex crystal of polylactide in high-speed melt spinning and subsequent drawing and annealing processes. *J Macromol Sci Part B: Phys* B42:403–420.

Tsuji H. 2000. In vitro hydrolysis of blends from enantiomeric poly(lactide)s. Part 1. Well-stereocomplexed blend and non-blended films. *Polymer* 41:3621–3630.

Tsuji H. 2002a. Poly(lactide). In: Doi Y, Steinbüchel A. editors. *Polyesters III* (Biopolymers, vol. 4) Weinheim: Wiley-VCH. p. 129–177.

Tsuji H. 2002b. Autocatalytic hydrolysis of amorphous-made polylactides: effects of L-lactide content, tacticity, and enantiomeric polymer blending. *Polymer* 43:1789–1796.

Tsuji H. 2003. In vitro hydrolysis of blends from enantiomeric poly(lactide)s. Part 4: well-homo-crystallized blend and nonblended films. *Biomaterials* 24:537–547.

Tsuji H. 2005. Poly(lactide) stereocomplexes: formation, structure, properties, degradation, and applications. *Macromol Biosci* 5:569–597.

Tsuji H. 2007. Degradation of poly(lactide)-based biodegradable materials. In: Albertov LB editor. *Polymer Degradation and Stability Research Developments*. New York: Nova Science Book Publishers, p. 11–59.

Tsuji H, Fukui I. 2003. Enhanced thermal stability of poly(lactide)s in the melt by enantiomeric polymer blending. *Polymer* 44:2891–2896.

Tsuji H, Ikada Y. 1993. Stereocomplex formation between enantiomeric poly(lactic acid)s. 9. Stereocomplexation from the melt. *Macromolecules* 26:6918–6926.

Tsuji H, Ikada Y. 1995. Properties and morphologies of poly(L-lactide): 1. Annealing condition effects on properties and morphologies of poly(L-lactide). *Polymer* 36:2709–2716.

Tsuji H, Ikada Y. 1999. Stereocomplex formation between enantiomeric poly(lactic acid)s. XI. Mechanical properties and morphology. *Polymer* 40:6699–6708.

Tsuji H, Miyauchi S. 2001a. Enzymatic hydrolysis of poly(lactide)s: Effects of molecular weight, L-lactide content, and enantiomeric and diastereoisomeric polymer blending. *Biomacromolecules* 2:597–604.

Tsuji H, Miyauchi S. 2001b. Poly(L-lactide): VI. Effects of crystallinity on enzymatic hydrolysis of poly(L-lactide) without free amorphous region. *Polym Degrad Stab* 71:415–421.

Tsuji H, Miyauchi S. 2001c. Poly(L-lactide): 7. Enzymatic hydrolysis of free and restricted amorphous regions in poly(L-lactide) films with different crystallinities and a fixed crystalline thickness. *Polymer* 42:4463–4467.

Tsuji H, Suzuki M. 2001. In vitro hydrolysis of blends from enantiomeric poly(lactide)s. 2. Well-stereocomplexed fiber and film. *Sen'i Gakkishi* 57:198–202.

Tsuji H, Tezuka Y. 2004. Stereocomplex formation between enantiomeric poly(lactic acid)s. XII. Spherulite growth of low-molecular-weight poly(lactic acid)s. *Biomacromolecules* 5:1181–1186.

Tsuji H, Hyon S-H, Ikada Y. 1991a. Stereocomplex formation between enantiomeric poly(lactic acid)s. 4. Differential scanning calorimetric studies on precipitates from mixed solutions of poly(D-lactic acid) and poly(L-lactic acid). *Macromolecules* 24:5657–5662.

Tsuji H, Horii F, Hyon S-H, Ikada Y. 1991b. Stereocomplex formation between enantiomeric poly(lactic acid)s. 2. Stereocomplex formation in concentrated solutions. *Macromolecules* 24:2719–2724.

Tsuji H, Hyon S-H, Ikada Y. 1991c. Stereocomplex formation between enantiomeric poly(lactic acid)s. 3. Calorimetric studies on blend films cast from dilute solution. *Macromolecules* 24:5651–5656.

Tsuji H, Hyon S-H, Ikada Y. 1992. Stereocomplex formation between enantiomeric poly(lactic acid)s. 5. Calorimetric and morphological studies in the stereocomplex formed in acetonitrile solution. *Macromolecules* 25:2940–2946.

Tsuji H, Ikada Y, Hyon S-H, Kimura Y, Kitao T. 1994. Stereocomplex formation between enantiomeric poly(lactic acid)s. VIII. Complex fibers spun from mixed solution of poly(D-lactic acid) and poly(L-lactic acid); *J Appl Polym Sci* 51:337–344.

Tsuji H, Miyase T, Tezuka Y, Saha SK. 2005a. Physical properties and crystallization, and spherulite growth of linear and 3-arm poly(L-lactide)s. *Biomacromolecules* 6:244–254.

Tsuji H, Tezuka Y, Saha SK, Suzuki M, Itsuno S. 2005b. Spherulite growth of L-lactide copolymers: Effects of tacticity and comonomers. *Polymer* 46:4917–4927.

Tsuji H, Nakano M, Hashimoto M, Takashima Katsura K, Mizuno A. 2006a. Electrospinning of poly(lactic acid) stereocomplex nanofibers. *Biomacromolecules* 7:3316–3320.

Tsuji H, Takai H, Saha SK. 2006b. Isothermal and non-isothermal crystallization behavior of poly(L-lactic acid): Effects of stereocomplex as nucleating agent. *Polymer* 47:3826–3837.

Tsuji H, Takai H, Fukuda N, Takikawa H. 2006c. Non-isothermal crystallization behavior of poly(L-lactic acid) in the presence of various additives. *Macromol Mater Eng* 291:325–335.

Tsuruno T, Tsuji H. 2007. Water vapor permeability of poly(lactic acid) stereocomplex films. *Polym Prepr Japan* 56:2273.

van Nostrum CF, Veldhuis TFJ, Bos GW, Hennink WE. 2004. Tuning the degradation rate of poly(2-hydroxypropyl methacrylamide)-graft-oligo(lactic acid) stereocomplex hydrogels. *Macromolecules* 37:2113–2118.

Vert M, Schwarch G, Coudane JJ. 1995. Present and future of PLA polymers. *J Macromol Sci Pure Appl Chem* A32:787–796.

Watanabe J, Ishihara K. 2003. Higher water intrusion property on novel porous matrix composed of bioinspired polymer stereocomplex for tissue engineering. *Chem Lett* 32:192–193.

Watanabe J, Eriguchi T, Ishihara K. 2002a. Stereocomplex formation by enantiomeric poly(lactic acid) graft-type phospholipid polymers for tissue engineering. *Biomacromolecules* 3:1109–1114.

Watanabe J, Eriguchi T, Ishihara K. 2002b. Cell adhesion and morphology in porous scaffold based on enantiomeric poly(lactic acid) graft-type phospholipid polymers. *Biomacromoleucles* 3:1375–1383.

Yamane H, Sasai K. 2003. Effect of the addition of poly(D-lactic acid) on the thermal property of poly(L-lactic acid). *Polymer* 44:2569–2575.

Yu L, Dean K, Li L. 2006. Polymer blends and composites from renewable resources. *Prog Polym Sci* 31:576–602.

Zhang J, Duan Y, Sato H, et al. 2005a. Crystal modifications and thermal behavior of poly(L-lactic acid) revealed by infrared spectroscopy. *Macromolecules* 38:8012–8021.

Zhang J, Sato H, Tsuji H, Noda I, Ozaki Y. 2005b. Infrared spectroscopic study of CH$_3$ $\cdots$ O $=$ C interaction during the poly(L-lactide)/poly(D-lactide) stereocomplex formation. *Macromolecules* 38:1822–1828.

Zhang J, Tashiro K, Tsuji H, Domb AJ. 2007. Investigation of phase transitional behavior of poly(L-lactide)/poly(D-lactide) blend used to prepare the highly-oriented stereocomplex. *Macromolecules* 40:1049–1054.

CHAPTER 8

Polyhydroxyalkanoate Blends and Composites

GUO-QIANG CHEN

Multidisciplinary Research Center (MRC), Shantou University, Shantou, Guangdong, China;
Department of Biological Sciences and Biotechnology, Tsinghua University, Beijing, China

RONG-CONG LUO

Multidisciplinary Research Center (MRC), Shantou University, Shantou, Guangdong, China

8.1 INTRODUCTION

Polyhydroxyalkanoates (PHAs) are a family of intracellular biopolymers synthesized by many bacteria as intracellular carbon and energy storage granules (Fig. 8-1). It is generally believed that PHA synthesis is promoted by unbalanced growth (Anderson and Dawes, 1990). The plastic-like properties and biodegradability of PHAs make them attractive as potential replacements for nondegradable polyethylene and polypropylene, as well as biodegradable and biocompatible biomaterials for implant purposes (Chen and Wu, 2005). Many efforts have been made to produce PHAs as environmentally degradable thermoplastics (Chen, 2003), including the large-scale production of poly-3-hydroxybutyrate (PHB) (Hrabak, 1992), copoly-esters of 3-hydroxybutyrate and 3-hydroxyvalerate (PHBV) (Chen et al., 1991;

Biodegradable Polymer Blends and Composites from Renewable Resources. Edited by Long Yu
Copyright © 2009 John Wiley & Sons, Inc.

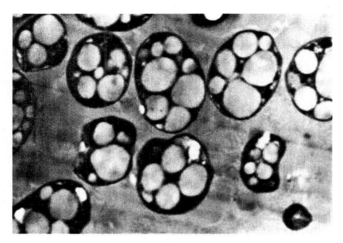

Fig. 8-1   Bacterial cells containing PHA granules imaged by scanning electron microscopy (Chen et al., 2000).

Byrom, 1992), copolyesters of 3-hydroxybutyrate and 3-hydroxyhexanoate (PHBHHx) (Chen et al., 2001), as well as medium-chain-length (mcl) PHA (Weusthuis et al., 2002) (Figs. 8-2 and 8-3).

PHB is the most common and lowest-cost PHA available but has the poorest properties for applications (Table 8-1) (Doi et al., 1995). Recently, a copolyester of 3-hydroxybutyrate and 4-hydroxybutyrate (P3HB4HB) was also produced by Tianjin Green Bioscience Co. Ltd in Tianjin (China) at one-tonne scale, allowing more investigation into this unique material.

As PHB is very brittle and prone to thermal degradation, it is necessary to improve its mechanical properties and processability through copolymerization or blending. In fact, since it is very difficult to develop new copolymers, blending with a second, already available, cheaper polymer has been attempted by many in the polymer research community.

There have been many reports on the blending of PHB with other biodegradable or nonbiodegradable polymers; these include poly(vinyl acetate) (PVAc) (Greco and Martuscelli, 1989; Kumagai and Doi, 1992a,b), poly(epichlorohydrin) (PECH) (Dubini et al., 1993; Sadocco et al., 1993), poly(vinyl alcohol) (PVA) (Azuma et al., 1992), atactic poly-[(R,S)-3-hydroxybutyrate] (Abe et al., 1995) and its block copolymer with poly(ethylene glycol) (P(R,S-HB-b-EG) (Kumagai and Doi,

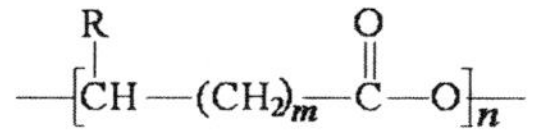

$$-\left[\overset{\displaystyle R}{\underset{\displaystyle |}{C}}H-(CH_2)_m-\overset{\displaystyle O}{\overset{\displaystyle \|}{C}}-O\right]_n-$$

Fig. 8-2   General molecular structure of polyhydroxyalkanoates (PHAs). $m = 1, 2, 3$, with $m = 1$ the most common; n can range from 100 to several thousands. R is variable. When $m = 1$, $R = CH_3$, the monomer structure is 3-hydroxybutyrate; with $m = 1$ and $R = C_3H_7$, the monomer is a 3-hydroxyhexanoate monomer.

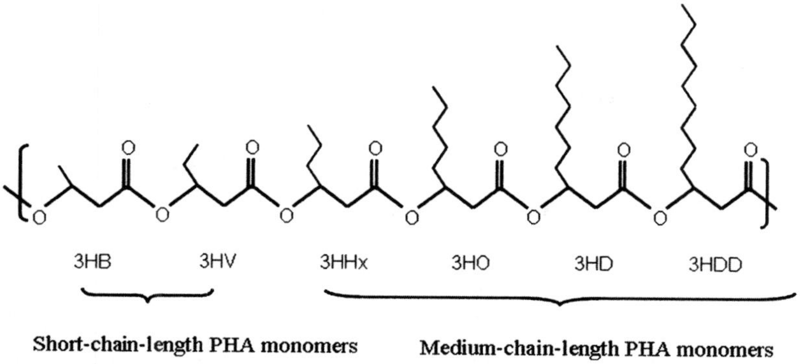

3HB 3HV 3HHx 3HO 3HD 3HDD

Short-chain-length PHA monomers Medium-chain-length PHA monomers

Fig. 8-3 Some commonly synthesized PHA monomers. Short-chain-length (scl) PHA monomers: 3HB, 3-hydroxybutyrate; 3HV, 3-hydroxyvalerate. Medium-chain-length PHA monomers: 3HHx, 3-hydroxyhexanoate; 3HO, 3-hydroxyoctanoate; 3HD, 3-hydroxydecanoate; 3HDD, 3-hydroxydodecanoate.

1993), poly(L-lactic acid-co-ethylene glycol-co-adipic acid) (Yoon et al., 1996), cellulose acetate butyrate (CAB) or cellulose acetate propionate (CAP) (Scandola et al., 1992), and poly(ethylene oxide) (PEO) (Avella and Martuscelli 1988; Kim et al., 1999). Poly(methyl methacrylate) (PMMA) is known to be immiscible with PHB at room temperature (Yoon et al., 1993); however, it is partially miscible in its molten state (Lotti et al., 1993). Immiscible blends have been prepared by mixing PHB with poly(1,4-butylene adipate) (PBA) (Kumagai and Doi, 1992a), PCL (Kumagai and Doi, 1992), poly(cyclohexyl methacrylate) (PCHMA) (Lotti et al., 1993), PHBV (Kumagai and Doi, 1992; Organ and Barham, 1993; Pearce and Marchessault, 1994; Gassner and Owen, 1996), and rubbers, such as

TABLE 8-1 Physical Properties of Various PHAs in Comparison with Conventional Plastics

Sample[a]	T_m (°C)	T_g (°C)	Tensile Strength (Mpa)	Elongation at Break (%)
PHB	177	4	43	5
P(HB-co-10% HV)	150	2	25	20
P(HB-co-20% HV)	135	−5	20	100
P(HB-co-10% HHx)	127	−1	21	400
P(HB-co-17% HHx)	120	−2	20	850
Polypropylene	170	–	34	400
Polystyrene	110	–	50	–
PET	262	69	56	7300
HDPE	135	–	29	–

[a]HV, 3-hydroxyvalerate; HHx, 3-hydroxyhexanoate; PET, poly(ethylene teraphthalate); HDPE, high-density polyethylene.

ethylene-propylene rubber (EPR) (Greco and Martuscelli, 1989; Abbate et al., 1991), ethylene-vinyl acetate (EVA) (Avella and Martuscelli, 1988; Abbate et al., 1991), modified EPR rubbers grafted with succinic anhydride (EPR-g-SA) or dibutyl maleate (EPR-g-DBM), or a modified EVA polymer containing OH groups (EVAL). Verhoogt and co-workers reviewed the properties and biodegradability of blends containing either PHB or PHBV (Verhoogt et al., 1994).

In this chapter, the blending of PHA with various polymers or additives is discussed.

8.2 PHA BLENDED WITH STARCH OR CELLULOSE

Starch is one of the cheapest biomaterials. If starch is mixable with PHB, it will help both reduction of the cost of PHB and improved PHB biodegradability. Godbole and co-workers (Godbole et al., 2003) studied the compatibility of PHB with starch. They indicated that blend films had a single glass transition temperature for all the proportions of PHB : starch tested; all combinations were found to be crystalline. The tensile strength was optimum for the PHB : starch ratio of 0.7 : 0.3 (wt/wt) (Table 8-2).

The results indicate that blending of starch with PHB in a ratio of 30 : 70 could be beneficial for cost reduction with improved properties over the virgin PHB. This blend material might also be used as a coating material on paper or cardboard used for food packaging etc. (Godbole et al., 2003) (Table 8-2).

The thermal behavior and phase morphology of PHB and starch acetate (SA) blends have been studied and they were found to be immiscible. The melting temperatures of PHB in the blends showed some shift with increase of SA content. The melting enthalpy of the PHB phase in the blend was close to the value for pure PHB. The glass transition temperatures of PHB in the blends remained constant at 9°C. The FTIR absorptions of hydroxyl groups of SA and carbonyl groups of PHB in the blends were found to be independent of the second component at 3470 cm^{-1} and 1724 cm^{-1}, respectively. The crystallization of PHB was affected by the addition of the SA component both from the melt on cooling and from the glassy state on heating. The temperature and enthalpy of nonisothermal crystallization of PHB in the blends were much lower than those of pure PHB. The crystalline morphology of PHB crystallized from the melt under isothermal conditions varied with SA content. The cold crystallization peaks of PHB in the blends shifted to higher temperatures compared with that of pure PHB. No mechanical study was performed by the authors for their property changes (Zhang et al., 1997) (Table 8-2).

When blends of PHB with cellulose acetate butyrate (CAB) were prepared by solution casting from chloroform solution at different compositions ranging from 20% to 100% PHB, the PHB/CAB blends were found to be miscible in the melt state as evidenced by a single glass transition (T_g) for each composition; a depression in the equilibrium melting point of PHB and a marked reduction in the spherulite growth rate of PHB in the PHB/CAB blends were also detected (Table 8-2). The phase structure of the blend in the solid state as revealed by SAXS (small-angle X-ray scattering) was

TABLE 8-2   Thermal and Mechanical Properties of Typical PHA-based Blends

Sample	Blend Ratio (w/w)	T_g (°C)	T_m (°C)	Young's Modulus (Mpa)	Tensile Strength (Mpa)	Elongation to Break (%)	References
PHB/Starch	70/30	7.1	167	949	19.2	9.4	Godbole et al. (2003)
PHB/SA	40/60	8.6	173	–	–	–	Zhang et al. (1997)
PHB/PIP-g-PVAc	80/20	6	175	711	13.8	13	Yoon et al. (1999)
PHB/BWF	90/10	–20	162	1200	–	–	Fernandes et al. (2004)
a-PHB/P(d,l)LA	50/50	23	/	100	–	300	Focarete et al. (2002)
PHB/LMWPLA	50/50	19	145; 167	–	–	–	Koyama and Doi (1997)
PHB/HMWPLA	50/50	5	177	1000	40	80	Park et al. (2004)
UHMWPHB/PLA	70/30	1	177	1600	100	110	Park et al. (2004)
PHBHHx/PLA	60/40	3.7; 65	137; 175	–	–	–	Furukawa et al. (2007)
PHB/P(d,l)LA	60/40	0.28; 44.6	175	274	–	27.7	Zhang LL et al. (1996)
PHB/PCL	77/23	–60; 4	60; 168	730	21	9	Kumagai and Doi (1992a)
PHB/PHO	75/25	–35	172	370	6.2	30	Dufresne and Vincendon (2000)
PLA/a-PHB	50/50	–1; 66	163	664	12	70	Dacko et al. (2006)
PHB/a-PHB	30/70	–1	168	170	3.4	51	Kunze et al. (2006)
PHB/PHBV	25/75	–	152; 163	150	2	7	Sombatmankhong et al. (2006)
PHBV/a-PHB	50/50	2	133	240	7	33	Scandola et al. (1997)
PHB/PHBHHx	40/60	0	150	–	18	140	Chen and Wu (2005)
PHB/P3HB4HB	50/50	–10	168	113	3	35	Luo et al. (2007)
PHB/PBA	75/25	–	–	1050	32	7	Ha and Cho (2002)
PHB/EPR	80/20	–	–	147	17	2	Abbate et al. (1991)
PHB/EVA	80/20	–	–	156	17	4	Abbate et al. (1991)
PHB/PEG	80/20	–	166; 152	–	13	40	Parra et al. (2006)

Notes: T_g, glass transition temperature; T_m, melting temperature; –, not determined; SA, starch acetate; PIP-g-PVAc, poly(*cis*-1,4-isoprene) grafted with poly(vinyl acetate); BWF, beech wood flour; a-PHB, atactic PHB; LMWPLA, low-molecular-weight PLA; HMWPLA, high-molecular-weight PLA; UHMWPHB, ultrahigh-molecular-weight PHB; PBA, poly(1,4-butylene adipate); EPR, ethylene-propylene rubber; EVA, ethylene-vinyl acetate.

characterized by the presence of a homogeneous amorphous phase situated mainly in the interlamellar regions of crystalline PHB and consisting of CAB molecules and noncrystalline PHB chains. Again, there was no study of mechanical property change (El-Shafee et al., 2001).

Pizzoli and co-workers investigated isothermal crystallization from the melt of blends of PHB with two cellulose acetobutyrates (CE1 and CE2) using hot-stage optical microscopy (Pizzoli et al., 1994). Space-filling spherulites were observed at all compositions (50–100% PHB) and crystallization temperatures (50–130°C) were explored. The spherulite radius increases linearly with time, and the radial growth rate (G) depends strongly on the cellulose ester content of the blend. The crystallization rate of the blend containing 50% CE2 is 2.5 orders of magnitude lower than that of pure PHB. The spherulites show banding whose spacing increases with increasing T_c for a given blend, while at constant T_c banding decreases with increasing cellulose ester concentration. In PHB/CE1 blends, owing to the ability of CE1 to crystallize from the melt concomitantly with PHB at certain compositions and crystallization temperatures, a very unusual deviation from linearity of the radial growth of PHB spherulites was observed. The rate was seen to increase with time, reflecting the compositional changes that occurred in the melt as a consequence of CE1 crystallization. Apart from the cases where crystallization of CE1 occurred, the morphology of melt-crystallized PHB/CE1 and PHB/CE2 blends was rather independent of the identity of the cellulose ester (Pizzoli et al., 1994).

Beech wood flour (BWF) composites were prepared to plasticize PHB. The type of plasticizer [tri(ethylene glycol) bis(2-ethylhexanoate)] (TEGB), poly(ethylene glycol) (PEG200)] and the amounts (5 and 20 wt%) were selected as independent variables in a factorial design. Thermal and mechanical properties of 90 wt% PHB composites were investigated (Table 8-2). Incorporation of PEG200 was found to compromise thermal stability of PHB as demonstrated by the greater decrease in the onset decomposition temperature (T_d) and the drop in its average molecular weight (M_w). The study showed that TEGB/PHB/BWF composites can be optimized to obtain new materials for disposable items (Fernandes et al., 2004).

Lignin fine powder as a new kind of nucleating agent for PHB was studied by Kai et al. (2004). The kinetics of both isothermal and nonisothermal crystallization processes from the melt for both pure PHB and PHB/lignin blend was studied. Lignin shortened the crystallization half-time $t_{1/2}$ for isothermal crystallization. The crystallization of the PHB/lignin blend was more favorable than that of pure PHB from a thermodynamic perspective. At the same time, according to polarized optical microscopy, the rate of spherulite growth from the melt increased with the addition of lignin. Polarized optical microscopy also showed that the spherulites found in PHB with lignin were smaller in size and greater in number than those found in pure PHB. The wide-angle X-ray diffraction (WAXD) indicated that addition of lignin caused no change in the crystal structure and degree of crystallinity. These results indicated that lignin is a good nucleating agent for the crystallization of PHB (Kai et al., 2004).

8.3 PHA BLENDED WITH PLA

The spherulitic structure, growth rates, and melting behavior of blends of bacterially produced PHB and poly(L-lactide) (PLLA) were investigated using polarized light microscopy. Results indicated that low-molecular-weight PLLA ($M_n = 1759$) was miscible in the melt over the whole composition range, whereas a blend of high-molecular-weight PLLA ($M_n = 159,400$) with PHB showed biphasic separation (Table 8-2). Two types of spherulite were formed on cooling, related to the crystallization of PHB and PLLA, respectively. In some blends, spherulites of opposite type interpenetrated when the growth fronts met. It is proposed that lamellae belonging to one type of spherulite continued to grow in the interlamellar regions of the other type of spherulite (Blümm and Owen, 1995).

Furukawa and co-workers investigated the miscibility and structure of PHB/PLLA and PHBHHx/PLLA blends using DSC, WAXD, and IR microspectroscopy (Table 8-2). The results are then compared between the two types of blends. They found that both PHB/PLLA and PHBHHx/PLLA blends are immiscible, but the PHBHHx/PLLA blends are somewhat more compatible. WAXD reflection patterns revealed that PHB component can be crystallized in the PHB/PLLA blends with any ratio and that the PHBHHx component can also be crystallized in the PHBHHx/PLLA blends except for the 20/80 blend. The a and b lattice parameters of each component in the blends are almost constant, suggesting that their crystalline structures are kept intact in the blends. The T_g values for PHBHHx and PLLA components in the PHBHHx/PLLA blends also do not change significantly, as well those for PHB and PLLA component in the PHB/PLLA blends. The observation of the T_c of PLLA in both blend systems suggests that each component in the PHB/PLLA and also in the PHBHHx/PLLA blends forms the mixed semicrystalline structure. These results indicate that the PHB/PLLA blends with decreasing T_c are totally immiscible, while the PHBHHx/PLLA blends with increasing T_c are somewhat compatible. Micro IR spectra from any spots of the 80/20 PHB/PLLA and PHBHHx/PLLA blends show crystalline bands due only to the PHB and PHBHHx components but not for PLLA. On the other hand, although the micro-IR spectra from any spots of the 20/80 PHB/PLLA blend also show the crystalline bands due to PHB, those from some spots of the 20/80 PHBHHx/PLLA blend show only the crystalline bands of the PLLA component. PHBHHx dispersed in a PLLA matrix at this low level does not crystallize (Furukawa et al., 2007) (Table 8-2).

JM Zhang and co-workers investigated the crystallization behaviors of the two components in their immiscible and miscible 50 : 50 blends by real-time infrared (IR) spectroscopy by adjusting the molecular weight of the PLLA component in PHB/PLLA blends (Zhang JM et al., 2006). In the immiscible PHB/PLLA blend, the stepwise crystallization of PHB and PLLA was realized at different crystallization temperatures. PLLA crystallizes first at a higher temperature (120°C). Its crystallization mechanism from the immiscible PHB/PLLA melt was not affected by the presence of the PHB component, while its crystallization rate was substantially depressed. Subsequently, in the presence of crystallized PLLA, the isothermal melt-crystallization of PHB took place at a lower temperature (90°C). It is interesting to find that there are two growth stages for PHB. At the early stage of the growth period the Avrami exponent is 5.0, which is unusually high, while in the late stage it is 2.5, which is very close to the reported value ($n \approx 2.5$) for the neat PHB system. In contrast to the stepwise crystallization of PHB and PLLA in the immiscible blends, the almost simultaneous crystallization of PHB and PLLA in the miscible 50:50 blend was observed at the same crystallization temperature (110°C). Detailed dynamic analysis by IR spectroscopy disclosed that the crystallization of PLLA actually occurs faster than that of PHB even in such apparently simultaneous crystallization. It has been found that in both the immiscible and miscible blends, the crystallization dynamics of PHB are heavily affected by the presence of crystallized PLLA (Zhang JM et al., 2006) (Table 8-2).

Furukawa and co-workers prepared four kinds of PHB/PLLA blends with a PLLA content of 20, 40, 60, and 80 wt% from chloroform solutions. Micro-IR spectra obtained at different positions of a PHB film are all very similar to each other, suggesting that there are no discernible segregated amorphous and crystalline parts on the PHB film at the resolution scale of micro-IR spectroscopy. On the other hand, the micro-IR spectra of two different positions of a PLLA film, where spherulite structures are observed and where they are not observed, are significantly different from each other. PHB and PLLA have characteristic IR marker bands for their crystalline and amorphous components. Therefore, it is possible to explore the structure of each component in the PHB/PLLA blends using micro-IR spectroscopy. The IR spectra of a position of blends except for the 20/80 blend are similar to that of pure PHB. DSC curves of the blend show that the heat of crystallization of PHB varies with the blending ratio of PHB and PLLA. The recrystallization peak is detected for PLLA and the 20/80 blend respectively at 106.5 and 88.2°C. The lowering of recrystallization temperature for the 20/80 blend compared with that of pure PLLA suggests that PHB forms small finely dispersed crystals that may act as nucleation sites of PLLA. The results for the PHB/PLLA blends obtained from IR microspectroscopy indicate that PHB crystallizes in any blends. However, crystalline structures of PHB in the 80/20, 60/40, and 40/60 blends are different from those of the 20/80 blend (Furukawa et al., 2005).

Park and co-workers prepared blends of PLLA with two kinds of PHB having different molecular weights, commercial-grade bacterial PHB (bacterial-PHB) and ultrahigh-molecular-weight PHB (UHMW-PHB), by the solvent-casting method and uniaxially drawn at two drawing temperatures, around the T_g of PHB (2°C) for PHB-rich blends and around the T_g of PLLA (60°C) for PLLA-rich blends. DSC analysis showed that this system was immiscible over the entire composition range (Table 8-2). Mechanical properties of all of the samples were improved in proportion to the draw ratio. Although PLLA domains in bacterial-PHB-rich blends remained almost unstretched during cold drawing, a good interfacial adhesion between two polymers and the reinforcing role of PLLA components led to enhanced mechanical properties proportional to the PLLA content at the same draw ratio (Park et al., 2004).

In contrast, in the case of UHMW-PHB-rich blends, the minor component PLLA was found to be also oriented by cold drawing in ice water due to an increase in the interfacial entanglements caused by the very long chain length of the matrix polymer.

As a result, the mechanical properties were considerably improved with increasing PLLA content compared with the bacterial-PHB system. Scanning electron microscopy observations on the surface and cross-section revealed that a layered structure with uniformly oriented microporous in the interior was obtained by selective removal of PLLA component after simple alkaline treatment (Park et al., 2004).

LL Zhang and co-workers investigated the miscibility, crystallization and morphology of PHB/poly(d,l-lactide) (PDLLA) blends. The results indicated that PHB/PDLLA blends prepared by casting a film from a common solvent at room temperature were immiscible over the range of compositions studied, while the melt-blended sample prepared at high temperature showed some evidence of greater miscibility. The crystallization of PHB in the blends was affected by the level of addition of PDLLA. The thermal history caused a depression of the melting point and a decrease in the crystallinity of PHB in the blends. Compared with plain PHB, the blends exhibited some improvement in mechanical properties (Zhang LL et al., 1996).

8.4 PHA BLENDED WITH PCL

Chee and co-workers demonstrated that PEO (poly[ethylene oxide]) is miscible with PHB, whereas poly-ε-caprolactone (PCL) is immiscible (Chee et al., 2002). In their biodegradation, Cara et al. 2003 showed that both pure PCL and PHB samples were degraded with strong erosion of the amorphous zones. In the PCL/PHB 70/30 blend, after only 20 days of incubation, spheres of PCL were bordering with spherulites of PHB, indicating complete degradation. The crystallinity content of homopolymers and blends were investigated at different degradation times: while PCL crystallinity remains constant, both PHB and the blend PHB-phase crystallinity increased. Data from differential scanning calorimetry fit well with those obtained by scanning electron microscopy, gel permeation chromatography, and weight loss analysis (Cara et al., 2003) (Table 8-2).

Two types of mixtures were prepared by solution blending: high molecular weight PHB/PCL and PHB/low-molecular-weight chemically modified PCLs (mPCL). Lovera and co-workers studied the morphology, crystallization, and enzymatic degradation of the blends by exposure to *Aspergillus flavus*. High-molecular-weight PHB/PCL blends were found to be immiscible in the entire composition range. Phenomena such as PCL fractionated crystallization and a decrease in PHB nucleation density were detected. When PHB was blended with mPCLs, the blends were partially miscible; two phases were formed, but the PHB-rich phase exhibited clear signs of miscibility through a depression of both the T_m and the T_g of the PHB component (which was stronger with lower-molecular-weight mPCL), and an increase in the growth rate of PHB spherulites in the blends as compared with neat PHB or the PHB component in the PHB/PCL blends. The biodegradation by exposure to *A. flavus* showed that the blends are synergistically attacked in comparison with the homopolymers. Two factors may influence the improved degradation rate of the blends: the dispersion of the components and their crystallinity, which was reduced in view of the fractionated crystallization and transfer of impurities. In the case of the PHB/mPCL blends, the increased miscibility between the components caused a reduction in the degradation rate (Lovera et al., 2007).

8.5 BLENDING OF DIFFERENT PHAs

Dufresne and Vincendon (2000) prepared blends of PHB with poly(3-hydroxyoctano-ate) (PHO) by co-dissolving the two polyesters in chloroform and casting the mixture. It has been observed that PHB shows no miscibility at all with PHO, resulting in two-phase systems in which the nature of the continuous phase is composition-dependent. The morphology of the blend strongly influences the mechanical behavior. The mechanical properties of these materials have been predicted from a model involving the percolation concept. It takes both linear and nonlinear mechanical behaviors into account and allows for the effect of the lack of adhesion between material domains and/or breakage of one of the components (Dufresne and Vincendon, 2000).

Scandola and co-workers blended synthetic atactic PHB (a-PHB) with a natural bacterial PHBV containing 10 mol% of 3HV using a simple casting procedure (Table 8-2). In the range of compositions explored (10–50% a-PHB), blends of PHBV and synthetic atactic a-PHB were miscible in the melt and solidified with spherulitic morphology. The degree of crystallinity decreased with increasing content of a-PHB in the film samples, and the elongation at break for a sample containing 50% of a-PHB was 30-fold that of pure PHBV. Degradation experiments using both hydrolytic (pH 7.4, 70°C) and enzymatic PHB depolymerase A from *Pseudomonas lemoignei* or Tris-HCl buffer (pH 8, 37°C) were performed for both polymers and polymer blends. The rate of enzymatic degradation of the blends was higher than that of PHBV and increased with a-PHB content in the blends studied, whereas pure a-PHB did not biodegrade under these conditions. 3-Hydroxybutyric acid (3HB) and its dimer were identified by HPLC as biodegradation products of both pure PHBV and its blends with a-PHB. Higher oligomers up to heptamer were detected as degradation products of the blends by APCI-MS and ESI-MS (Scandola et al., 1997).

Saito and co-workers showed that the PHB/PHBV (9% HV) blends exhibited almost perfect cocrystallization, the PHB/PHBV (15% HV) blends form PHB-rich crystalline phase, and the PHB/PHB (21% HV) blends showed phase segregation and formation of the crystalline phases of component polymers as well as the cocrystalline phase. These results indicate that the degree of phase segregation changes depending on the HV content of PHBV. As the HV content increases, phase segregation proceeds to higher degree before cocrystallization and, as a result, the copolymer content in the cocrystalline phase decreases and/or the crystalline phases of the component polymers are formed. The phase structure of the blends is determined by the competition between cocrystallization and phase segregation. The necessary conditions for cocrystallization are supposed to be miscibility in the melt state, similarity of the crystalline structures, similarity of the crystallization rates of the component polymers, and large crystallization rates. The miscibility prevents phase segregation.

The similarity of the crystalline structure lowers the free energy of cocrystallization. The similarity of the crystallization rates allows the simultaneous crystallization of two components. The high crystallization rate does not give sufficient time for phase segregation. It has been reported that the crystalline lattice of PHBV containing 40% or less HV unit is the same as that of PHB. The fact that PHB and PHBV with high HV content are immiscible indicates that the extent of miscibility between PHB and PHBV gradually decreases with the increase of HV content of PHBV. The fact that the crystallization rate of PHBV gradually decreases with the increase of HV content shows that the difference in the crystallization rate of component polymers gradually increases with the increase of HV content of PHBV. Both of these factors promote phase segregation with the increase of the HV content of PHBV for PHB/PHBV blends. Therefore, as the HV content of PHBV for PHB/PHBV blends increases, the PHBV content in the cocrystalline phase gradually decreases and the crystalline phases of the individual component polymers are formed (Saito et al., 2001).

The hydrolytic degradation of PHBV and PHO blends with low- and high-molecular-weight additives was examined. DSC and atomic force microscopy (AFM) results revealed that hydrolyzable PLA and hydrophilic PEG, selected as additives that might accelerate hydrolytic degradation, were immiscible with PHBV or PHO over the range of compositions studied. The results show that the presence of a second component, whatever its chemical nature, is sufficient to perturb the crystallization behavior of highly crystalline PHBV, and increase hydrolytic degradation. In contrast, the degradation of PHO was unaffected by blending with PLA or PEG. PHO degradation is a very slow process, requiring several months of incubation. However, the introduction of polar carboxylic groups in side-chains led to an increase in the degradation rate. Carboxylic groups promote water penetration into the polymer (Renard et al., 2004) (Table 8-2).

8.6 PHA BLENDED WITH OTHER POLYMERS

De Lima and Felisberti showed that PHB/PEP (poly[epichlorohydrin]) and PHB/ECO (poly[epichlorohydrin-co-ethylene oxide]) blends are immiscible. The apparent melting temperature (T_m) of PHB in the blends decreases slightly with increasing elastomeric content and the melting point depression cannot be associated with miscibility, because the blends are immiscible. Thus, T_m of the blends are affected by morphological effects. There is an influence of PEP and ECO on the crystallization that occurs upon cooling of PHB, even when the blends are immiscible due to an expressive decrease of the intensity of the peak at crystallization, that is completed during the second heating. The degree of crystallinity of blends with PEP was found to decrease with an increase in PEP content. PHB/ECO blends exhibit degrees of crystallinity that can be considered nearly independent of the ECO content. Study of the morphology of blends showed that the presence of elastomer influences the ratio of the growth rate and the nucleation rate. The elastomer component, probably resides in the intraspherulitic zones (de Lima and Felisberti, 2006).

Xing and co-workers made a systematic study of the miscibility, crystallization, and morphology of PHB/PVPh [poly(p-vinylphenol)] blends (Xing et al., 1997). The single glass transition temperatures of the blends suggest that PHB and PVPh will form miscible blends in the whole composition range in melt. After quenching, the melting enthalpy of PHB in the blend is substantially lowered and approaches zero at about 40% PVPh content, which is due to the high glass transition temperature of the blend. The equilibrium melting point of PHB in the blends, which was obtained from DSC results using the Hoffman–Weeks equation, decreases with the increase in PVPh content. The large negative values of the interaction parameter determined from the equilibrium melting point depression support the miscibility and strong hydrogen-bonding interactions between the components. They found that the rate of isothermal crystallization of PHB is strongly affected by blending it with PVPh. The rate of spherulite growth decreases with the increase in PVPh content. Isothermally crystallized blends of PHB with PVPh were examined by WAXD and SAXS. The long period increases with the addition of amorphous PVPh, which is a strong indication of interlamellar segregation (Xing et al., 1997).

Qiu and co-workers prepared four blends of PHB and poly(butylene succinate) (PBS), both biodegradable semicrystalline polyesters, with the ratio of PHB/PBS ranging from 80/20 to 20/80 by co-dissolving the two polyesters in N,N-dimethylformamide and casting the mixture. Results indicated that PHB showed some limited miscibility with PBS for PHB/PBS 20/80 blend as evidenced by the small change in the glass transition temperature and the depression of the equilibrium melting point temperature of the high-melting-point component PHB. However, PHB showed immiscibility with PBS for the other three blends as shown by the existence of unchanged composition-independent glass transition temperature and the biphasic melt. During the nonisothermal crystallization of PHB/PBS blends, two crystallization peak temperatures were found for PHB/PBS 40/60 and 60/40 blends, corresponding to the crystallization of PHB and PBS, respectively, whereas only one crystallization peak temperature was observed for PHB/PBS 80/20 and 20/80 blends. However, it was found that after the nonisothermal crystallization the crystals of PHB and PBS actually co-existed in PHB/PBS 80/20 and 20/80 blends from the two melting endotherms observed in the subsequent DSC melting traces, corresponding to the melting of PHB and PBS crystals, respectively. The subsequent melting behavior was also studied after nonisothermal crystallization. In some cases, double melting behavior was found for both PHB and PBS, which was influenced by the cooling rates used and the blend composition (Qiu et al., 2003).

Parra and co-workers investigated the thermal properties, tensile properties, water vapor transmission rate, enzymatic biodegradation and mass retention of blends of PHB with PEG (PHB/PEG), in proportions of 100/0, 98/2, 95/5, 90/10, 80/20, and 60/40 wt%, respectively. They found that the addition of plasticizer did not alter the thermal stability of the blends, although an increase in the PEG content reduced the tensile strength and increased the elongation at break of pure PHB (Parra et al., 2006) (Table 8-2).

8.7 PHA COMPOSITES

It has been demonstrated that the incorporation of a small amount of α-cyclodextrin can greatly enhance the nucleation and crystallization of PHB. This natural and environmentally safe compound, may thus be used as a nucleation agent for biodegradable polyester PHB (He and Inoue, 2003).

The di-*n*-butyl phthalate (DBP) diluent is miscible at molecular level with the natural polyester PHB. In PHB/DBP mixtures both the glass transition and melting temperatures decrease with increasing diluent content, the shift of T_g being more pronounced than that of T_m. The whole crystallization window moves to lower temperatures and broadens, as reflected by the changes of the cold crystallization process in dynamic DSC scans and by the shifts of the isothermal spherulite growth rate curves. It is interesting to note that the maximum growth rate (G_{max}) does not change in the range of diluent contents explored and that all isothermal crystallization data fall on the same master curve when the crystallization temperatures are normalized to account for the T_g and T_m changes with composition. The constancy of G_{max} indicates that in the polymer–diluent mixtures investigated, opposite "dilution" and "chain mobility" effects that should respectively decrease and increase the crystal growth rates tend to compensate (Pizzoli et al., 2002).

Ceccorulli and co-workers found blends obtained by melt compounding of PHB with CAB are miscible over the whole composition range. In the range of PHB contents from 0% to 50% the blend glass transition temperature (T_g) depends strongly on composition, while a much less substantial dependence is found when the amount of PHB exceeds 50%. In the former composition range, in addition to the strongly composition-dependent T_g, another relaxation associated with mobilization of the low-T_g component is observed at a lower temperature (Ceccorulli et al., 1993).

The DBP is miscible in all proportions with both CAB and PHB. Analogously to the polymeric CAB/PHB blends, the two polymer/diluent systems investigated (CAB/DBP and PHB/DBP) show a dual dependence of T_g on composition. In binary mixtures such behavior appears to be independent of the macromolecular or low-molecular-weight nature of the low-T_g component. Addition of a fixed amount of DBP plasticizer to CAB/PHB blends with varying composition (PHB content from 0% to 100%) causes a significant decrease of T_g of the binary polymer blends; the T_g depression is larger the higher the amount of DBP in the ternary blend. Concomitant with the expected "plasticizing" effect on T_g, the presence of DBP also induces a decrease in the characteristic temperature of the additional low-temperature transition observed in CAB/PHB blends. In the ternary blends, the temperature of such a transition is a function of DBP content only, being independent of the relative amount of the two polymers (CAB and PHB) (Ceccorulli et al., 1993).

In conclusion, the many efforts that have been made to improve the properties of PHA, including blending and composite preparations, have achieved various degrees of success. In the future, blends or composite preparation should be tailor made to suite a particular application. With increasing costs of petrochemicals, more and more effort is being directed to the development of useful biobased materials for various application for PHA; some of these effects will be rewarded sooner or later.

REFERENCES

Abbate M, Martuscelli E, Ragosto G, Scarinzi G. 1991. Tensile properties and impact behaviour of poly(D(−)3-hydroxybutyrate)/rubber blends. *J Mater Sci* 26:1119–1125.

Abe H, Matsubara I, Doi Y. 1995. Physical properties and enzymic degradability of polymer blends of bacterial poly[(R)-3-hydroxybutyrate] and poly[(R,S)-3-hydroxybutyrate] stereoisomers. *Macromolecules* 28:844–853.

Anderson AJ, Dawes EA. 1990. Occurrence, metabolism, metabolic role and industrial uses of bacterial polyhydroxyalkanoates. *Microbiol Rev* 45:450–472.

Avella M, Martuscelli E. 1988. Poly-D-(−)(3-hydroxybutyrate)/poly(ethylene oxide) blends: phase diagram, thermal and crystallization behavior. *Polymer* 29:1731–1737.

Azuma Y, Yoshie N, Sakurai M, Inoue Y, Chu JR. 1992. Thermal behaviour and miscibility of poly(3-hydroxybutyrate)/poly(vinyl alcohol) blends. *Polymer* 33:4763–4767.

Blümm E, Owen AJ. 1995. Miscibility, crystallization and melting of poly(3-hydroxybutyrate)/poly(L-lactide) blends. *Polymer* 36:4077–4081.

Byrom D. 1992. Production of poly-β-hydroxybutyrate and poly-β-hydroxyvalerate copolymers. *FEMS Microbiol Rev* 103:247–250.

Cara FL, Immirzi BE, Mazzella A, Portofino S, Orsello G, De Prisco PP. 2003. Biodegradation of poly-ε–caprolactone/poly-β-hydroxybutyrate blend. *Polym Degrad Stab* 79:37–43.

Ceccorulli G, Pizzoli M, Scandola M. 1993. Effect of a low molecular weight plasticizer on the thermal and viscoelastic properties of miscible blends of bacterial poly(3-hydroxybutyrate) with cellulose acetate butyrate. *Macromolecules* 26:6722–6726.

Chee MJK, Ismail J, Kummerlöwe C, Kammer HW. 2002. Study on miscibility of PEO and PCL in blends with PHB by solution viscometry. *Polymer* 43:1235–1239.

Chen GQ. 2003. Production and application of microbial polyhydroxyalkanoates. In: Chiellini E, Solaro R, editors. *Biodegradable Polymers and Plastics* (Proceedings of the 7th World Conference on Biodegradable Polymers & Plastics). Kluwer Academic/Plenum. p. 155–166.

Chen GQ, Wu Q. 2005. Polyhydroxyalkanoates as tissue engineering materials. *Biomaterials* 26:6565–6578.

Chen GQ, Koenig KH. Lafferty RM. 1991. Production of poly-D(−)-3-hydroxybutyrate and poly-D(−)-3-hydroxyvalerate by strains of *Alcaligenes latus*. *Antonie van Leeuwenhoek* 60:61–66.

Chen GQ, Wu Q, Zhao K, Yu HP, Chan A. 2000. Chiral Biopolyesters-Polyhydroxyalkanoates Synthesized by Microorganisms. Chinese. *J Polym Sci* 18:389–396.

Chen GQ, Zhang G, Park SJ, Lee SY. 2001. Industrial production of poly(hydroxybutyrate-co-hydroxyhexanoate). *Appl Microbiol Biotechnol* 57:50–55.

Dacko P, Kowalczuk M, Janeczek H, Sobota M. 2006. Physical properties of the biodegradable polymer compositions containing natural polyesters and their synthetic analogues. *Macromol Symp* 239:209–216.

de Lima JA, Felisberti MI. 2006. Poly(hydroxybutyrate) and epichlorohydrin elastomers blends: Phase behavior and morphology. *Eur Polym J* 42:602–614.

Doi Y, Kitamura S, Abe H. 1995. Microbial synthesis and characterization of poly(3-hydroxybutyrate-co-3-hydroxyhexanoate). *Macromolecules* 28:4822–4828.

Dubini PE, Beltrame PL, Canetti M, Seves A, Marcandalli B, Martuscelli E. 1993. Crystallization and thermal behaviour of poly-D(−)3-hydroxybutyrate)/poly(epichlorohydrin) blends. *Polymer* 34:996–1001.

Dufresne A, Vincendon M. 2000. Poly(3-hydroxybutyrate) and poly(3-hydroxyoctanoate) blends: morphology and mechanical behavior. *Macromolecules* 33:2998–3008.

El-Shafee E, Saad GR, Fahmy SM. 2001. Miscibility, crystallization and phase structure of poly(3-hydroxybutyrate)/cellulose acetate butyrate blends. *Eur Polym J* 37:2091–2104.

Fernandes EG, Pietrini M, Chiellini E. 2004. Bio-based polymeric composites comprising wood flour as filler. *Biomacromolecules* 5:1200–1205.

Focarete ML, Scandola M, Dobrzynski P, Kowalczuk M. 2002. Miscibility and mechanical properties of blends of (L)-lactide copolymers with atactic poly(3-hydroxybutyrate). *Macromolecules* 35:8472–8477.

Furukawa T, Sato H, Murakami R, et al. 2005. Structure, dispersibility, and crystallinity of poly(hydroxybutyrate)/poly(L-lactic acid) blends studied by FT-IR microspectroscopy and differential scanning calorimetry. *Macromolecules* 38:6445–6454.

Furukawa T, Sato H, Murakami R, et al. 2007. Comparison of miscibility and structure of poly(3-hydroxybutyrate-co-3-hydroxyhexanoate)/poly(L-lactic acid) blends with those of poly(3-hydroxybutyrate)/poly(L-lactic acid) blends studied by wide angle X-ray diffraction, differential scanning calorimetry, and FTIR microspectroscopy. *Polymer* 48:1749–1755.

Gassner F, Owen AJ. 1996. Some properties of poly(3-hydroxybutyrate)-poly(3-hydroxyvalerate) blends. *Polym Int* 39:215–219.

Godbole SS, Gote M, Latkar T, Chakrabarti S. 2003. Preparation and characterization of biodegradable poly-3-hydroxybutyrate–starch blend films. *Bioresource Technol.* 86:33–37.

Greco P, Martuscelli E. 1989. Crystallization and thermal behaviors of poly(D-3-hydroxybutyrate)-based blends. *Polymer* 30:1475–1483.

Ha CS, Cho WJ. 2002. Miscibility, properties and biodegradability of microbial polyester containing blends. *Prog Polym Sci* 27:759–809.

He Y, Inoue Y. 2003. α-Cyclodextrin enhanced crystallization of poly(3-hydroxybutyrate). *Biomacromolecules* 4:1865–1867.

Hrabak O. 1992. Industrial production of poly-β-hydroxybutyrate. *FEMS Microbiol Rev* 103:251–256.

Kai WH, He Y, Asakawa N, Inoue Y. 2004. Effect of lignin particles as a nucleating agent on crystallization of poly(3-hydroxybutyrate). *J Appl Polym Sci* 94:2466–2474.

Kim MN, Lee AR, Lee KH, Chin IJ, Yoon JS. 1999. Biodegradability of poly(3-hydroxybutyrate) blended with poly(ethylene-co-vinyl acetate) or poly(ethylene oxide). *Eur Polym J* 35:1153–1158.

Koyama N, Doi Y. 1997. Miscibility of binary blends of poly[(R)-3-hydroxybutyric acid] and poly[(S)-lactic acid]. *Polymer* 38:1589–1593.

Kumagai Y, Doi Y. 1992a. Enzymatic degradation and morphologies of binary blends of microbial poly(3-hydroxy butyrate) with poly(ε-caprolactone), poly(1,4-butylene adipate and poly(vinyl acetate). *Polym Degrad Stab* 36:241–248.

Kumagai Y, Doi Y. 1992b. Enzymatic degradation of binary blends of microbial poly(3-hydroxybutyrate) with enzymatically active polymers. *Polym Degrad Stab* 37:253–256.

Kumagai Y, Doi Y. 1993. Synthesis of a block copolymer of poly(3-hydroxybutyrate) and poly(ethylene glycol) and its application to biodegradable polymer blends. *J Environ Polym Degrad* 1:81–87.

Kunze C, Bernd HE, Androsch R, et al. 2006. In vitro and in vivo studies on blends of isotactic and atactic poly(3-hydroxybutyrate) for development of a dura substitute material. *Biomaterials* 27:192–201.

Lotti N, Pizzoli M, Ceccorulli G, Scandola M. 1993. Binary blends of microbial poly(3-hydroxybutyrate) with polymethacrylates. *Polymer* 34:4935–4940.

Lovera D, Marquez L, Balsamo V, Taddei A, Castelli C, Müller AJ. 2007. Crystallization, morphology, and enzymatic degradation of polyhydroxybutyrate/polycaprolactone (PHB/PCL) blends. *Macromo. Chem Phys* 208:924–937.

Luo RC, Xu KT, Chen GQ. 2007. Miscibility, crystallization, mechanical properties and thermal stability of blends of poly(3-hydroxybutyrate) and poly(3-hydroxybutyrate-co-4-hydroxybutyrate). *J Appl Polym Sci* 105:3402–3408.

Organ SJ, Barham PJ. 1993. Phase separation in a blend of poly(hydroxybutyrate) with poly(hydroxybutyrate-co-hydroxyvalerate). *Polymer* 34:459–467.

Park JW, Doi Y, Iwata T. 2004. Uniaxial drawing and mechanical properties of poly[(R)-3-hydroxybutyrate]/poly(L-lactic acid) blends. *Biomacromolecules* 5:1557–1566.

Parra DF, Fusaro J, Gaboardi F, Rosa DS. 2006. Influence of poly (ethylene glycol) on the thermal, mechanical, morphological, physical-chemical and biodegradation properties of poly(3-hydroxybutyrate). *Polym Degrad Stab* 91:1954–1959.

Pearce RP, Marchessault RH. 1994. Melting and crystallization in bacterial poly(β-hydroxyvalerate), PHV, and blends with poly(β-hydroxybutyrate-co-hydroxyvalerate). *Macromolecules* 27:3869–3874.

Pizzoli M, Scandola M, Ceccorulli G. 1994. Crystallization kinetics and morphology of poly(3-hydroxybutyrate)/cellulose ester blends. *Macromolecules* 27:4755–4761.

Pizzoli M, Scandola M, Ceccorulli C. 2002. Crystallization and melting of isotactic poly(3-hydroxy butyrate) in the presence of a low molecular weight diluent. *Macromolecules* 35:3937–3941.

Qiu Z, Ikehara T, Nishi T. 2003. Poly(hydroxybutyrate)/poly(butylene succinate) blends: miscibility and nonisothermal crystallization. *Polymer* 44:2503–2508.

Renard E, Walls M, Guerin P, Langlois V. 2004. Hydrolytic degradation of blends of polyhydroxyalkanoates and functionalized polyhydroxyalkanoates. *Polym Degrad Stab* 85: 779–787.

Sadocco P, Canetti M, Seves A, Marcandalli B, Martuscelli E. 1993. Small-angle X-ray scattering study of the phase structure of poly(D-(−)-3-hydroxybutyrate) and atactic poly(epichlorohydrin) blends. *Polymer* 34:3368–3375.

Saito M, Inoue Y, Yoshie N. 2001. Cocrystallization and phase segregation of blends of poly(3-hydroxybutyrate) and poly(3-hydroxybutyrate-co-3-hydroxyvalerate). *Polymer* 42: 5573–5580.

Scandola M, Ceccorulli G, Pizzoli M. 1992. Miscibility of bacterial poly(3-hydroxybutyrate) with cellulose esters. *Macromolecules* 25:6441–6446.

Scandola M, Focarete ML, Grazyna AG, et al. 1997. Polymer blends of natural poly(3-hydroxybutyrate-*co*-3-hydroxyvalerate) and a synthetic atactic poly(3-hydroxybutyrate). Characterization and biodegradation studies. *Macromolecules* 30:2568–2574.

Sombatmankhong K, Suwantong O, Waleeforncheepsawat S, Supaphol P. 2006. Electrospun fiber mats of poly(3-hydroxybutyrate), poly(3-hydroxybutyrate-co-3-hydroxyvalerate), and their blends. *J Polym Sci Part B Polym Phys* 4:2923–2933.

Verhoogt H, Ramsay BA, Favis BD. 1994. Polymer blends containing poly(3-hydroxyalkanoate)s. *Polymer* 35:5155–5169.

Weusthuis RH, Kessler B, Dielissen MPM, Witholt B, Eggink G. 2002. Fermentative production of medium-chain-length poly(3-hydroxyalkanoate). In: Doi Y, Steinbüchel A, editors. *Biopolymers (Polyesters I)*. New York: Wiley–VCH. p. 291–316.

Xing P, Dong L, An Y, Feng Z, Avella M, Martuscelli E. 1997. Miscibility and crystallization of poly(α-hydroxybutyrate) and poly(p-vinylphenol) blends. *Macromolecules* 30: 2726–2733.

Yoon JS, Choi CS, Maing SJ, Choi HJ, Lee HS, Choi SJ. 1993. Miscibility of poly-D(−) (3-hydroxybutyrate) in poly(ethylene oxide) and poly(methyl methacrylate). *Eur Polym J* 29:1359–1364.

Yoon JS, Chang MC, Kim MN, Kang EJ, Kim C, Chin IJ. 1996. Compatibility and fungal degradation of poly[(R)-3-hydroxybutyrate]/aliphatic copolyester blend. *J Polym Sci Part B Polym Phys* 34:2543–2551.

Yoon JS, Lee WS, Jin HJ, Chin IJ, Kim MN, Go JH. 1999. Toughening of poly(3-hydroxybutyrate) with poly(cis-1,4-isoprene). *Eur Polym J* 35:781–788.

Zhang JM, Sato H, Furukawa T, Tsuji H, Noda I, Ozaki Y. 2006. Crystallization behaviors of poly(3-hydroxybutyrate) and poly(L-lactic acid) in their immiscible and miscible blends. *J Phys Chem B* 110:24463–24471.

Zhang LL, Xiong CD, Deng XM. 1996. Miscibility, crystallization and morphology of poly(p-hydroxybutyrate)/poly(D,L-lactide) blends. *Polymer* 37:235–241.

Zhang LL, Deng XM, Zhao SJ, Huang ZT. 1997. Biodegradable polymer blends of poly(3-hydroxybutyrate) and starch acetate. *Polym Int* 44:104–110.

PART III

HYDROPHOBIC AND HYDROPHILIC POLYMERIC BLENDS

CHAPTER 9

Starch–Poly(hydroxyalkanoate) Composites and Blends*

RANDAL SHOGREN

National Center for Agricultural Utilization Research, USDA/ARS, Peoria, IL 61604, USA

9.1 SUMMARY OF STARCH AND PHA STRUCTURE AND PROPERTIES

9.1.1 Starch

The structure and properties of starch have been described in a number of reviews (Whistler et al., 1984; Parker and Ring, 2001; Tester et al., 2004). Starch is composed

---

*Product names are necessary to report factually on available data; however, the USDA neither guarantees nor warrants the standard of the product, and the use of the name by the USDA implies no approval of the product to the exclusion of others that may also be suitable.

Biodegradable Polymer Blends and Composites from Renewable Resources. Edited by Long Yu
Copyright © 2009 John Wiley & Sons, Inc.

Amylose

Amylopectin

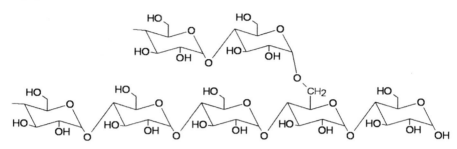

Fig. 9-1   Simplified chemical structures of amylose and amylopectin.

of the polysaccharides amylose and amylopectin (Fig. 9-1) along with small ($<1\%$) amounts of protein and lipid. Amylose is a mostly linear polymer of α-(1,4) linked glucose residues with molecular weights on the order of 10^5–10^6. Amylopectin is highly branched and consists of short, α-(1,4) linked chains of 10–40 glucose residues connected by α-(1,6) bonds. Molecular weights for amylopectin are typically $>10^8$.

Starch is biosynthesized as small granules 1–60 μm in diameter in plants and serves as an energy storage medium. Sources of starch include cereal seeds such as corn, wheat, and rice and tubers such as potato and tapioca. Starch granules are semicrystalline, with amylopectin forming thin crystalline and amorphous layers. As a result, native granular starch is insoluble in cold water but can absorb water and other small molecules into the granule due to the hydrophilic character of the numerous hydroxyl groups as well as the presence of pores in cereal starches (Huber and Bemiller, 2000). The packing characteristics of starch granules in different fluids have been described (Willett, 2001). Maximum volume fractions of different types of starches are typically in the range 0.58–0.63 and depend on granular shape, size distribution, and degree of interparticulate adhesion.

On heating in water to 60–150°C, starch granules swell, lose their crystallinity (gelatinize) and are dispersed or solubilized in water. At low water contents (10–30%) and moderate temperatures (100–200°C), gelatinized starch possesses rheological properties similar to those of polymer melts (Willet et al., 1995) and can be processed by extrusion, injection molding, compression molding, etc. Other plasticizers such as glycerol, urea, and sorbitol can be substituted in

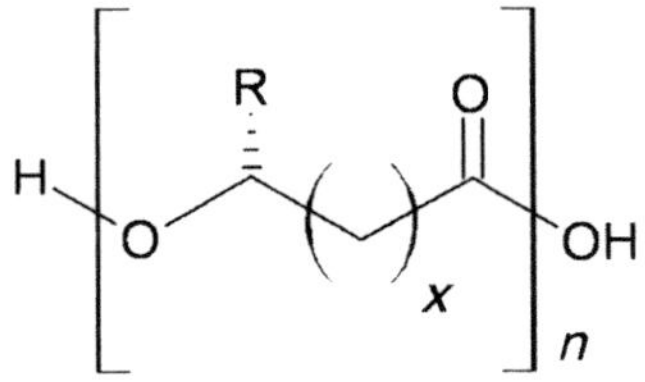

Fig. 9-2 General chemical structure of PHAs. Typically, $x = 1-8$, $n =$ several thousand.

part for water to aid in starch destructurization and to add flexibility to starch (Shogren, 1993).

9.1.2 Poly(hydroxyalkanoate)s

Poly(hydroxyalkanoate)s (PHAs) are aliphatic polyesters biosynthesized by bacteria and function as energy storage reservoirs (Inoue and Yoshi, 1992; Marchessault, 1996; Sudesh et al., 2000). The general structure of PHAs is given in Fig. 9-2. Poly(3-hydroxybutyrate) (PHB) is the most common PHA and is a rather rigid, crystalline polymer having a T_g of 4°C and T_m of 180°C. PHAs can be made with a variety of other hydroxyalkanoate monomers, however. In particular, copolymers of 3-hydroxybutyrate with 3-hydroxyvaleric acid, 3-hydroxyhexanoic acid, and 4-hydroxybutyric acid yield materials that are less crystalline, more flexible, and more easily processable. PHBV, for example, is a copolyester of 3-hydroxybutyrate and 3-hydroxyvalerate. Such copolymers have mechanical properties similar to those of polyethylene and polypropylene and have been commercialized, although production volumes are still fairly low due to high cost.

9.2 WHY BLEND STARCH WITH PHAs?

The main reason for adding starch to PHAs has been to reduce the overall cost, since starch is relatively inexpensive ($0.15/lb for corn starch) while PHAs typically cost several dollars per pound. Of course, other cheap fillers such as ground minerals could also be used. The advantage of starch is that, like PHAs, it is completely biodegradable and is already in the form of a fine, white powder. Starch could also potentially serve as a reinforcing filler, increasing the modulus and strength of the blend. Since starch biodegrades very rapidly, it may also have an influence on the overall degradation rate of starch–PHA blends. This could be important to degradation rates of starch–PHA plastics both in the environment and also for in-vivo medical devices.

9.3 PROBLEMS WITH STARCH–PHA BLENDS

The main problems with starch–PHA blends are the poor compatibility of these very different materials as well as the water sensitivity of starch. Since starch is quite

TABLE 9-1   Surface Energy Data for Starch and PHAs

	Starch		PHA			
Type	Surface Energy (mN/m)	Type	Surface Energy (mN/m)	Method	Reference
Corn	56	–	–	Drop shape	Shogren and Biresaw (2007)
Corn	35–42	–	–	Dynamic contact angle	Lawton (1995)
–	–	PHBV	41.5	Contact angle	Lawton (1997)
Corn	43	PHBV-12	42.2	Contact angle	Biresaw and Carriere (2001)
Not specified	48–56	–	–		Odidi et al. (1991)

hydrophilic and PHAs are hydrophobic, there is poor adhesion between these two polymers and hence mechanical properties of composites will be poor. Also, since starch readily absorbs water, composite starch–PHA materials will be less water resistant than pure PHA or other commodity plastics. Much of the research on starch–PHA composites has therefore focused on improving adhesion or compatibility between the components.

Surface energies of starch (Lawton, 1995; Shogren and Biresaw, 2007) and PHBV (Lawton, 1997) as well as work of adhesion (Biresaw and Carriere, 2001) have been estimated. As shown in Table 9-1, there is a rather wide range in experimental estimates for the surface energy of starch. This is likely due to contamination of the surfaces of starch films with hydrophobic compounds (lipids, hydrocarbons, silicon oils, etc.) (Russell et al., 1987; Shogren and Biresaw, 2007), giving lower than expected values. The value of 56 mN/m (Shogren and Biresaw, 2007) may be the most accurate, since a new surface was created during the formation of the pendant drop and the result agrees well with a value of 59 mN/m determined by group contribution theory (Shogren and Biresaw, 2007). These values for the surface energy of starch are significantly higher than for the more hydrophobic PHBV (42 mN/m) and thus give rise to a large interfacial tension.

9.4   GRANULAR STARCH–PHA COMPOSTIES

Ramsay and co-workers first reported on the mechanical properties and biodegradation of starch-PHBV composites (Ramsay et al., 1993). Tensile strength and elongation (Table 9-2) both decreased dramatically with increasing starch loading, reflecting the poor adhesion between phases. Scanning electron micrographs also showed starch granules separating from the PHA matrix. Elastic modulus,

TABLE 9-2 Mechanical Property Data for Selected Starch–PHA Composites

PHA	Starch Type	Starch (wt %)	Additive (wt %)	Tensile Strength (MPa)	Elongation at Break (%)	Modulus (GPa)	Reference
PHBV12	None	0	None	17.7 ± 3.9	25 ± 8	1.5 ± 0.1	Ramsay et al. (1993)
	Wheat	25	None	8.6 ± 2.6	5.1 ± 1.0	2.1 ± 0.1	
	Wheat	50	None	7.7 ± 1.0	1.0 ± 0.2	2.5 ± 0.1	
PHBV12	None	0	Triacetin, 10%	24 ± 1.0	38 ± 2.5	0.18 ± 0.04	Shogren (1995)
	Corn	50	Triacetin, 10%	10 ± 1.0	11 ± 2.5	0.30 ± 0.04	
	PEO-coated corn	50	Triacetin, 10%	19 ± 1.0	21 ± 2.5	0.17 ± 0.04	
PHB	None	0	Triacetin, 30%	11.3 ± 0.7	15.5 ± 1.3	0.19 ± 0.01	Innocentini-Mei et al. (2003)
	Corn	20	Triacetin, 30%	3.0 ± 0.3	2.3 ± 0.2	0.17 ± 0.02	
	Corn-PU	20	Triacetin, 30%	7.5 ± 0.1	6.4 ± 0.2	0.20 ± 0.01	
PHBV8	Corn	25	Acetyltributyl citrate, 5%	17.1	15.6	0.46	Willett et al. (1998)
	Corn-g-PGMA	25	Acetyltributyl citrate, 5%	23.6	13	0.54	
PHBV12	Corn propionate/maleate, DS 1.4	50	Triacetin, 15%	14.0 ± 2.5	6.9 ± 0.6	0.15 ± 0.01	Bloembergen (1995)
PHBV	Corn propionate, DS 2.3	70	None	10.3	0.5	0.90	Rimsa and Tatarka (1994)
PHBV12	Corn	30	None	9.3	3.6	0.93	Swanson, unpublished data (1991)
	Corn	30	0.4% dicumyl peroxide	20.3	3.9	1.1	

however, increased with increasing starch content due to the rigidity of the starch granules. Other authors have reported similar results (Kotnis et al., 1995; Shogren, 1995; Koller and Owen, 1996; Rosa et al., 2003). The effects of added plasticizer and mineral filler have also been studied (Kotnis et al., 1995). Mechanical properties of starch-filled polymers have been related to various theoretical models (Willett, 1994; Owen and Koller, 1996; St. Lawrence et al., 2001). To improve adhesion, investigators have taken two approaches: (1) adding a coupling agent and (2) chemically modifying the starch and/or PHA.

In the first approach, Shogren (1995) showed that coating starch with polyethylene oxide (PEO) greatly increased strength and elongation of starch–PHBV composites (Table 9-1). Values were, however, still less than for pure PHBV. Presumably, PEO has favorable interactions with both starch and PHBV and therefore can serve as a binding or interfacial agent. Other amphiphilic polymers such as hydroxyl functional polyester-ethers (Willett and Doane, 2001), and copolymers such as styrene and maleic anhydride (Krishnan and Narayan, 1996) were claimed to improve the mechanical properties of starch–PHBV blends.

In the second approach, Avella et al. (2002) showed that addition of a free-radical former (2% bis[*tert*-butylperoxyisopropyl]benzene) to PHBV/starch 70/30 caused a two-fold increase in impact resistance to $1.9 \, kJ/m^2$, similar to $1.8 \, kJ/m^2$ for pure PHBV. Presumably, some starch–PHBV graft copolymer was formed by free radical combination reactions and acted as an interfacial binding agent. Unpublished work by Swanson from 1991 at USDA/ARS/NCAUR also indicated that addition of a free-radical initiator (0.4% dicumyl peroxide) to a 70/30 PHBV/starch composite gave tensile strengths similar to PHBV-12 alone (20.3 MPa). The same study also indicated that orientation of PHBV and PHBV/starch ($<20\%$ starch) blends by drawing resulted in very high elongations (200–400%). Innocentini-Mei et al. (2003) showed that copolymerization of starch with a diisocyanate and propylene glycol resulted in improved tensile strengths and elongations in blends with PHB (Table 9-1), but values were significantly lower than for pure PHB. Composites of starch-*g*-poly(glycidyl methacrylate) ($>7\%$ PGMA) and PHBV had significantly higher tensile strengths than unmodified starch–PHBV (Table 9-1) (Willett et al., 1998). Composites of maleic anhydride-grafted starch with PHBV and a CO_2-epoxypropane copolymer were also reported (Li and Liu, 2004). Composites of starch (50%) with acrylic acid-grafted PHB had higher tensile strengths (14 MPa) than unmodified starch–PHB (7 MPa) (Liao and Wu, 2007).

Willett and co-workers also studied the effect of water immersion for 28 days on water absorption and mechanical properties (Willett et al., 1998). Weight gains for PHBV–starch bars containing 25% starch were about 4–5% compared with 0.9% for PHBV alone and 40–50% for starch. After soaking, tensile strengths did not change, elongations increased, and elastic modulus values were reduced to less than half of their dry values. Water vapor permeabilities have not been reported for starch–PHA composites but would be expected to be significantly higher than for pure PHAs. Water vapor permeabilities of PHBVs alone are 20–40 times higher than for polyethylene (Shogren, 1997) and this needs to be considered for certain applications where long-term water resistance is important.

Blends of other chemically modified starches and PHAs have also been considered. Blends of starch acetates with PHB (Zhang et al., 1997) and PHBV (Koenig and Huang, 1995) were found to be incompatible and rather brittle. Blends of starch valerate and PHBV were judged to be compatible for starch valerate contents <20%) (Seves, 1998). Starch propionates, maleates and other mixed esters of DS 1.2–1.8 were also found to be partially compatible with PHBV (Bloembergen and Narayan, 1995) and were reasonably strong (Table 9-1). Mechanical properties were rather insensitive to humidity, as expected due to the more hydrophobic nature of the starch esters. Similarly, Rimsa and Tatarka (1994) found that tensile strength and elongation of blends of starch propionate (DS 2.3) changed little with relative humidity. Trimethylsilylstarch acetates were also judged to be partially miscible with PHBV (Choi et al., 1994).

9.5 GELATINIZED STARCH–PHA BLENDS

There have been several studies of blends of PHAs with starch that has been gelatinized or melted in the presence of water and glycerol or other nonvolatile plasticizer (Verhoogt et al., 1995; Thire et al., 2006; Godbole et al., 2003; Lai et al., 2006). These blends are generally incompatible as evidenced by separation of large phase sizes (Verhoogt et al., 1995). The advantage of using plasticized over using granular starch is that plasticized starch has significant flexibility with elongations of typically 10–100% (Shogren et al., 1992). Strength properties of the blends tend to be poor both because of poor interfacial adhesion and because of the rather low strength of the starch–plasticizer phase (1–8 MPa) (Thire et al., 2006; Lai et al., 2006; Shogren, 1993). In addition, the presence of hydrophilic plasticizers such as glycerol increases the water absorption of thermoplastic starch at high humidities (Shogren et al., 1992), so that strength and stiffness would be expected to decline as humidity increased. As with granular starch blends, compatibilizers such as ethylene-vinyl acetate copolymer (Dabi and Kataria, 1994) or poly(vinyl alcohol) (Deng et al., 2002) could be added to improve properties. Properties of blends of plasticized starch with other polyesters have been reviewed (Averous, 2004).

9.6 THERMOPLASTIC STARCH/PHA LAMINATES AND FOAMS

In these types of structures, thermoplastic starch makes up the bulk of the material with the PHA being a small component, typically 5–20% or less. The PHA acts as a water resistant outer coating and/or aids the foam expansion process. Lawton (1997), Shogren and Lawton (1998), and Doane et al. (2000) have described coating starch-based foams and films with PHBV. Solvent-applied PHBV coatings ordinarily have low adhesion to starch (Lawton, 1997), but much improved adhesion was found by first coating starch foam plates with natural resins such as shellac or rosin (Shogren and Lawton, 1998), zein, or hydroxyl-functional polyester (Doane et al., 2000). These materials have both hydrogen bonding capability and hydrophobic content and thus have good adhesion to both starch and PHAs. Martin et al. (2001) prepared glycerol-plasticized starch films laminated with PHBV by coextrusion. They also found low values of peel strength for PHBV (0.01 N/mm), especially compared with other polyesters such as polyester amide (0.16 N/mm). Wang et al. (2000) found that peel strengths of polyesters on thermoplastic starch could be enhanced by creating a rough interface during the coextrusion process. Adhesion of PHBV to thermoplastic starch using only water as a plasticizer was higher than when glycerol was added. Laminates of thermoplastic starch with PHAs have also been claimed in the patent literature (Buehler, 1994; Wnuk, 1995; Bastioli et al., 1996; Bond and Noda, 2003), but little information on the properties of the materials were provided.

Water vapor permeabilities of PHBVs (for 25 μm thick film) ranged from 1.8–3.5 g/m^2/day at 6°C, 13–26 g/m^2/day at 25°C, and 124–245 g/m^2/day at 50°C (Shogren, 1997). Therefore, a coating of PHBV/shellac on an 8-in. (20 cm) diameter starch foam plate for instance would transmit <0.8 g water per day at room temperature. Such a small amount of water would not be expected to significantly change the mechanical properties of a starch plate weighing 10–20 g (Shogren and Lawton, 1998). Martin et al. (2001) also found that thermoplastic starch films extrusion laminated with polyesters did not swell significantly after soaking in water for a few days. Thus, thermoplastic starch foams and films laminated with a

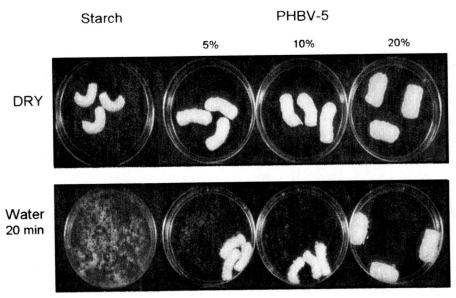

Fig. 9-3 Photographs of extruded foam peanuts containing corn starch and starch/PHBV-5 (5–20% PHBV) before and after soaking in water for 20 min. Peanuts were prepared as described in Willett and Shogren (2002).

thin layer of PHA appear to be suitable for applications requiring short-term exposure to water.

Willett and Shogren (2002) have described properties of extruded starch/PHBV foams containing 5–20% PHBV. Addition of PHBV enhanced foam expansion considerably, giving foam densities of 25 kg/m^3 with 20% PHBV-5 compared with 61 kg/m^3 for starch alone. Addition of other linear polymers also gave higher foam expansions, suggesting that perhaps the viscoelastic properties of the added polymer are important for expansion. Most of the PHBV existed as separate elongated inclusions about 1–5 μm in length within the starch matrix. Foam surfaces were, however, enriched with PHBV, probably due to the lower surface energy of PHBV than starch. As a result, the starch/PHBV foams had much greater water resistance than starch foams and fragmentation on impact (friability) was reduced. Photographs of starch foam peanuts prepared with 0–20% PHBV-5 are shown in Fig. 9-3. After soaking in water for 20 min., the pure starch peanuts have largely dispersed in water, whereas the peanuts containing 20% PHBV-5 have similar size and shape as the dry ones.

9.7 BIODEGRADABILITY, RECYCLING, AND SUSTAINABILITY

There have been several studies of the biodegradation of starch–PHA blends in different environments (Yasin et al., 1989; Tanna et al., 1992; Lauzier, 1993; Ramsay et al., 1993; Yasin and Tighe, 1993; Imam et al., 1995; Willett and O'Brien, 1997; Imam et al., 1998; Imam et al., 1999; Gordon et al., 2000; Avella et al., 2002; Rosa et al., 2003) as well as studies of the individual components (Allen et al., 1994; Vikman et al., 1995). Selected data from these studies are summarized in Table 9-3.

TABLE 9-3 Biodegradation of Starch–PHA Composites

Starch (%)	PHA	Thickness (mm)	Environment	Time (day)	Weight loss (%)	Reference
0	PHBV19	0.8	Activated sludge	30	30	Ramsay et al.
25		0.8	Activated sludge	30	85	(1993)
50		0.8	Activated sludge	30	100	
0	PHBV7	0.5	Compost	20	60	Tanna et al.
30		0.5	Compost	20	100	(1992)
0	PHBV12	3.2	Soil	125	7	Imam et al.
30		3.2	Soil	125	25	(1998)
50		3.2	Soil	125	49	
0	PHBV12	0.5	Marine	150	10–20	Imam et al.
30		0.5	Marine	150	50–90	(1999)
50		0.5	Marine	150	90–100	
30	PHBV5	Not given	Compost	20	100	Avella et al. (2002)

In all cases starch–PHA blends were found to be biodegradable over a period of weeks to months in all environments tested including soil, compost, activated sludge under aerobic and anaerobic conditions, and marine environments. This was expected since both starch and PHAs are natural products that are hydrolyzed by amylases and esterases. A number of factors influence biodegradation rates including sample thickness, moisture content, temperature, microbial activity, presence of starch/PHA degraders, presence of plasticizer, and molecular weight and crystallinity of the polymers. Rates of biodegradation tended to be high in compost and activated sludge, probably due to the high numbers of PHA depolymerase-producing microorganisms as well as the high temperatures used in composting. In the marine environment, microbial counts can vary significantly depending on location (near shore vs. deep water) (Imam et al., 1999). Imam and co-workers also found that starch-degrading microorganisms were about 10 times more abundant than PHA degraders. As a result, biodegradation and removal of starch from the composite samples largely preceded biodegradation of PHBV. In most of the studies, rates of biodegradation increased with starch content, probably because more surface area for microbial/enzyme attack was created as the more rapidly degraded starch was removed. Rates of degradation also increased with addition of $CaCO_3$ filler (Willett and O'Brien, 1997). These aspects allow the tailoring of biodegradation rates of starch–PHA composites to suit a particular application.

Slow but significant rates of hydrolysis of starch/PHBV and PHBV occur on exposure to aqueous environments (Yasin et al., 1989; Yasin and Tighe, 1993). This may be an important factor if PHA blends are to be considered for very long-term applications such as consumer durables.

Much of the initial interest in biodegradable plastics was the result of the perceived shortage of landfill space, hazards of nondegradable plastics to wildlife, and the MARPOL treaty, which forbids disposal of plastics from ships at sea. The costs of landfilling, which are particularly high in some areas of coastal U.S.A., Europe and Asia, continue to provide impetus for the development and use of biodegradable plastics. For many applications such as agricultural mulch film, planting pots, and military meals-ready-to-eat packaging, it is difficult or expensive to collect and dispose of these plastic articles after use. Nondegradable plastics in the marine environment including old netting, fishing line, and garbage washed into the sea from coastal areas continue to threaten marine life.

More recently, with growing worldwide concern about global warming and shortages of oil, there is interest in PHBV, starch and other polymers made from renewable resources such as agricultural commodities. Although biodegradation is still an option for disposal of starch–PHBV materials, more consideration is being given to recycling back into the monomeric hydroxyacids (Kaihara et al., 2005; Reddy et al., 2003). This can easily be accomplished by enzymatic depolymerization. Presumably, new PHAs could then be biosynthesized from the hydroxyacids and glucose from depolymerized starch. As the prices of agricultural commodities of all kinds grow rapidly due to increases in global population, economic growth, and increased biofuel production, this may become a more attractive option than mineralization back to CO_2 and water. Recycling of polyesters such as PHAs by the depolymerization/repolymerization route should be much easier than for petrochemical addition polymers like polyethylene and this bodes well for the future of PHA–starch composites.

The sustainability of PHAs and starch blends has been reviewed recently by Patel and Narayan (2005). They conclude that energy use and CO_2 emissions for the production of PHAs are currently similar to or higher than those for petrochemical polymers. This is likely to change in the future, however, as production becomes more efficient. They did note, however, that a Japanese study found CO_2 emissions to be much lower for PHB production than for commodity synthetic polymers. They also summarized findings that energy use and CO_2 emissions are significantly lower for starch, thermoplastic starch, and starch blends than for polyethylene or polystyrene.

9.8 APPLICATIONS AND PRODUCTION

Applications for PHA/starch blends can be diverse since properties of PHAs run the gamut from highly flexible, mimicking polyethylene, to rigid, similar to polypropylene or poly(ethylene terephthalate). These can include film and sheet for packaging, disposable items, and agriculture; injection molded articles for food serving, housewares, electronics, and automotive uses; foams for cushioning and insulation; and fibers for textiles, wipes, filtration and hygiene, and coatings/adhesives. The degree of market penetration for starch/PHA blends will depend on the price of PHA and the amount of PHA needed in the blend. Initially, blends that require PHAs to be only a small fraction of the material such as PHA-laminated starch films and foams may be preferred. Applications such as biodegradable garbage bags and mulch films (Halley et al., 2001) are attractive targets since they are single-use items. The starch component of the films can serve as a good barrier against permeation of oxygen or various organic molecules (Rankin, 1958), a property lacked by PHAs or other hydrophobic polymers such as polyethylene. Currently, microbial PHAs are produced on a small scale by Metabolix/ADM but a 110 million pound per year plant is set to start in late 2008. Later, production of PHAs in plants is expected at prices comparable to or lower than petrochemical polymers.

9.9 FUTURE RESEARCH NEEDS AND DIRECTIONS

From previous studies, it appears that compatibilization of granular starch/PHA blends by free radical generation works well. This work needs to be confirmed, studied in greater depth and extended to gelatinized starch blends. In addition, other methods such as grafting onto starch, that have proved valuable for blends of starch with other polyesters (Mani and Bhattacharya, 2001; Dubois and Narayan, 2003; Kalambur and Rizvi, 2006) could be investigated. Starch–PHA graft copolymers could have some additional interesting applications as surfactants and as self-assembled nanostructures. Starch-based nanocomposites have been prepared with other polyesters and nanoclays and these show improved mechanical properties (Dufresne, 1998; Kalambur and Rizvi, 2004). The compatibility of moderately and highly derivatized starches with PHAs is a guessing game at present and would benefit from molecular modeling approaches. PHAs biosynthesized with functional or reactive side-chains are another avenue for improving compatibility that can be pursued. The effects of starch modification or reactive compatibilization on the biodegradation rates have not been characterized extensively. Starch has the potential to form inclusion complexes with ionic polymers (Shogren et al., 1991), so similar complexes might form between starch and a carboxy-modified PHA, for instance. Few studies have addressed the problem of increased water sensitivity of starch–PHA composites. Competition of starch/PHA blends with natural fiber–PHA composites is apparent (Hodzic, 2005) and the advantages and disadvantages of both need to be compared and analyzed. Finally, natural PHA latex and blends with starch have potential as adhesives (Lauzier et al., 1993) and further investigation in this area seems warranted.

REFERENCES

Allen AL, Mayer J, Stote R, Kaplan DL. 1994. Simulated marine respirometry of biodegradable polymers. *J Environ Polym Degrad* 2:237–244.

Avella M, Errico ME, Rimedio R, Sadocco P. 2002. Preparation of biodegradable polyesters/high amylose starch composites by reactive blending and their characterization. *J Appl Polym Sci* 83:1432–1442.

Averous L. 2004. Biodegradable multiphase systems based on plasticized starch: a review. *J Macromol Sci* 44:231–274.

Bastioli C, Romano M, Scarati M, Toscin M. 1996. Biodegradable starch-based articles. U.S. Patent 5,512,378.

Biresaw G, Carriere CJ. 2001. Correlation between mechnical adhesion and interfacial properties of starch/biodegradable polyester blends. *J Polym Sci B: Polym Phys* 39:920–930.

Bloembergen S, Narayan R. 1995. Biodegradable moldable products and films comprising blends of starch esters and polyesters. U.S. Patent 5,462,983.

Bond EB, Noda I. 2003. Polyhydroxyalkanoate copolymer/starch compositions for laminates and films. U.S. Pat. Appl. 20030108701.

Buehler FS, Schmid E, Schultz HJ. 1994. Starch/polymer mixture, process for the preparation thereof, and products obtainable therefrom. U.S. Patent 5,346,936.

Choi Y, Kim S, Chang W, Seo K, Kim W. 1994. Preparation of biodegradable alloys 3. *Polymer (Korea)* 18:376–383.

Dabi S, Kataria RL. 1994. Melt-processable biodegradable compositions containing polyesters and starch. Eur. Pat. Appl. EP 606923.

Deng P, Ran X, Hu J, Dong L. 2002. Preparation of biodegradable starch-polyvinyl alcohol–polyester films. *Faming Zhuanli Shenqing Gongkai Shuomingshu* CN 1357563.

Doane WM, Lawton JW, Shogren R. 2000. Biodegradable polyester and natural polymer laminates. U.S. Patent 6,040,063.

Dubois P, Narayan R. 2003. Biodegradable compositions by reactive processing of aliphatic polyester/polysaccharide blends. *Macromol Symp* 198:233–243.

Dufresne A. 1998. High performance nanocomposite materials from thermoplastic matrix and polysaccharide fillers. *Recent Res Dev Macromol Res* 3:455–474.

Godbole S, Gote S, Latkar M, Chakrabarti T. 2003. Preparation and characterization of biodegradable poly-3-hydroxybutyrate-starch blend films. *Bioresource Technol* 86:33–37.

Gordon SH, Imam SH, Shogren RL, Govind NS, Greene RV. 2000. A semiempirical model for predicting biodegradation profiles of individual polymers in starch–poly(β-hydroxybutyrate-co-β-hydroxvalerate) bioplastic. *J Appl Polym Sci* 76:1767–1776.

Halley P, Rutgers R, Coombs S, et al. 2001. Developing biodegradable mulch films from starch-based polymers. *Starch/Staerke* 53:362–367.

Hodzic A. 2005. Bacterial polyester-based biocomposites: a review. In: Mohanty AK, Misra M, Drzal LT, editors. *Natural Fibers, Biopolymers and Biocomposites*. Boca Raton, FL: Taylor & Francis. p. 597–616.

Huber KC, Bemiller JN. 2000. Channels of maize and sorghum starch granules. *Carbohydr Polym* 41:267–276.

Imam SH, Gordon SH, Shogren RL, Greene RV. 1995. Biodegradation of starch-poly(β-hydroxybutyrate-co-valerate) composites in municipal activated sludge. *J Environ Polym Degrad* 3:205–213.

Imam SH, Chen L, Gordon SH, Shogren RL, Weisleder D, Greene RV. 1998. *J Environ Polym Degrad* 6:91–98.

Imam SH, Gordon SH, Shogren RL, Tosteson TR, Govind NS, Greene RV. 1999. Degradation of starch-poly(β-hydroxybutyrate-co-β-hydroxvalerate) bioplastic in tropical coastal waters. *Appl Environ Microbiol* 65:431–437.

Innocentini-Mei LH, Bartoli JR, Baltieri RC. 2003. Mechanical and thermal properties of poly(3-hydroxybutyrate) blends with starch and starch derivatives. *Macromol Symp* 197:77–87.

Inoue Y, Yoshie N. 1992. Structure and physical properties of bacterially synthesized polyesters. *Prog Polym Sci* 17:571–610.

Kaihara S, Osanani Y, Nishikawa K, Toshima K, Doi Y, Matsumura S. 2005. Enzymatic transformation of bacterial polyhydroxyalkanoates into repolymerizable oligomers directed towards chemical recycling. *Macromol Biosci* 5:644–652.

Kalambur SB, Rizvi SSH. 2004. Starch-based nanocomposites by reactive extrusion processing. *Polym Int* 53:1413–1416.

Kalambur S, Rizvi SSH. 2006. An overview of starch-based plastic blends from reactive extrusion. *J Plastic Film Sheet* 22:39–58.

Koenig MF, Huang SJ. 1995. Biodegradable blends of composites of polycaprolactone and starch derivatives. *Polymer* 36:1877–1882.

Koller I, Owen AJ. 1996. Starch filled PHB and PHB/HV copolymer. *Polym Int* 39:175–181.

Kotnis MA, O'Brien GS, Willett JL. 1995. Processing and mechanical properties of biodegradable poly(hydroxybutyrate-co-valerate-starch compositions. *J Environ Polym Deg* 3:97–105.

Krishnan M, Narayan R. 1996. Biodegradable multi-component polymeric materials based on unmodified starch-like polysaccharides. U.S. Patent 5,500,465.

Lai SM, Don TM, Huang YC. 2006. Preparation and properties of biodegradable thermoplastic starch/poly(hydroxybutyrate) blends. *J Appl Polym Sci* 100:2371–2379.

Lauzier CA, Monasterios CJ, Saracovan I, Marchessault RH, Ramsay BA. 1993. Film formation and paper coating with poly(β-hydroxyalkanoate), a biodegradable latex. *Tappi J* 76:71–77.

Lawton JW. 1995. Surface energy of extruded and jet cooked starch. *Starch/Staerke* 47:62–67.

Lawton JW. 1997. Biodegradable coatings for thermoplastic starch. In: Campbell GM, Webb C, McKee SL, editors. *Cereals: Novel Uses and Processes*. New York: Plenum. p. 43–47.

Li J, Liu J. 2004. Preparing poly-3-hydroxyalkanoate/carbon dioxide-epoxypropane compolymer/starch ternary composite. *Faming Zhuanli Shenquing Gongkai Shuomingshu* CN 1470553.

Liao HT, Wu CS. 2007. Performance of an acrylic-acid-grafted poly(3-hydroxybutyric acid)/starch bio-blend: characterization and physical properties. *Designed Monomers Polym* 10:1–18.

Mani R, Bhattacharya M. 2001. Properties of injection moulded blends of starch and modified biodegradable polyesters. *Eur Polym J* 37:515–526.

Marchessault RH. 1996. Tender morsels for bacteria. Recent developments in microbial polyesters. *Trends Polym Sci* 4:163–168.

Martin O, Schwach E, Averous L, Couturier Y. 2001. Properties of biodegradable multilayer films based on plasticized wheat starch. *Starch/Staerke* 53:372–380.

Odidi IO, Newton JM, Buckton G. 1991. The effect of surface treatment on the values of contact angles measured on a compressed powder surface. *Int J Pharm* 72:43–49.

Owen AJ, Koller I. 1996. A note on the Young's modulus of isotropic two-component materials. *Polymer* 37:527–530.

Parker R, Ring SG. 2001. Aspects of the physical chemistry of starch. *J Cereal Sci* 34:1–17.

Patel M, Narayan R. 2005. How sustainable are biopolymers and biobased products? The hope, the doubts and the reality. In: Mohanty AK, Misra M, Drzal LT, editors. *Natural Fibers, Biopolymers and Biocomposites*. Boca Raton, FL: Taylor & Francis. p. 597–616.

Ramsay BA, Langlade V, Carreau PJ, Ramsay JA. 1993. Biodegradability and mechanical properties of poly(β-hydroxybutyrate-co-β-hydroxyvalerate)-starch blends. *Appl Environ Microbiol* 59:1242–1246.

Rankin JC, Wolff IA, Davis HA, Rist CE. 1958. Permeability of amylose film to moisture vapor, selected organic vapors, and the common gases. *Ind Eng Chem* 3:120–123.

Reddy CSK, Ghai R, Rashmi, Kalia VC. 2003. Polyhydroxyalkanoates: an overview. *Bioresource Technol* 87:137–146.

Rimsa S, Tatarka P. 2004. Starch ester blends with linear polyesters. PCT Int. Appl. WO 94/07953.

Rosa D, Rodrigues TC, Guedes C. 2003. Effect of thermal aging on the biodegradation of PCL, PHB-V, and their blends with starch in soil compost. *J Appl Polym Sci* 89:3539–3546.

Russell PL, Gough BM, Greenwell P, Fowler A, Munro HS. 1987. A study by ESCA of the surface of native and chlorine-treated wheat starch granules: the effects of various surface treatments. *J Cereal Sci* 5:83–100.

Seves A, Beltrame PL, Selli E, Bergamasco L. 1998. Morphology and thermal behavior of poly(3-hydroxybutyrate-co-hydroxyvalerate)/starch valerate blends. *Angew Makromol Chem* 260:65–70.

Shogren RL. 1993. Effects of moisture and various plasticizers on the mechanical properties of extruded starch. In: Ching C, Kaplan DL, Thomas EL, editors. *Biodegradable Polymers and Packaging*. Lancaster, PA: Technomic. p. 141–150.

Shogren RL. 1995. Poly(ethylene oxide) coated granular starch - poly(hydroxybutyrate-co-hydroxyvalerate) composite materials. *J Environ Polym Degrad* 3:75–80.

Shogren R. 1997. Water vapor permeability of biodegradable polymers. *J Environ Polym Degrad* 5:91–95.

Shogren R, Biresaw G. 2007. Surface properties of water soluble maltodexrin, starch acetates and starch acetates/alkenylsuccinates. *Colloids Surf A: Phys Eng Aspects* 298:170–176.

Shogren RL, Lawton JW. 1998. Enhanced water resistance of starch-based materials. U.S. Patent 5,756,194.

Shogren RL, Greene, RV, Wu YV. 1991. Complexes of starch polysaccharides and poly(ethylene co-acrylic acid): structure and stability in solution. *J Appl Polym Sci* 42:1701–1709.

Shogren RL, Swanson CL, Thompson AR. 1992. Extrudates of cornstarch with urea and glycols: structure/mechanical property relations. *Starch/Staerke* 44:335–338.

St. Lawrence S, Walia PS, Willett JL. 2001. Mechanical properties of granular starch filled blends. *Polym Mater Sci Eng* 85:369–370.

Sudesh K, Abe H, Doi Y. 2000. Synthesis, structure and properties of polyhydroxyalkanoates: biological polyesters. *Prog Polym Sci* 25:1503–1555.

Tanna ST, Gross R, McCarthy SP. 1992. Biodegradation of blends of bacterial polyester and starch in a compost environment. *Polym Mater Sci Eng* 67:290–291.

Tester RF, Karkalas J, Qi X. 2004. Starch-composition, fine structure and architecture. *J Cereal Sci* 39:151–165.

Thire RM, Rebeiro TA, Andrade CT. 2006. Effect of starch addition on compression-molded poly(3-hydroxybutyrate)/starch blends. *J Appl Polym Sci* 100:4338–4347.

Verhoogt H, St-Pierre N, Truchon FS, Ramsay BA, Favis BD, Ramsay JA. 1995. Blends containing poly(hydroxybutyrate-co-12%-hydroxyvalerate) and thermoplastic starch. *Can J Microbiol* 41:323–328.

Vikman M, Itavaara M, Poutanen K. 1995. Measurement of the biodegradation of starch-based materials by enzymatic methods and composting. *J Environ Polym Degrad* 3:23–29.

Wang L, Shogren RL, Carriere C. 2000. Preparation and properties of thermoplastic starch–polyester laminate sheets by coextrusion. *Polym Eng Sci* 40:499–506.

Whistler RL, BeMiller JN, Paschall EF. 1984. *Starch: Chemistry and Technology*, 2nd ed. Orlando, FL: Academic Press.

Willett JL. 1994. Mechanical properties of LDPE/granular starch composites. *J Appl Polym Sci* 54:1685–1695.

Willett JL. 2001. Packing characteristics of starch granules. *Cereal Chem* 78:64–68.

Willett JL, Doane WM. 2001. Biodegradable polymer compositions, methods for making same and articles therefrom. U.S. Patent 6,191,196.

Willett JL, O'Brien GS. 1997. Biodegradable composites of starch and polyhydroxybutyrate-co-valerate) copolymers. In: Campbell GM, Webb C, McKee SL editors. *Cereals: Novel Uses and Processes*. New York: Plenum. p. 35–41.

Willett JL, Shogren, RL. 2002. Processing and properties of extruded starch/polymer foams. *Polymer* 43:5935–5947.

Willett JL, Jasberg BK, Swanson CL. 1995. Rheology of thermoplastic starch: effects of temperature, moisture content and additives on melt viscosity. *Polym Eng Sci* 35:202–210.

Willett JL, Kotnis MA, O'Brien GS, Fanta GF, Gordon SH. 1998. Properties of starch-graft-poly(glycidyl methacrylate)-PHBV composites. *J Appl Polym Sci* 70:1121–1127.

Wnuk AJ, Koger TJ, Young TA. 1995. Biodegradable, liquid impervious multilayer film compositions. U.S. Patent 5,391,423.

Yasin M, Tighe BJ. 1993. Strategies for the design of biodegradable polymer systems: manipulation of polyhydroxybutyrate-based materials. *Plast Rubber Compos Process Appl* 19:15–27.

Yasin M, Holland SJ, Jolly AM, Tighe BJ. 1989. Polymers for biodegradable medical devices. VI. Hydroxybutyrate-hydroxyvalerate copolymers: accelerated degradation of blends with polysaccharides. *Biomaterials* 10:400–412.

Zhang L, Deng X, Zhao S, Huang Z. 1997. Biodegradable polymer blends of poly(3-hydroxybutyrate) and starch acetate. *Polym Int* 44:104–110.

CHAPTER 10

Biodegradable Blends Based on Microbial Poly(3-hydroxybutyrate) and Natural Chitosan

CHENG CHEN and LISONG DONG

State Key Laboratory of Polymer Physics and Chemistry, Changchun Institute of Applied Chemistry, Chinese Academy of Sciences, Changchun, P. R. China

10.1   INTRODUCTION

Biodegradable blends have been attracting increasing interest because of their potential applications, which range from environmentally degradable resins through biomedical implants to absorbable surgical sutures. Among the various biodegradable materials, microbial poly(3-hydroxybutyrate) (PHB) and chitosan (CS) have been a new focus of research as important candidates for biomaterials because they are renewable polymers from sustainable natural resources. PHB is a natural occurring crystalline polyester, which is accumulated as intracellular energy reserves (Doi 1990; Reddy et al., 2003). CS is a natural polysaccharide produced by the

Biodegradable Polymer Blends and Composites from Renewable Resources. Edited by Long Yu
Copyright © 2009 John Wiley & Sons, Inc.

deacetylation of chitin, which is the second most abundant biopolymer in nature after cellulose. Although both PHB and CS are biocompatible and biodegradable *in vivo* and *in vitro*, they are difficult to apply individually because of some inherent drawbacks (Avella et al., 2000; Chen et al., 2002).

The blending of PHB and CS is a convenient and effective approach to improve their physicochemical properties and to develop low-cost and high-performance biodegradable composites. Because there are many functional groups on its molecular chains, CS is very useful in the development of composite materials such as blends or alloys with other polymers (Society, 1988; Muzzarelli, 1977). Blending of PHB with CS offers the possibility of new materials for medical applications due to their low cost and excellent biocompatibility and biodegradability, especially in tissue engineering. NIH 3T3 fibroblasts can be cultured on scaffolds made from the PHB/CS blends (Cao et al., 2005). The presence of PHB enhances cell attachment to the scaffolds and the cells spread and grow well on PHB/CS scaffolds. In addition, the PHB and CS conjugates may form strong and elastic films (Yalpani et al., 1991) and copolymers of PHB grafting CS also show high antibacterial activity, which can be used for wound dressings (Hu et al., 2003). By incorporating PHB with CS, the functional groups on the CS chains can be available for further modification such as conjugation of biomolecules, which will widen the application of the blend system. Nanofiber technology represents an important direction for material research studies, and material performance of chitosan nanofibers in the medical field is very promising (Ohkawa et al., 2006). An elegant method for nanofiber production is now well known as "spinning," which allows fabrication of a fine and dense meshwork of polymer fibers directly from solution in the presence of an electric field (Doshi and Reneker, 1995). Application of the pure chitosan nanofibers in medical technology has been reported as a guided bone regeneration material (Shin et al., 2005) and as a scaffold for regenerating nerve tissue. Poly(L-lactide) (PLLA)/CS fabrics are produced by the blend spinning technique. The compatibility between PLLA/CS fabrics and osteoblasts under in vitro degradation has been investigated for the potential application of PLLA/CS fabrics as supporting materials for chest walls and bones. Excellent adhesion between osteoblasts and PLLA/CHS fabrics is observed, indicating good biocompatibility of the fabrics with osteoblasts (Zhang et al., 2007). PLLA/CS fabrics may thus be an ideal osteoblast carrier for the repair of damaged chest wall and bones over a large area, since the pore size and porosity of the PLLA/CS braid can be adjusted according to the cell characteristics. Given that PHB is usually hydrolyzed by extracellular depolymerases but is hardly degraded by intracellular depolymerases, fabrics based on PHB and CS are likely to have wider application as biomedical materials than PLLA/CS fabrics because PHB shows longer biodegradation times in vivo than does PLLA (Sang, 1996).

So far there have been few studies on the blending of PHB and CS because they are difficult to mix due to their natural properties. Generally, PHB and CS are blended by solution blending rather than melt blending because CS hardly melts and its glass transition temperature (T_g) is very close to the thermal decomposition temperature of PHB (Sakurai et al., 2000). The properties of melt-blended CS with aliphatic polyesters such as poly(lactic acid) (PLA) show that the CS and polyesters form phase-separated systems (Correlo et al., 2005). The general properties of the blends are usually strongly relevant to the preparation methods, and preparation with different methods will exhibit major differences in terms of properties, particularly in the PHB/CS blend system. Three major methods for preparing PHB/CS blends are presented below, together with the corresponding properties of the blends.

10.2 PREPARATION AND PROPERTIES

10.2.1 Solution-Casting Method (Ikejima et al., 1999)

10.2.1.1 Preparation of PHB/CS and PHB/Chitin Blends 1,1,1,3,3,3-Hexafluoro-2-propanol (HFIP) was used as a common solvent. PHB, CS, and chitin were dissolved in HFIP and cast on a poly(tetrafluoroethylene) (PTFE) dish. After drying, PHB/CS and PHB/chitin films with different compositions were obtained. Although this method is very simple, the common solvent HFIP is very expensive, which limits its application.

10.2.2 Characterization and Properties

10.2.2.1 Miscibility In general, differential scanning calorimetry (DSC) is not a good method to accurately measure the T_g values of CS and its blends. The molecular structure of CS consists of rigid β-1,4-linked D-glucosamine units, so changes in heat capacity corresponding to changes in specific volume (or molecular mobility) at T_g are correspondingly small. This gives rise to a very small observed baseline step change in the DSC curves (Sakurai et al., 2000), which will introduce a large error in the DSC measurement of the T_g of CS. Because the conventional method of T_g measurement using DSC could not be applied in the PHB/CS blend system, dynamic mechanical thermal analysis (DMTA) was used to detect the miscibility of the PHB/CS blends (Ikejima and Inoue, 2000). However, the temperatures of the tan δ transition detected in the PHB/CS blends were not greatly different from that detected in pure PHB (around 15°C, which corresponded to the T_g of the PHB amorphous region).

There was an exception, in that a single T_g was found in the PHB/CS blend system where the T_g of CS was found to be 103°C (Cheung et al., 2002), implying that the PHB/CS blends were miscible. For all different compositions, the T_g values were higher than those of the calculated weight-averaged values. This was attributed to the strong intermolecular hydrogen bonding between the PHB and CS components. The phenomenon is completely different from the common miscible blends with specific interactions in which the T_g values of the blends are intermediate to those of two components (Zhang et al., 1997), whose values can be described by several equations such as the Fox, Gordon–Taylor, and Couchman equations. In order to further investigate the phase structure of the PHB/CS blends, ^{1}H CRAMPS (combined rotation and multiple pulse spectroscopy) was used. [1]H T_1 was measured with a modified BR24 sequence that yielded an intensity decay to zero mode rather than the traditional inversion recovery mode (Cheung et al., 2002). Single exponential T_1 decay was observed for the PHB/CS blends and their T_1 values were either faster than or intermediate to those of the plain polymers. The $T_{1\rho}$ decay of β-hydrogen belonging to PHB was biexponential. The slow $T_{1\rho}$ decay component was explained as a result of the crystalline phase of PHB. The fast $T_{1\rho}$ of: β-hydrogen and the $T_{1\rho}$ of CS in the blends either followed the same trend as or were faster than the weight-averaged values based on the $T_{1\rho}$ of the plain polymers. Together with the result that single T_g was detected by DSC, evidence from solid-state ^{13}C NMR strongly suggested that CS was miscible with PHB at all compositions.

10.2.2.2   *Thermal and Crystallization Behavior*   In the PHB/CS blend system investigated by Ikejima (Ikejima et al., 1999), the melting curves of the blends exhibited two melting peaks in the range of PHB melting. The peak appearing at a lower temperature was attributed to the melting of the crystalline film as cast. The other melting peak at a higher temperature was attributed to the melting of the recrystallized component. Although the PHB/CS blends were prepared and investigated under the same conditions, a single melting peak was found in the same blend system (Cheung et al., 2002); in particular, the peak at a lower temperature appeared in a different temperature region. This is possibly due to the differences in the materials used.

DSC also revealed that the crystallization of the PHB component in the blends was suppressed with increasing CS content. Taking advantage of the FTIR crystalline-sensitive peak of the PHB carbonyl stretching band, the crystallinity of the PHB component was determined quantitatively, and the results was in agreement with that obtained from the DSC measurement. Compared with α-chitin, CS exhibited a stronger ability to suppress the crystallization of PHB, due to the presence of the amino groups (Ikejima et al., 1999). From the solid-state ^{13}C NMR results, PHB in the blends was found to be trapped in the "glass" environment of CS. The corresponding X-ray diffraction (XRD) data also confirmed the results (Ikejima and Inoue, 2000). Because the CS resonances in the solid-state ^{13}C NMR spectra were significantly broadened after blending with PHB, it was suggested that there were intermolecular hydrogen bonds between the carbonyl groups of PHB and the amino groups of CS. The specific interaction between PHB and the highly rigid CS molecules surrounding the PHB molecules would make the PHB molecules in the blends inflexible, and would induce insufficient crystallization relative to pure PHB. Hence, the lamellar thickness of the PHB crystalline component became thinner, which was large enough to show detectable XRD peaks but was too small to show an observable melting endotherm in the DSC thermogram and the crystalline band absorption in the FTIR spectrum (Ikejima et al., 1999; Ikejima and Inoue, 2000).

10.2.2.3   *Environmental biodegradation*   Environmental biodegradation of PHB/CS blends was investigated using the biochemical oxygen demand (BOD), whose values were determined by the amount of oxygen consumption (Ikejima and Inoue, 2000). Compared with the low biodegradability of pure CS and chitin,

the biodegradability was clearly improved by blending with PHB, especially for CS. This was attributed to the low crystallinity of the PHB component.

10.2.3 Precipitation Blending Method (Chen et al., 2005)

10.2.3.1 Preparation of PHB/CS and Maleated PHB/CS Blends PHB and maleated PHB (PHB-*g*-MA) dissolved in dimethyl sulfoxide (DMSO) were mixed with the CS solution in acetic acid–DMSO. After being well homogenized, mixtures with different compositions were precipitated in excess acetone, and the precipitates were filtered off and dried to obtain the PHB/CS blends and the PHB-*g*-MA/CS blends.

Because few common solvents are known for PHB and CS, precipitation blending allows on to obtain PHB/CS blends more easily and cheaply than the common solution-casting. The method may also be extended to similar blend systems consisting of hydrophilic CS and hydrophobic aliphatic polyesters (Chen et al., 2005).

10.2.4 Characterization and Properties

10.2.4.1 Intermolecular Hydrogen Bonds In the PHB/CS blend system, the detection of intermolecular hydrogen bonds depends on the preparation methods of the samples and the related measurements. When the PHB/CS blends were prepared with the solution-casting method, intermolecular hydrogen bonds between the two components could be detected by ^{13}C solid-state NMR, while no detectable changes were found in the FTIR spectra. However, obvious changes were seen in the FTIR spectra of PHB/CS blends that were prepared by the precipitation blending method. It was interesting that in the precipitation blending method the intermolecular hydrogen bonds between PHB and CS components depended on the compositions in the blends. In the FTIR spectra of the PHB/CS blends with compositions of 20/80 and 40/60, the CS amino band at 1596 cm^{-1} disappeared, and the PHB amorphous carbonyl vibration at 1740 cm^{-1} became very clear, which indicated that the intermolecular hydrogen bonds were caused by the PHB carbonyls and the CS amino groups in the amorphous phase. Compared with the PHB/CS blends, the blends containing PHB-*g*-MA and CS showed intermolecular hydrogen bonds up to a higher composition of PHB-*g*-MA/CS = 60/40, indicating that the MA groups grafted onto PHB chains could further form hydrogen bonds with amino groups on the CS chains and then intensify the interaction between two components (Chen et al., 2005). When the PHB content amounted to 60% or above and the PHB-*g*-MA content was 80%, the intermolecular interaction between both components disappeared. X-ray photoelectron spectroscopy (XPS) provides a very sensitive way to detect changes in the molecular interactions in the blends. In PHB/CS blends, it was found that the N1s binding energy (BE) of CS in the blends with the compositions of 20/80 and 40/60 was increased, and the C1s BE values of the PHB carbonyl bonds in the blends with the same compositions decreased. These BE shifts were exhibited more obviously in the PHB-*g*-MA/CS blends, besides the changes of the PHB C1s peak shapes. This was because the

chemical environments of carbon or nitrogen atoms in the blends were perturbed by specific intermolecular hydrogen bonds.

10.2.4.2 Thermal Behavior of PHB/CS and PHB-g-MA/CS Blends

PHB/CS blends prepared by the solution-casting method can gradually crystallize as the solvent slowly evaporates, whereas crystallization is difficult during the precipitation process. Hence, PHB/CS blends prepared by the precipitation blending method exhibited a single melting peak in the heating process, which resulted mainly from the melting of the crystals formed during both the thermal treatment and heating process. As the PHB content in the blends decreased, the melting temperatures, the melting enthalpies, and the crystallinity of the PHB/CS blends gradually decreased, which suggests that the introduction of CS into PHB could effectively hinder its crystallization. The MA groups grafted onto the PHB chains might further suppress the crystallization of the PHB component.

10.2.4.3 Crystallization of PHB/CS and PHB-g-MA/CS Blends

When PHB was blended with CS, the crystallization of the PHB component was clearly suppressed. Moreover, the CS diffraction peaks disappeared in all compositions (Fig. 10-1a), indicating that the crystallization of CS was obviously changed. The suppression was more remarkably in the PHB-g-MA/CS blends. It should be noted that the crystallite growth of the PHB and PHB-g-MA components was different in different directions. Relative to that of the (020) diffraction, the intensity of the (110) diffraction in the PHB and PHB-g-MA components was always suppressed in the blends with compositions for which there were specific interactions (Fig. 10-1b), indicating that the hydrogen bonds in the blends not only affected the crystallinity of the blends but also disturbed the original crystal structures of the two components.

10.2.5 Emulsification Casting Method (Cao et al., 2005)

10.2.5.1 Preparation of PHB/CS Films

CS dissolved in acetic acid was mixed with PHB solution in chloroform. After being cast into films, the mixtures were neutralized in 0.5 M NaOH aqueous solution, washed thoroughly with deionized water, and then dried at room temperature. It is noted in the method that the viscosity of the CS solution should be high enough to keep the stability of the PHB droplets dispersed in the CS solution. It is very difficult to form homogeneous solutions with water and chloroform, and serious phase separation will take place in the PHB/CS blend as the solvents fully evaporate. Because of this solubility limitation, the method can only prepare blend films with a low PHB content (the highest PHB content in the blends is 30%) (Cao et al., 2005).

In order to improve the emulsification effect, the emulsifying agent poly(vinyl alcohol) (PVA) was used (Shih et al., 2007). PHB solution in methylene chloride was mixed with CS solutions in acetic acid in the presence of 1% (g/ml) PVA. The emulsified microspheres (shown in Fig. 10-2) were finally collected by a freeze-drying method.

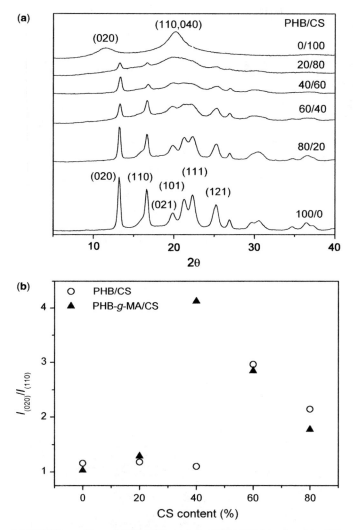

Fig. 10-1 (a) WAXD profiles of PHB, CS, and the PHB/CS blends with different compositions prepared by the precipitation blending method. (b) The relationship of the values of $I_{(020)}/I_{(110)}$ with CS content in the blends.

10.2.6 Characterization and Properties

10.2.6.1 Physical Properties of PHB/CS Films   In the PHB/CS films prepared with the emulsification casting method, PHB remained as microspheres in the blend films. There was no interfacial adhesion between the PHB microspheres and the CS matrices due to a lack of intermolecular interaction. After blending PHB with CS, the PHB/CS films exhibited a higher tensile strength than the pure CS film. With

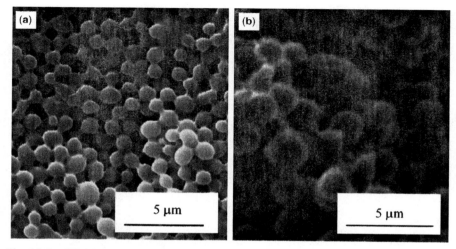

Fig. 10-2   SEM images of PHB/CTS microspheres at different PHB/CTS ratios: (a) 1 : 1; (b) 5 : 1.

increase in the PHB content, the elastic modulus of the PHB/CS films decreased and the elongation-at-break increased, which was attributed to the toughening effect of the PHB microspheres by a crack-pinning mechanism. In addition, the swelling capability of the PHB/CS films was be lowered because of the introduction of hydrophobic PHB. The same method was used to prepare the PLA/CS blends. The incorporation of PLA into CS improved the water barrier properties and decreased

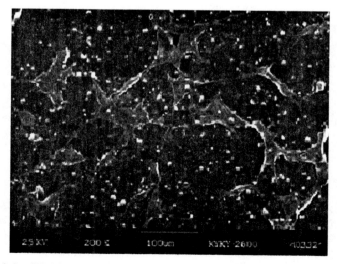

Fig. 10-3   SEM image of NIH 3T3 cells on PHB/CS films after 24 h in culture.

the water sensitivity of CS. However, the tensile strength and elastic modulus of CS clearly decreased with the addition of PLA (Suyatma et al., 2004).

10.2.6.2 Cytocompatibility of PHB/CS Films
NIH 3T3 fibroblasts were used to examine the cytocompatibility of PHB/CS films. The fibroblasts were attached to the blend films, and exhibited normal flat morphology even if the cell were grown on PHB microspheres (shown in Fig. 10-3). These results indicated that the PHB/CS films could support the attachment and growth of fibroblasts. The experiments on cell adhesion and cell proliferation further showed that the PHB/CS films had better cytocompatibility than the CS film, probably due to the better biocompatibility of PHB and the rough surface of the blend films.

10.3 CONCLUSIONS

New kinds of biomaterials based on PHB and CS can be produced using different effective blending methods. PHB/CS blends are expected to possess excellent biocompatibility and biodegradability and are especially suitable for application as biomedical materials. When prepared by the solution-casting method, the PHB/CS blends are miscible at all compositions, which means that the blends should synergistically possess the properties of both components. The introduction of CS hinders the crystallization of PHB because of the rigid CS surrounding environment and intermolecular hydrogen bonds between the components. The precipitation blending method offers an easy and convenient approach to prepare the blends or composites including hydrophilic CS and hydrophobic polyesters. The ductility of CS can be improved by decreasing self-hydrogen bonds among CS chains and lowering its crystallinity (Alexeev et al., 2000; Kolhe and Kannan, 2003). Thus, by manipulating the formulation conditions, the composition, and the thermal treatment, one may use the precipitation blending method as an effective way to improve the ductility of CS. The emulsification casting method differs from these two methods. By taking advantage of suitable solvents and emulsifying agents, polymeric blends with complete different solubility can be achieved. By adjusting the types of solvents and emulsifying agents, and changing the evaporation rate of the solvents, the phase morphology and size of the blends may be effectively controlled to satisfy the end-use demand.

REFERENCES

Alexeev VL, Kelberg EA, et al. 2000. Improvement of the mechanical properties of chitosan films by the addition of poly(ethylene oxide). *Polym Eng Sci* 40(5):211–1215.

Avella M, Martuscelli E, et al. 2000. Review—Properties of blends and composites based on poly(3-hydroxy)butyrate (PHB) and poly(3-hydroxybutyrate-hydroxyvalerate) (PHBV) copolymers. *J Mater Sci* 35(3):523–545.

Cao W, Wang A, et al. 2005. Novel biodegradable films and scaffolds of chitosan blended with poly(3-hydroxybutyrate). *J Biomater Sci Polym Ed* 16(11):1379–1394.

Chen C, Fei B, et al. (2002). The kinetics of the thermal decomposition of poly(3-hydroxy-butyrate) and maleated poly(3-hydroxybutyrate). *J Appl Polym Sci* 84(9):1789–1796.

Chen C, Dong, Cheung, MK. 2005a. Preparation and characterization of biodegradable poly (L-lactide)/chitosan blends. *Eur Polym J* 41(5):958–966.

Chen C, Zhou XS, et al. 2005b. Thermal behavior and intermolecular interactions in blends of poly(3-hydroxybutyrate) and maleated poly(3-hydroxybutyrate) with chitosan. *J Polym Sci Part B Polym Phys* 43(1):35–47.

Cheung MK, Wan KPY, et al. 2002. Miscibility and morphology of chiral semicrystalline poly-(R)-(3-hydroxybutyrate)/chitosan and poly-(R)-(3-hydroxybutyrate-co-3-hydroxy-valerate)/chitosan blends studied with DSC, ^{1}H T_1 and $T_{1\rho}$ and CRAMPS. *J Appl Polym Sci* 86(5):1253–1258.

Correlo VM, Boesel LF, et al. 2005. Properties of melt processed chitosan and aliphatic poly-ester blends. *Mater Sci Eng A* 403:57–68.

Doi Y. 1990. *Microbial Polyesters*. New York: VCH.

Doshi J, Reneker DH. 1995. Electrospinning process and applications of electrospun fibers. *J Electrostatics* 35(2 and 3):151–160.

Hu SG, Jou CH, et al. 2003. Antibacterial and biodegradable properties of polyhydroxyalkano-ates grafted with chitosan and chitooligosaccharides via ozone treatment. *J Appl Polym Sci* 88(12):2797–2803.

Ikejima T, Inoue Y. 2000. Crystallization behavior and environmental biodegradability of the blend films of poly(3-hydroxybutyric acid) with chitin and chitosan. *Carbohydr Polym* 41(4):351–356.

Ikejima T, Yagi K, et al. 1999. Thermal properties and crystallization behavior of poly (3-hydroxybutyric acid) in blends with chitin and chitosan. *Macromol Chem Phys* 200(2):413–421.

Kolhe P, Kannan RM. 2003. Improvement in ductility of chitosan through blending and copolymerization with PEG: FTIR investigation of molecular interactions. *Biomacromolecules* 4(1):173–180.

Muzzarelli RAA. 1977. *Chitin*. Oxford: Pergamon.

Ohkawa K, Minato KI, et al. 2006. Chitosan nanofiber. *Biomacromolecules* 7(11):3291–3294.

Reddy CSK, Ghai R, et al. 2003. Polyhydroxyalkanoates: an overview. *Bioresource Technol* 87(2):137–146.

Sakurai K, Maegawa T, et al. 2000. Glass transition temperature of chitosan and miscibility of chitosan/poly(N-vinyl pyrrolidone) blends. *Polymer* 41(19):7051–7056.

Sang YL. 1996. Bacterial polyhydroxyalkanoates. *Biotechnol Bioeng* 49(1):1–14.

Shih WJ, Chen YH, et al. 2007. Structural and morphological studies on poly(3-hydroxy-butyrate acid) (PHB)/chitosan drug releasing microspheres prepared by both single and double emulsion processes. *J Alloys Compd* 434–435:826–829.

Shin SY, Park HN, et al. 2005. Biological evaluation of chitosan nanofiber membrane for guided bone regeneration. *J Periodontol* 76(10):1778–1784.

Society JCC. 1988. *Chitin and Chitosan*. Tokyo: Gihodo.

Suyatma N, Copinet A, et al. 2004. Mechanical and barrier properties of biodegradable films made from chitosan and poly(lactic acid) blends. *J Polym Environ* 12(1):1–6.

Yalpani MR, Marchessault H, et al. 1991. Synthesis of poly(3-hydroxyalkanoate) (PHA) conjugates: PHA-carbohydrate and PHA-synthetic polymer conjugates. *Macromolecules* 24:6046–6049.

Zhang L, Deng X, et al. 1997. Miscibility, thermal behaviour and morphological structure of poly(3-hydroxybutyrate) and ethyl cellulose binary blends. *Polymer* 38(21):5379–5387.

Zhang X, Hua H, et al. 2007. In vitro degradation and biocompatibility of poly(L-lactic acid)/chitosan fiber composites. *Polymer* 48(4):1005–1011.

PART IV

NATURAL FIBER—REINFORCED COMPOSITES

CHAPTER 11

Starch–Cellulose Fiber Composites

ANALÍA VÁZQUEZ

INTECIN, Facultad de Ingenería-Universidad Nacional de Buenos Aires, Consejo Nacional de Investigaciones Científicas y Técnicas (CONICET)

VERA ALEJANDRA ALVAREZ

Research Institute on Science and Technology of Material Science (INTEMA), Universidad Nacional de Mar del Plata, J.B. Justo 4302, Mar del Plata, Argentina

Biodegradable Polymer Blends and Composites from Renewable Resources. Edited by Long Yu
Copyright © 2009 John Wiley & Sons, Inc.

11.1 INTRODUCTION

Starch and cellulose are some of the most abundant biopolymers. These materials are renewable, biodegradable, abundant, have low cost, and can be used in various applications (Vilpoux and Averous, 2004). The use of agricultural products is considered an important way to reduce surplus farm products and to develop nonfood applications.

The incorporation of starch into different types of biodegradable and nonbiodegradable polymers has been motivated by growing interest in enhancing the biodegradability of materials and in the reduction of the amount of waste. Since, 1970, starch has been used as a particulate filler or destructurized starch in commodity polymers (van Soest, 1997). Although such polymers were termed "biodegradable," only a part of them was starch and the rest of the polymer was not biodegradable. In order to prepare a strictly "biodegradable polymer" based on starch; the starch granules would be mixed with plasticizers or additives to make the starch thermoplastic.

The hydrophilic and moisture sensitivity characteristics of starch limit its application; for this reason starch is blended with other biodegradable polymer such as poly(hydroxyalkanoates), polycaprolactone, poly(vinyl alcohol), among others (Averous, 2004).

Starch reinforced with cellulose is one case of a natural fully biodegradable composite. The use of cellulose fiber to reinforce starch is improves the mechanical properties of starch. Cellulose fibers commonly used are sisal, wood, cotton, jute, and kenaf.

11.2 STARCH POLYMERS

11.2.1 Structure

Starch is made up of glucose repeat units. Native starch is based on two polysaccharides, the linear D-glucan amylose at 20–30 wt% and the highly branched amylopectin at 70–80 wt%. Figure 11-1 shows the structure of the polysaccharide component of starch (van Soest and Vliegenthart, 1997). Starch is a semicrystalline polymer: amylopectin is the major crystalline constituent; amylose and the amylopectin branches comprise the amorphous part. Figure 11-1 shows the double helices of the amylopectin outer chains arranged as thin laminar domains that represent the crystalline part. The co-crystallization of amylopectin and crystallization into single-helical structures can lead to additional crystallinity.

Starch is available from numerous resources such as potatoes and such grains as corn, wheat, etc. The proportion of amylopectin determines the character of the crystalline region and the strength (Placket and Vazquez, 2004).

11.2.2 Chemical Composition

Tables 11-1 and 11-2 show the chemical composition of different starches derived from maize, wheat, rice, potato, and tapioca. The values published in the literature

(a)

(b)

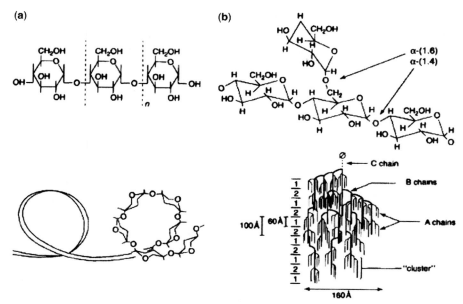

Fig. 11-1 (a) Structure of amylose and schematic representation of an amylose single helix; (b) Structure of amylopectin and schematic representation of the crystalline regions. (Reproduced with permission from *Trends in Biotechnology* **1997**, 15(6):208–213. Copyright 2008 Elsevier Editorial.)

present a range and the differences can be due to different growing conditions and the methods used in the chemical composition analysis.

The crystalline structure of starch granules can be disordered by heating in the presence of large quantities of water. During this process, the hydrogen bonds are broken

TABLE 11-1   Composition, Size, and Diameter of Different Starches[a]

Starch Source	Amylose Content (%)	Amylopectin Content (%)	Source	Diameter (μm)	Shape
Dent corn	25	75	Cereal	5–30	Polygonal, round
Waxy corn	<1	>99	Cereal	5–30	Polygonal, round
Tapioca	17	83	Root	4–35	Oval, truncated, "kettle drum"
Potato	20	80	Tuber	5–100	Oval, spherical
High-amylose corn	55–75 (or higher)	45–30 (or lower)	Cereal	5–30	Polygonal, round irregular
Wheat	25	75	Cereal	1–45	Round, lenticular
Rice	19	81	Cereal	1–3	Polygonal, spherical compound granules

[a]*Source*: Thomas DJ, William AA. 1999. Practical guides for the food industry. In: *Starches*. American Association of Cereal Chemists. Chapter 1, p. 6–9. ISBN 1–891127–01–2.

TABLE 11-2 Characteristics of Different Starches

Starch Source	Amylose (%)	Swelling Power (g/g) (°C)	Solubility (%)(°C)	Gelatinization Temperature (°C)	Reference
Maize	29.3	–	–	70–81	
Chayote	12.9			64–75	Jiménez-Hernandez et al. (2007)
Cassava tapioca	18.6–23.6	51 (95)	26 (95)	57–84.1	Hoover (2001), Freitas et al. (2004), Chang et al. (2006)
Yam starch	30–36				Mali et al. (2004), Freitas et al. (2004)
Normal potato	21.1–31.0	1159 (95)	82 (95)	57.0–80.3	Hoover (2001)
Normal corn	23–27	22 (95)	22 (95)	62.3–84.3	Singh et al. (2003)
High-amylose corn	42.6–67.8	6.3 (95)	12.4 (95)	66.8–73.3	
Normal rice	5–28.4	23–30 (95)	11–18 (95)	57.7–97.5	
Waxy rice[a]	0–2.0	45–50 (95)	2.3–3.2 (95)	66.1–78.8	
High-amylose rice	25–33	–	–	–	
Normal wheat	18–30	18.3–26.6 (100)	1.55 (100)	46.0–76	
Normal wheat A-granules	28.4–27.8	–	–	–	
Normal wheat B-granules	27.5–24.5	–	–	–	
Waxy wheat	25.1–29.5	–	–	–	
Normal soybean	19.8			51.8–55.8	Stevenson et al. (2007)
Black bean	35–39			61.2–81.2	Zhou et al. (2004)

[a]Waxy starch: starch particularly rich in amylopectin.

and from exposed hydroxyl groups; new water–amylose and water–amylopectin hydrogen bonds are generated. The starch chain interactions in the amorphous and crystalline regions can be inferred from the swelling power and solubility. The higher content of phosphate groups on amylopectin, which produces repulsion between phosphate groups on adjacent chains in potato starch, increasing the hydration due to the weakness of bonding in the crystalline domains, gives higher swelling power and solubility (Hoover, 2001).

The mechanical properties—tensile strength, modulus, and elongation at break—of starch films can be directly correlated with the amylase content (Thunwall et al., 2006). The higher the amylase content, the stronger the film. The branches of amylopectin commonly produce films with reduced mechanical properties (Thranathna, 2003).

11.2.3 Gelatinization

Gelatinization, which is an order–disorder phase transition of starches when they are heated in the presence of water, is assumed to involve several steps: (i) diffusion of water into the granule; (ii) uptake of water by the amorphous regions and hydration; (iii) radial swelling of the starch granules; (iv) loss of optical birefringence; (v) absorption of heat; (vi) loss of crystalline order due to uncoiling and dissociation of double helices in the crystalline regions; and (vii) amylose leaching (Lelieve and Mitchell, 1975; Stevens, 1981; Hoover, 2001).

11.2.4 Thermoplastic Starch

By gelatinization of starch with a plasticizer, it is possible to obtain *thermoplastic starch* (TPS). This can be done by application of mechanical, thermal, or thermomechanical energy in the presence of water and plasticizer [glycerol, sorbitol, xylose, sucrose, poly(ethyleneglycol)] and leads, as described earlier, to starch destructurization and loss of the organization of the intermolecular hydrogen bonding combined with partial depolymerization.

The *degree of gelatinization* will depend on the plasticizer content and processing parameters (shear stress, melt viscosity, and temperature) (Röper and Koch, 1990; van Soest et al., 1996a).

Starches showed two relaxations, which appeared between $-75°C$ and $-40°C$ and between $70°C$ and $150°C$. Some authors related this behavior to starch-rich and starch-poor regions caused by partially separation of phases. The high-temperature relaxation corresponds to the *glass transition temperature* and the other to a plasticizer-rich phase (Curvelo et al., 2001; Mathew and Dufresne, 2002; Forssell et al., 1997; Mylläinen et al., 2002). In the absence of glycerol, amylopectin showed only the upper glass transition.

The *type and content of plasticizer* influences the glass transition temperature, the mechanical properties (especially the modulus), and water absorption (Lourdin et al., 1997). An example of the effect of glycerol and water on the glass transition temperature is given in Fig. 11-2. Mali et al. (2006) have also determined the glass transition as a function of the glycerol content for corn, cassava, and yam starches, showing a decreasing trend with increasing glycerol content.

The *mechanical behavior* of TPS depends on the degree of destructurization reached and the plasticization effect (i.e., glass transition temperature and modulus) (Asa Rindlav-Westling et al., 1998; Rindlay et al., 1998; Chang et al., 2000; Myllarinen et al., 2002). The presence of intact or nondisrupted granular starch leads to inferior mechanical properties. In the case of high-amylose materials; the creation of a network leads to a stiffer and stronger product (van Soest and Borger, 2006). Thus, it is important to know how the mechanical properties are affected. Da Róz et al. (2006) showed that increasing the ethylene glycol content from 15 to

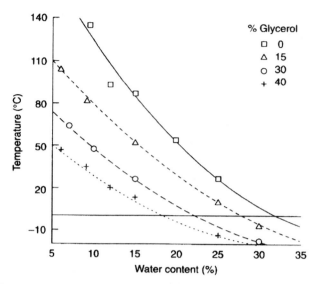

Fig. 11-2 Glass transition temperature vs. glycerol and water content. (Reproduced with permission from *Trends in Biotechnology* **1997**, 15(6):208–213. Copyright 2008 Elsevier Editorial.)

30 wt% in corn starch produces an increase on the tensile modulus from 63 MPa to 126 MPa due to the increase of crystallinity. However, when sorbitol or diethyleneoxide glycol was added to corn starch, the modulus decreased from 219 MPa to 59 MPa with the increase of sorbitol from 15 to 30 wt%, due to the plasticization effect. Glycerol can produce strong hydrogen bonding with starch, increasing the strength and toughness of the finished material (van Soest, 1997). Some authors also studied the relation between water and glycerol and the phenomena of plasticization and antiplasticization (Chang et al., 2006). The flexibility and workability of the rigid neat polymer can be improved by the incorporation of low-molecular-mass compounds or diluents, which act as plasticizers. Conversely, when they are incorporated in low quantities they can stiffen the material and behave as mechanical antiplasticizers (Sears and Darby, 1982). Chang et al. (2006) found that water acts an antiplasticizer but glycerol behaved as antiplasticizer only when the tapioca starch films were dried. Water and glycerol have synergic effects as plasticizers. The tensile modulus of the film increased at low glycerol content (close to 2.5%) but only when the humidity was lower than $0.22a_w$ (water activity).

There have been several studies on the preparation of TPS; in the majority of these the films were obtained by *casting*, but *melt processing* seems to be a more realistic technique for industrial application. The temperature, time, and content and type of plasticizer, as well as shear stress (i.e., viscosity × rpm) produce different degrees of destructurization (Shogren et al., 1992; Da Róz et al., 2006). During processing an increase in the shear stress (higher extrusion screw speed or viscosity) causes an increase in the amount of single-helix-type crystallinity. High temperature or shear stress can also degrade the polymer (Warburton et al., 1993; Sagar and Merril, 1995), so that additives or plasticizer are also necessary to improve the flow behavior.

11.2.5 Retrogradation

Starches suffer aging by *retrogradation* during storage and this has a tendency to change their structure. During aging; starch molecules reassociate in more ordered structures by forming simple juncture points and entanglements, helices, and

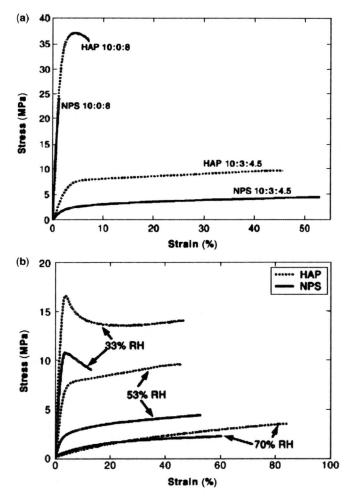

Fig. 11-3 Stress–strain behavior of HAP (higher amylose-content potato starch) and NPS (normal potato thermoplastic starch) (10:3:4.5 and 10:0:8): (a) conditioned at 21°C and 53% RH; (b) at 33% RH, 53% RH, and 70% RH. (Reproduced with permission from *Biomacromolecules* **2006**, 7:981–986. Copyright 2008 American Chemical Society.)

crystal structures. An X-ray diffraction pattern of the B-type develops slowly with time (Katz, 1930). The formation of B-type crystallinity in wheat starch gel polymers was explained using the crystalline growth theory by Marsh and Blanshard (1988). Miles et al. (1985) have proposed that, if the solution concentration is sufficiently high, an interconnected gel network can be formed in the polymer-rich region, with crystallization taking place subsequently. In another paper, Gidley (1989) suggested that the gelation behavior is based on the formation and subsequent aggregation of B-type double interchain helices of amylose. It was proposed that the gel contains rigid, crystalline, double-helical junction zones linked by single-chain segments which are amorphous and more mobile (Asa Rindlav-Westling et al., 1998). Amylose is responsible for the short-term changes and the amylopectin is responsible for all long-term rheological and structural changes (Gudmundsson, 1994). Starches from different botanical sources, despite similar amylase/amylopectin ratios, can retrograde to different extents, indicating that the structure of amylopectin is of importance. Other compounds such as lipids and surfactants can retard the retrogradation. Because of retrogradation, the prepared samples should be kept under controlled atmosphere before measuring their final mechanical properties. Since starch is a hygroscopic polymer, the relative humidity determines the mechanical properties. As an example, Fig. 11-3 shows the different behavior in tensile test of different potato starch films (Thunwall et al., 2006).

The mechanical properties are clearly influenced by the moisture content; a lower value leads to an increased modulus and a reduced elongation at break. Mali et al. (2006) also demonstrated different mechanical properties at initial time and after storage of samples of cassava, corn, and yam starch.

11.3   STARCH–CELLULOSE FIBER BASED COMPOSITES

11.3.1   Natural Fibers

During the past decades there has been an increasing interest in the use of natural fibers as an alternative to synthetic fibers such as those of glass, carbon, and aramide. This interest is based on the potential advantages of such fibers, mainly weight reduction, lower raw material price, and the ecological advantages of using renewable resources.

Several cellulose-based products and wastes have been incorporated as reinforcement, mainly to achieve cost savings (Joseph et al., 1993) and, in the case of biodegradable polymers, to maintain their biodegradability. Natural fibers also have numerous advantages compared to man-made fibers (Rozman et al., 1998): they are inexpensive, abundant and renewable, lightweight, biodegradable, and nonabrasive to processing equipment. The use of these fibers in biodegradable matrices is primarily for ecological reasons. However, there is only limited literature regarding the mechanical properties of natural fibers with thermoplastic biodegradable polymers (Hanselka and Herrman, 1994; Bastioli, 1995; Dufresne and Vignon, 1998; Kuruvilla and Mattoso, 1998; Iannace et al., 1999). Nevertheless, natural fibers also have some disadvantages such as lower durability and lower strength, though their specific properties are comparable.

Fig. 11-4 SEM image of sisal fiber.

Each natural fiber (Fig. 11-4) is a composite material in which soft lignin and hemicellulose acts as matrix and rigid cellulose microfibrils act as reinforcement. Microfibrils form hollow cells and are helically wound along the fiber axis. Uncoiling of these spirally oriented fibrils consumes large amounts of energy and is one of the predominant failure modes.

In biocomposites, the biofibers (natural fibers) serve as reinforcement to improve the strength, stiffness, and toughness of the neat matrix. The origin, source, and nature of different fibers together with their properties (physical, chemical, and mechanical) have been studied by several authors (McGovern, 1987; Batra et al., 1998). Whereas the properties of conventional fibers are found in a defined range, in the case of natural fibers they depend on the factors previously described and also (Bledzki et al., 1996) on the extraction method, the quality of the soil in which they grew (Barkakaty et al., 1976), the age of the plant from which they came (Chand et al., 1993), and the preconditioning used (Ray et al., 1976; Chawa et al., 1979).

Cellulose is a natural polymer with high strength and stiffness per unit weight, and is the building material of long fibrous cells. These cells can be found in the stem, the leaves, or the seeds of plants. Natural fibers can be classified accordingly:

- *Bast fibers* (*flax, hemp, jute, kenaf, ramie*). These consist of a wood core surrounded by a stem. Within the stem there are a number of fiber bundles, each containing individual fiber cells or filaments. The filaments are made of cellulose and hemicellulose, bonded together by a matrix, which can be lignin or pectin. The pectin surrounds the bundles, thus holding them onto the stem. The pectin is removed during the retting process. This enables separation of the bundles from the rest of the stem (scutching).
- *Leaf fibers* (*sisal, abaca* (*banana*), *palm*). These are coarser than bast fibers. Applications include ropes, and coarse textiles. Within the total production of leaf fibers, sisal, obtained from the agave plant, is the most important. Its stiffness is relatively high and it is often used in binder twines.

• *Seed fibers* (*cotton, coir, kapok*). Cotton is the commonest seed fiber and is used for textiles all over the world. Coir is the fiber of the coconut husk, it is a thick and coarse but durable fiber.

The strength and stiffness of natural fibers can be correlated with the spiral angle (the lower the angle, the higher the mechanical property); also, the chemical composition and the complex structure of natural fibers strongly affect their properties. Although fibers with higher cellulose content exhibit higher strength and fibers with lower cellulose content have lower strength, because of their complex nature there is no strict correlation between the cellulose content and the strength of the fiber. The lignin content influences the structure, properties, and morphology of the fibers. Waxy components generally affect the wettability and the adhesion characteristics of the fibers.

Tables 11-3 and 11-4 summarize the composition and properties of several natural fibers.

Natural fibers also contribute to the enhancement of material toughness. Different energy dissipation mechanisms can be identified depending on the fiber length. In thermoplastics reinforced with short fibers, fibers of subcritical length are pulled out rather than broken, as they are too short to reach their strength. In this case, the relevant energy dissipation mechanisms are debonding, sliding, restricted pull-out, and brittle or ductile matrix fracture Other failure mechanisms (i.e., fiber splitting into ultimate cells, stretching and uncoiling of microfibrils in the cells of fibers, transverse microcracking, and multiple ultimate cell fracture) have also been described in polymers reinforced with natural fibers (Fig. 11-5).

11.3.2 Starch/Natural Fiber Composites

Various reinforcing materials are mixed with starch-based polymers to increase the modulus or impact toughness. For preparing starch/natural fiber composites, different processing techniques have been used including kneading, extrusion, and postcompression molding or injection molding. When a biodegradable thermoplastic polymer is blended with natural fibers, good dispersion and distribution of the fibers are necessary.

Kneading, intensive mixing, and twin-screw extrusion produce a high shear stress, which is the main factor in improving the quality of mixing. Figure 11-6 shows schematically the influence of shear and chemical treatment on natural fibers in terms of dispersion and distribution. Intensive or dispersive mixing is related to reduction of phase domains, breaking the solid agglomerations, and produces a good distribution. Such a mixing mechanism is associated with shear stress levels developed in the equipment. Agglomerates break into individual particles when internal stresses exceed a threshold value. This value will depend on the nature of the bonds holding the particles of the agglomerate together, as well as the sizes of the fibers.

Extensive or distributive mixing or blending is related to decrease of the concentration of the agglomerated phase. Such a situation can be improved by

TABLE 11-3 Composition of Different Natural Fibers

Natural Fiber	Cellulose (%)	Lignin (%)	Hemicellulose (%)	Pectin (%)	Wax (%)	Spiral Angle (°)	Humidity (%)	Reference
Jute	61–71.5	12–13	13.6–20.4	0.2	0.5	8.0	12.6	Bledzky (1996), Hon (1992), Ugbolue (1990)
Flax	71	2.2	18.6–20.6	2.3	1.7	10.0	10.0	Bledzky (1996), Hon (1992), Ugbolue (1990)
Hemp	70.2–74.4	3.7–5.7	17.9–22.4	0.9	0.8	6.2	10.8	Bledzky (1996), Hon (1992), Ugbolue (1990)
Ramie	68.6–76.2	0.6–0.7	13.1–16.7	1.9	0.3	7.5	8.0	Bledzky (1996), Hon (1992), Ugbolue (1990)
Kenaf	31–39	15–19	21.5	–	–	–	–	Hon (1992), Rowell (1992),
Sisal	67–78	8.0–11.0	10.0–14.2	10.0	2.0	20.0	11.0	Bledzky (1996), Hon (1992), Ugbolue (1990)
Henequen	77.6	13.1	4–8	–	–	–	–	Hon (1992)
Cotton	82.7	–	5.7	–	0.6	–	–	Bledzky (1997)

TABLE 11-4 Properties of Various Fibers

Fiber	Density (g/cm³)	Diameter (μm)	Strength (MPa)	Tensile Modulus (GPa)	Elongation at Break (%)	Price (US$/kg)	Reference
			Natural Fibers				
Cotton	1.5–1.6	–	287–800	5.5–12.6	7.0–8.0	0.40	Chand (1994, Bledzky (1999)
Jute	1.3–1.45	25–200	393–773	13–26.5	1.16–1.5	0.30	Bledzky (1996/99), Ugbolue (1990)
Flax	1.50	–	345–1100	27.6	2.7–3.2	0.26	Bledzky (1996/99), Ugbolue (1990)
Hemp	–	–	690	–	1.6	–	Bledzky (1996), Ugbolue (1990)
Ramie	1.50	–	400–938	61.4–128	1.2–3.8	0.29	Bledzky (1996/99), Chand (1994)
Sisal	1.45	50–200	468–640	9.4–22.0	3–7	0.36	Bledzky (1996/99), Chand (1994)
			Synthetic fibers				
Glass E	2.5	–	2000–3500	70	2.5	3.20	Bledzky (1996)
Glass S	2.5	–	4570	86	2.8	3.25	Saechttling (1987)
Aramide	1.4	–	3000–3150	63–67	3.3–3.7	–	
Carbon	1.7	–	4000	230–240	1.4–1.8	500	

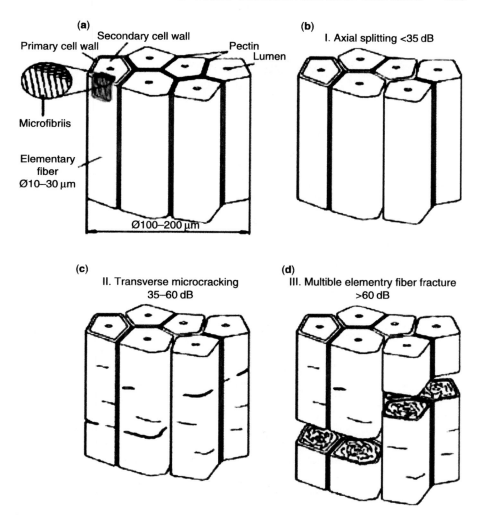

Fig. 11-5 Energy-dissipation mechanisms of natural fibers. (Reproduced with permission from *Macromolecular Materials and Eng.* **2003**, 288(9):699–707. Copyright 2008, Wiley Ed.)

good compatibility between matrix and fibers, because the agglomerate will be more easily separated and a greater interfacial area will be exposed to the polymer matrix.

Carvalho et al. (2003) studied the effect of processing conditions on the degradation of starch/cellulose composites, demonstrating that the degree of degradation could be evaluated by size-exclusion chromatography. Starch from waxy maize and bleached kraft pulp from *Eucalyptus urograndis* was used as raw material. The effect of chain scission is more important on the high-molecular-weight fraction of

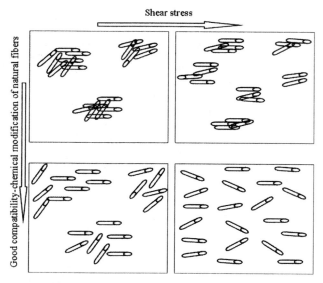

Fig. 11-6   Schematic of natural fiber mixing.

starch, mainly the amylopectin fraction. Increase of glycerol content reduced chain degradation, but increase of fiber content produced an increase of degradation, possibly due to the higher viscosity and as consequence higher shear stresses.

Curvelo et al. (2001) prepared pulp fibers with thermoplastic corn starch (with glycerol). Starch and 30 wt% of glycerol were premixed in a bags to obtain a powder. Then the composites were prepared in a Haake Rheomix 600 batch mixer at 170°C. Composites with low quantities of fiber (16 wt%) showed an significant increase in the tensile properties (tensile modulus and tensile strength). These composites exhibited good adhesion between the fibers and the matrix. Moisture sorption was reduced with the incorporation of fiber. These results are attributed to the fact that starch is more hydrophilic than cellulose and the fibers absorb some of the glycerol.

Gáspar et al. (2005) studied the effect of the addition of cellulose, hemicellulose, and zein (protein) to thermoplastic starch. The best mechanical properties were obtained with hemicellulose and zein; the addition of cellulose fibers produced a decrease in the mechanical properties of the starch. These authors also analyzed the water absorption behavior of these materials.

Averous et al. (2001) studied the effect of the addition of cellulose fibers to TPS matrix, and focused especially on the interaction between the fibers and the matrix. They observed an increase in the main transition temperature and correlate this behavior with the matrix–fiber interaction and a decrease in the mobility of the starch chains. Mechanical properties (an increment with fiber incorporation) and SEM observations clearly support the suggestion.

A series of papers were published by Gómez et al. (2006). They studied biocomposites of starch and natural fibers, processed by compression molding, using different starch sources: potato, sweet potato, and corn. Natural fibers, including jute, sisal, and cabuya, in the range of 2.5–12.5 wt%, were used as reinforcements and two different plasticizers (water and glycerol) were employed. The mechanical properties increased with fiber content. Whereas the tensile strength was particularly improved with 10 wt% of sisal, the higher values of impact strength were obtained with cabuya fibers.

A very interesting material was studied by Duanmu et al. (2007), who synthesized and prepared composites obtained from allylglycidyl-ether (AGE) modified potato starch as matrix, ethyleneglycol dimethacrylate (EGDA) as crosslinker, and wood fiber as reinforcement, showing that the tensile properties (modulus and strength) were higher (by approximately one order of magnitude) than cellulosic fiber-reinforced natural polymer composites previously reported. They also studied the effect of humidity on the mechanical properties of these materials, showing a clear detriment of mechanical behavior in wet environments.

Despite the mechanical properties of nanocellulose fibers (Young's modulus 140 GPa and tensile strength 1.76 GPa), the mechanical properties of nanocomposites based on nanofibers have not shown much better properties than natural fiber-based composites. This may be due to the fiber–matrix compatibility and processing techniques used in the fabrication of such nanocomposites. Lu et al. (2006) studied the effect of ramie nanocrystallites (0–40 wt%) on the behavior of plasticized starch. The nanocrystals were prepared by acid hydrolysis and had average length of 538.5 nm and diameter of 85.4 nm. Both the tensile strength and modulus were improved by the incorporation of nanocrystals. The authors observed the homogeneity of the biocomposites by SEM and also by studying the tensile stress–strain curves. Their results differ from those of Anglés and Dufresné (2001), who studied waxy maize starch reinforced with tunicin cellulose whiskers. The differences may be related to the plasticizer accumulation and to the strong interaction between the components and a plasticizer in the cellulose–amylopectin interface.

Takagi and Asano (2008) have obtained high reinforcing ratios ($E_{composite}/E_{matrix}$) using a stirrer and compression molding. They observed that when the molding pressure was increased, the density and the mechanical properties also increased. As consequence, low void content was obtained and the flexural strength increased from 10 to 65 MPa, and the flexural modulus increased from 1 to 6.5 GPa.

Table 11-5 shows the tensile properties of different starch/cellulose fiber composites; Fig. 11-7 shows the reinforcing ratio for the tensile strength of the composites with respect to each starch matrix.

It is clear that the incorporation of cellulose-based fibers into starch matrices produces an increase in the mechanical properties. The degree of reinforcement depends on the kind of starch (source), the plasticizer used (type and content), the fibers (type and aspect ratio), the processing technique, and the relative humidity of the environment. All of these parameters have to be taken into account in order to obtain a material with desired properties.

TABLE 11-5 Tensile Properties of Composites Based on Starch/Cellulose Fibers

Starch Type	Reinforcement	l/d	Glycerol (%)	E (GPa)	σ (MPa)	Reference
Corn	None	–	30	0.9	9.0	Gáspár et al. (2005)
	10 wt% cellulose	–		0.6	3.0	
Corn	None	–	30	0.125	5.0	Curvelo et al. (2001)
	16 wt% *Eucalyptus grandis* pulp-wood	–		0.320	11.0	
Corn	None	–	30(U/F)	0.042	4.1	Ma et al. (2005)
	5 wt% micro winceyette fiber	–		0.045	7.0	
	10 wt% micro winceyette fiber	–		0.051	8.2	
	15 wt% micro winceyette fiber	–		0.090	11.2	
	20 wt% micro winceyette fiber	–		0.140	15.2	
Wheat	None	–	18	0.087	3.6	Averous et al. (2001)
	10 wt% leaf-wood cellulose	3		0.348	7.2	
		15		0.435	10.8	
		45		0.510	14.4	
Wheat	None	0	3	0.056	2.8	Lu et al. (2006)
	5 wt% ramie cellulose nanocrystallites	6		0.07	4.0	
	10 wt% ramie cellulose nanocrystallites			0.15	5.15	
	20 wt% ramie cellulose nanocrystallites			0.23	5.35	
	30 wt% ramie cellulose nanocrystallites			0.41	6.15	
	40 wt% ramie cellulose nanocrystallites			0.475	6.9	

Material	Treatment				Reference
Wheat	None	–	0.052	3	Averous et al. (2001)
	15 wt% leaf-wood cellulose	3	0.296	7	
		15	0.350	10	
		45	0.430	13	
	20 wt% leaf-wood cellulose	3	0.757	13	
		15	0.630	15	
		45	0.670	22	
Waxy maize (amylomaize)	None	33		3	Funke et al. (1998)
	2 wt% Cellunier F			5.25	
	7 wt% Cellunier F			7	
	15 wt% Cellunier F			4.50	
	2 wt% Temming 500			4.75	
	7 wt% Temming 500			6.25	
	15 wt% Temming 500			7	
Potato (allyglycidylether modified)	None (LDS: low degree of substitution) dry	100	0.313	13	Duanmu et al. (2007)
	LDS 40 wt% wood dry		2.285	56	
	LDS 60 wt% wood dry		2.823	59	
	LDS 60 wt% wood wet		0.138	4	
	None (HDS: high degree of substitution) dry		0.359	15	
	HDS 40 wt% wood dry		2.875	77	
	HDS 60 wt% wood dry		3.130	90	
	HDS 60 wt% wood wet		0.428	26	
	HDS 70 wt% wood		3.777	146	

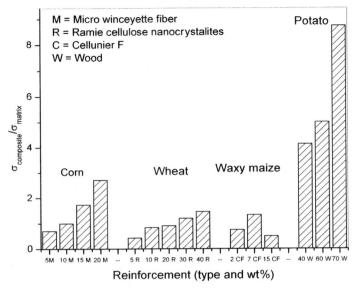

Fig. 11-7   Ratio of composite and matrix tensile strengths of different starches with degree of natural fiber incorporation.

11.4   STARCH-BASED POLYMER BLENDS AS POLYMER MATRIX

Another possibility for improving the behavior of starches is blending them with another polymer to produce starch-based blends. An number of important biodegradable polymers are derived from both synthetic and natural sources (Kaplan et al., 1993; Chiellini and Solaro, 1996; Amass et al., 1998), but most of these are quite costly. Starch is a potentially useful material for biodegradable plastics because of its natural abundance and low cost. However, starch-based materials, such as thermoplastic starch, have some drawbacks including poor long-term stability caused by water absorption, poor mechanical properties, and processability.

The development of low-cost biopolymers such as starch-based materials obtained from renewable resources has become crucial. For maintaining the biodegradability of the blend, known biopolymer components include aliphatic polyesters like polycaprolactone (PCL), poly(lactic acid) (PLA), poly(hydroxybutyrate-co-valerate) (PHBV), and poly(ester amide) (PEA) (Huang et al., 1993; Ramsay et al., 1993; Bastioli et al., 1995; Verhoogt et al., 1995; Chiellini and Solaro, 1996; Amass et al., 1998; Lorcks, 1998; Myllymäki et al., 1998; Averous et al., 2000; Bastioli, 2001). Some starch-based blends have been commercialized, such as MaterBi[®] (Novamont, Italy) and Bioplast[®] (Biotec, Germany).

Novamont, based in Italy, is a prominent European company in the business of manufacturing and supplying starch-based products. It markets its product under

TABLE 11-6 Biodegradable Starch-Based Polymers Commercially Used

Product	Company	Components
Amipol	Japan Cornstarch	Starch (100%)
Biofil	Samyang Genex Co.	Starch/polystyrene
Greenpol	Yukong Ltd.	Starch/polycaprolactone
MaterBi	Novamont	Starch/PVA–starch/PCL–starch/ cellulose derivatives
Novon	Chisso Warner Lambert	Starch (90–95%)/additives

the name MaterBi. Biotec GmbH in Germany has three product lines manufacturing Bioplast® granules for injection molding, Bioflex® film, and Biopur® foamed starch. Other manufacturers of starch-based polymer presently include Avebe, Earth-Shell, Groen Granulaat, Hayashibara Chemical Labs, Midwest Grain Products, National Starch, Rodensburg Biopolymers, Starch Tech, Supor, and Vegemat (Placket and Vazquez, 2004). Some commercial blends are summarized in Table 11-6.

Polycaprolactone (PCL)/starch blends and also cellulose derivative/starch blends are among the most widely used biodegradable polymers. Whereas the first of these is mainly used for films, the second is commonly processed by extrusion or injection. The properties of both matrices are summarized in Table 11-7.

PCL is commercially available and widely produced. The addition of PCL to starch produces an increase in the tensile strength as a function of PCL content (Koening and Huang, 1995; Pranamuda et al., 1996). Similar results were obtained with poly(hydroxyalkanoates) (PHA) and polylactides (PLA) (Ramsay et al., 1993; Koening and Huang, 1995; Kotnis et al., 1995; Ke and Sun, 2000).

Several workers have improved the mechanical behavior of starch-based blends by chemical treatment of one of the phases in the blend. The free hydroxyl groups in starch can be chemically modified to produce starch esters, starch ethers, and other derivatives. The C-2, C-3, and C-6 hydroxyl groups can be tailored, the C-6 primary hydroxyl being preferred due to its exposed position. Another possibility is to crosslink the starch, including starch modified by epoxidation and starch

TABLE 11-7 Physical and Mechanical Properties of Some Biodegradable Commercial Blends

Property	PCL/Starch Blend	Cellulose Derivative/Starch Blend
T_g (°C)	−60	105
T_m (°C)	66	–
MFI (g/10 ml)	2–4	6–30
Tensile strength (MPa)	20–50	15–35
Tensile modulus (MPa)	100–600	600–5000
Elongation at break (%)	200–600	20–150
Flexural modulus (MPa)	400–600	2000–2500

phosphates or diphosphates. For economic reasons, the last treatments are not often used in commercial products (Placket and Vazquez, 2004). Willett et al. (1998) reported improvements in tensile strength by grafting poly(glycidyl methacrylate) to starch granules. The size and shape of the granules as well as the chemical composition depend on the type of starch. For example, Willett and Felker (2005) obtained different results when potato or corn starch granules were used in the blend. The starch granules can be more easily debonded (i.e., at lower stress) from PEA matrix in potato/PEA blends than in corn/PEA blends.

However, the best properties were obtained by the addition of TPS (no granules) into other polymers because the matrix is more coherent. Averous et al. (2001) prepared a TPS/PEA blend in which PEA was the minor phase and reported that the blend had good interphase compatibility.

Chemical treatment such as grafting of one of the polymers in the blend does not improve only the interphase—the size of the separated phase decreases and the mechanical properties are also improved.

11.5 STARCH-BASED BLEND/NATURAL FIBER COMPOSITES

Starch-based blends have limited mechanical properties for several applications, but an increase of tensile strength can be achieved by compounding them with fibers. The fibers have a higher tensile strength and Young's modulus and a lower elongation at break than the matrix. As already mentioned, discontinuous fibers have been widely used as reinforcement (Mouzakis et al., 2000) of thermoplastic matrices in order to obtain better mechanical properties and processability. Several authors have studied the effect of incorporation of natural fibers on the behavior of starch-based blends, focusing on different aspects. In the following, we analyze each aspect separately.

11.5.1 Mechanical Properties

The mechanical properties of short-fiber-reinforced polymer composites depend on many factors: the mechanical properties of each component (the matrix and the reinforcement); the fiber content (volume fraction); fiber aspect ratio (length to diameter ratio) and fiber orientation; and the fiber-matrix interaction/adhesion, which is related to the compatibility between them. Fiber aspect ratio should be above a critical value for maximum stress in the fiber before composite failure. In addition, dissipation mechanisms, which determine the toughness, are also affected by the fiber aspect ratio: fibers with subcritical aspect ratio will be pulled out rather than broken. Similarly, fiber orientation has a significant influence on the mechanical properties of the composites, the stress value being maximum along the axis of orientation of the fiber. Processing conditions also have an important effect on the mechanical properties in terms not only of fiber orientation but also of fiber breakage, length reduction and defibrillation, and diameter reduction (Joseph et al., 1999).

11.5.1.1 Tensile Properties (Table 11-8; Fig. 11-8) One of the most studied starch-based blends is starch–polycaprolactone. Wollerdorfer and Bader (1998) have studied the effect of incorporation of flax and ramie as reinforcements of MaterBi, type ZI01U which consists of corn starch and a biodegradable polyester and a starch blend produced by compounding thermoplastic wheat starch with 40% poly-ε-caprolactone (PCL). In the case of MaterBi, the tensile strength clearly increased when fibers were incorporated; nevertheless, there was no additional rise with fiber contents higher than 15 wt%. For noncommercial blends, while no reinforcement effect was observed with ramie, the TPS/PCL compounds displayed a similar effect to those based on MaterBi by the addition of flax.

Cyras et al. (2001) studied MaterBi ZF03 reinforced with short sisal fibers. They have demonstrated that the initial fiber bundles are separated during processing into smaller fibers that have a rough surface and that the reduction of fiber diameter contributes to the enhancement on the fiber–matrix contact due to higher available surface. Nevertheless, the aspect ratio (l/d) did not show not an important increase with the fiber content, because processing also produces a shortening of the fibers. Regarding mechanical properties, the authors shows that fibers operate as reinforcement: both the elastic modulus and the tensile strength were higher with incorporation of sisal fibers despite the damage occurred during processing.

Ali et al. (2003) reinforced MaterBi Z with short sisal fibers. They found an important increase in both the tensile strength and the tensile modulus of this matrix on the incorporation of 20 wt% of sisal fibers.

As can be seen from Table 11-8, the incorporation of around 20 wt% of sisal fibers into this kind of matrix produced an >600% increase in the tensile modulus. However, analysis of the tensile strength shows that the increment is related to the type of fiber (sisal > ramie > flax) and also to the fiber content (being higher for higher fiber content).

Another important starch-based blend is *starch/cellulose derivatives*. Within blends of this kind, Wollerdorfer and Bader (1998) studied Bioplast GS 902, which is a blend consisting of potato starch, modified cellulose, and synthetic polymers. The observation that resistance did not change with fiber incorporation or fiber content was probably related to incomplete disintegration of starch grains.

Ali et al. (2003) have also studied the effect of 20 wt% of short sisal fibers on the mechanical properties of MaterBi Y101, showing a clear increase in tensile modulus and a decrease in tensile strength.

Lanzillota et al. (2002) analyzed the effect of flax fiber addition on the tensile behavior of MaterBi Y101, showing that tensile strength as well as tensile modulus increased when fibers were incorporated into the neat matrix, but the fiber content did not show any important additional effect.

Alvarez et al. (2003) have analyzed the tensile properties of MaterBi Y/short sisal fiber composites. They determined that the elastic modulus and mechanical strength of the composites increase with fiber content, confirming the reinforcing

TABLE 11-8 Tensile Properties of Starch-Based Blends/Natural Fiber Composites

Matrix	Reinforcement[a]	σ (MPa)	E (MPa)	Reference
	Starch/Biodegradable Polyester			
MaterBI® ZI01U (corn starch and a biodegradable polyester)	None	15.26		Wollerdorfer and Bader (1998)
	15 wt% ramie	25.10		
	15 wt% flax	20.83		
	20 wt% flax	21.64		
	25 wt% flax	20.51		
MaterBi ZF03 (corn starch and a biodegradable polyester)	None	4.03 ± 0.68	28 ± 8	Ali et al. (2003)
	20 wt% sisal	10.4 ± 1.44	222 ± 24	
MaterBi-Z ZF03 (corn starch and a biodegradable polyester)	None	7.3 ± 1.3	37 ± 0.9	Cyras et al. (2001)
	10 wt% sisal	10.9 ± 0.9	138 ± 13	
	20 wt% sisal	12.7 ± 0.5	257 ± 17	
	30 wt% sisal	14.4 ± 1.6	687 ± 119	
TPS : PCL [thermoplastic wheat starch with 40% poly-ε-caprolactone (PCL)]	None	20.84		Wollerdorfer and Bader (1998)
	15 wt% ramie	19.95		
	15 wt% flax	27.61		
MaterBi® LF01U (potato, corn, and wheat starch)	None	24	95	Romhány et al. (2003b)
	20 wt% flax UD	48 ± 6	2600 ± 300	
	40 wt% flax UD	73 ± 3	5900 ± 600	
	60 wt% flax UD	78 ± 4	9300 ± 1400	
	20 wt% flax CP	30 ± 6	1800 ± 400	
	40 wt% flax CP	53 ± 5	4500 ± 500	
	60 wt% flax CP	55 ± 3	5900 ± 30	
Bioplast GS 902 (potato starch, modified cellulose, and synthetic polymers)	None	27.72		Wollerdorfer and Bader (1998)
	15 wt% ramie	24.77		
	15 wt% flax	24.23		
	25 wt% flax	29.37		
	35 wt% flax	27.15		
	Starch/Cellulose Derivatives			
MaterBi Y101(starch and cellulose derivatives)	None	17.6 ± 3.32	704.6 ± 16.9	Ali et al. (2003)
	20 wt% sisal	14.2 ± 0.92	1032.2 ± 39.7	
MaterBi Y101 (starch and cellulose derivatives)	None	12.6 ± 0.7	945 ± 90	Alvarez et al. (2003)
	5 wt% sisal	14.4 ± 0.7	1390 ± 150	
	10 wt% sisal	15.7 ± 0.9	1870 ± 169	
	15 wt% sisal	16.8 ± 0.9	2220 ± 178	
MaterBi Y101 (starch and cellulose derivatives)	None	28	1000	Lanzillota et al. (2002)
	20 wt% flax	48	3900	
	30 wt% flax	49	3900	
	40 wt% flax	51	4300	
SCA (starch and cellulose acetate)	None	31.5	1300	Cunha et al. (2004)
	20 wt% wood flour	19.0	3000	
	40 wt% wood flour	22.5	4200	
	50 wt% wood flour	33.0	5700	
	60 wt% wood flour	15.0	5700	

[a]UD, unidirectional; CP, cross-ply.

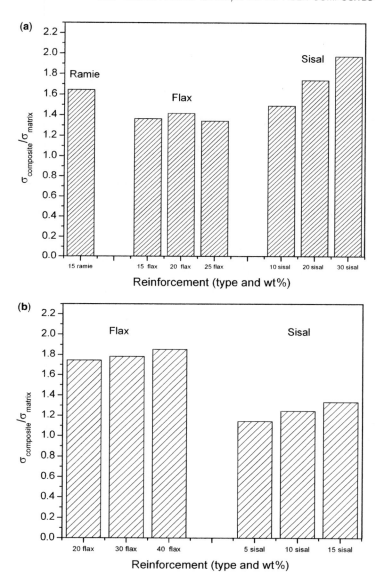

Fig. 11-8 Ratio of composite and matrix tensile strength as a function of fiber content for different fibers in (a) starch/polycaprolactone blends and (b) starch/cellulose derivative blends.

action of the fibers. They also established (Alvarez et al., 2006a) the effect of fiber orientation on the tensile properties, showing higher values for parallel-oriented fibers, intermediate values for random orientation, and lower values for perpendicular orientation. Also, the fiber orientation was correlated with the processing technique (injection, oriented at the skin and random at the core; and compression molding, random).

Cunha et al. (2004) analyzed the mechanical properties of starch–cellulose acetate blend/wood flour composites. They observed that the tensile strength and the modulus improve considerably up to 50 wt% of reinforcement. In contrast, the toughness decreases slowly. The material with 50 wt% wood flour displayed the highest tensile strength and tensile modulus but is difficult to process. However, Cunha and co-workers have demonstrated that the processability can be improved by using glycerol as plasticizer. They also analyzed the effect of orientation (radial and tangential) with respect to injection point.

Comparing the results of different authors (Table 11-8), it is clear that the tensile modulus increase with natural fiber incorporation (up to 330% for 40 wt% of flax fibers). On the other hand, dissimilar results have been found in relation to the tensile strength; whereas some authors (Wollerdorfer and Bader, 1998; Ali et al., 2003) have found no differences between the matrix and the biocomposites, others (Lanzillota et al., 2002; Alvarez et al., 2004a) have shown an increasing trend with fiber incorporation and fiber content. The differences, working with the same kind of matrices and fibers, could be related to the processing technique and the fiber aspect ratio (l/d), as will be explained later.

The starch based blends described exhibit important differences, first of all, in terms of the crystallinity: whereas starch/polycaprolactone blend is a semicrystalline polymer, starch/cellulose derivative blend is an amorphous polymer. The glass transition temperature of the first is below room temperature (around $-60°C$), whereas that of the second one is above room temperature (near to $100°C$). These characteristics make the polymers very different (see Table 11-7) and the initial mechanical properties (without reinforcement) are also very different, especially the tensile modulus, which is 8 times higher for cellulose derivative/starch blends. Accordingly, the ratio of fiber modulus to matrix modulus (E_f/E_m) is also very different, being around 45 for polycaprolactone/starch blends and 6 for cellulose derivative/starch blends.

Romhany et al. (2003a) have analyzed the effect of flax fibers (unidirectional and crossed-ply arrangements) on the behavior of thermoplastic starch-based composites (MaterBi LF01U) obtained by hot pressing using the film stacking method. The mechanical performance and also the mode of failure of the composites depended strongly on the fiber content and the flax fiber lay-up. The tensile strength increased with higher flax fiber content up to 40 wt% and then remained almost constant. In particular, for MaterBi Y/40 wt% unidirectional fiber, the tensile strength was three times greater than that of the pure matrix. In addition, the fibers increased the tensile modulus of the neat matrix by several orders of magnitude. It should be noted that the strength depends on the interphase generated between fiber and matrix, and is related to compatibility but not the modulus.

11.5.1.2 *Flexural Properties* (Table 11-9)

Shibata et al. (2005) have studied composites prepared with a biodegradable resin, CP-300, which is blend of corn starch and PCL reinforced with kenaf and bagasse fibers. For kenaf, the flexural modulus increased as a function of fiber volume fraction up to 60% and then decreased. A similar trend was observed for flexural strength. They showed that on

TABLE 11-9   Flexural Properties of Starch-Based Blend/Natural Fiber Composites

Matrix	Reinforcement	σ (MPa)	E (GPa)	Reference
CP-300 (corn-starch and PCL)	None	18.75	0.49	Shibata et al. (2005)
	17 vol% kenaf	31.56	1.97	
	30 vol% kenaf	41.25	2.94	
	50 vol% kenaf	43.44	3.40	
	60 vol% kenaf	49.38	3.83	
	68 vol% kenaf	40.00	3.67	
	20 vol% bagasse	28.44	1.00	
	33 vol% bagasse	38.13	1.41	
	50 vol% bagasse	48.44	2.18	
	64 vol% bagasse	49.06	2.48	
	74 vol% bagasse	48.13	2.28	
MaterBi Y101 (starch and cellulose derivatives)	None		2.28 ± 0.25	Alvarez et al. (2004a)
	5 wt% sisal		2.32 ± 0.20	
	10 wt% sisal		2.86 ± 0.25	
	15 wt% sisal		3.42 ± 0.29	

the surface of the composites with fiber contents higher than 60%, many fibers were not sufficiently wetted by the resin; this was also observed for bagasse.

Alvarez et al. (2004a) studied the flexural modulus of MaterBi Y/short sisal fibers biocomposites. They demonstrated that incorporation of short sisal fibers notably improved the flexural stiffness of this matrix and that higher fiber content produced a higher enhancement.

11.5.1.3   *Impact Properties* (Table 11-10)

Johnson et al. (2003) studied the effect of incorporation of fiber from *Miscanthus* fiber on the impact performance of MaterBI Y101U. The miscanthus-filled MaterBi exhibited up to 30% higher impact load than the neat matrix.

Alvarez et al. (2005a) analyzed the out-of-plane impact fracture of MaterBi Y/short sisal fiber biocomposites. They established an increasing trend of the total fracture energy and related it to an increase on circumferential shear-cracking mechanism observed in impacted samples. The active mechanisms in these biocomposites were axial splitting, uncoiling of microfibrils, microcracking, and fiber pull-out. These authors have also demonstrated that under quasi-static loading the biocomposites exhibited higher resistance to crack initiation than did the neat matrix and attributed the differences to the new energy dissipation mechanisms derived from the fibers.

From the results (Table 11-10), a decrease in the disk strength in comparison to the neat matrix is evident for composites with low fiber content. A possible explanation for such behavior is that the fibers spread through the matrix can act as crack initiation points during impact (Mouzakis et al., 2000). However, an increasing tendency of

TABLE 11-10 Impact Properties of Starch-Based Blend/Natural Fiber Composites

Matrix	Reinforcement	σ (MPa)	E (kJ/m)	Reference
MaterBI® YI01U	None	106.4 ± 9.2		Johnson et al.
(starch, cellulose	10 vol% miscanthus	125.9 ± 10.1		(2003)
derivatives,	20 vol% miscanthus	138.6 ± 1.9		
additives)				
MaterBi-Y YI01U	None	111.0 ± 5.4	0.17 ± 0.03	Alvarez et al.
(starch, cellulose	5 wt% sisal	80.1 ± 1.4	0.40 ± 0.13	(2005a)
derivatives,	10 wt% sisal	97.1 ± 1.4	0.43 ± 0.04	
additives)	15 wt% sisal	106.4 ± 9.2	0.45 ± 0.03	
	20 wt% sisal	129.8 ± 7.6	0.61 ± 0.03	

strength with fiber content was observed. Indeed, there were enough fibers at some fiber content (20 wt% of sisal or 10 wt% of miscanthus) to enhance load transfer. For the same matrix, 20 wt% of sisal fibers gave an increase on the impact strength of around 17% whereas the same amount of miscanthus fiber produced increases up to nearly 30%. In addition, the energy values from Table 11-10 suggest that the incorporation of sisal fibers into the biodegradable matrix is an efficient way to improve impact fracture properties.

11.5.1.4 Creep Properties (Table 11-11) Creep behavior is one of the most important mechanical properties that needs to be known for the structural design of thermoplastic composites where the dimensional stability of materials is important. Additionally, creep measurements are of primary interest in any application where the polymer must sustain loads for long periods. In spite of the importance of this subject, little information can be found in the scientific literature about creep of thermoplastic composites with natural fibers, and even less related to biodegradable polymers (Park and Balatinecz, 1998; Vázquez et al., 1999; Cyras et al., 2001; Alvarez et al., 2004a).

TABLE 11-11 Creep Properties of Starch-Based Blend/Natural Fiber Composites

Matrix	Reinforcement	Q (kJ/mol)	E_1 (MPa)	τ_2 (s)	Reference
MaterBi-Z ZF03	None	318.7	320	250.0	Cyras et al. (2002)
(corn starch and a	10 wt% sisal	343.3	590	269.0	
biodegradable	20 wt% sisal	381.0	1020	236.4	
polyester)	30 wt% sisal	382.0	950	326.9	
	40 wt% sisal	350.3	805	232.6	
MaterBI-Y YI01U	None	102	2000	2.84	Alvarez et al.
(starch, cellulose	5 wt% sisal	114	2700	1.83	(2004a)
derivatives,	10 wt% sisal	124	3400	1.05	
additives)	15 wt% sisal	147	4000	0.81	

Cyras et al. (2002) analyzed the flexural creep behavior of MaterBi Z/short sisal fiber composites at different temperatures, demonstrating that creep performance was enhanced as a function of the fiber content but also that the fragmentation of the polymer chains and natural fibers clearly influenced the creep behavior.

Alvarez et al. (2004a) studied the creep performance of MaterBi Y/short-sisal fibers and analyzed the creep behavior at different temperatures and fiber content. They showed that the highest fiber content produces the highest creep resistance and flexural modulus.

Several models have been used to predict behavior and analyze properties. For MaterBi Y, the test temperature was below the T_g value. The elastic modulus E_1 increases with the fiber content. These results are in agreement with those obtained for flexural modulus. The viscosity η_1 increased with the fiber content, and lower flow occurred at the dashpot and the permanent deformation decreased. The relaxation time τ_2 decreased with the fiber content; this is an indication that the viscous creep decreases and adding fibers enhances the elastic part. For MaterBi Z, measurements were made above T_g value. The elastic modulus E_1 increases with fiber content up to 20 wt%. These results agree with those for flexural modulus. The viscosity η_1 increased with the fiber content, and lower flow occurred at the dashpot and the permanent deformation decreased. The relaxation time τ_2, increased with the fiber content; this is an indication that the viscous creep increases and adding fibers produces a smaller elastic part, perhaps due to some plasticization of the matrix related to with some kind of degradation of the natural fibers.

11.5.2 Effect of Processing Conditions (See Tables 11-12 and 11-13)

When composites are processed by extrusion, shear stresses are developed during mixing and extrusion and these stresses cause fiber damage. The shear stresses generated in extrusion process, used to disperse the fibers into the polymer, produce a great reduction in fiber length and diameter. The extent of fiber breakage depends on the residence time, the temperature, viscosity and speed of rotation, and the characteristics of the matrix (Joseph et al., 1993; Ali et al., 2003).

Several authors (e.g., Johnson, 2003) have stated that shear force in an extruder could cause fiber breakage. It was contended that as the rotational speed of the screw increases, the shear forces will increase and fiber breakage will increase, producing a greater degree of fiber damage.

Processing conditions can also influence the extraction of lignin from natural fibers and/or promote some chemical or physical modification of components during the mixing process. This effect can result from higher processing temperature, speed of rotation, or time of mixing employed (Ali et al., 2003). Thus, the appropriate selection of processing parameters is of crucial importance for the final behavior of the biocomposites.

Ali et al. (2003) studied the effect of processing conditions on the tensile and creep performance of MaterBi Y and MaterBi Z with short sisal fibers. They have shown a clear increase in the aspect ratio with increasing speed of rotation, whereas only a moderate change was seen by extending the time of mixing. The changes in aspect

TABLE 11-12   Effect of Processing Conditions on the Final Natural Fiber Dimensions

(a) MaterBi Y 20 wt% sisal, 180°C[a]

Conditions	l/d
20 rpm, 2 minutes	60.6
20 rpm, 4 minutes	66.4
20 rpm, 6 minutes	74.7
50 rpm, 2 minutes	73.4

(b) MaterBi Y 15 wt% sisal, double-screw extruder at 60 rpm[b]

T (°C)	l (mm)	d (mm)	l/d
180	2.23	0.17	14.48
185	1.57	0.13	14.64
190	1.19	0.09	16.40

(c) MaterBi Y 15 wt% sisal, double-screw extruder at 180°C[b]

SR (rpm)	l (mm)	Reduction (%)	d (mm)	Reduction (%)	l/d
25	2.16	58.2	0.173	49.0	13.41
60	1.53	70.4	0.130	57.0	14.66
80	1.47	71.0	0.098	67.3	21.07

[a]Ali et al. (2003).
[b]Alvarez et al. (2005b).

ratio were greater for MaterBi Z. These authors found an increase in the tensile properties when the speed of rotation was higher due to the higher final aspect ratio of the sisal fibers, related to higher shear forces. They have observed similar tendencies for creep performance.

Johnson et al. (2003) studied the effect of processing parameters on the impact performance of MaterBi Y101U/mischanthus fiber biocomposites, finding that the temperature of the barrel and the rotational speed of the screw affect the performance. They demonstrated that materials molded at higher temperature (190°C) have better impact properties than those molded at lower temperature (170°C). In addition, at higher speed of rotation (300 rpm) the composite molded at lower temperature resisted higher impact loads than that molded at higher temperature; they related this behavior to the degree of heating imparted by the screw to the polymer melt due to the shear forces, which can degraded the natural fibers and affect the impact performance of MaterBi Y/miscanthus biocomposites. On the other hand, at lower temperature, the speed of rotation can homogenize the material, improving the impact performance.

Alvarez et al. (2005b) evaluated the effect of the processing conditions on the final properties of MaterBi Y/short sisal fiber composites, using three different temperature profiles in a twin-screw extruder and varying the speed of rotation.

TABLE 11-13 Effect of Processing Conditions on the Mechanical Properties of Starch-Based Blends/Natural Fiber Composites

Matrix	Reinforcement	σ (MPa)	E (MPa)	Reference
(a) Tensile Properties				
MaterBi ZF03 (corn starch and a biodegradable polyester)	20 wt% sisal; 120°C; 110 rpm			Ali et al. (2003)
	2 minutes	9.03 ± 0.49	198.6 ± 27	
	6 minutes	10.39 ± 1.44	222.0 ± 24.2	
	20 wt% sisal; 120°C; 6 minutes			
	60 rpm	6.67 ± 1.65	116.7 ± 9.1	
	110 rpm	10.39 ± 1.44	222.0 ± 24.2	
	20 wt% sisal; 110 rpm; 6 minutes			
	120°C	10.39 ± 1.44	222.0 ± 24.2	
	140°C	10.09 ± 2.32	275.8 ± 58.1	
MaterBi Y101 (starch and cellulose derivatives)	20 wt% sisal; 180°C; 20 rpm			Ali et al. (2003)
	2 minutes	14.15 ± 0.92	1032.2 ± 39.7	
	4 minutes	12.00 ± 5.80	1081.6 ± 93.2	
	6 minutes	18.60 ± 1.98	1183.3 ± 25.8	
	20 wt% sisal; 20 rpm; 2 minutes			
	140°C	12.30 ± 0.57	958.0 ± 5.9	
	160°C	12.60 ± 2.83	995.9 ± 59.5	
	180°C	14.15 ± 0.92	1032.2 ± 39.7	
	20 wt% sisal; 180°C; 2 minutes			
	20 rpm	14.15 ± 0.92	1032.2 ± 39.7	
	50 rpm	22.35 ± 0.21	1249.5 ± 13.6	
MaterBI-Y YI01U (starch, cellulose derivatives, additives)	15 wt% sisal; 60 rpm			Alvarez et al. (2005b)
	180°C	15.2 ± 1.6	880 ± 79	
	185°C	16.8 ± 0.9	1552 ± 160	
	190°C	16.4 ± 1.8	1601 ± 91	
	15 wt% sisal; 185°C			
	25 rpm	14.8 ± 1.1	1172 ± 158	
	60 rpm	16.8 ± 0.9	1552 ± 160	
	80 rpm	16.1 ± 1.5	1443 ± 99	

(Continued)

TABLE 11-13 *Continued*

(b) Impact Properties

Matrix	Reinforcement	σ (MPa) 170°C		E (MPa) 190°C		Reference
		150 rpm	300 rpm	150 rpm	300 rpm	
MaterBI® YI01U (starch, cellulose derivatives, additives)	10 vol% miscanthus	117.0 ± 13	122.3 ± 10	135.5 ± 21	125.9 ± 10	Johnson et al. (2003)
	20 vol% miscanthus	107.5 ± 33	113.4 ± 22	135.5 ± 11	138.6 ± 2	
		σ (MPa)		E (kJ/m)		
MaterBI-Y YI01U (starch, cellulose derivatives, additives)	15 wt% sisal; 60 rpm					Alvarez et al. (2005b)
	180°C	101.0 ± 1.6		0.74 ± 0.01		
	185°C	134.0 ± 0.9		0.84 ± 0.10		
	190°C	116.7 ± 1.8		0.82 ± 0.05		
	15 wt% sisal; 185°C					
	25 rpm	130.2 ± 0.6		0.68 ± 0.01		
	60 rpm	134.0 ± 0.9		0.84 ± 0.10		
	80 rpm	128.8 ± 1.2		0.74 ± 0.04		

The mechanical properties increased when rotation speed changed from 25 to 60 rpm; they then decreased because fiber breakage increased with rotational speed. Aspect ratio also increases (Table 11-12), but at high speed degradation of the matrix occurs. Impact properties increased with temperature but decreased at temperatures higher than 185°C, again related to the thermal degradation of the matrix. The changes in the aspect ratio were evidently greater as a function of the speed of rotation than as a function of the temperature.

As a general result, we can say that higher temperature produces a decrease in the viscosity of the matrix and better mixing, but it is also possible that the viscous dissipation increases and the matrix may experience some thermal degradation; the viscosity of the matrix increases as a result of this effect, especially in the case of cellulose derivative/starch blends whose processing temperatures are higher (around 180°C) than those of polycaprolactone/starch blends (around 100–120°C). Shear stresses increase due to the increase in viscosity, and the fiber aspect ratio increases. The tensile stress transfer to the fibers increases when the aspect ratio of the fibers increases. In addition, lignin extraction could lead to different surface properties of the fibers, which are responsible for the fiber–matrix adhesion. Carvalho et al. (2003) found a similar tendency for matrix degradation studied by changes in the molecular weight and molecular weight distribution during the mixing of thermoplastic starch (TPS) compounds of conventional corn starch and glycerol reinforced with cellulosic fibers.

Fiber breakage increases with rotational speed, but aspect ratio also increases; this effect was observed independently of the other blend component (polycaprolactone or cellulose derivatives) but the changes are more important in the case of polycaprolactone/starch blends for which the viscosity is clearly lower and the processing temperature is far from the degradation temperature, avoiding other possible consequences.

11.5.3 Rheological Behavior

The design of the most suitable processing conditions is guided mainly by the rheological behavior of the composites (Jayamol et al., 1996). A number of investigations on the rheological behavior of short natural fiber-reinforced thermoplastic and elastomers have reported (Wang and Lee, 1987; Fujiyama and Kawasaki, 1991) that incorporation of fillers in thermoplastics and elastomers will increase the melt viscosity, which may result in unusual rheological effects.

Alvarez et al. (2004b) studied the rheological behavior of MaterBi Y/short sisal fiber composites, demonstrating that the viscosity of all materials decreased with frequency (pseudoplastic behavior) but increased with fiber content. In addition, the enlargement of viscosity and storage modulus with fiber content was not linear, showing a saturation effect at higher fiber concentrations. They also analyzed the effect of processing technique, showing that compression-molded composites displayed higher viscosities than injected ones and were more difficult to process.

Cunha et al. (2004) analyzed the rheological properties of starch-cellulose acetate blend/wood fiber composites. They determined that the effect of the wood flour content on the shear viscosity was complex and a linear relationship was not established. The shear viscosity decreases with shear rate, but for composites with wood flour contents higher than 40 wt% there was evidence of quasi-newtonian behavior independently of the temperature.

11.5.4 Effect of Fiber Dimensions (See Table 11-14)

Transference of stress from the matrix to the fibers takes place when the experimental aspect ratio is higher than the critical value. For fibers with an aspect ratio lower than the critical value, pull-out occurred more than fiber breakdown. A weak interface between fiber and matrix leads to a lower reinforcement effect. The critical aspect ratio can be estimated from the interfacial strength (obtained from pull-out or micro-droplet test) and fiber strength (obtained form the single-fiber test) by use of the Kelly–Tyson equation:

$$\frac{l}{d} = \frac{\sigma_f}{2\tau} \qquad (11.1)$$

where σ_f is the fiber strength and τ is the fiber–matrix interfacial strength.

Shibata et al. (2005) have also studied the effect of natural fiber length on the flexural properties of CP-300, determining that the flexural modulus rapidly decreased for fiber lengths less than 2.8 mm for kenaf and 3.2 mm for bagasse, and founding a similar trend for the flexural strength. They related this result to the critical aspect ratio (l/d) because, according to results of Hsueh (2002), the elastic modulus decreases sharply below a fiber aspect ratio of 15. The authors calculated the critical length of kenaf (3.4 mm) and bagasse (3.6 mm): fibers below these values are possibly pulled out, not broken, in the flexural test.

It is clear that the reinforcement effect depends on the experimental aspect ratio (in relation to the critical value), which is directly correlated with the fiber and matrix properties and the interfacial adhesion, which depends on the fiber–matrix compatibility (wettability of the fibers by the matrix, etc). Whereas the critical aspect ratio for sisal–starch/cellulose derivatives was around 20, in the case of sisal–starch/polycaprolactone it was near to 79; this is related to the higher fiber–matrix compatibility in the former case, which is also linked to the matrix polarity (8.7 vs. 6.7). Given that natural fibers have polar character, the higher the polarity, the higher the fiber–matrix compatibility and also the higher the interfacial adhesion. It is also necessary to take into account that in some cases the additives in the blends can occupy the interphase and make it less compatible.

11.5.5 Influence of Fiber Treatments (See Tables 11-11, 11-15–11-17)

One of the most important factors for achieving good fiber reinforcement in composites is the fiber–matrix adhesion, which depends on the structure and polarity of both

TABLE 11-14 Effect of Fiber Dimensions on the Mechanical Properties of Starch-Based Blend/Natural Fiber Composites

Matrix	Reinforcement	l (mm)	σ (MPa)	l (mm)	E (GPa)	Reference
		(a) Flexural Properties				
CP-300 (corn-starch and PCL)	60 vol% kenaf	2	38.0	2	2.80	Shibata et al. (2005)
		3	40.0	3	2.95	
		6	41.5	6	3.90	
		11	49.5	9	4.00	
		17	47.0	15	3.90	
	66 vol% bagasse	1	23.0	2	1.50	
		4	40.0	3	2.30	
		9	50.0	9	2.50	
		16	48.0	15	2.50	

(b) Impact Properties

Matrix	Reinforcement	σ (MPa)		Reference
		<1 mm	>3 mm	
MaterBI® YI01U (starch, cellulose derivatives, additives)	10 vol. % Mischantus	125.9 ± 10.1	140.2 ± 5.6	Johnson et al. (2003)
	20 vol. % Mischantus	138.6 ± 1.9	130.8 ± 7.7	

TABLE 11-15   Effect of Chemical Treatments on the Diameter and Mechanical Properties of Sisal Fibers

Sisal Fibers	E (MPa)	σ (MPa)	ε (%)	d (mm)
Untreated	7.5	190	3.2	0.30
Alkali-treated	8.7	399	4.9	0.15
Acetylated	3.7	53	2.1	0.24

TABLE 11-16   Interfacial Shear Strength and Estimated Critical Aspect Ratio for Untreated and Treated Sisal–Starch/PCL Blend and Sisal–Starch/Cellulose Derivative Blend Composites

	Starch/Polycaprolactone		Starch/Cellulose Derivatives	
Sisal Fibers	τ (MPa)	$(l/d)_c$	τ (MPa)	$(l/d)_c$
Untreated	1.2 ± 0.6	79.2	3.3 ± 1.4	20.2
Alkali-treated	1.5 ± 0.6	95.0	4.4 ± 2.5	26.6
Acetylated	1.6 ± 0.4	13.3	3.7 ± 1.8	5.9

components. In order to improve this parameter, a number of fiber treatments can be carried out on the natural fibers to modify, not only the interphase but also the morphology of the fibers (Vázquez et al., 1999; Cyras et al., 2001; Rong et al., 2001; Plackett and Vázquez, 2004; Alvarez and vázquez, 2006b). Of these processes, the most used is the alkaline treatment (Ray and Sarkar, 2001) in which impurities such as waxes, pectins, hemicelluloses, and mineral salts are removed; the change from cellulose I to cellulose II takes place and the texture of the fibers, principally their accessibility in aqueous media, is modified, which also changes the morphology of natural fibers. The other commonly used method is termed acetylation, which makes the fiber surface more hydrophobic.

The equations for both treatments are summarized here:

Alkali treatment:

$$\text{Fiber—OH} + \text{NaOH} \longrightarrow \text{Fiber—O}^-\text{Na}^+ + \text{H}_2\text{O} \tag{11.2}$$

Acetylation:

$$\text{Fiber—OH} + \text{CH}_3\text{COOH} \xrightarrow[\text{H}^+]{(\text{CH}_3\text{CO})_2\text{O}} \text{Fiber—O—C}\!\!=\!\!\text{O(CH}_3) + \text{H}_2\text{O} \tag{11.3}$$

Previous treatments influence the dimensions of the natural fibers and also their mechanical properties. Table 11-15 shows the changes in the fiber dimensions, the mechanical properties, and interfacial adhesion due to chemical treatments.

The interfacial strength increases with alkali treatment, but in the case of acetylated fibers it decreases with treatment, showing the lower adhesion between fibers and matrix.

TABLE 11-17 Effect of Fiber Treatments on the Tensile Properties of Starch Based Blend/Natural Fiber Composites

Matrix	Reinforcement	σ (MPa) Untreated	σ (MPa) Treated	E (MPa) Untreated	E (MPa) Treated	Reference
MaterBi ZF03 (corn starch and a biodegradable polyester)	20 wt% sisal; 120°C; 110 rpm; 6 minutes					Ali et al. (2003)
	Untreated	10.39 ± 1.44		222.0 ± 24.2		
	Treated	13.03 ± 1.68		292.2 ± 31.8		
MaterBi ZF03 (corn starch and a biodegradable polyester)	10 wt% sisal	10.9 ± 0.9	6.4 ± 0.8	138 ± 13	112 ± 16	Cyras et al. (2006)
	20 wt% sisal	12.7 ± 0.5	8.8 ± 0.9	257 ± 17	221 ± 24	
	30 wt% sisal	14.4 ± 1.6	10.9 ± 2.1	410 ± 36	687 ± 119	
MaterBi YI01U (starch, cellulose derivatives, additives)	20 wt% sisal; 180°C; 20 rpm; 6 minutes					Ali et al. (2003)
	Untreated	18.60 ± 1.98		1183.3 ± 25.8		
	Treated	22.35 ± 0.21		1249.5 ± 13.6		
MaterBi YI01U (starch, cellulose derivatives, additives)	5 wt% sisal	14.4 ± 0.7	15.2 ± 1.0	1.39 ± 0.2	1.41 ± 0.2	Alvarez et al. (2003)
	10 wt% sisal	15.7 ± 0.9	16.8 ± 1.2	1.87 ± 0.2	1.91 ± 0.2	
	15 wt% sisal	16.8 ± 0.9	18.1 ± 0.8	2.20 ± 0.2	2.16 ± 0.2	

Note: For the last group (MaterBi YI01U), the E values are in GPa.

For alkaline-treated fibers, the critical aspect ratio becomes higher. On the other hand, acetylated fibers show the lowest value of the critical aspect ratio, but this may be due to the low strength of that fiber that appears in equation (11.1).

Ali et al. (2003) have studied the effect of fiber treatment on the tensile properties of MaterBi Y and MaterBi Z with 20 wt% of short sisal fibers. They showed that alkaline treatment favored an increased fiber aspect ratio and improved mechanical properties of the MaterBi Z and MaterBi Y/short sisal fiber composites.

Vázquez et al. (1999) and Cyras et al. (2001) observed that the alkaline treatment used on the sisal fibers produced fibrillation and collapse of the cellular structure due to the removal of the cementing material, which leads to a better packing of cellulose chains. The higher fiber density allowed them to increase the fiber content in the extruder, limited by the fiber volume.

For MaterBi Y/short sisal fiber composites Alvarez et al. (2003) found that the water-uptake rate decreased on alkaline treatment of the fibers due to the network formed by the high content of fibers, which impairs the diffusion of the water through the matrix. However, alkaline treatment makes the sisal fibers more hydrophilic and the composites absorb more water at equilibrium in comparison with untreated fiber composites. These authors also showed that the critical strain energy release rate G_{Ic} of the treated sisal fiber composite was higher than that of the untreated ones, probably due to the fibrillar morphology, which increases the toughening and the fiber energy dissipation mechanisms of the composite.

Other results are quite dissimilar; in the case of starch/PCL matrix the modulus clearly increased (up to 70% for 30 wt% of fibers) when fibers were alkali-treated, whereas the tensile strength can either increase or decrease with respect to the composites with untreated fibers. In the case of starch/cellulose derivatives, the change in the modulus of the composite induced by fiber treatment is almost negligible.

In addition, it must be taken into account that the fiber treatment combined with the processing technique can lead to damaged fibers with properties different from those of the treated fibers alone (measured by the single-fiber test).

Alvarez et al. (2004b) have also analyzed the effect of fiber treatment on the rheological behavior of MaterBi Y/short sisal fiber composites, showing that treated fiber composites exhibited higher viscosities than untreated ones. The fiber aspect ratio increased from 17.9 for untreated fibers to 19.5 for the treated ones, so the effects can be attributed to fiber fragmentation (Cyras et al., 2004). The results argue for more severe processing conditions in the case of treated fibers.

11.5.6   Aqueous and Soil-Burial Degradation (See Table 11-18)

Information on the environmental biodegradability composites with starch-based blends is still limited. Bastioli (1998) have reported that the presence of starch influences the biodegradation rate of the intrinsically biodegradable synthetic component during composting of MaterBi Y. On the other hand, the same material exposed to a respirometric test simulating soil burial conditions was only partially degraded up to about 18% (Solaro et al., 1998). The effect of natural fibers on the biodegradation of starch-based polymers and blends is still under study.

TABLE 11-18 Components of Starch Based Blend/Natural Fiber Composites Before and After Soil-Burial Degradation

Matrix	Reinforcement	Starch (%)	Other (%)	Additives (%)	Cellulose (%)	Reference
MaterBi-Z ZF03 (corn starch and a biodegradable polyester)	*Initial*					di Franco et al. (2004)
	None	16	75 (PCL)	9	–	
	15 wt% sisal	25 (starch + hemicellulose)	45	8	21	
	Final					
	None	7	93 (PCL)	0	–	
	15 wt% sisal			0		
MaterBi-Y Y101 (starch and cellulose derivatives)	*Initial*					Alvarez et al. (2006c)
	None	38	38 (cellulose derivatives)	22	–	
	15 wt% sisal	41 (starch + hemicellulose)	47 (cellulose + cellulose derivatives)	12	47 (cellulose + cellulose derivatives)	
	Final					
	None	2	98	0	–	
	15 wt% sisal	0	100 (cellulose + cellulose derivatives)	0	100 (cellulose + cellulose derivatives)	

di Franco et al. (2004) have also evaluated the susceptibility of the MaterBi Z/sisal fiber composites to different degrading environments. In hydrolytic tests, both the matrix and the composites displayed stability at pH 7.2 at 25°C and 40°C. In addition, sisal fibers support water access, which swells the material and produces hydrolysis of the starch (the most bioavailable component). When fiber content increased, the material became more hydrolytically stable, which was related to the presence of a fiber–fiber physical network. Microbial attack was evidence by the presence of a biofilm, particularly on the fiber surface. In soil burial, the matrix was degraded to about 50%, the weight loss pattern of the composite being associated with the presence of strong fiber–fiber and fiber–matrix interactions.

When Alvarez et al. (2006c) studied the degradation in soil of MaterBi Y/short sisal fiber composites, they observed that water sorption was predominant during the first month, followed by weight loss; the composites absorbed less water than the pure MaterBi-Y due to the fiber–fiber and fiber–matrix interactions. Also, the amorphous nature of the matrix favored the preferential abstraction of starch, as was determined by thermogravimetry. Only an insignificant difference in weight loss between the matrix and the composites were observed. The authors also analyzed the decrease in mechanical properties as a function of the exposure time.

The same authors have also studied the effect of water on the mechanical behavior of such materials (Alvarez et al., 2004c, d, 2007), determining that the rate of water absorption (measured by the diffusion coefficient) increased with temperature; the water acts as plasticizer and increases flexibility, and the flexural modulus was related to the water content. The decrease of modulus was slightly greater with higher fiber content, but the effect of temperature was more pronounced.

The results obtained for MaterBi Y (cellulose derivative/starch blend) composites differ from those for composites based on MaterBi Z (polycaprolactone/starch blend) where fibers were found to promote the biodegradation and then the preferential removal of starch. MaterBi Z is a semicrystalline blend ($T_m = 60°C$), whereas MaterBi Y is amorphous. Crystalline regions are more difficult to degrade and may act as preference points for microorganism intake. Fibers may act as channels and facilitate microbial ingress in natural-fiber semicrystalline matrix composites. In the case of MaterBi Y, the amorphous structure favors microbial access to the matrix (mainly to the destructurized starch) and fibers have a minor role, as can be concluded from the slight difference in weight loss suffered by the matrix and the composites.

11.6   CONCLUSIONS

Starch is one of the most abundant natural polymers and can be obtained from various botanical sources. However, its drawbacks are water absorption and mechanical instability with humidity, and fragility when dry. One of the most common uses is in packaging films and the possibility blending with other biodegradable polymers with better and more stable mechanical properties has been studied and is currently studied. Another possibility for improving the final properties is to mix starch with natural fibers or nanocellulose fibers. The potential improvement in mechanical properties is greater with nanocellulose, due to the high mechanical properties of the nanocellulose fiber, the higher aspect ratio, and better compatibility between starch and cellulose. However, the cost of the nanofibers will limit the application of these nanofillers. Natural fibers have the problem that they have lower mechanical properties, but they could be used for common applications. A future direction of study to reveal about the final material is the use of nanocellulose fibers as reinforcement and their distribution in the composite. Another important area of study needed on natural fibers is the modification of starch and the fibers themselves in order to produce a more water-resistant material. Time-dependent properties (creep, fatigue) also need to be studied, and nondestructive methods to evaluate the behavior of the composites need to be developed. There is some work in the literature on starch foams and reinforced foam, and this is an interesting application of the material; this subject was not included here for reasons of space restrictions. The use of starch and cellulose in multilayers is another interesting area, which is treated in other chapter.

ACKNOWLEDGMENTS

The authors gratefully acknowledge CONICET (Consejo Nacional de Investigaciones Científicas y Ténicas) and Ministerio de Ciencia y Technología for their financial support and SECYT.

REFERENCES

Ali R, Iannace S, Nicolais L. 2003. Effect of processing conditions on mechanical and viscoelastic properties of biocomposites. *Journal of Applied Polymer Science* 88:1637–1642.

Alvarez VA, Ruseckaite RA, Vázquez A. 2003. Mechanical properties and water absorption of MaterBi-Y sisal fibres composites: Effect of alkaline treatment. *Journal of Composite Materials* 37(17):1575–1588.

Alvarez VA, Kenny JM, Vázquez A. 2004a. Creep behaviour of biocomposites based on sisal fibre reinforced MaterBi-Y. *Polymer Composites* 25(3):280–288.

Alvarez VA, Terenzi A, Kenny JM, Vázquez A. 2004b. Melt rheological behaviour of starch based matrix composites reinforced with short sisal fibres. *Polymer Engineering and Science* 44(10):1907–1914.

Alvarez VA, Fraga A, Vázquez A. 2004c. Effect of the moisture and fibre content on the mechanical properties of biodegradable polymer and sisal fibre biocomposites. *Journal of Applied Polymer Science* 91(6):4007–4016.

Alvarez VA, Vázquez A. 2004d. Effect of water sorption on the flexural properties of a fully biodegradable composite. *Journal of Composite Materials* 38(3):1165–1182.

Alvarez VA, Vázquez A. 2004e. Thermal degradation of cellulose derivatives/starch blends and sisal short fibres composites. *Polymer Degradation and Stability* 84:13–21.

Alvarez VA, Vázquez A, Bernal CR. 2005a. Fracture behaviour of sisal fibre reinforced starch based composites. *Polymer Composites* 26(3):316–323.

Alvarez VA, Ianonni A, Kenny JM, Vázquez A. 2005b. Influence of the twin-screw processing conditions on the mechanical properties of biocomposites. *Journal of Composite Materials* 39:2023–2038.

Alvarez VA, Vázquez A, Bernal CR. 2006a. Effect of microstructure on the tensile and fracture properties of sisal fibre/starch based composites. *Journal of Composites Materials* 40(1):21–35.

Alvarez VA, Vázquez A. 2006b. Influence of fibre chemical modification procedure on the mechanical properties of MaterBi-Y/sisal fibre composites. *Composite Part A* 37:1672–1680.

Alvarez VA, Ruscekaite RA, Vázquez A. 2006c. Degradation of sisal fibre/Mater Bi-Y biocomposites buried in soil. *Polymer Degradation and Stability* 91(12):3156–3162.

Alvarez VA, Ruseckaite RA, Vázquez A. 2007. Aqueous degradation of Mater Bi Y–sisal fibre biocomposites. *Journal of Thermoplastic Composite Materials* 20(5):291–303.

Amass W, Amass A, Tighe B. 1998. A review of biodegradable polymers: uses, current developments in the synthesis and characterization of biodegradable polyester, blends of biodegradable polymers and recent advances in biodegradation studies. *Polymer International* 47:89–144.

Anglès MN, Dufresne A. 2001. Plasticized starch/tunicin whiskers nanocomposite materials. 2: Mechanical behaviour. *Macromolecules* 34(9):2921–2931.

Avérous L, Fringant C. 2001. Association between plasticized starch and polyesters: processing and performances of injected biodegradable systems. *Polymer Engineering and Science* 41(5):727–734.

Averous L, Fauconnier N, Moro L, Fringant C. 2000. Blends of thermoplastic starch and polyesteramide: Processing and properties. *Journal of Applied Polymer Science* 76(7):1117–1128.

Averous L, Fringant C, Moro L. 2001. Plasticized starch-cellulose interactions in polysaccharide composites. *Polymer* 42:6565–6572.

Averous L. 2004. Biodegradable multiphase systems based on plasticized starch: A review. *Journal of Macromolecular Science Part C – Polymer Reviews* C44(3):231–274.

Bastioli C. 1998. Properties and applications of MaterBi starch-based materials. *Polymer Degradation and Stability* 59:263–272.

Biastioli C. 1995. Starch-polymer composites. In: Scott G, Gilead D, editors. *Degradable Polymers: Principles and Applications*. 2nd edn. London: Chapman & Hall.

Bastioli C. 2001. Global status of the production of biobased packaging materials. *Starch/Staerke* 53(8):351–355.

Batra SK, 1981. Other long vegetable fibres. In: Lewin M, Pearce EM. editors. *Handbook of Fibre Science and Technology*. New York: Marcel Dekker. p. 727–808.

Bastioli C, Cerrutti A, Guanella I, Romano GC, Tosin M. 1995. Physical state and biodegradation behaviour of starch-polycaprolactone systems. *Journal of Environmental Polymer Degradation* 3(2):81–95.

Bledzki AK, Gassan J. 1997. Natural fiber reinforced plastics. In: Cheremisinoff NP, editor. *Handbook of Engineering Polymeric Materials*. New York: Marcel Dekker.

Bledzki AK, Gassan J. 1999. Composites reinforced with cellulose based fibres. *Progress in Polymer Science (Oxford)* 24(2):221–274.

Bledzki AK, Reihmane S, Gassan J. 1996. Properties and modification methods for vegetable fibres for natural fibre composites. *Journal of Applied Polymer Science* 29:1329–1336.

Carvalho AJF, Zambon MD, Curvelo AAS, Gandini A. 2003. Size exclusion chromatography characterization of thermoplastic starch composites. 1. Influence of plasticizer and fibre content. *Polymer Degradation and Stability* 79:133–138.

Chand N, Hashmi SAR. 1993. *Metals Materials and Processes* 5:51.

Chand N, Rohatgi, PK. 1994. Natural fibres and their composites. Periodical Experts Publishers: Delhi. p. 55.

Chang YP, Abd Karim A, Scow CC. 2006. Interactive plasticizing-antiplasticizing effects of water and glycerol on the tensile properties of tapioca starch films. *Food Hydrocolloids* 20:1–8.

Chang YP, Cheah PB, Seow CC. 2000. Plasticizing and antiplasticising effects of water on physical properties of tapioca starch films in the glassy state. *Journal of Food Science* 65(3):445–451.

Chawa KK, Bastos AC. 1979. Mechanical properties of jute fibres and polyester-jute composites. *International Conference on the Mechanical Behavior of Materials* 3:191.

Chiellini E, Solaro R. 1996. Biodegradable polymeric materials. *Advanced Materials* 8(4):305–313.

Chiellini E, Corti A, D'Antone S, Solaro R. 2003. Biodegradation of poly(vinyl alcohol) based materials. *Progress in Polymer Science* 28(6):963–1014.

Cunha AM, Liu ZQ, Feng Y, Yi X-S, Bernardo CA. 2004. Preparation, processing and characterization of biodegradable wood flour/starch-cellulose acetate compounds. *Journal of Materials Science* 36(20):4903–4909.

Curvelo AAS, de Carvalho AJF, Agnelli JAM. 2001. Thermoplastic starch/cellulosic fibres composites: preliminary results. *Carbohydrate Polymers* 45:183–188.

Cyras VP, Iannace S, Kenny JM, Vázquez A. 2001. Relationship between processing and properties of biodegradable composites based on PCL/starch matrix and sisal fibres. *Polymer Composites* 22(1):104–110.

Cyras VP, Martucci JF, Iannace S, Vázquez A. 2002. Influence of the fibre content and the processing conditions on the flexural creep behaviour of sisal–PCL–starch composites. *Journal of Thermoplastic Composite Materials* 15:253–265.

Cyras VP, Vallo C, Kenny JM, Vázquez A. 2004. Effect of chemical treatment on the mechanical properties of starch-based blends reinforced with sisal fibre. *Journal of Composite Materials* 38(16):1387–1399.

Cyras VP, Zenklusen MCT, Vazquez A. 2006. Relationship between structure and properties of modified potato starch biodegradable films. *Journal of Applied Polymer Science* 101(6):4313–4319.

Da Róz AL, Carvalho AFJ, Gandini A, Curvelo AAS. 2006. The effect of plasticizers on thermoplastic starch compositions obtained by melt processing. *Carbohydrate Polymers* 63:417–424.

di Franco CR, Cyras VP, Busalmen JP, Ruseckaite RA, Vázquez A. 2004. Degradation of polycaprolactone/starch blendsand composites with sisal fibre. *Polymer Degradation and Stability* 86:95–103.

Duanmu J, Gamstedt EK, Rosling A. 2007. Hygromechanical properties of composites of crosslinked allylglycidyl-ether modified starch reinforced by wood fibres. *Composites Science and Technology* 67(15–16):3090–3097.

Dufresne A, Vignon MR. 1998. Improvement of Starch film performances using cellulose microfibrils. *Macromolecules* 31(8):2693–2696.

Forssell P, Mikkilä J, Moates G, Parker R. 1997. Phase and glass transition behaviour of concentrated barley starch-glycerol-water mixtures, a model for thermoplastic starch. *Carbohydrate Polymers* 34:275–282.

Freitas RA, Paula RC, Fitosa JPA, Rocha S, Sierakowski MR. 2004. Amylose contents, rheological properties and gelatinization kinetics of yam (*Dioscorea alata*) and cassava (*Manihot utilissima*) starches. *Carbohydrate Polymers* 55:3–8.

Fujiyama M, Kawasaki Y. 1991. Rheological properties of polypropylene/high-density polyethylene blend melts. II. Dynamic viscoelastic properties. *Journal of Applied Polymer Science* 42(2):481–488.

Funke U, Bergthaller W, Lindhauer MG. 1998. Processing and characterization of biodegradable product based on starch. *Polymer Degradation and Stability* 59:293–296.

Gáspár M, Benkó Zs, Dogossy G, Réczey K, Cigány T. 2005. Reducing water absorption in compostable starch-based plastics. *Polymer Degradation and Stability* 90:563–569.

Gómez C, Torres FG, Nakamatsu J, Arroyo OH. 2006. Thermal and structural analysis of natural fibre reinforced starch-based biocomposites. *International Journal of Polymeric Materials* 55(11):893–907.

Gidley MJ. 1989. Molecular mechanism underlying amylase aggregation and gelation. *Macromolecules* 22:351–358.

Gudmundsson R. 1994. Retrogradation of starch and the role of its components. *Thermochimica Acta* 246:329–41.

Hanselka H, Herrman AS. 1994. Fibre properties and characterization. In: Internationales Techtexil Symposium. Frankfurt: Germany. p. 20–22.

Hon DN-S. 1994. Degradative effects of ultraviolet light and acid rain on wood surface quality. *Wood and Fiber Science* 26(2):185–191.

Hoover R. 2001. Composition, molecular structure, and physicochemical properties of tuber and root starches: a review. *Carbohydrate Polymers* 45(3):253–267.

Hsueh H. 2000. Young's modulus of unidirectional discontinuous-fibre composites. *Composites Science and Technology* 60:2671–2680.

Huang SJ, Koening MF, Huang M. 1993. Design, synthesis, and properties of biodegradable composites. In: Ching C, Kaplan DL, Thomas EL, editors. *Biodegradable Polymers and Packaging*. Basel: Technomic. p. 97–110.

Iannace S, Nocilla L, Nicolais L. 1999. Biocomposites based on sea algae fibres and biodegradable thermoplastic matrices. *Journal of Applied Polymer Science* 73:583–592.

Jayamol G, Janardhan R, Anand JS, Bhagawan SS, Sabu T. 1996. Melt rheological behaviour of short pineapple fibre reinforced low density polyethylene composites. *Polymer* 37(24):5421–5431.

Jiménez-Hernández J, Salazar-Montoya JA, Ramos-Ramírez EG. 2007. Physical, chemical and microscopic characterization of a new starch from chayote tuber (*Sechium edule*) and its comparison with potato and maize starches. *Carbohydrate Polymers* 68:679–686.

Johnson RM, Tucker N, Barnes S. 2003. Impact properties of Miscanthus/Novamont MaterBI biocomposites. *Polymer Testing* 22:209–215.

Joseph K, Thomas S, Pavithran C, Brahmakumar M. 1993. Tensile properties of short sisal fibre-reinforced polyethylene composites. *Journal of Applied Polymer Science* 47:1731–1739.

Joseph PV, Joseph K, Thomas S. 1999. Effect of processing variables on the mechanical properties of sisal-fibre-reinforced polypropylene composites. *Composites Science and Technology* 59:1625–1640.

Kaplan DJ, Mayer JM, Ball D, McMassie J, Allen AL, Stenhouse P. 1993. Fundamentals of biodegradable polymers chap 1.In: Ching C, Kaplan DL, editors. *Biodegradable Polymers and Packaging*. Basel: Technomic. 1–42.

Katz JR. 1930. Adhandlungen zur physikalschen Chemie der Starke und der Brotbereitung. *Zeitschrift für Physikalische Chemie* 150:37–59.

Ke T, Sun X. 2000. Physical properties of poly(lactic-acid) and starch. Composites with various blending ratios. *Cereal Chemistry* 77(6):761–768.

Koening MF, Huang SI. 1995. Biodegradable blends and composites of polycaprolactone and starch derivatives. *Polymer* 36(9):1877–1882.

Kotnis MA, O'Brien GS, Willett JL. 1995. Processing and mechanical properties of biodegradable Poly(hydroxybutyrate-co-valerate)-starch compositions. *Journal of Environmental Polymer Degradation* 3(2):97–105.

Kuruvilla J, Mattoso LHC. 1998. Sisal fibre reinforced polymer composites: status and future. In *Natural Polymers and Composites*.p. 333.

Lörcks J. 1998. Properties and applications of compostable starch-based plastic material. *Polymer Degradation and Stability* 59:245–249.

Lanzillota C, Pipino A, Lips D. 2002. New functional biopolymer natural fiber composites from agricultural resources. In: *Proceedings of the Annual Technical Conference – Society of Plastics Engineers*, San Francisco, California,Vol. 2. p. 2185–2189.

Lelieve J, Mitchell J. 1975. A pulsed NMR study of some aspects of starch gelatinization. *Stärke* 27:113–115.

Lourdin D, Coignard L, Bizot H, Colonna P. 1997. Influence of equilibrium relative humidity and plasticizer concentration on the water content and glass transition of starch materials. *Polymer* 38:5401–5406.

Lu Y, Weng L, Cao, X. 2006. Morphological, thermal and mechanical properties of ramie crystallites—reinforced plasticized starch biocomposites. *Carbohydrate Polymers* 63(2):198–204.

Mali S, Karam LB, Pereira Ramos L, Grossman MVE. 2004. Relationships among the composition and physicochemical properties of starches with the characteristics of their films. *Journal of Agricultural and Food Chemistry* 52:7720–7725.

Mali S, Grossman MVE, García MA, Martino MN, Zaritzky NE. 2006. Effect of controlled storage on thermal, mechanical and barrier properties of plasticized films form different starch sources. *Journal of Food Engineering* 75:453–460.

Marsh RDL, Blanshard JMV. 1988. The application of polymer crystal growth theory to kinetics of formation of the β-amylose polymorph in a 50% wheat-starch gel. *Carbohydrate Polymers* 27:261–270.

Mathew AP, Dufresne A. 2002. Morphological investigation of nanocomposites from sorbitol plasticized starch and tunicin whiskers. *Biomacromolecules* 3(3):609–617.

Mc. Govern JN. 1987. "Other Fibers" Chapter 9 of Pulp and Paper Manufacture. In: Hamilton F, Leopold B, Kocurek MI, editors. *Secondary Fibers and Non-Wood Pulping*, 3rd ed. TAPPI, Atlanta: GA. Vol. 3. p. 110–121.

Miles MJ, Morris VJ, Ring SG. 1985. Gelation of amylase. *Carbohydrate Research* 135:257–269.

Mouzakis DE, Harmia T, Karger-Kocsis J. 2000. Fracture behaviour of discontinuous long glass fibre reinforced injection molded polypropylene. *Polymers and Polymer Composites* 8:167–175.

Myllärinen P, Partanen R, Seppala J, Forsell P. 2002. Effect of glycerol on behaviour of amylose and amylopectin films. *Carbohydrate Polymers* 50:355–361.

Myllymäki O, Myllärinen P, Forssell P, et al. 1998. Mechanical and permeability properties of biodegradable extruded starch/polycaprolactone films. *Packaging Technology and Science* 11:265–274.

Park B-D, Balatinecz JJ. 1999. Short term flexural creep behaviour of wood-fibre/polypropylene composites. *Polymer Composites* 19:377–382.

Plackett D, Vázquez A. 2004. Biopolymers and biocomposites. In: Baille C, editor. *Green Composites: Polymer Composites and the Environment.* Cambridge: Woodhead Publishers. Chapter 5. ISBN 1 85573 739 6.

Pranamuda H, Tokiwa J, Tanaka H. 1996. Physical properties and biodegradability of blends containing poly(ε-caprolactone) and tropical starches. *Journal of Environmental Polymer Degradation* 4:1–7.

Röper H, Koch H. 1990. The role of starch in biodegradable thermoplastic materials. *Starch* 42:123–130.

Ramsay BA, Langlade V, Carreau PJ, Ramsay JA. 1993. Biodegradability and mechanical properties of poly(hydroxyubutyrate-co-hydroxyvalerate)-starch blends. *Applied Environmental Microbiology* 59:1242–1246.

Ray D, Sarkar BK. 2001. Characterization of alkali-treated jute fibres for physical and mechanical properties. *Journal of Applied Polymer Science* 80:1013–1020.

Ray PK, Chakravarty AC, Bandyopadhay SB. 1976. Fine structure and mechanical properties of jute differently dried after retting. *Journal of Applied Polymer Science* 20:1765–1766.

Rindlav-Westling A, Stading M, Hormansson A-M, Gatenholm P. 1998. Structure and mechanical properties of amylase and amylopectin films. *Carbohydrate Polymers* 36(2–3):217–224.

Rindlay A, Stading M, Hermansson A, Gatenholm P. 1998. Structure, mechanical and barrier properties of amylase and amylopectin films. *Carbohydrate Polymers* 36:217–224.

Romhány G, Karger-Kocsis J, Czigány T. 2003. Tensile fracture and failure behavior of thermoplastic starch with unidirectional and cross-ply flax fiber reinforcements. *Macromolecular Materials and Engineering* 288(9):699–707.

Romhány G, Karger-Kocsis J, Czigány T. 2003a. Tensile fracture and failure behaviour of technical flax. *Journal of Applied Polymer Science* 90(3):3638–3645.

Romhány G, Karger-Kocsis J, Czigány T. 2003b. Tensile fracture and failure behaviour of thermoplastic starch with unidirectional and cross-ply flax fibre reinforcements. *Macromolecular Engineering* 288(9):699–707.

Rong MZ, Zhang MQ, Lui Y, Yang GC, Zeng HM. 2001. The effect of fibre treatment on the mechanical properties of unidirectional sisal reinforced epoxy composites. *Composites Science and Technology* 61:1437–1447.

Rowell RM. 1992. Opportunities for lignocellulosic materials and composites. In: *Materials and Chemicals from Biomass.* American Chemical Society, Chapter 2. p. 12–27.

Rozman HD, Peng GB, Mohd Ishak ZA. 1998. The effect of compounding techniques on the mechanical properties of oil palm empty fruit bunch-polypropylene composites *Journal of Applied Polymer Science* 70:2647–2655.

Saechtling H. 1987. Starch-polymer composites (Chapter 6). In: Munich, editor. *International Plastics Handbook for the Technologist, Engineer and User.* 2nd ed. Hansen. p. 12.

Sagar AD, Merril EW. 1995. Starch fragmentation during extrusion processing. *Polymer* 36(9):1883–1995.

Sears JK, Darby JR. 1982. *The Technology of Plasticizers.* New York: Wiley.

Shibata S, Cao Y, Fukumoto I. 2005. Press forming of short natural fibre-reinforced biodegradable resin: Effects of fibre volume and length on flexural properties. *Polymer Testing* 24:1005–1011.

Shogren RL, Swason CL, Thompson AR. 1992. Extrudates of cornstarch with urea and glycols: Structure/mechanical properties relations. *Starch* 44:335–338.

Singh N, Singh J, Lovedeep Kaur L, Singh Sodhi N, Singh Gill B. 2003. Morphological, thermal and rheological properties of starches from different botanical sources. *Food Chemistry* 81(2):219–231.

Solaro R, Corti A, Chiellini E. 1998. A new respirometric test simulating soil burial conditions for the evaluation of polymer biodegradation. *Journal of Environmental Polymer Degradation* 4:203–208.

Stevens DJ, Elton GA. 1981. Thermal properties of the starch/water system. I Measurement of heat of gelatinization by differential scanning calorimetry. *Starke* 23:8–11.

Stevenson DG, Jane JL, Inglett GE. 2007. Structure and physicochemical properties of starch from immature seeds of soybean varieties (glycine max) exhibiting normal, low-linoleic or low-saturated fatty acid oil profiles at maturity. *Carbohydrate Polymers* 70:149–159.

Takagi H, Asano A. 2008. Effect of processing conditions on flexural properties of cellulose nanofiber reinforced "green" composites. *Composite Part A* 39(4):685–689.

Thranathna RN. 2003. Biodegradable films and composite coatings: past, present and future. *Trends in Food Science and Technology* 14:71–78.

Thunwall M, Boldizar A, Rigdah M. 2006. Compression molding and tensile properties of thermoplastic potato starch materials. *Biomacromolecules* 7:981–986.

Torres FG, Arroyo OH, Grande C, Esparza E. 2006. Bio-Photo-degradation of natural fibre reinforced starch-based biocomposites. *International Journal of Polymeric Materials* 55(12):1115–1132.

Ugbolue SCO. 1990. Structure-property relationships in textile fibres. *Text Inst* 20(4):1–12.

Vázquez A, Domínguez V, Kenny JM. 1999. Bagasse-fibre-polypropylene based composites. *Journal of Thermal Composite Materials* 12:477–497.

van Soest JJG, Borger DB. 2006. Structure and properties of compression-molded thermoplastic starch materials from normal and high-amylose maize starches. *Journal of Applied Polymer Science* 64(4):631–644.

van Soest JJG, De Wit D, Vliegenthart JFG. 1996a. Mechanical properties of thermoplastic waxy maize starch. *Journal of Applied Polymer Science* 61(11):1927–1937.

van Soest JJG, Hulleman SHD, Vliegenthart JFG. 1996b. Crystallinity in starch bioplastics. *Industrial Crops and Products* 5:11–22.

van Soest JJG, Vliegenthart JFG. 1997. Crystallinity in starch plastics: consequence for material properties. *TIBTECH* 15:208–213.

van Soest JJG. 1995. In: van Soest JJG, editor. *Starch Plastics: Structure–Properties Relationship.* p. 1–168.

Verhoogt H, St-Pierre N, Truchon FS, Ramsay BA, Favis BD, Ramsay JA. 1995. Blends containing poly(hydroxybutyrate-co-12%-hydroxyvalerate) and thermoplastic starch. *Canadian Journal of Microbiology* 41(1):323–328.

Vilpoux O, Averous L. 2004. Starch-based plastics. In: Cereda MP, editor. *Technology, Use and Potentialities of Latin American Starchy Tubers.* Chapter 18. p. 551–553.

Wang K, Lee LJJ. 1987. Rheological and extrusion behaviour of dispersed multiphase polymeric systems. *Journal of Applied Polymer Science* 33:431–453.

Warburton SD, Donald AM, Smith AC. 1993. The deformation of thin films made from extruded starch. *Carbohydrate Polymers* 21(1):17–21.

Willett JL, Felker FC. 2005. Tensile yield properties of starch filled poly(ester amide) materials. *Polymer* 46:3035–3042.

Willett JL, Kotnis MA, O'Brien GS, Fanta GF, Gordon SH. 1998. Properties of starch-graft-poly(glycidyl methacrylate)-PHBV composites. *Journal of Applied Polymer Science* 70:1121–1127.

Wollerdorfer M, Bader H. 1998. Influence of natural fibres on the mechanical properties of biodegradable polymers. *Industrial Crops and Products* 8:105–112.

Yongshang Lu, Lihui Weng, Xiaodong Cao. 2006. Morphological, thermal and mechanical properties of ramie crystallites-reinforced plasticized starch biocomposites. *Carbohydrate Polymers* 198–204.

Zhou Y, Hoover R, Liu Q. 2004. Relationship between amylase degradation and the structure and physicochemical properties of legume starches. *Carbohydrate Polymers* 57:200–317.

CHAPTER 12

Poly(Lactic Acid)/Cellulosic Fiber Composites

MITSUHIRO SHIBATA

Department of Life and Environmental Sciences, Faculty of Engineering, Chiba Institute of Technology, Narashino, Chiba 275-0016, Japan

12.1   INTRODUCTION

Biocomposites composed of polymers from renewable resources (PFRR) and cellulosic fibers from plants have been gathering much attention from the standpoint of protection of the natural environment and saving of petroleum resources (Bledzki and Gassan, 1999; Saheb and Jog, 1999; Mohanty et al., 2000, 2002; Wool and Sun, 2005; Yu et al., 2006). Poly(lactic acid) (PLA) is one of the most promising PFRR, because it is derived from abundant agricultural products such as corn, sugar cane, and sugar beat via fermentation and chemical processes, and can be used as a structural material with sufficient lifetime to maintain mechanical properties without rapid hydrolysis even under humid conditions, as well as showing good compostability (Garlotta, 2001; Warmington, 2001; Inoue, 2003; Ohara, 2003; Sawyer, 2003; Scott and Sissell, 2003; Vink et al., 2003). Lignocellulosic natural fibers are grouped into leaf, bast, leafstalk, stalk, seed, and fruit origins. Well-known examples include (i) *Leaf:* sisal, pineapple leaf fiber, and henequene; (ii) *Bast:* flax, ramie, jute,

---

Biodegradable Polymer Blends and Composites from Renewable Resources. Edited by Long Yu
Copyright © 2009 John Wiley & Sons, Inc.

hemp, and kenaf; (iii) *Leafstalk:* abaca (manila hemp) and banana; (iv) *Stalk:* bamboo and wood fiber; (iv) *Seed:* cotton and kapok; and (v) *Fruit:* coconut (coir) (Bledzki and Gassan, 1999; Mohanty et al., 2000). Most of them have until now been investigated as reinforcing materials for PLA. As examples, PLA composites using flax (Oksman et al., 2003; Wong et al., 2003, 2004; Shanks, 2006a, b), jute (Plackett et al., 2003), kenaf (Nishino et al., 2003; Serizawa et al., 2006), abaca (Shibata et al., 2003; Teramoto et al., 2004), bamboo (Lee and Ohkita, 2004; Lee and Wang, 2006), and wood flour (Plackett, 2004; Mathew et al., 2005, 2006; Gamstedt et al., 2006; Yu et al., 2007) as cellulosic natural fibers have been reported by several groups for improvement of mechanical properties. As a whole, it is relatively easy to improve the rigidity of PLA by natural fiber reinforcement. However, the improvement of mechanical properties related to toughness, such as tensile and flexural strength and impact strength, is very difficult. Recently, it has been reported that the impact strength of PLA composites is improved by use of kenaf fibers form which fine particles have been eliminated. The PLA/kenaf composite can be used as casing materials for electronic products such as mobile phones (Serizawa et al., 2006). Recently, wood pulp, microcrystalline cellulose, and cellulose whisker, which are derived from plant-based lignocellulose fibers (Mathew et al., 2005, 2006; Hou et al., 2006; Oksman et al., 2006; Petersson and Oksman, 2006), and rayon (Fink and Ganster, 2006) and lyocell (Shibata et al., 2004), which are known as cellulosic man-made fibers, have also been investigated as reinforcing materials for PLA. This section describes the improvement of rigidity by reinforcement of PLA with short abaca fiber and wood flour, and the improvement of toughness by the reinforcement of PLA with lyocell fabric.

12.2 PLA/ABACA COMPOSITES

Abaca fiber is produced from leafstalk of a banana-shaped plant, *Musa textiles* Née (Manila hemp), which is mainly supplied from Philippines. Abaca fiber has relatively high tensile strength (810 MPa) and modulus (34 MPa) among various natural fibers (Table 12-1) (Shibata et al., 2002, 2003). The abaca fiber was used as natural fiber reinforcing PLA (Shibata et al., 2003). Because strongly polarized lignocellulosic fibers are inherently incompatible with hydrophobic polymers (Luo and Netravali, 1999a, b), the esterification of abaca fiber with acetic anhydride (AA-abaca) or butyric anhydride (BA-abaca), alkali treatment (Alk-abaca), and cyanoethylation (AN-abaca) with acrylonitrile were carried out as shown in Fig. 12-1 (Shibata et al., 2002).

The reactions were confirmed by FT-IR spectral analysis of the modified fibers. The AA-, BA-, and Alk-abaca fibers exhibited slightly lower tensile strength and slightly higher tensile modulus than untreated abaca fiber (Shibata et al., 2003). The tensile strength and modulus of AN-abaca were considerably lower than those of the untreated abaca fiber. The untreated and modified abaca fibers chopped into ~5 mm length (fiber content: 0, 5, 10, 15, 20 wt%) and PLA were mixed at 190°C and subsequently injection-molded to give PLA/abaca composites.

TABLE 12-1   Mechanical Properties of Natural Fibers[a]

Fiber	Tensile Strength (MPa)	Tensile Modulus (GPa)	Elongation at Break (%)	Density (g/cm³)
Flax	345–1100	27.6	2.7–3.2	1.50
Jute	393–773	13–26.5	1.2–1.5	1.3–1.45
Ramie	400–938	61.4–128	1.2–3.8	1.45
Sisal	468–640	9.4–22.0	3–7	1.5
Abaca	756–813	31.1–33.6	2.9	1.5
Cotton	287–800	5.5–12.6	7.0–8.0	1.5–1.6
Coir	131–175	4–6	15–40	1.2
Lyocell	450–630	16–18	4–10	1.5
E-glass	2000–3500	70	2.5	2.5

[a]Data for fibers except for abaca and lyocell are cited from Mohanty et al. (2000). Data for abaca are cited from Shibata et al. (2002, 2003). Data for lyocell are cited from Mieck et al. (2002).

Flexural properties of the PLA/abaca composites as a function of fiber content are shown in Fig. 12-2. Flexural moduli of the PLA composites increased with fiber content. Although the PLA composites with BA-abaca or AA-abaca showed slightly higher modulus, the influence of fiber treatment was not so large. On the other hand, flexural strength did not increase regardless of the fiber treatment in the fiber content range below 20 wt%.

It is generally considered that the biodegradation of PLA in soil at room temperature takes longer time than that of other biodegradable aliphatic polyesters such as poly(3-hydroxybutyrate-*co*-3-hydroxyvalerate) (PHBV), poly(ε-caprolactone) (PCL), and poly(butylene succinate) (PBS). Control PLA and PLA/AA-abaca composite with fiber content of 10 wt% showed no weight loss after burial for 6 months in 1 : 1 mixture of black soil and leaf mold for gardening (Teramoto et al., 2004). It is thought that the penetration of water or microorganism through the fiber–matrix interface is restricted for PLA/AA-abaca composite because the interfacial adhesion is improved due to the surface modification of the fiber. On the other hand, $\sim$10% of the original weight of PLA/untreated abaca composite with fiber content of 10 wt%

Fiber-OH + (RCO)$_2$O $\longrightarrow$ Fiber-OCOR + RCOOH

AA-abaca (R = CH$_3$)
BA-abaca (R = CH$_2$CH$_2$CH$_3$)

Fiber-OH + NaOH $\rightleftharpoons$ Fiber-O$^-$ Na$^+$ + H$_2$O

Alk-abaca

Fiber-OH + CH$_2$=CHCN $\xrightarrow{\text{NaOH}}$ Fiber-OCH$_2$CH$_2$CN

AN-abaca

Fig. 12-1 Surface modifications of abaca fiber.

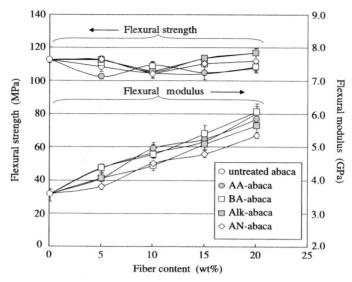

Fig. 12-2 Flexural properties of PLA composites as a function of fiber content.

was lost at 60 days; the weight loss did not increase thereafter. The slowdown of the weight loss around 10% is presumably related to the preferential decomposition of the fiber in the early stage and a much slower biodegradability of the matrix PLA in the next stage. The buried specimen of PLA/untreated abaca was very fragile and some parts were broken in the washing process. The PLA near the fiber–matrix interface of PLA/untreated abaca composite may be somewhat degraded by the action of water absorbed by the fiber. However, the surface of the matrix PLA of PLA/untreated abaca and PLA/AA-abaca composites appeared unchanged after the burial test. For PLA/untreated abaca composite, some cracks around the interface were observed. The cracks are thought to be formed by interfacial delamination and the shrinkage of matrix PLA due to the crystallization of PLA. The degradation of abaca fiber in the composite may occur through the cracks. Such cracks were not observed in PLA/AA-abaca composite. This result is thought to be related to the improvement of interfacial adhesion due to the surface modification of abaca fiber.

12.3 PLA/WOOD FLOUR COMPOSITES

Wood is an abundant and cheap natural resource, composed of cellulose, lignin, and hemicellulose. For composites of PLA and chopped abaca fibers of ~5 mm, the preparation of PLA/abaca composites with fiber content more than 25 wt% by injection molding was difficult because the injection molding gate becomes stopped up. In order to prepare PLA composites with higher fiber content by injection molding, wood flour (WF) which is much finer than the chopped abaca fiber, was used.

Wood pulverized using a grinder was passed through successive sieves of different mesh sizes. SEM images of the separated WFs are shown in Fig. 12-3. The aspect ratio of all the WF particles was ~5. Because the WF of 635 mesh contains all the particles passing through a 635 mesh, some fine particles appeared. Figs. 12-4 and 12-5 show the tensile and flexural properties of PLA/WF composites.

The tensile and flexural moduli increased with increasing WF content. The composites with the WF of 635 mesh showed slightly lower modulus because of the

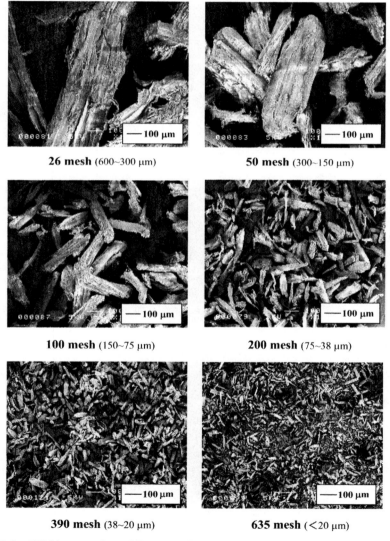

26 mesh (600~300 μm)   **50 mesh** (300~150 μm)

100 mesh (150~75 μm)   **200 mesh** (75~38 μm)

390 mesh (38~20 μm)   **635 mesh** (<20 μm)

Fig. 12-3   SEM images of wood flour particles separated with sieves of different mesh sizes.

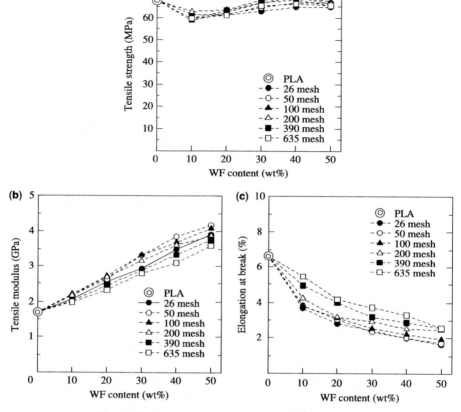

Fig. 12-4 Tensile properties of PLA/WF composites.

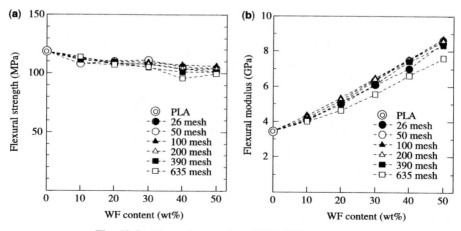

Fig. 12-5 Flexural properties of PLA/WF composites.

presence of the fine particles. The tensile strength of all the composites was somewhat lower than that of control PLA. The tensile strength was lowest for the composite with WF content of 10 wt% and then increased gradually with increasing WF content. When composites with the same WF content are compared, the composites with mesh size between 50 and 200 mesh showed higher strength and moduli in both the tensile and flexural tests. Elongation at break decreased with increasing WF content. The strength decrease is due to the lowering of flexibility. The PLA/abaca composites had slightly higher modulus than PLA/WF composites with the same fiber content (Fig. 12-5 vs. Fig. 12-2). This is attributed to the fact that the abaca fiber has higher aspect ratio than WF.

Figure 12-6 shows dynamic viscoelastic curves of PLA/WF composites with WF content of 20 wt%. For control PLA, the storage modulus (E') dropped around 50°C due to the glass transition, and rose again around 120°C due to crystallization of PLA. For PLA/WF composites, although the temperature at which E' starts to decrease is almost the same, that at which it starts to increase due to PLA crystallization is shifted to ~75°C, which is considerably lower than that for PLA. This suggests that cold crystallization of PLA is promoted by the presence of WF. The tan δ peak temperature corresponding to the glass transition temperature was almost unchanged, indicating that WF does not affect the mobility of PLA chains.

Figure 12-7 shows dynamic viscoelastic curves of PLA composites with the WF of 100 mesh as a function of WF content. There was little difference in the temperature

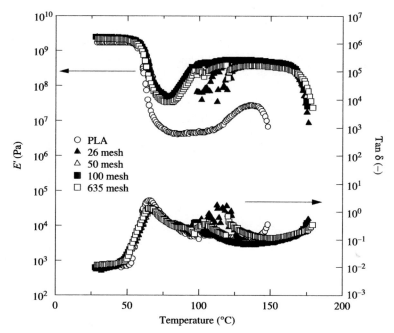

Fig. 12-6 Dynamic viscoelastic curves of PLA and PLA/WF composites with WF content of 20 wt%.

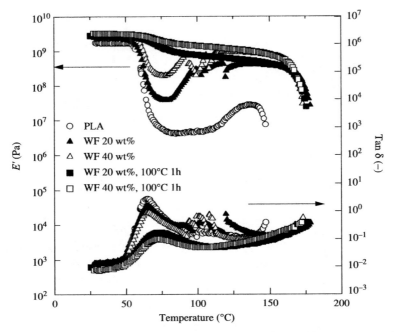

Fig. 12-7 Dynamic viscoelastic curves of PLA and PLA/WF (100 mesh) composites.

at which E' starts to increase due to PLA crystallization between the composites with WF contents of 20 and 40 wt%. When the PLA/WF composites were annealed at 100°C for 1 h, the decrease of E' due to the glass transition of the PLA component became much smaller because of the increase in crystallinity. The annealed PLA composite with WF content of 40 wt% showed a higher storage modulus at 100–150°C than the annealed composite with WF content of 20 wt%.

When the esterification of the WF surface with acetic anhydride/pyridine was investigated in an attempt to improve the interfacial adhesiveness, there was slight improvement of tensile strength and modulus. The use of acetic anhydride-treated WF (AA-WF) was rather effective in diminishing the water absorption of the PLA composites. Thus, PLA/AA-WF with WF content of 40 wt% showed lower water absorption (2.4%) after dipping in water for 24 h than the PLA/WF with the same WF content (3.1%). Composites with a higher WF content showed higher water absorption, and the use of finer WF resulted in a slight decrease in water absorption.

12.4 PLA/LYOCELL COMPOSITES

The man-made fiber lyocell is environmentally benign because it is manufactured from wood pulp by dissolution of cellulose in N-methylmorpholine N-oxide, which is used repeatedly by recycling (Firgo, 1995). Compared with flax and abaca, lyocell fiber has a considerably greater elongation at break (see Table 12-1). In addition, variations in the mechanical properties as well as in the shape between different batches of lyocell fiber are less than for natural fibers. Very tough and flexible fabrics are manufactured from lyocell fibers. Although lyocell fiber/ biodegradable polymer composites are very interesting eco-friendly composites ("green composites"), little is reported in the literature (Mieck et al., 2002). PLA/ lyocell composites were prepared by sandwiching lyocell fabric (300 dtex, 600 dtex, 2/2 twill, thickness ~0.35 mm) between two layers of PLA sheet at 160–190°C and 3–10 MPa pressure (Shibata et al., 2004). The PLA/lyocell composites obtained were also annealed at 100°C for 3 h in order to enhance the crystallinity. For PLA-based composites, multilayered laminate composites for Izod impact tests were also prepared by sandwiching 6–8-ply lyocell fabrics between alternating 7–9-ply PLA sheets.

Figure 12-8 shows the tensile properties of PLA/lyocell composites with various fiber contents. PLA sheets prepared by pressure molding were used for measurements at fiber content of 0%, Tensile strength and moduli of the composites increased with increasing fiber content. PLA composites showed higher elongation at break (6.4–10.4%) than pure PLA sheet (2.1%). This are attributed to the fact that lyocell fiber has a higher elongation than PLA.

The crystallization of PLA at room temperature is very slow because of a high T_g value (60°C). The original degree of crystallinity (χ_c) of the composite can be evaluated from the value of ($\Delta H_m - \Delta H_{g,c}$), where ΔH_m and $\Delta H_{g,c}$ are respectively the heats of melting and of crystallization from the glassy state of the composite in the first heating DSC scan, respectively. Taking the heat of melting of 100% crystalline PLA as 93 J/g (Fisher et al., 1973), χ_c values of the PLA component of the original and annealed PLA/lyocell composites (fiber content ~40 wt%) were evaluated as 15% and 35%. Figure 12-8 also shows the comparison of tensile properties between the original and annealed PLA/lyocell composites. Despite the increase of χ_c by annealing, the annealed composites showed significantly lower tensile strength and moduli than the original composites.

Figure 12-9 shows SEM images of the surface of the original and annealed PLA/ lyocell composites. The annealed composite had much more cracking than the original sample, which should be why the annealed PLA composite showed a lower tensile strength than the original composite.

Figure 12-10 shows SEM images of the fractured surface of the original PLA/ lyocell composite at different magnifications. It is obvious that PLA permeates among the lyocell fibers in the cloth, indicating that interfacial delamination between lyocell cloth and PLA is difficult. It is therefore thought that the shrinkage of PLA adhering to lyocell cloth caused the cracks after annealing. The occurrence of microcracks is assumed to be the reason for the lowering of tensile properties.

Izod impact testing of multilayered PLA/lyocell laminate composites with fiber content ~50 wt% compared with PLA sheet prepared by pressure molding was done to evaluate the improvement of toughness in the PLA/lyocell composites. The PLA composites with 6-ply (28.2 kJ/m^2) and 8-ply (40.8 kJ/m^2) lyocell fabrics showed approximately two and three times respectively higher impact strength

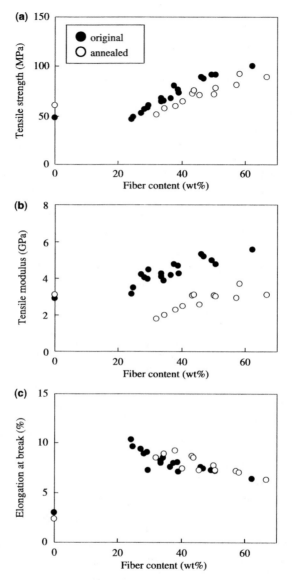

Fig. 12-8 Tensile properties of native and annealed PLA/lyocell composites as a function of fiber content.

than pure PLA (14.1 kJ/m^2) (Fig. 12-11). Although a simple comparison between a laminate composite and an injection-molded short-fiber composite may be inappropriate, the short abaca fiber-reinforced PLA composite (fiber length 5 mm; fiber content 10 wt%) prepared by injection molding (Shibata et al., 2003) showed a

Fig. 12-9   SEM images of the surface of the original and annealed PLA/lyocell composites: (a) original sample, and (b) annealed sample.

considerably lower Izod impact strength $(10-12\,\text{kJ/m}^2)$. Mieck and co-workers found that the impact strength of lyocell fiber composites was much higher than that of natural fiber composites when the composites were prepared with the same fiber length and by the same method (Benevolenski et al., 2000). The improvement of impact strength for the PLA/lyocell composite may be attributed to the use of lyocell fabric and the fact that lyocell fiber has greater elongation than plant-based fibers such as abaca and flax.

The lyocell fabric itself biodegraded after 30 days, as shown in Fig. 12-12. The lyocell fabric in PLA/lyocell composite also clearly degraded after 60 days (Fig. 12-12). Because pure PLA film was almost unchanged after 120 days, degradation of the PLA in the composite is not thought to occur. Microcracking or delamination in the PLA/lyocell composite may be responsible for the direct biodegradation of the lyocell fabric in the composite.

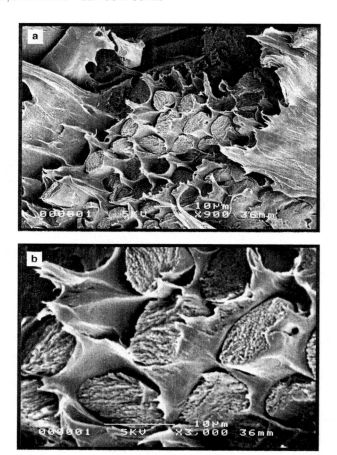

Fig. 12-10 SEM images of the fractured surface of the original PLA/lyocell composites at different magnifications: (a) ×900, (b) ×3000.

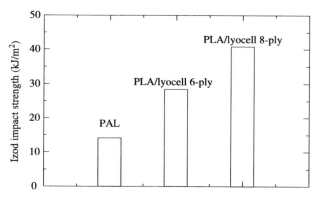

Fig. 12-11 Izod impact strength of PLA and multi-layered PLA/lyocell composites.

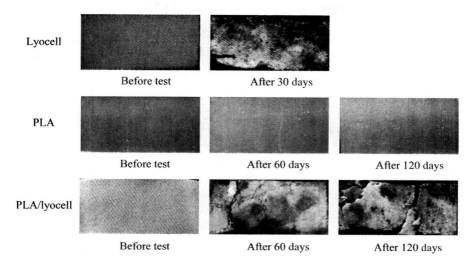

Lyocell

Before test   After 30 days

PLA

Before test   After 60 days   After 120 days

PLA/lyocell

Before test   After 60 days   After 120 days

Fig. 12-12 Photographs of lyocell fabric, PLA, and PLA/lyocell composite before and after the soil-burial test.

12.5 CONCLUSION

PLA/abaca and PLA/WF composites were prepared by melt mixing and subsequent injection molding. For both composites, tensile and flexural moduli increased with increasing fiber content, but the strength of the composites was somewhat lower than that of control PLA. The influence of fiber surface treatment on the mechanical properties of the PLA composites was less marked. PLA/lyocell fabric composites with fiber content 40–67 wt% prepared by pressure molding showed considerably higher tensile strength and elongation at break than PLA. Multilayered PLA/lyocell laminate composites showed considerably higher Izod impact strength than PLA. These results are attributed to the fact that lyocell fiber itself has higher strength and elongation than PLA, and that interfacial adhesion is good. Consequently, reinforcement with natural plant fibers is effective in improving the rigidity of the composites, while improvement of toughness this way is very difficult. Reinforcement with lyocell fabric was very effective in improving the toughness of PLA. In soil burial tests on the PLA composites, fiber degradation occurred with interfacial delamination and/or microcracking of matrix PLA.

REFERENCES

Benevolenski OI, Karger-Kocsis J, Mieck KP, Reussmann T. 2000. Instrumented perforation impact response of polypropylene composites with hybrid reinforcement flax/glass and flax/cellulose fibers. *J Thermoplast Compos Mater* 13:481–496.

Bledzki AK, Gassan J. 1999. Composites reinforced with cellulose based fibres. *Prog Polym Sci* 24:221–274.

Fink HP, Ganster J. 2006. Novel thermoplastic composites from commodity polymers and man-made cellulose fibers. *Macromol Symp* 244:107–118.

Firgo H. 1995. Lyocell process. *Lenzinger Berichte* 75:47–50.

Fisher EW, Sterzel HJ, Wegner G. 1973. Investigation of the structure of solution grown crystals of lactide copolymers by means of chemical reactions. *Kolloid Z Z Polym* 251:980–990.

Gamstedt EK, Bogren KM, Neagu RC, Akerho LMM, Lindstrom M. 2006. Dynamic-mechanical properties of wood-fiber reinforced polylactide: Experimental characterization and micromechanical modeling. *J Thermoplast Compos Mater* 19:613–637.

Garlotta D. 2001. A literature review of poly(lactic acid). *J Polym Environ* 9:63–84.

Hou QX, Chai XS, Yang R, Ragauskas AJ, Elder T. 2006. Characterization of lignocellulosic-poly(lactic acid) reinforced composites. *J Appl Polym Sci* 99:1346–1349.

Inoue Y. 2003. Environmentally safe biodegradable plastics from oil-alternative renewable plant resources. *Foods & Food Ingred J Jpn* 208:648–654.

Lee SH, Ohkita T. 2004. Eco-composite from poly(lactic acid) and bamboo fiber. *Holzforschung* 58:529–536.

Lee SH, Wang S. 2006. Biodegradable polymers/bamboo fiber biocomposite with bio-based coupling agent. *Composites Part A* 37:80–91.

Luo S, Netravali AN. 1999a. Interfacial and mechanical properties of environment-friendly "green" composites made from pineapple fibers and poly(hydroxybutyrate-co-valerate) resin. *J Mater Sci* 34:3709–3719.

Luo S, Netravali AN. 1999b. Mechanical and thermal properties of environment-friendly "green" composites made from pineapple leaf fibers and poly(hydroxybutyrate-co-valerate) resin. *Polym Compos* 20:367–378.

Mathew AP, Oksman K, Sain M. 2005. Mechanical properties of biodegradable composites from poly lactic acid (PLA) and microcrystalline cellulose (MCC). *J Appl Polym Sci* 97:2014–2025.

Mathew AP, Oksman K, Sain M. 2006. The effect of morphology and chemical characteristics of cellulose reinforcements on the crystallinity of polylactic acid. *J Appl Polym Sci* 101:300–310.

Mieck KP, Luetzkendorf R, Reussmann T, Nechwatal A, Eilers M, Biehl D. 2002. Lyocell fiber reinforced plastics composites—state-of-the-art and prospects. *Technical Textiles* 45, E58–60.

Mohanty AK, Misra M, Hinrichsen G. 2000. Biofibres, biodegradable polymers and biocomposites: An overview. *Macromol Mater Eng* 276/277:1–24.

Mohanty AK, Misra M, Drzal LT. 2002. Sustainable bio-composites from renewable resources: Opportunities and challenges in the green materials world. *J Polym Environ* 10:19–26.

Nishino T, Hirao K, Kotera M, Nakamae K, Inagaki H. 2003. Kenaf reinforced biodegradable composites. *Compos Sci Technol* 63:1281–1286.

Ohara H. 2003. Polylactate produced from biomass. *J Appl Glycosci* 50:405–410.

Oksman K, Skrifvars M, S Eli JF. 2003. Natural fibres as reinforcement in polylactic acid (PLA) composites. *Compos Sci Technol* 63:1317–1324.

Oksman K, Mathew AP, Bondeson D, Kivien I. 2006. Manufacturing process of cellulose whiskers polylactic acid nanocomposites. *Compos Sci Technol* 66:2776–2784.

Petersson L, Oksman K. 2006. Biopolymer based nanocomposites: Comparing layered silicates and microcrystalline cellulose as nanoreinforcement. *Compos Sci Technol* 66:2187–2196.

Plackett D. 2004. Maleated polylactide as an interfacial compatibilizer in biocomposites. *J Polym Environ* 12:131–138.

Plackett D, Andersen TL, Pedersen WB, Nielsen L. 2003. Biodegradable composites based on L-polylactide and jute fibres. *Compos Sci Technol* 63:1287–1296.

Saheb DN, Jog JP. 1999. Natural fiber polymer composites: A review. *Adv Polym Technol* 18:351–363.

Sawyer DJ. 2003. Bioprocessing—no longer a field of dreams. *Macromol Symp* 201:271–282.

Scott A, Sissell K. 2003. Green chemistry starts to take root. *Chem Week* 165:18.

Seriwzawa S, Inoue K, Iji M. 2006. Kenaf-fiber-reinforced poly(lactic acid) used for electronic products. *J Appl Polym Sci* 100:618–624.

Shanks RA, Hodzic A, Ridderhof D. 2006a. Composites of poly(lactic acid) with flax fibers modified by interstitial polymerization. *J Appl Polym Sci* 99:2305–2313.

Shanks RA, Hodzic A, Ridderhof D. 2006b. Composites of poly(lactic acid) with flax fibers modified by interstitial polymerization. *J Appl Polym Sci* 101:3620–3629.

Shibata M, Takachiyo K, Ozawa K, Yosomiya R, Takeishi H. 2002. Biodegradable polyester composites reinforced with short abaca fiber. *J Appl Polym Sci* 85:129–138.

Shibata M, Ozawa K, Teramoto N, Yosomiya R, Takeishi H. 2003. Biocomposites made from short abaca fiber and biodegradable polyesters. *Macromol Mater Eng* 288:35–43.

Shibata M, Oyamada S, Kobayashi S, Yaginuma D. 2004. Mechanical properties and biodegradability of green composites based on biodegradable polyesters and lyocell fabric. *J Appl Polym Sci* 92:3857–3863.

Teramoto N, Urata K, Ozawa K, Shibata M. 2004. Biodegradation of aliphatic polyester composites reinforced by abaca fiber. *Polym Degrad Stab* 86:401–409.

Vink ETH, Rabago KR, Glassner DA, Gruber PR. 2003. Applications of life cycle assessment to NatureWorks polylactide (PLA) production. *Polym Degrad Stab* 80:403–419.

Warmington A. 2001. Green progress. *Eur Plast News* 28:49–50.

Wong S, Shanks RA, Hodzic A. 2003. Poly(L-lactic acid) composites with flax fibers modified by plasticizer absorption. *Polym Eng Sci* 43:1566–1575.

Wong S, Shanks RA, Hodzic A. 2004. Mechanical behavior and fracture toughness of poly(L-lactic acid)-natural fiber composites modified with hyperbranched polymers. *Macromol Mater Eng* 289:447–456.

Wool RP, Sun XS. 2005. *Bio-based Polymers and Composites*. Burlington: Elsevier Academic Press.

Yu L, Dean K, Li L. 2006. Polymer blends and composites from renewable resources. *Prog Polym Sci* 31:576–602.

Yu L, Petinakis S, Dean K, Bilyk A, Wu D. 2007. Green polymeric blends and composites from renewable resources. *Macromol Symp* 249/250:535–539.

CHAPTER 13

Biocomposites of Natural Fibers and Poly(3-Hydroxybutyrate) and Copolymers: Improved Mechanical Properties Through Compatibilization at the Interface

SUSAN WONG and ROBERT SHANKS

Co-operative Research Centre for Polymers, School of Applied Sciences, Science, Engineering and Technology, RMIT University, Melbourne, Victoria, Australia

Biodegradable Polymer Blends and Composites from Renewable Resources. Edited by Long Yu
Copyright © 2009 John Wiley & Sons, Inc.

13.1 TRADITIONAL COMPOSITES AND NOVEL BIODEGRADABLE COMPOSITES

Composites are comprised of two discrete phases: one is the dispersed reinforcing phase such as fibers and organic and mineral fillers; the other is the continuous binding phase (referred to as the matrix) such as thermoplastics, thermosetting resins, and concrete. These have been widely used in many diverse fields over the past few decades. Their uses range from automotive interior parts, sporting goods, electronic components, artificial joints, insulation materials, and building structures, to furniture. The idea of introducing fillers such as fibers into a matrix is for them to act as a reinforcing component to yield improved properties compared with the individual components. The most common fibers used in conventional composites are glass, carbon, or aramid fibers with polypropylene, epoxy resins, unsaturated polyester resins, and polyurethanes. Although conventional composites give high strength and modulus, one major disadvantage is the difficulty of disposal after use. Synthetic composites are well interconnected and relatively stable and therefore separation for recycling is difficult.

With increasing environmental interest, the idea of a "green" environment is of great concern; thus many manufacturers are forced to seek alternative materials to produce many products to address environmental and recycling problems. An innovative idea that began in the late 1980s was to use natural fibers as an alternative to conventional fibers in consumer products. To keep to the idea of green materials, the use of biodegradable polymers as a matrix to replace synthetic matrices was introduced. These materials combined together yielded new materials known as biocomposites.

At present, composites comprised of natural fibers are not commonly used and have been restricted to uses in automotive interior parts where high load-bearing properties are not of great importance. Natural fibers such as flax, sisal, or kenaf have being introduced for automotive interior trim parts such as door panels, parcel shelves, and roofing (Pou et al., 2001). Vehicle manufactures such as Opel, Daimler Chrysler, BMW, Audi, and Ford have already introduced this type of technology into their vehicles (Karus, 1999). The use of natural fibers offers the advantages of weight reduction of 10–30%, good mechanical and manufacturing properties, good performance in accidents (high stability, no splintering), no emission of toxic substances, and overall reduction in costs.

However in the light of these advantages, the use of biocomposites is limited in further applications due to their properties being less than those in conventional composites such as epoxy resin/polypropylene-glass/carbon fiber composites. If possible, the production of these composites should employ existing equipment and technologies that are used to fabricate conventional composites.

The properties of the biocomposites need to be improved to increase the range of applications so that, ideally, the composites exhibit high modulus and strength and have good thermal and dimensional stability and resistance to impact (high toughness). The main controlling factor of these properties identified in literature is the degree of bonding or adhesion between the fiber and the matrix at the interface.

The interface is extremely important as the modulus, strength, and toughness of a composite relies heavily on the chemical interactions or mechanical interlocking between the fiber and the matrix. The strength and toughness are largely dependent on the ability of the matrix to transfer the applied stress to the reinforcing fibers. The inferior fiber–matrix adhesion arises from the hydrophilic nature of natural fibers, which have limited compatibility with the hydrophobic nature of thermoplastics. In brief, methods such as surface modification of the fibers and the use of additives have been investigated to combat such problems, with some success. However, in order to produce a composite with superior properties for specific applications, knowledge of the properties of the individual constituents is important so that conflicting characteristics can be identified and altered accordingly.

13.2 NATURAL FIBERS

Natural fibers with good mechanical performance are found in many natural sources such as varieties of plants. There are many different types of natural fibers as they can be extracted from many types and parts of plants. Generally, they are sourced primarily from the structural components such as the stems, leaves, and seeds. There are various classifications of these fibers, but the widely accepted one is based on their location in the plant. Accordingly, they have been subgrouped into three categories such as seed, bast, and leaf fibers. Some examples are cotton and coir (seed), flax and hemp (bast) and sisal and abaca (leaf). The bast and leaf fiber families are the most commonly used for polymer composites as they generally exhibit higher strength than seed fibers.

13.2.1 Types of Natural Fibers

In brief, leaf fibers found in the leaves of the plant have a coarser texture than bast fibers. Sisal and abaca fibers are among the most common leaf fibers. Sisal fibers can be obtained from three different zones of a leaf, and two kinds of fibers— "mechanical" and "ribbon"—can be extracted. The characteristics of the fibers depend largely on the maturity of the leaves. The fibers from mature leaves are generally coarser, longer and stronger than those from immature leaves. The extraction processes are by retting (controlled decay), hand-scraping, or decortication (mechanical separation). These fibers exhibit moderately high stiffness, high tensile strength (Table 13-1) and durability and have been used as wall coverings, floor mats, floor coverings and upholstery padding.

Coir fibers are categorized as seed fibers that are extracted from the husk of coconuts, and two types of fibers are classed as white or brown. The white variety is finer and lighter in color and is obtained from immature coconuts. They are flexible and therefore can be spun into yarn or twine, tufted to make floor mats and woven in carpets. The brown variety, obtained from mature coconuts, is coarse and stiff and is used in mattresses, brooms, nets, and air filters. These applications are those where high strength is not required as their tensile strength is generally weaker (see

TABLE 13-1 Typical Properties of Natural Fibers[a]

Properties	Flax	Coir	Sisal	E-glass
Density (g/cm^3)	1.4	1.25	1.33	2.55
Tensile strength (MPa)	800–1500	220	600–700	2400
Tensile modulus (GPa)	60–80	6	38	73
Specific modulus	26–46	5	29	29
Elongation at break (%)	1.2–1.6	15–25	2–3	3
Spiral angle (°)	10.0	41–45	20.0	–

[a]Adapted from Brouwer (2000).

Table 13-1). The extraction processes are wet milling or defibering on special machines.

The most common bast fibers are flax; these are extracted from the flax plant and have many uses depending on their quality. Finer flax is used mainly for linen, sheeting, lace, and apparel, while coarser fibers are used in twines, canvases, bags, fishnets, sewing threads, fire hoses, and sail cloth. The oil extracted from the flax seeds are used as linseed oil. Flax fibers will be discussed in some detail as they have been chosen as the fibers for the following studies.

13.2.2 Plant and Bast Fiber Structure (Flax)

The cross-section of the flax plant stem has five distinct regions, identified as the outer surface (epidermis layer), the intermediate layer (cortical parenchyma), the bast layer containing the fibers, the cambium layer, and woody tissue. Their regions are illustrated in Fig. 13-1.

In the outer surface layer (epidermis layer), there is a thin layer of wax to prevent excessive evaporation of moisture and to protect the plant from environmental conditions. The next, cortex, layer consists of circular cortical cells that contain pectin substances and coloring matter. In the third, bast, layer, the fiber bundles are found and they are surrounded by parenchyma. The fourth layer, the cambium layer, contains the tender growth tissue composed of thin-walled cells that separate the fiber layer from the fifth layer. The fifth layer is composed of woody tissue consisting of thick-walled cells and thin-walled cells surrounding the pith cavity, which is an air chamber throughout the length of the stalk. The fiber bundle located in the third layer of the stem can be further divided into subgroups according to the type of fibers.

As the schematic of the structure of the flax shows in Fig. 13-2, the fiber bundles are comprised of technical fibers with diameters of 50–100 μm and finer fibers within the technical fibers called elementary fibers (or ultimates) with diameters of 10–20 μm (Singleton et al., 2003). The technical fibers are bonded together with weak pectin and lignin interphase, and within the technical fiber 40 or more elementary fibers are found. The elementary fibers are bonded together by a stronger pectin interphase. Generally the elementary fibers have a higher tensile modulus of

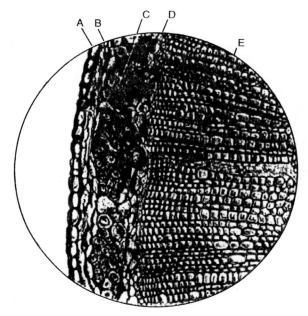

Fig. 13-1 Cross-section of flax plant stem: (A) outer layer (epidermis), (B) intermediate layer, (C) bast layer, (D) cambium layer, (E) woody tissue. (Reprinted from Rouette, 2001, *Encyclopedia of Textile Finishing*, Woodhead Publishing Limited, Cambridge, UK.)

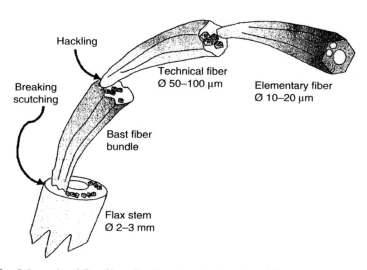

Fig. 13-2 Schematic of flax fiber showing the sub-grouping of fibers. (Reprinted from Van Den Oever et al., 2000, Influence of the physical structure of flax fibers on the mechanical properties of flax fiber reinforced polypropylene composites, *Applied Composite Materials* 7:387–402. With kind permission of Springer Science and Business Media.)

up to 80 GPa and tensile strength of approximately 1.5 GPa compared with the technical fibers with a typical tensile modulus of 50–60 GPa and a tensile strength of 0.6–0.7 GPa (Singleton et al., 2003). Further discussion of fiber properties will be presented in a later section. In practice, most fibers used in many studies are technical fibers or fiber bundles.

The separation techniques or processes that separate the fibers into their individual filaments can be retting and scutching, and the steam explosion technique. The fiber extraction is thought to have some influence on the fiber properties as it alters the chemical composition of the fibers.

The retting process is performed after harvesting and it can be done through biological action (use of enzymes) or chemical action. In brief it is basically a method of controlled rotting/decay and the harvested stalks are left in the fields under conditions of moderate humidity and warmth. Biological retting can remove some chemical components that are detrimental to the fiber strength such as pectic substances, proteins, sugars, starch, fats, and waxes. Lignin is usually resistant to biological degradation, and chemical methods of extraction can be employed.

After retting, the process called scutching is used to extract the fibers from the stalks. The stalks are first mechanically broken from the woody matter into small pieces of shrives that are then scraped to loosen the fibers from the shrives. Fluted rollers are used to break the core into pieces of shrive and the stalks are fed into a machine with rotating bladed-wheels that scutch the fibers clean. During this process, some fibers are broken and the longer fibers are sorted from the short. The long fibers are termed "scutched flax" or "line" and they can be spun after a special combing technique known as "hackling" which further separates the fibers according to length.

In the steam explosion method, the fibers are subjected to steam and additives (if necessary) at high pressures and elevated temperatures so that the steam can penetrate the space between the fibers of the fiber bundles. The middle lamella and the substances adhering on the fibers are dissolved, become water soluble, and can subsequently be removed by washing.

13.2.3 Chemical Components of Bast Fibers

Natural fibers are considered as lignocellulosic since they are composed of cellulose, hemicellulose, and lignin. The compositions of natural fibers depend on the type of fiber as well as the age, origin, and mode of extraction. These are shown in Table 13-2, which shows that most natural fibers consist of major components of cellulose, hemicellulose, and lignin. Other minor components are wax and pectin that function as protective barriers in plants.

Cellulose is the major component of natural fibers and is a polymer of hydrophilic glucan with linear chains of β-1,4-bonded anhydroglucose units containing hydroxyl groups (Mohanty et al., 2000b, 2001). The degree of polymerization of native cellulose is as high as 14,000, but purification usually reduces this to the order of 2500 (Bledzki and Gassan, 1999). Cellulose is chemically defined as poly(1,4-β-D-anhydroglucopyranose). The structure of cellulose is illustrated in Fig. 13-3.

TABLE 13-2 Chemical Components of Some Natural Fibers[a]

Type	Cellulose (wt%)	Lignin (wt%)	Hemicellulose (wt%)	Pectin (wt%)	Wax (wt%)	Moisture (wt%)
Bast						
Jute	61–71.5	12–13	13.6–20.4	0.2	0.5	12.6
Flax	71	2.2	18.6–20.6	2.3	1.7	10.0
Hemp	60.2–74.4	3.7–5.7	17.9–22.4	0.9	0.8	10.8
Ramie	68.6–76.2	0.6–0.7	13.1–16.7	1.9	0.3	8.0
Kenaf	31–39	15–19	21.5	–	–	–
Leaf						
Sisal	67–78	8.0–11.0	10.0–14.2	10.0	2.0	11.0
PALF[b]	70–82	5–12	–	–	–	–
Henequen	77.6	13.1	4–8	–	–	–
Seed						
Cotton	82.7	–	5.7	–	0.6	–

[a] Adapted from Mohanty et al. (2000b).
[b] Pineapple leaf fiber.

The function of cellulose in nature is to impart strength and rigidity to plants. The organization of cellulose chains within the fiber cell is depicted in Fig. 13-4. The structure of natural fibers is quite complex and each fiber could be considered as essentially a composite in itself. Crystalline cellulose is located in many cell walls including the primary and secondary walls and can be oriented or disordered. Disordered crystalline cellulose can be found in the primary wall, while the oriented crystalline cellulose (crystallites) is mainly found in the secondary walls. Elementary fibers (or ultimates) are made of oriented crystalline cellulose in a helical formation wound along the fiber axis at some angle (called the spiral angle), which differs depending on the type of natural fiber. The helical cellulose fibrils exist in a series of layers in the secondary walls and the direction of the spiral changes from Z-spiral to S-spiral successively. The cellulose microfibrils are generally embedded in a soft lignin and hemicellulose matrix in the amorphous regions of the fiber walls. When the helical cellulose chains are fully extended, they form a flat ribbon

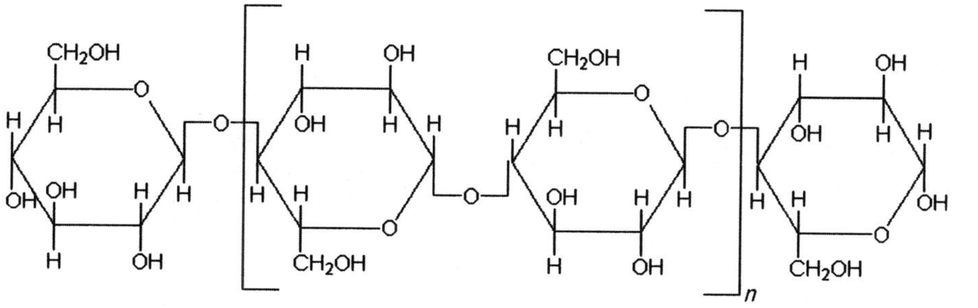

Fig. 13-3 Structure of cellulose.

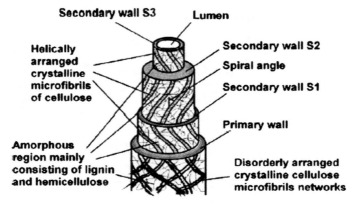

Fig. 13-4   Organization of cellulose chains in fiber cells. (Reprinted from Rong et al., 2001, The effect of fibre treatment on the mechanical properties of unidirectional sisal-reinforced epoxy composites, *Composites Science and Technology* 61:1437–1447. With permission from Elsevier.)

with the hydroxyl groups protruding laterally. These hydroxyl groups form intramolecular and intermolecular hydrogen bonds that give the fibers their hydrophilic nature. Moisture content ranges from 8% to 13%.

Hemicellulose is amorphous and consists of short-chained isotropic polysaccharides. It is chemically linked or partly intermingled and oriented with cellulose molecules as depicted in Fig. 13-5. Hemicellulose is not a form of cellulose as the name would suggest. There are three major differences between hemicellulose and cellulose. Hemicellulose contains several different sugar units, whereas cellulose contains only β-1,4-D-glucopyranose units. It also exhibits a considerable degree of chain branching, whereas cellulose is strictly linear. In addition, the degree of polymerization of native cellulose is about 10–100 times greater than that of hemicellulose. Although the chemical bonds between cellulose and hemicellulose are not covalent, they are still very strong and difficult to separate even through stringent extraction processes.

Lignin is a component that acts as a structural support material in plants by binding the other components together. Lignin is found between cellulose and hemicellulose

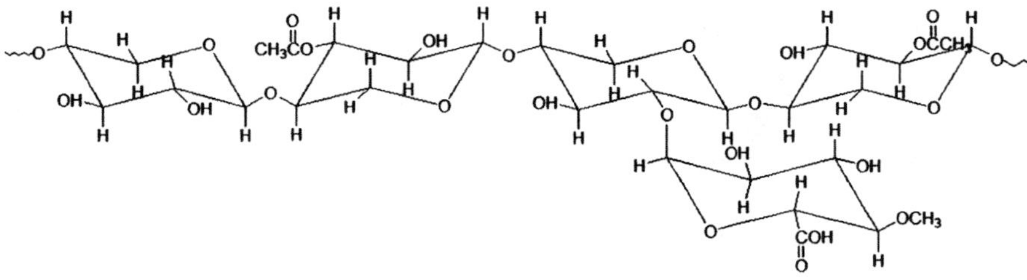

Fig. 13-5   Structure of hemicellulose.

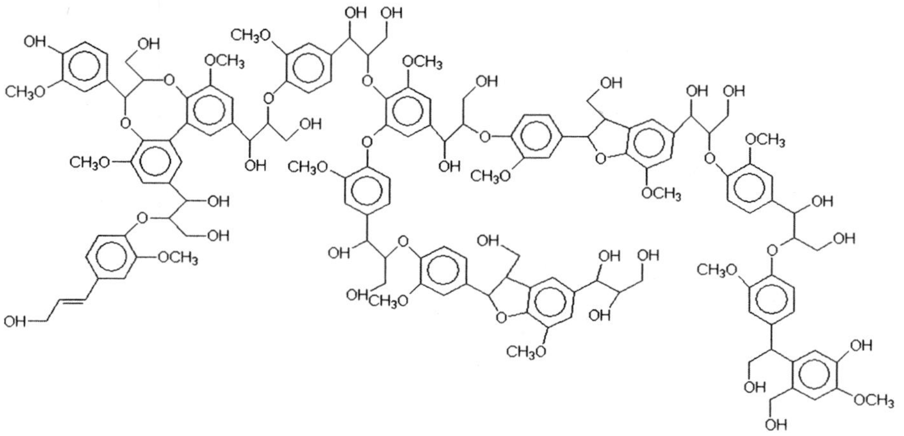

Fig. 13-6   Structure of lignin.

and on walls of cells. The exact chemical structure of lignin has not been established, but through identification of its functional groups it is said to be a phenolic polymeric material (Mohanty et al., 2001). It is found that lignin contains high carbon and low hydrogen content, which suggests that it is highly unsaturated or aromatic in nature. A probable structure of lignin is given in Fig. 13-6.

Although the chemistry of lignin in natural fibers is not well known, it is believed that two types of linkages form between the carbohydrate groups and lignin. One is an ester-type, which is formed between the hydroxyl of lignin and the carboxyl of hemicellulose uronic acid. The second is of an ether-type that occurs between hydroxyls of lignin and hydroxyls of cellulose. The ester-type linkage is sensitive to alkali and the ether-type is insensitive to alkali. Because lignin exists in combination with more than one neighboring cellulose/hemicellulose chain, a crosslinked structure is formed.

Pectins are polygalacturonosyl-containing polysaccharides that have other polysaccharides covalently associated with them. The branched, hydrated pectin molecules are capable of forming semirigid gels in the presence of calcium ions. The calcium ions bind neighboring galacturonan chains, which provide crosslinks that enhance the rigidity of the plant walls. During biological retting, acetic and butyric acids are formed. Pectin can be readily removed by boiling with alkali.

13.3 MECHANICAL PROPERTIES OF NATURAL FIBERS

Extensive studies on the mechanical properties of natural fibers are presented in numerous literature reports as the final composite properties are largely dependent on the reinforcing component. It has been established that natural fibers can compete with glass fibers in terms of mechanical properties. However, the mechanical properties of the fibers can vary depending on the quality of the fibers, which in turn is strongly affected by factors such as the different growing conditions, methods of fiber extraction, and maturity of the plant at harvest. With these sources of variability in mind, a general comparison of the natural fibers can still be made with careful control to minimize the variability in the determination of these properties. The mechanical properties of different types of natural fibers in reported studies are summarized and comparisons between the properties of glass fibers and natural fibers are presented.

It has been established that properties such as density, ultimate tensile strength, and modulus are related to the internal structure as well as the chemical composition of the fibers. More specifically, the strength and stiffness (modulus) of the natural fibers have been found to be dependent on the cellulose content and the spiral angle of the cellulose microfibrils in the inner secondary cell walls along the fiber axis (Li et al., 2000). Some typical properties of selected natural fibers are presented in Table 13-1.

E-glass fibers exhibited higher tensile modulus compared with all the natural fibers as expected, and consequently exhibited higher tensile strength and elongation at break. If the fibers are viewed in relation to the density of the fibers, that is, the specific modulus which takes account of the density, the specific modulus of flax and sisal are comparable with those of E-glass fibers. This is one advantage of natural fibers over conventional fibers when weight is important, since composites comprised of natural fibers are of much lower density. The elongation at break of natural fibers is similar to that of glass fibers, with the exception of coir fibers. As coir fibers have a much lower modulus than the other fibers shown, the elongation at break is expected to be higher than that of materials with higher modulus.

Of the natural fibers shown, higher tensile strength and modulus are exhibited by those with high cellulose content and lower content of amorphous chemical components such as hemicellulose, lignin, and wax (see Table 13-1). The strengths of the individual components of natural fibers are thought to be greatest for cellulose and the strength decreases in the following order for the other components: noncrystalline cellulose, hemicellulose, and lignin (Mohanty et al., 2000b).

The higher tensile strength and modulus of the natural fibers has been explained in terms of the fiber failure mechanism. In relation to the organization of cellulose chains within the fiber presented earlier, the cellulose chains are arranged in a helical formation in the secondary walls and are bound together by amorphous hemicellulose and lignin within the secondary walls and in the primary wall. When under tension, the primary wall consisting of amorphous materials fractures in a brittle manner, while the secondary walls are bridged by relatively thick cellulose fibrils. It has been reported that the higher the cellulose content, the greater the number of cellulose chains that can bridge a crack, resulting in greater strength for fracture to occur (Joffe et al., 2003). Another approach to the cause of breakage of native cellulose under tension was discussed on a molecular level. It has been stated that tensile breakage could be due to rupture of covalent bonds in the cellulose molecules or between secondary valence bonds (primarily hydrogen bonds) between the cellulose molecules (Gassan and Bledzki, 2001).

It is well recognized that the spiral angle of the cellulose chains has some influence on the tensile strength and modulus. It was established that the fibers with smaller spiral angles (such as flax and sisal) have higher strength and modulus compared with fibers with larger angles (coir). It was postulated that the chains with smaller spiral angles consume more energy to uncoil the helically oriented fibrils to their flat state (when subjected to tension) compared with chains with larger angles.

As mentioned earlier, flax fibers are the chosen fibers in the following studies of composites because they are natural fibers with exceptionally good mechanical properties. Comparison of the properties of flax with other natural fibers and E-glass is illustrated in Fig. 13-7.

Breaking length is a term for tenacity (breaking stress), a specific measure used in the textile industry, and is given in kilometers. It is the length at which the fiber breaks due to its own weight (Mohanty et al., 2000b). From the graph, flax fibers exhibit the highest stiffness along with hemp and ramie. It is worth noting that the stiffness of hemp even exceeds that of E-glass. The breaking length of flax is one of the highest compared with sisal, spruce, and cotton, and is comparable with that of E-glass. When subjected to tension, flax fibers have been known to undergo some amount of strain hardening, which is due to the progressive reorientation of the microfibrils with off-axis orientation in the unstrained fiber (Joffe et al., 2003). The elongation values are similar, with breakage occurring at about 3% strain. Based

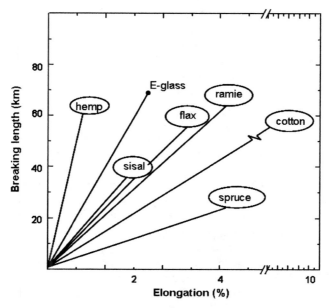

Fig. 13-7 Breaking length versus elongation of some natural fibers. (Reprinted from Herrmann et al., 1998, Construction materials based upon biologically renewable resources—from components to finished parts, *Polymer Degradation and Stability* 59:251–261. With permission from Elsevier.)

on these findings, flax fibers proved to be an appropriate fiber for the manufacture of composites compared with E-glass fibers.

13.3.1   Factors that Affect Fiber Mechanical Properties

As well the variability of natural fibers that depends on the quality (e.g., environmental growing conditions, maturity of plant at harvest), the mechanical properties of natural fibers can be affected by factors that may be introduced during manufacturing, during its service-life, and when subjected to different chemical treatments to improve their compatibility with polymeric matrices. These can alter the chemical structure or chemical composition of the fibers, which determine their strength. Such factors are vital and have been identified in order to make allowances for changes in the composites. Change in the chemical structure of the fibers can be brought about by thermal treatments, surface modifications of the fibers, and the effects of moisture. These are discussed in detail in the following.

13.3.1.1   Thermal Treatments   During manufacture of fiber composites, it is often necessary to subject the thermoplastic matrix and fibers to heat to form the composites. Natural fibers are sensitive to thermal treatments as degradation can occur that weakens the fibers. The inherent moisture that is naturally bound to the hemicellulose component in the fibers can change depending on temperature and exposure time. Several publications have investigated these issues and have generally found that exposure to temperatures at or below 170°C did not have a significant effect on the strength of the fibers (Coutinho et al., 1999; Sharma et al., 1999; Gassan and Bledzki, 2001; Van de Velde and Baetens, 2001). However, exposure to higher temperatures (up to 247°C) can cause a significant reduction in the mechanical properties and change in chemical composition (Joffe et al., 2003). The strain was found to be more limited than the fiber strength. Removal of water by drying the fibers before fabrication of composites was sufficient to control the amount of moisture within the composites (Bledzki and Gassan, 1999).

The thermal stability of natural fibers is reported to depend on their structure and chemical composition. In general, the thermal stability of the individual components of natural fibers (in the absence of oxygen) is reported to show higher decomposition temperatures in the order lignin $<$ α-cellulose $<$ hemicellulose (Gassan and Bledzki, 2001). When the fibers are subjected to heat, the physical and/or chemical structural changes that can occur are depolymerization, hydrolysis, oxidation, dehydration, decarboxylation, and recrystallization (Gassan and Bledzki, 2001). Correlation of the degree of polymerization (X_n) and the degree of crystallinity (X_c) with the strength of flax and jute fibers (measured as tenacity) was presented in a study by Gassan and co-workers (Gassan and Bledzki, 2001). They concluded that when the fibers were exposed to temperatures below 170°C for a maximum of 120 min, a decrease in the X_n value resulted in a slight decrease in tenacity. For temperatures above 170°C, the tenacity showed a rapid decrease along with X_n, which depended on the exposure time and temperature. As jute fibers exhibited a lower value of X_n than flax, the decrease in tenacity was also greater.

It was found that chain scission of cellulose led to an increase in X_c upon heating of the fibers. It was thought that crystallization results from an increase in the size of preexisting crystallites by realignment of chains on the crystallite surface and at their ends, or by forming completely new crystallites within the amorphous regions of the fibers. However, an increase in X_c on exposure to high temperatures did not have any profound influence on the tenacity.

13.3.1.2 Effect of Moisture

Removal of the moisture from the fibers before manufacture of composites is important for dimensional stability and elimination of voids within the composite, which would compromise its mechanical properties. However, exposure of fibers to mild humidity did not have a profound affect on the strength of the fibers as the moisture in fact had a plasticizing effect (Joffe et al., 2003). The effect of humidity on the fiber properties was found to be highly dependent on the exposure time, relative humidity, and any treatment of the fibers. Stamboulis and co-workers (Stamboulis et al., 2000, 2001) found that exposure of flax fibers up to a maximum humidity of 66% did indeed increase the strength of the fibers; this was attributed to plasticization due to "free" water molecules. It was thought that water was able penetrate into the cellulose network of the fiber in the capillaries and spaces between the fibrils. At relatively mild humidity, free water molecules force the cellulose molecules apart, destroying some rigidity of the cellulose structure. Due to this effect, cellulose molecules can move more freely because of plasticization and can reorientate to yield higher X_c and thus higher strength. However, at higher humidity, an increase in the absorbed bound water on the fibers and a decrease in free water molecules decreases the plasticizing effect, resulting in reduced fiber strength. On prolonged exposure, this may lead to degradation of the mechanical properties caused by fungus growth on the fiber surface.

Another explanation at a molecular level was presented by Stamboulis (Stamboulis et al., 2001). It was established that the strength was generally dependent on the spiral angle of the cellulose network. At high angles (flatter chains), the strength and modulus were reduced compared with a smaller angle. When water molecules penetrated into the fiber, the spiral angle was thought to be reduced (became steeper) resulting in higher tensile strength.

Along with marginal improvements in tensile strength with mild humidity, it is also important to note that the amorphous components of the fibers were dissolved, separating the filaments from the bundle. At high humidity, some damage to the fibers due to swelling resulted, which contributed to a reduction in strength. Swelling of the fibers resulted in formation of kink bands, determined to be defects along the fiber.

13.3.1.3 Surface Treatments of Fibers

Surface treatments are often applied to fibers to enhance adhesion and durability under environmental conditions (temperature and moisture). As treatments usually change the fiber surface, they may alter the fiber strength. Surface treatments will be mentioned only briefly in this section as details will be discussed later.

Briefly, the mechanical properties of fibers after various treatments do not always produce an effect on the fiber strength. For example, acetylation of natural fibers brought about an increase in hydrophobicity so that compatibility with hydrophobic polymers was achieved. This was attained by reacting hydroxyl groups of the fiber constituents with acetyl groups through esterification (Zafeiropoulos et al., 2002). However, according to Joffe et al. (2003) no appreciable difference in the fiber strength was found upon acetylation of dew-retted flax, though for green flax a pronounced increase in the strength was observed. Treatment with stearic acid increased the hydrophobic nature of the fibers by reducing the number of free hydroxyl groups. Furthermore, the long hydrocarbon chain of stearic acid provided extra protection from moisture. The carboxyl groups of stearic acid reacted with the hydroxyl groups through esterification (Zafeiropoulos et al., 2002). This treatment was found to have no effect on the mechanical properties for low treatment times, but deterioration of the properties was observed with longer intervals. Ultimately, the effects on the mechanical properties depended on how the chemical or physical treatments changed the fiber structure and the chemical composition, so that no general trends can be given.

13.4 BIODEGRADABLE POLYMERS

There is a large range of biodegradable polymers that have been under extensive research. Most are produced synthetically from monomers of natural origin (biodegradable polymers) but some are produced by bacteria (biopolymers). These polymers have many potential applications, ranging from consumer packaging to medical applications (such as sutures, surgical implants) to automotive components. There are a number of types of biodegradable polymers and biopolymers and many kinds of classifications have been presented in the literature. In this study, these polymers are categorized by the origin of the monomers (synthetic or natural). A schematic representation of this classification is presented in Fig. 13-8.

The main function of the polymer in composites is to bind the fibers together, thus forming and stabilizing the shape of the composite structure. But most importantly, the matrix is responsible for transmitting the applied shear forces from the matrix to the fibers, resulting in high-strength composites. The selection of the matrix is important for the intended application in terms of the temperature range to which the composite will be subjected during use, the magnitudes of the mechanical loads, and the flexibility and stiffness required. From the range of biodegradable polymers available, the most promising for future use in place of synthetics such as poly (propylene) are the aliphatic polyesters. These are the most widely used polymers in the literature surveyed. As this study concerns the use of aliphatic polyesters as the matrices for biodegradable composites, their structure and synthesis and some properties of these polymers are presented.

As depicted in Fig. 13-8, the family of aliphatic polyesters can be produced from natural or synthetic monomers. From the category of natural monomers, these consists of poly(α-hydroxy acid)s and poly(β-hydroxyalkanoate)s. The classes of poly(ω-hydroxyalkanoate)s and poly(alkylene dicarboxylate)s are from synthetic monomers. These classes are summarized below.

13.4.1 Poly(α-Hydroxy Acid)s

Poly(lactic acid) and poly(glycolic acid) are two polymers from the class of poly(α-hydroxy acid)s. As poly(lactic acid) has been a primary concern in this study, it is discussed in detail.

Poly(lactic acid) (PLA) is a biodegradable, linear, aliphatic, thermoplastic polyester derived from completely renewable resources such as corn and sugar beets. It is a poly(α-hydroxy acid) and is also known as poly(lactate) or poly(lactide); its chemical structure is given in Fig. 13-9.

PLA exhibits good mechanical properties that are comparable with those of polystyrene (Martin and Averous, 2001). Its major drawback is its brittleness, but plasticization with various citrate esters (Labrecque et al., 1995, 1997; Martin and Averous, 2001; Ljungberg and Wesslen, 2002) or polyglycols (Sheth et al., 1997), and blending with other polymers (Kopinke and MacKenzie, 1997; Park et al., 2000; Ke and Sun, 2001) can improve its ductility. PLA is used primarily in medical applications such as sutures, drug delivery, and orthopedic implants because it is biocompatible (nontoxic for humans) (Labrecque et al., 1997). There is potential for large-scale uses in packaging and many consumer goods such as hygiene products.

PLA can be prepared by direct condensation of lactic acid and by ring-opening polymerization of the cyclic lactide dimer. Many different catalysts are employed such as complexes of aluminum, zinc, tin, and lanthanides. Stereochemically

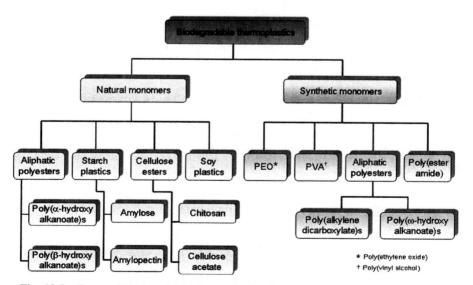

Fig. 13-8 Types of biodegradable thermoplastics from natural and synthetic monomers.

different PLAs can be produced from the two stereoisomers of lactide, L-lactide and D-lactide, and a third that exists as a mixture of D,L-lactide. The stereochemical compositions of the polymer have a dramatic affect on the melting temperature, the rate of crystallization, and the ultimate extent of crystallization. The glass transition temperature (T_g) is mainly affected by the molecular weight (M_w), which can range from "high" (425,000 g/mol) to "low" (12,000 g/mol) (Drumright et al., 2000).

The maximum melting temperature of stereochemically pure poly(lactide) (either D or L) is about 180°C with an enthalpy of melting of 40–50 J/g (Drumright et al., 2000). Higher enthalpy of melting in the range 76–78 J/g was reported (Perego et al., 1996). By introducing different stereochemical defects (D,L- or D-) into poly(L-lactic acid), reductions in the melting temperature, the rate of crystallization, and the magnitude of crystallinity can be observed with little effect on the T_g. The rate of crystallization is relatively slow even at the optimum crystallization temperature of 105–115°C for poly(L-lactic acid) (Drumright et al., 2000).

The properties of PLAs that make them suitable for many applications are good crease-retention and crimp properties, excellent grease and oil resistance, easy low-temperature heat sealability, and good barrier properties to flavors and aromas. They generally show a high modulus and strength but low toughness. The mechanical properties of PLA are largely dependent on the molecular weight and its processing conditions. After annealing of poly(L-lactic acid) (PLLA) with varying molecular weights, an increase in molecular weight showed only a slight increase in the tensile properties, but a more marked affect was noticed for the flexural properties. This was attributed to the crystallinity being lower at higher molecular weights as the lower-molecular-weight PLLA allowed crystallization to be more complete due to higher chain mobility (Perego et al., 1996). Another study, by Huda and co-workers (Huda et al., 2002), showed the importance of processing conditions for the dynamic mechanical properties. By using two different methods of processing—solution casting and heat pressing—the dynamic storage modulus (E'), which indicates the stiffness, they showed that the heat-pressed PLA retained its high stiffness to a higher temperature than did the solution cast sample.

Although PLA shows poor melt strength and is not very shear-sensitive, branching can be introduced by treating with peroxide or through addition of multifunctional initiators or monomers. Branching is proven to increase the viscosity even at low shear rates, making PLA applicable for use in extrusion coating, extrusion blow-molding, and foaming. Further modifications can make PLA suitable for injection molding, sheet extrusion, blow molding, thermoforming, film forming, or fiber spinning.

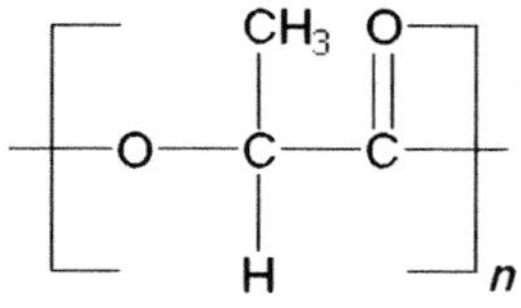

Fig. 13-9 Chemical structure of poly(lactic acid).

13.4.2 Poly(β-Hydroxyalkanoate)s

The polymers of poly(β-hydroxybutyrate) (PHB) and its copolymer, poly(β-hydroxy-butyrate-*co*-hydoxyvalerate) (PHB-HV) come under the class of poly(β-hydroxyalkanoate)s. These are synthesized biochemically by microbial fermentation and bacteria accumulate them in granular form in their body as an energy storage material (Ha and Cho, 2002). PHB can be fermented from a variety of sources such as sugars and molasses or from hydrogen and carbon dioxide, depending on the bacteria used. Upon degradation under certain environmental conditions and with enzymes, these polymers become nontoxic residues of carbon dioxide and water.

Generally, poly(β-hydroxyalkanoate)s are linear polyesters that are semicrystalline, hydrophobic, and biocompatible and have good mechanical strength and modulus, resembling poly(propylene). Their structures are given in Fig. 13-10.

Pure PHB is highly crystalline (about 80%) (Janigova et al., 2002), resulting in its brittle nature and low elongations as drawing is not possible. PHB is frequently referred to as a hard, brittle plastic. It exhibits a narrow processing window as it readily undergoes degradation via chain-scission to form crotonic acid and oligomers (Kopinke and MacKenzie, 1997) along with a reduction in M_w and melt viscosity. Plasticizers such as glycerol, triacetin, citrate esters, and tributyrin are employed to reduce the processing temperature as well as the brittleness (El-Hadi et al., 2002; Janigova et al., 2002). The brittle nature of PHB comes from large volume-filling spherulites from few nuclei due to their high purity, accompanied by a number of interspherulitic cracks. Two types of cracks can be present, circular breaks around the center of the spherulite and splitting between the crystal interfaces. The crystallization rate depends on sample preparation but is generally faster with precipitation from solution than with solution casting. The mechanical properties of PHB depend largely on the crystallization process. It was found that during rapid cooling, the degree of crystallinity was reduced, producing smaller spherulites, which consequently exhibited better mechanical properties compared with PHB with high crystallinity and large spherulites. The T_g of PHB is around $0-5°C$ and it has a melting temperature (T_m) of about $178°C$ (Janigova et al., 2002). Young's modulus of PHB is usually about $3.5 GPa$, which is comparable with poly(propylene) or poly (ethylene terephthalate) (Ha and Cho, 2002).

The copolymer of PHB with hydroxyvalerate (HV) units was produced in order to reduce the brittleness of PHB. The HV content in the copolymer can range from 0 to 95 mol% and the ductility depends on the relative content of HV. The Young's modulus generally decreases with greater HV content; thus properties from rigid

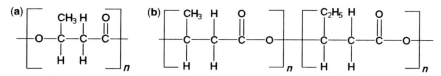

Fig. 13-10 Chemical structures of (a) poly(β-hydroxybutyrate) and (b) poly(β-hydroxy-butyrate-*co*-hydroxyvalerate).

(with high crystallinity) to elastomeric can be obtained depending on the composition. The copolymers generally have lower crystallinity and thus reduced T_m compared with pure PHB. The T_m can be reduced to as low as 75°C with 40 mol% of HV and greater HV content results in a gradual increase in T_m to 108°C at 95 mol% HV (Mohanty et al., 2000b). The T_g of copolymers was reduced to about -10 to -20°C. The crystallization rate of copolymers was found to decrease with greater HV content and the HB and HV units cocrystallize.

13.4.3 Poly(ω-Hydroxyalkanoate)s

Poly(ε-caprolactone) (PCL) is a typical poly(ω-hydroxyalkanoate) that is a partially crystalline linear polyester with a very low T_g of -60°C and a low T_m of 60°C (Mohanty et al., 2000b). PCL is produced from the cyclic ester lactone monomer by ring-opening reactions with stannous octanoate as a catalyst in the presence of an initiator with an active hydrogen atom. Its structure is shown in Fig. 13-11.

PCL is described as a tough and semirigid polymer at room temperature, exhibiting a tensile modulus between 520 and 600 MPa depending on its molecular weight (Corden et al., 1999). The molecular weights can range from 200 to 100,000 g/mol and the polymers exist from liquids to hard waxes. PCL is flexible at room temperature due to its low T_g. The crystallinity of PCL is quite high (about 45–60%) as its T_g is significantly low, which allows crystallization to proceed easily. It is miscible with a range of polymers and organic materials and therefore is widely used as a compatabilizer. From the literature surveyed, PCL is not used as a polymeric matrix for composites because the mechanical properties are generally lower than those of the other types of aliphatic polyesters. PCLs are mainly used in blend formulations as a plasticizer for other brittle polymers to enhance their mechanical properties.

13.4.4 Poly(Alkylene Dicarboxylate)s

The types of polyesters known as poly(alkylene dicarboxylate)s include poly(butylene succinate) (PBS), poly(butylene succinate-*co*-butylene adipate) (PBSBA), and poly(ethylene succinate) (PES). These are produced through polycondensation reactions of glycols (e.g., ethylene glycol and 1,4-butanediol) with aliphatic dicarboxylic acids (such as succinic acid and adipic acid). The general structure of poly(alkylene dicarboxylate)s is depicted in Fig. 13-12.

The average molecular weights of poly(alkylene dicarboxylate)s range from 40,000 to 1,000,000 g/mol (Fujimaki, 1998). Their T_m values are about 116°C but

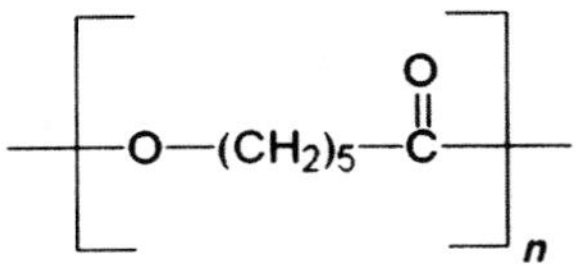

Fig. 13-11 Chemical structure of poly(ε-caprolactone).

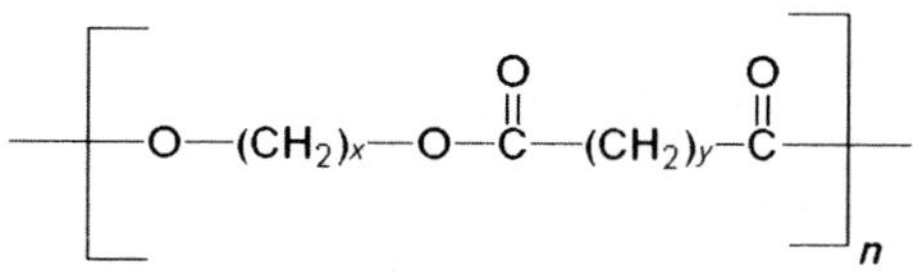

Fig. 13-12 Chemical structure of poly(alkylene dicarboxylate)s.

T_m values of their copolymers are slightly lower and their T_g values range from -45 to $-10°C$. The tensile modulus for PBS is 0.54 MPa with a yield strength of 32.9 MPa and a large elongation of 560%. The tensile modulus of PES is similar to that of PBS but its elongation and yield strength are lower than for PBS (200% and 20.5 MPa, respectively) (Fujimaki, 1998). Depending on the polymer structure, these polymers are suitable for processing with conventional methods and equipment. The linear types are suitable for injection molding and fabrication of filaments, while the comb types with short branching are applicable to film casting, foaming, and sheet extrusion. The star types with few long branches are suitable for tubular films, foamed fibrils, and manufacture of bottles. Like PCL, aliphatic polyesters of this type are not used for manufacture of natural fiber composites because of their low mechanical properties relative to the poly(α-hydroxy acid)s and poly(β-hydroxyalkanoate)s.

13.5 MAJOR PROBLEMS ASSOCIATED WITH HIGH-STRENGTH COMPOSITES

The problem of adhesion between fibers and matrices is well recognized in the field of composites. It is highly desirable to have a strong fiber–matrix interface as it is thought that the composite strength stems from the ability of the matrix to transfer its stress to the fibers. The fiber–matrix interface is generally the weakest part of the composite. The degree of interaction between the fibers and the matrix is often called the "interfacial bonding" and is related to the "interfacial strength." Many studies have found that good interfacial bonding is crucial for high-strength and high-modulus composites.

In natural fiber composites the polymers used as the matrix are often hydrophobic (nonpolar), while the natural fibers are hydrophilic (polar). This leads to limited compatibility and gives inherently inferior mechanical properties. The presence of natural waxy compounds on the fiber surface also limits the fiber–matrix bonding due to poor surface wetting.

The importance of the interfacial bonding for the mechanical properties was revealed in a study by Lui and co-workers (Lui and Netravali, 1999). This entailed unidirectional biocomposites comprised of pineapple fibers and poly(hydroxybutyrate-co-hydroxyvalerate) (PHB-HV). The variables of interest were fiber loading and fiber orientation. It was found that as the fiber loading increased, the tensile and flexural strength and modulus increased accordingly in the longitudinal direction compared with the pure polymer. An increase in tensile strength from 26 MPa (pure polymer) to 80 MPa was exhibited with 29% volume of fibers. However, the tensile and flexural strength of the composites in the transverse direction decreased as a function of fiber content. A reduction in tensile strength from 26 MPa (pure polymer) to 8 MPa resulted with the inclusion of 29% volume of fibers. The fractured surfaces of these composites revealed that failure was initiated at the fiber–matrix interface, as observed by SEM. No polymer was observed on the fiber surface and fiber pull-out from the matrix was evident, indicative of poor interfacial adhesion. Further evidence of poor adhesion between the fibers and the matrix was implied in the crystallization kinetics. No change in the kinetics during crystallization of PHBV was detected regardless of the fiber loading. Accordingly, the melting temperature and enthalpy remained constant. If there is interaction between the matrix and the fiber existed, the fiber surface would be expected to act as a nucleating agent, affecting the crystallization kinetics.

Various physical and chemical methods have been investigated to improve the interfacial bonding and gain some improved levels of adhesion. These methods usually alter the surface energy of the fibers and can alter the surface structure. The ultimate interfacial bonding is complex and is difficult to achieve as there are many parameters that contribute to the interfacial properties. These include the surface energies of the fibers and matrix, matrix morphology, fiber surface condition, presence of residual stresses, moduli of the fiber and the matrix, and the presence of reactive functionalities (Gao and Kim, 2000). Roughness of the fiber surface is thought to contribute to a good interface as it can cause mechanical interlocking between the matrix and the fiber.

Pathways to improvement of the interface have taken directions toward the use of physical methods to alter the fiber surface (with no change to chemical composition) as well as chemical methods such as chemical surface modification and the use of additives (i.e., coupling agents). Chemical methods have proved to be the most popular and convenient for chemists. The various methods are summarized in sections 13.5.1 and 13.5.2 with particular emphasis on the use of additives.

13.5.1 Physical Modification of Natural Fibers

Physical methods do not involve changes in the chemical composition of the fibers. Some of these methods include stretching, calendering, thermal treatment, and production of hybrid yarns, all of which change the structural and surface properties of the fibers and influence the mechanical bonding between the fibers and the polymer.

Electric discharge methods (corona and cold plasma) are physical treatments. Corona treatment is effectively surface oxidation and it has been found to alter the surface energy of cellulose fibers (Bledzki and Gassan, 1999). Cold plasma treatment basically functions the same way as the corona discharge method. Depending on the choice of gases, a variety of surface modifications could be achieved such as surface crosslinking, increasing or decreasing the surface energies, and production of reactive free radicals and groups. Electric discharge methods are very effective for use with polymer substrates that are "nonactive" such as polystyrene, polyethylene, and polypropylene and they have been found to be successful with cellulose-based fibers.

13.5.2 Chemical Modifications of Natural Fibers

Chemical modifications are applied to natural fibers in an attempt to improve the matrix–fiber adhesion by bringing about changes in the chemical composition of the fibers as well as the surface properties. Some chemical modifications could lead to reduced water absorption. Since the use of natural fibers as reinforcements was introduced, numerous chemical modifications have been published in the literature such as dewaxing (Gassan and Bledzki, 1999; Mohanty et al., 2000a), alkalization (mercerization) (Gassan and Bledzki, 1999; Mwaikambo and Ansell, 1999; Mohanty et al., 2000a; Ray and Sarkar, 2001), acetylation (Mwaikambo and Ansell, 1999; Hill and Abdul Khalil, 2000; Singh et al., 2000), coating with coupling agents (Singh et al., 2000; Rozman et al., 2001), cyanoethylation (Mohanty et al., 2000a; Rout et al., 2001), esterification and etherification (Kazayawoko et al., 1997; Baiardo et al., 2002), and grafting (acrylonitrile and methyl methacrylate) (Samal, 1994, 1997; Saha et al., 2000; Mishra et al., 2001). Other modifications include crosslinking with formaldehyde, *p*-phenylenediamine, and phthalic anhydride; nitration; dinitrophenylation; benzoylation; and transesterification (Samal, 1995, 1997, 2001). Schematic presentations of some of these chemical reactions are shown in Fig. 13-13.

As there are many chemical modification methods, only the most popular methods, alkalization, acetylation, treatment with anhydrides, and the use of silane coupling agents are discussed in detail. Some novel innovations that have been developed recently are also presented. Excellent review papers on the other surface treatments of natural fibers can be found in Bledzki and Gassan (1999) and Mohanty et al. (2001).

13.5.2.1 Alkalization   Alkalization (mercerization) is commonly used as a pretreatment to improve the polymer–fiber composite properties through treatment of the fibers with an alkali (KOH, LiOH, and NaOH). The most effective and commonest alkali used is NaOH. The chemical reaction is shown as reaction (1) in Fig. 13-13. The use of an alkali produce a change to a new cellulose form, from cellulose I to cellulose II, partial dissolution of noncellulosic components (such as hemicellulose, lignin, and pectin), and a rougher surface topography. All these factors combined lead to polymer–fiber composites with improved modulus and strength compared with their untreated counterparts. However, careful consideration must be exercised toward the alkali concentration and treatment times as extremely harsh treatments will degrade the cellulose crystal structure and inhibit maximum conversion of cellulose I to cellulose II forms.

Native cellulose found in flax and other natural fibers exhibits a monoclinic crystalline lattice of cellulose I (Van de Weyenberg et al., 2006). When subjected to NaOH, the cellulose fiber swells, widening the small pores between the lattice plane, and NaOH penetrates into them. Sodium ions displace hydrogen ions on the hydroxyl groups, forming Na-cellulose I. After removal of the excess NaOH by

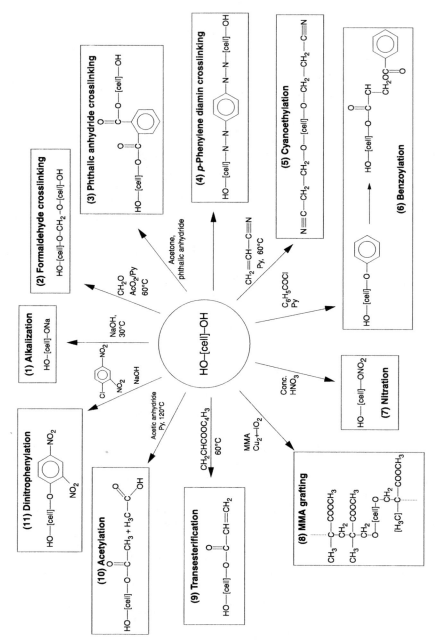

Fig. 13-13 Schematic presentation of chemical modifications of natural fibers.

324

washing with distilled water, sodium ions are removed and conversion occurs from cellulose I to cellulose II, which is more thermodynamically stable. Subsequent to caustic treatment, it was found that the crystallinity increased along with improved crystalline packing due to the removal of noncellulosic components (Mwaikambo and Ansell, 1999; Das and Chakraborty, 2006; Van de Weyenberg et al., 2006). The dissolution of the noncellulosic constituents allowed the cellulose fibrils to rearrange, exhibiting a reduced spiral angle (the angle between the fibrils and fiber axis) and increased the molecular orientation. Ray and Sarkar (2001) showed that after alkali treatment of coir fibers, the fiber modulus and the tenacity increased while the strain at break was reduced.

The partial removal of noncellulosic components (hemicellulose, pectin, and lignin) caused the fiber bundles to separate into technical and elementary fibers of smaller diameters. This led to an increase in the fiber aspect ratio (the ratio of length of the fiber to the diameter) and accordingly to surface area. The surface topography was markedly altered, whereby the surface roughness was greatly enhanced. A number of large holes or pits on the fiber surface were visible (Fig. 13-14). Rout and co-workers attributed those holes to the removal of fatty deposits of tyloses that lie hidden inside the surface of the untreated fiber. The increasing surface roughness was thought to be responsible for better fiber–matrix adhesion, giving rise to additional sites for mechanical interlocking (Rout et al., 2001).

Shrinkage of jute fibers upon alkalization was reported by Gassan and co-workers (Gassan and Bledzki, 1999), which led to a beneficial change in the mechanical properties. It was found that the higher the alkali concentration (up to 28%), the greater the degree of shrinkage from a greater loss of lignin and hemicellulose. As a result, improved yarn modulus and strength were observed due to removal of hemicellulose, which caused a change in the orientation of the amorphous cellulose and some regions of the crystalline cellulose.

Composites comprised of alkali-treated jute fibers and poly(3-hydroxybutyrate-*co*-8%-3-hydroxyvalerate) (tradename Biopol®) were studied by Mohanty and

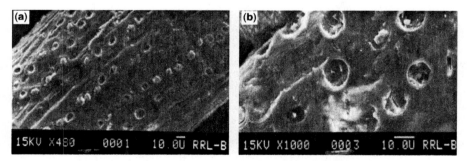

Fig. 13-14 SEM images of coir fibers before (a) and after (b) alkali treatment. Scale bar given in 10 μm. (Reprinted from Rout et al., 2001, Scanning electron microscopy study of chemically modified coir fibers, *Journal of Applied Polymer Science* 79, 1169–1177. Copyright 2001 John Wiley & Sons. Reprinted with permission of John Wiley & Sons, Inc.)

co-workers (Mohanty et al., 2000c). The fibers were oriented and were prepared by a prepreg method. For the composites tested in the parallel direction to the fiber orientation, the tensile, flexural, and impact strengths were significantly improved relative to the pure polymer and to those comprised of defatted fibers (surface waxes removed with a solution of 1 : 2 alcohol and benzene).

The tensile strength of the alkali-treated composite was about 2.4 times greater than that of the pure polymer, while that of the defatted fiber composite was about 1.9 times higher than that of the pure polymer. Similar incremental improvements in flexural strength were obtained of 1.5 times and 1.25 times for alkali and defatted composites, respectively. The authors attributed the improved composite properties seen in the alkali fiber composites to the rough fiber surface topography, enabling mechanical interlocking. Fiber fibrillation increased the effective contact surface area between the fiber and the matrix, and was suspected to contribute to the enhanced mechanical properties.

13.5.2.2 *Acetylation*   Acetylation involves replacing the hydroxyl groups of the natural fibers with acetyl moieties. It is done to plasticize the fibers, improve the dimensional stability, and improve dispersion of fibers into polymeric matrices, traditionally nonbiodegradable polymers such as polyalkenes. After acetylation, the moisture regain was considerably reduced as the fibers became more hydrophobic due to the substitution of hydroxyl groups with acetyl groups (Khalil et al., 1997). It has been found that acetylation of hydroxyl groups from lignin and hemicellulose occurs more readily than of those from cellulose due to the tightly packed, hydrogen-bonded crystalline nature of cellulose (Tserki et al., 2005). Acetylated fibers were obtained by immersing the plant fibers in acetic anhydride as illustrated in Fig. 13-13, reaction (10) in the presence of pyridine (Bauer and Owen, 1988). Any unreacted reagents and acid by-products were removed by Soxhlet extraction with acetone. Alternatively, acetylation could occur in the presence or absence of an acid catalyst. In the presence of a catalyst, the untreated fibers were initially soaked in acetic acid to cause them to swell so that a faster reaction could occur before acetylation with acetic anhydride (Mwaikambo and Ansell, 1999).

Interestingly, the type of natural fibers has proved to be paramount to the degree of acetylation. Because the substitution of hydroxyl groups with acetyl moieties occurs mostly within lignin and hemicellulose, fibers that contain greater amounts of these components generally show greater degrees of esterification. The reactivity of acetic anhydride is greatest with lignin, followed by hemicellulose and lastly cellulose. Teserki and co-workers quantified the ester content after acetylation of flax, hemp, and wood fibers for various reaction times (Tserki et al., 2005). Wood fibers that are rich in lignin and hemicellulose resulted in the greatest ester content (17.2 wt%) after 120 min of treatment. Flax and hemp fibers that are abundant in cellulose displayed lower ester content of 6.9 and 5.8 wt%, respectively.

The effect of acetylation on the surface topography of the fibers was inspected by SEM (Fig. 13-15). For all untreated fibers, uneven layers of wax on the surface can be clearly seen. Upon acetylation, the esterified fibers were smooth compared with the untreated fibers. This indicated that some of the waxy surface substances

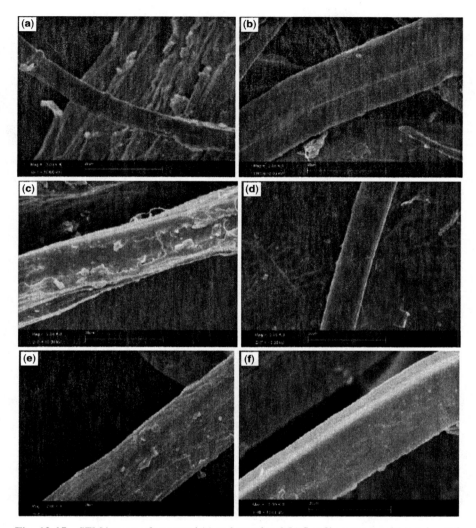

Fig. 13-15   SEM images of untreated (a) and acetylated (b) flax fiber, untreated (c) and acetylated (d) hemp fiber and untreated (e) and acetylated (f) of wood fiber. (Reprinted from Tserki et al., 2005, A study of the effect of acetylation and propionylation surface treatments on natural fibers, *Composites Part A: Applied Science and Manufacturing* 36(8):1110–1118. Reproduced with permission from Elsevier.)

were removed and it was postulated that a smooth fiber surface resulted from the acetyl groups.

Acetylated fibers have proved to be effective in many nonbiodegradable polymeric matrices. The tensile strength and modulus of acetylated fiber/polyester composites were found to be higher, probably due to better resin wetting. Compared with the untreated fiber composites, the increase in hydrophobicity of the acetylated fibers prevented debonding at the interface as covalent bonds could be formed between the matrix and the fibers. The interfacial shear strength (IFSS) (a measure of adhesion) of oil palm acetylated fibers with polystyrene was higher than that of unmodified fibers as the matrix had a greater ability to wet the fiber surface, increasing the work of adhesion. Similar results were found with various commercial epoxy resins. Overall, the improvement in the IFSS was attributed to an increase in compatible surface energies and formation of chemical, physical, and mechanical bonds.

Zini and co-workers studied the effects of acetylation of flax fibers on the tensile properties in poly(3-hydroxybutyrate-co-3-hydroxyhexanoate) (P[3HB-*co*-3HH]) composites. In the unmodified short fiber composite, the tensile strength was lower than in the pure polymer (1.3 GPa and 1.4 GPa, respectively). This was thought to be due to a lack of adhesion between the fibers and the polymer (Zini et al., 2007).

The tensile strength of the resultant composite was greater than that of the unmodified composite upon acetylation of the fibers. The tensile strength increased from 1.3 GPa to 1.8 GPa. With the aid of SEM (Fig. 13-16), a strong interface was observed between the acetylated fibers and the matrix compared with the unmodified fiber system.

It can be clearly seen that large cavities at the interface were present in the unmodified composite, indicative of inferior bonding between the fibers and the matrix. This resulted in poor transfer of stress from the fibers to the matrix, giving lower strength than the pure polymer. Conversely, no cavities were observed at the interface of the acetylated composite, as consequence of a stronger interface.

Due to the difference in surface morphology after acetylation, polarized optical microscopy revealed that transcrystallinity—the phenomenon whereby the fiber behaves as a nucleating agent, inducing spherulite growth along the perpendicular of the fiber—was absent during isothermal crystallization compared with the unmodified fiber. The authors could not definitively explain why transcrystallinity did not develop. However, they did note that their results supported earlier findings that transcrystallinity did not enhance the strength of the fiber composites as the acetylated composites displayed improved strength (without transcrystallinity) compared with the unmodified composites, which exhibited transcrystallization.

13.5.2.3 *Treatment with Anhydrides*

Anhydrides used in composites are by definition considered as organic coupling agents. Coupling agents are described as "substances that are used in small quantities to treat a surface so that bonding occurs between it and another surface" (Lu et al., 2000). Anhydrides that have been successfully used as coupling agents are acetic anhydride (AA), alkyl succinic anhydride (ASA), succinic anhydride (SA), phthalic anhydride (PA), butyric anhydride (BA) and maleic anhydride (MA). The most common of this variety is MA. The carboxylate groups in AA, SA and PA can covalently bond to cellulose through esterification or through hydrogen bonding. The reaction is shown in Fig. 13-17.

MA on the other hand is an α,β-unsaturated carbonyl compound with one carbon–carbon double bond and two carboxylate groups. The conjugated structure significantly enhanced the reactivity of the carbon–carbon double bond with the

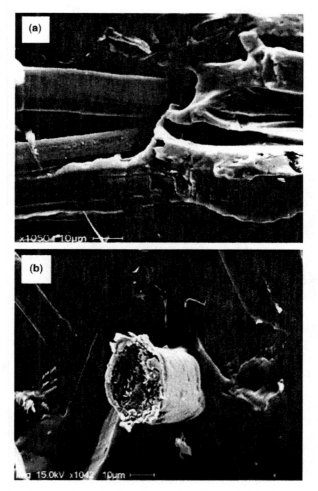

Fig. 13-16 SEM images of fractured surfaces of (a) unmodified fiber composite and (b) acetylated fiber composite. (Reprinted from Zini et al., 2007, Biocomposite of bacterial poly(3-hydroxybutyrate-*co*-3-hydroxybutyrate) reinforced with vegetable fibers, *Composite Science and Technology*, 67(10):2085–2094. Reproduced with permission from Elsevier.)

polymeric matrix using a radical initiator. Crosslinking that occurred caused a strong adhesion at the interface.

BA was used to treat abaca leaf fibers in PHB-HV composites (Shibata et al., 2002). The level of fiber loading ranged from 0 to 20%wt and was randomly orientated. Fibers

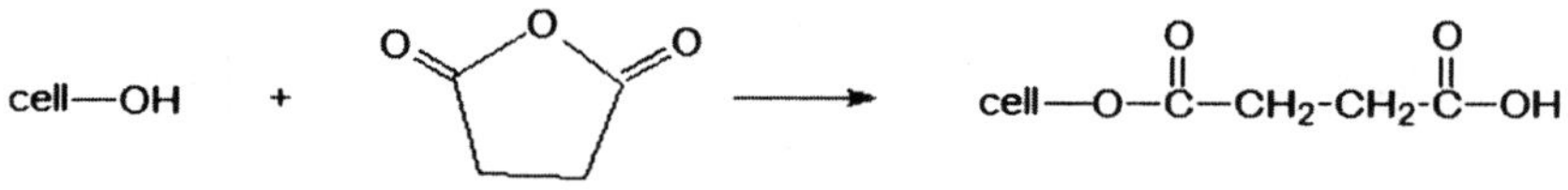

$$\text{cell—OH} \quad + \quad \text{(succinic anhydride)} \quad \longrightarrow \quad \text{cell—O—\overset{O}{\overset{\|}{C}}—CH}_2\text{-CH}_2\text{-\overset{O}{\overset{\|}{C}}—OH}$$

Fig. 13-17 Reaction between cellulose fibers and succinic anhydride (SA).

were treated by soaking in a solution of BA and pyridine (molar ratio of 1 : 1) at room temperature for various prescribed times (0.5 to 24 h). In this study, an improvement in flexural strength and modulus was observed relative to the unmodified composite, but no improvements in the tensile properties were detected.

As expected, an increase in the fiber loading (up to 20 %wt) of the treated fibers in the composites resulted in a substantial rise in the flexural modulus and flexural strength compared with the pure polymer and the untreated composites. The authors of the study found that the anhydride treated fiber composites were comparable with PHB-HV/glass fiber composites at the same fiber loading. The length of the anhydride treatment times of the fibers did not display a significant effect on the flexural properties. The composites with fibers that were treated for 5 h exhibited a marginal rise in flexural strength and modulus compared with the composites with 12 h treatment.

On SEM it was found that there were no appreciable differences in surface topography between the unmodified and the anhydride-treated fibers. The treated fiber surface was only marginally rougher than that of the unmodified fibers, implying that the improvement in flexural properties was not due to a physical effect. However, micrographs of the fractured surface of the unmodified composite and the treated fiber composite revealed that strong fiber–matrix adhesiveness was attained, as was implied by the absence of cavities between the fiber–matrix interfaces in the anhydride-treated composite.

Grafting of anhydrides onto polymers can be effective for enhancing the compatibility between polymeric matrices and natural fibers. It is well known that maleated polypropylene (PP-g-MA) is very effective in polypropylene/natural fiber composites. The mechanism of PP-g-MA is similar to that of anhydrides with cellulose. The reaction scheme is presented in Fig. 13-18. A covalent bond is formed between the cellulose fiber and PP-g-MA, while the PP chain can form chain entanglements with the matrix to give strong adhesion (Mohanty et al., 2001).

In recent publications it was revealed that anhydride grafting onto PHB-HV can be achieved by reactive blending and is an effective coupling agent in natural fiber composites (Avella et al., 2007). The anhydride group can bond with the hydroxyl group of cellulose, like the PP-g-MA, and have the same surface energy/polarity of PHB-HV to enhance the interfacial strength.

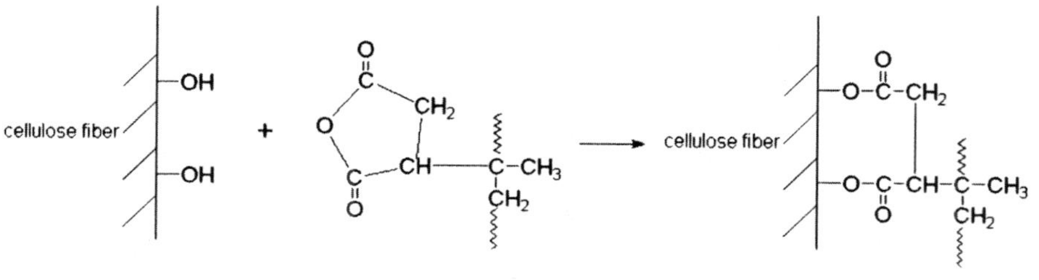

Fig. 13-18 Esterification of cellulose and PP-g-MA.

Anhydride grafting onto PHB-HV involved dibenzoyl peroxide (DBPO), PHB-HV and maleic anhydride mixed in a Brabender-like apparatus at 170°C. Details of the synthesis of maleated PHB are given in Chen et al. (2003). The reaction is shown in Fig. 13-19.

Different amounts of maleic anhydride (3, 5, and 7 wt%) were grafted onto PHB-HV through thermal decomposition of DBPO. It was found that attachment of MA mainly occurred onto the hydroxybutyrate units rather than hydroxyvalerate. It is speculated that this was due to steric hindrance of the ethyl group on the β-carbon of the hydroxyvalerate unit. The reduced acidity from the presence of the ethyl group could also reduce the availability of hydrogen to radical attack.

Maleated PHB-HV was used at 5 wt% in PHB-HV/kenaf fiber composites (Avella et al., 2007). Dynamic mechanical analysis showed that for these composites, the storage modulus (E') was higher than for the uncompatibilized composites and the pure PHB-HV. That is, the stiffness or hardness of the maleated composites is high and can be subjected to higher stress before permanent deformation. The damping factor (tan δ), which is the ratio of the loss modulus (viscous flow component or energy loss) to the storage modulus, was influenced by the presence of maleated PHB-HV. The damping factor of composites with maleated PHB-HV was significantly reduced compared with the unmodified composites with identical fiber loading. An increase in fiber loading also resulted in a decline in the damping factor. A reduction in tan δ indicated that the polymer chains were immobilized to some degree, which in this case could be due to the maleated PHB-HV in composites of the same fiber loading. A reduction in the damping factor also coincided with a rise in the glass transition temperature (T_g), which was measured from the maximum of the damping factor. The increase in T_g supported the idea that the PHB-HV matrix chains were immobilized in the presence of the fibers and even further with maleated PHB-HV.

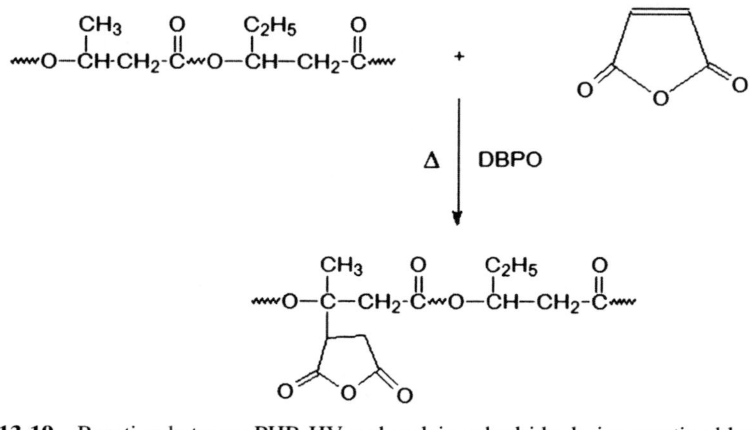

Fig. 13-19 Reaction between PHB-HV and maleic anhydride during reactive blending.

To quantify the strength of the interfacial adhesion, Avella et al. (2007) proposed an adhesion parameter, A, for the fiber–matrix interface, using the T_g values measured from the damping factor:

$$A = \frac{1}{1 - V_f} \frac{\tan \delta_c(T)}{\tan \delta_m(T)} - 1 \qquad (13.1)$$

where V_f is the volume fraction of fillers in the composite, and $\tan \delta_c(T)$ and $\tan \delta_m(T)$ are the $\tan \delta$ values measured at temperatures T of the composite and the pure matrix, respectively. With a higher fiber loading, with and without maleated PHB-HV, A was found to increase, indicating that the interface was weaker. The authors explained this observation by considering that at 20 wt% of kenaf fibers, the maximum surface contact between the matrix and the fibers may be reached; thus higher fiber content did not promote higher adhesion. In the presence of maleated PHB-HV, A was reduced compared with the composites of the same fiber content, implying that the interface was much improved.

Aside from the tensile properties, the flexural and impact properties also increased in a corresponding manner. With the incorporation of kenaf fibers alone, the flexural modulus increased as much as 108% compared with the pure PHB-HV (at 30 vol% fiber loading). Further improvements were observed for composites with maleated PHB-HV due to a stronger interface. An increase as great as 161% was observed at 30 vol% fiber loading. In a corresponding manner, the maximum flexural stress increased by 39% upon incorporation of unmodified kenaf fibers. In the presence of maleated PHB-HV, the composites increased by 46% compared with the pure PHB-HV. A similar rise in the impact strength was observed.

13.5.2.4 *Treatment with Silane Coupling Agents*

Silanes are classified as organic–inorganic coupling agent and were introduced primarily for use with glass fiber-reinforced composites, but they have been found also to be effective with natural fibers. They can couple with virtually any polymer or mineral in composites. Of all modification techniques, the use of silanes to improve the interfacial adhesion of natural fibers and polyolefins is by far the most popular. Fiber surface silanization is thought to enhance the interfacial adhesion thus the interfacial load transfer efficiency by improving the chemical affinity of the fibers with the polymer matrix. The other positive aspects brought about by silanization are enhanced tensile strengths of the composites, reduced effects of moisture on the composite properties, and composite strength.

For most coupling agents, the chemical structure can be represented by the formula $R\text{-}(CH_2)_n\text{-}Si(OR')_3$, where $n = 0$ to 3, OR' is the hydrolyzable alkoxy group, and R is the functional organic group. Some common silanes are vinyltris(2-methoxyethoxy) silane (A-172), γ-methacryloxypropyltrimethoxysilane (A-174), and (3,4-epoxy-cyclohexyl)ethyltrimethoxysilane (A-187). The organo functional group is responsible for bonding with the polymer via copolymerization and/or formation of an interpenetrating network. The attachment of the silane onto the fiber is accomplished after hydrolyzing the silane as depicted in Fig. 13-20.

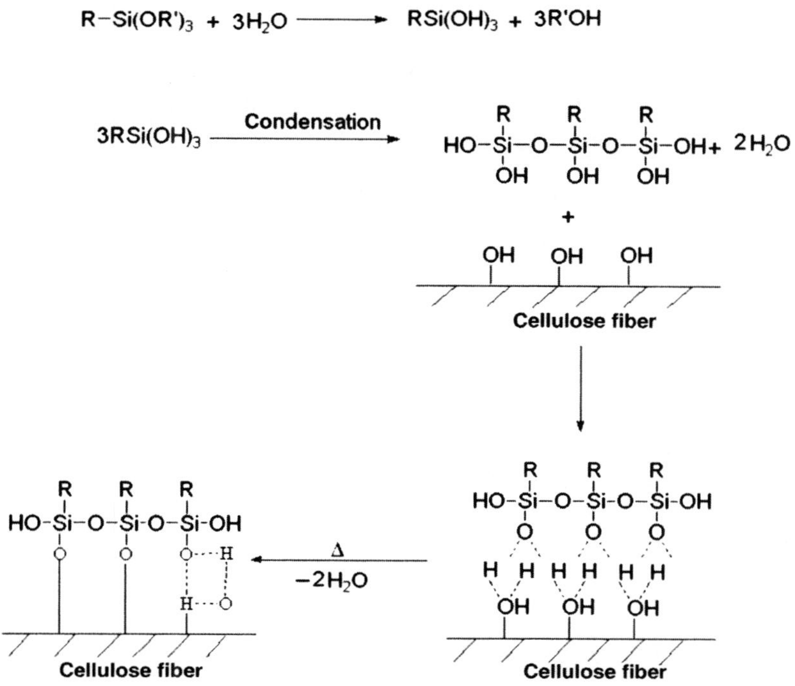

$$R-Si(OR')_3 + 3H_2O \longrightarrow RSi(OH)_3 + 3R'OH$$

Fig. 13-20 Reaction of silane with cellulose.

From the above reactions, the alkoxysilanes first undergo hydrolysis, followed by condensation, and finally formation of bonds. Along with the bond formation, polysiloxanes could be formed. Valandex-Gozalez et al. (1999) found that the presence of polysiloxanes inhibited the adsorption of silanes on the fiber. In the same study, the efficiency of silane treatments on the fiber–polyester composite was found to be better for fibers that had been pretreated with alkali than for those with no pretreatment. The adsorption of silane on alkalized fibers was much greater than on the untreated fibers as there were a larger amounts of exposed cellulose on the surface due to partial removal of lignin, waxes, and hemicellulose. For fibers that were rich in lignin and waxes, the silane had to diffuse through these compounds before it could interact with cellulose, which had a limiting effect on the adsorption. Upon alkalization, the fiber surface became rougher, which increased the effective surface area for adsorption.

Silane coupling agents in anisotropic composites comprised of flax fibers and PHB and copolymers (PHB-HV) were used with some success (Shanks et al., 2004). Flax fibers were treated with an aqueous solution of trimethoxymethacrylsilane prior to composite fabrication. The dynamic mechanical and thermal properties were investigated. Interestingly, it was found in this study that for the untreated fiber composites, the introduction of hydroxyvalerate units into the PHB matrix had indeed increased the stiffness, as indicated by E'. Hydroxyvalerate units are usually introduced into PHB polymers to reduce its brittle nature; it could therefore be expected that the copolymer composites would be less brittle than the PHB composite if no interactions existed between the fibers and the matrix. In agreement with the rise in storage modulus, the T_g values detected by the loss modulus (E'') of the copolymer-unmodified composites were generally greater than for the PHB composites. This suggested that even without the use of silanes, the hydroxyvalerate units provided better chemical affinity between the matrix and the fibers. This was supported by optical microscopy, whereby nucleation of the matrix along the fiber surface was observed.

The storage moduli of silane-treated composites generally exhibited higher values compared with their untreated equivalents, indicating that a greater stiffness was attained, possibly due to greater interfacial strength. The increase in fiber–matrix adhesion observed was in parallel with the enhanced nucleating ability of the fibers, which was observed by optical microscopy. For the silane-treated fibers, many sporadic crystalline regions nucleated from the fiber surface were observed. This was most evident when PHB was the matrix, as there was no crystal growth from the fiber surface for the untreated fibers but such phenomena occurred with silane-treated fibers.

13.5.2.5 *Modification of Natural Fibers by Plasticizer Absorption*   A novel technique to overcome the problem of moisture variability in natural fibers was presented by absorption of plasticizers into the fibers after removal of water (Wong et al., 2002). It was anticipated that fabricated composites would result in better interfacial adhesion by enhancing the compatibility between the hydrophilic fibers with the hydrophobic matrix. It was expected that the dimensional stability of the fibers and composites under different humidity levels would show less variability due to the fiber treatments with less hydrophilic liquids.

The plasticizers chosen in this study were mostly derived from natural sources, such as tributyl citrate (TBC) and glyceryl triacetate (GTA) and one that was commonly used to treat wood, poly(ethylene glycol) (PEG). The plasticizers were introduced into the fibers by impregnation at elevated temperatures after removal of water. Excess plasticizer was removed by washing with acetone and then dried prior to composite fabrication with PHB. The amount of plasticizer present was determined to be between 7 and 8 vol%, which was similar to 8 vol% of moisture typically present in flax. The fiber loading was fixed at 50 vol%.

Dynamic flexural testing was performed to determine the mechanical properties. The storage modulus (G') values of pure PHB and the composites of interest are shown in Fig. 13-21. Upon the addition of unmodified fibers, the stiffness, as indicated by G' was significantly reduced relative to the pure PHB below temperatures of about 70°C. This showed that flax fibers in the unmodified form failed to act as reinforcement to the PHB matrix. Conversely, the composites containing GTA- and PEG-treated fibers displayed higher stiffness throughout the entire temperature range tested, while those treated with TBC exhibited higher G' only below about 0°C. This enhancement in the presence of GTA and PEG (and to a limited degree

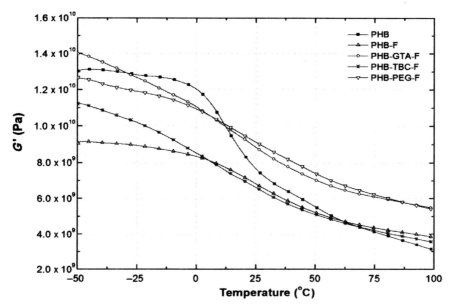

Fig. 13-21   Storage moduli (G') of PHB and composites.

with TBC) was suspected to be due to stronger interfacial adhesion, as suggested by the damping factor tan δ, shown in Fig. 13-22.

The tan δ peak intensity can indirectly indicate the magnitude of the interfacial adhesion in composites. A lower intensity generally indicates that the mobility of the polymer chains is hindered, most probably from the interactions with the fibers or fillers. In this case, the intensity of the unmodified PHB composite relative to pure PHB was much greater, indicating that the motion of the polymer chains was not affected by the fibers (little to no interactions). The reduced degree of crystallinity (i.e., larger amount of the amorphous phase), as measured by DSC, would certainly contribute to the higher tan δ intensity.

The composites containing GTA and PEG generally displayed the lowest tan δ intensity throughout the temperature range investigated. Their degree of crystallinity (as determined by DSC) was lower than that of the unmodified composite, hence the decline in the intensity could be due to greater interfacial adhesion. The adhesion parameter A, as proposed by Avella et al. (2007), was calculated for each of the composites at three temperatures that were most applicable to the service temperature, shown in Table 13-3 (see equation 13.1 above).

At all three temperatures of interest, A was lowest for the composite treated with GTA followed by PEG, as anticipated. Lower values of A correspond to a higher degree of adhesion. It appeared that the presence of TBC had a negligible effect on the interfacial adhesion as the value was not significantly different from that of the unmodified composite. Interestingly, it appeared that the degree of interfacial adhesion was a function of temperature, as A for any composite was observed to

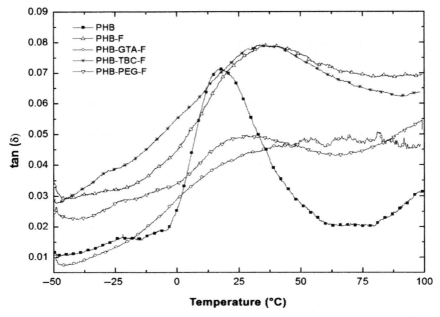

Fig. 13-22 Damping factor (tan δ) of PHB and composites.

be the lowest at 25°C regardless of the fiber treatment. At 0°C, the adhesion was reduced by as much as 3 times. At 50°C, the adhesion was most adversely affected, showing a reduction as great as 7 times relative to that at room temperature. As the systems are complex, a definitive reason cannot easily be determined but it would most likely involve the temperature-sensitive nature of PHB and surface energetic; the situation is further complicated by the presence of plasticizers.

Further evidence to support the effectiveness of GTA and PEG in improving the interfacial adhesion was provided by SEM on inspection of the cross-sections of the fractured composites. These are shown in Fig. 13-23.

The most obvious feature between the different composites was the magnitude of the cavities located at the interface. The largest cavities were seen for the unmodified composite and that with TBC (Fig. 13-23a and c, respectively), these exhibiting

TABLE 13-3 Adhesion Parameter *A* of Various PHB Composites at Three Temperatures

	A Value at Temperature		
Composite	0°C	25°C	50°C
PHB-flax	2.49	1.28	4.56
PHB-GTA-flax	1.28	0.31	2.47
PHB-PEG-flax	1.67	0.50	2.32
PHB-TBC-flax	3.32	1.33	4.43

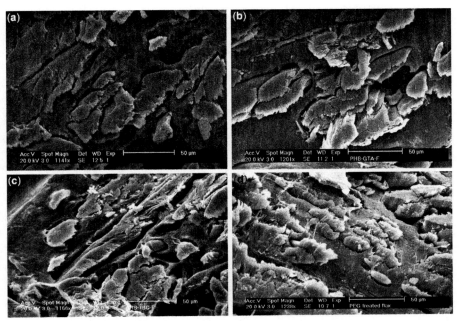

Fig. 13-23 SEM images of PHB–plasticizer–flax systems: (a) PHB–flax, (b) PHB–GTA–flax, (c) PHB–TBC–flax, and (d) PHB–PEG–flax. (Reprinted from Wong et al., 2002, Properties of poly(3-hydroxybutyric acid) composites with flax fibers modified by plasticiser absorption, *Macromolecular Materials and Engineering* 287:647–655. Copyright Wiley-VCH Verlag GmbH & Co. KGaA. Reproduced with permission.)

higher A and lower G' values. Conversely, the composites with GTA and PEG showed smaller cavities, corresponding to lower A and higher G' values.

Polarized optical microscopy was employed to observe the nucleating ability of the fibers in the PHB matrix with various treatments (Fig. 13-24). A dense transcrystalline layer resulted from the untreated fibers, whereas lower-density transcrystals were attained in the presence of the plasticizers. Plasticizer located at the fiber surface would inhibit nucleation at the fiber surface, hence the density of transcrystallinity. Coincidently, a lower degree of transcrystallinity corresponded to a reduced degree of crystallinity as measured by DSC.

13.5.2.6 Novel Dihydric Phenols as Interfacial Bonding Additives

Another novel approach to improving the interfacial bonding to yield stable composites with high mechanical strength was the use of hydrogen-bonding additives. A suitable dihydric phenol is 4,4′-thiodiphenol (TDP). It has been shown with Fourier transform infrared (FT-IR) spectroscopy that hydrogen bonding existed between the hydroxyl groups of cellulose and the hydroxyl and carbonyl groups of aliphatic polyesters (Wong et al., 2004). Another advantage of such

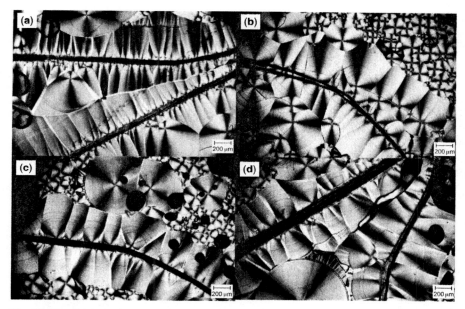

Fig. 13-24 Optical micrographs of PHB–plasticiser–flax systems: (a) PHB–flax, (b) PHB–GTA-flax, (c) PHB–TBC–flax, and (d) PHB–PEG–flax. (Reprinted from Wong et al., 2002, Properties of poly(3-hydroxybutyric acid) composites with flax fibers modified by plasticiser absorption, *Macromolecular Materials and Engineering* 287:647–655. Copyright Wiley-VCH Verlag GmBH & Co. KGaA. Reproduced with permission.)

treatment of the fibers was a noticeable increase in the fiber thermal stability. It was postulated that during processing of TDP and flax fibers, a more ordered cellulose structure could have been obtained due to some disruption of hydrogen bonds of cellulose.

The fibers were pre-treated by dewaxing with acetone via Soxhlet extraction for 24 h prior treatment with TDP. This was intended to remove the surface waxes to maximize the amount of TDP bonding onto the fiber surface. The success of dewaxing was observed via SEM as the presence of fibers with a rugged surface. After drying of any residual solvent and moisture, the washed fibers were immersed in a solution of TDP dissolved in 1,4-dioxane and the solvent was evaporated to leave a "coating" of TDP onto the fibers. The TDP concentration was varied from 0 to 20 vol% based on the fibers at 5% increments. The fiber loadings of the composites were fixed at 50 vol% and the fibers were randomly orientated.

Dynamic mechanical analysis in 3-point bend mode was performed on the composites to observe the improvement in the mechanical properties. The values of storage modulus G' are depicted in Fig. 13-25.

Upon addition of unmodified fibers (no TDP) to PHB, the stiffness markedly reduced due to poor interfacial adhesion arising from the difference in polarity of the constituents. For all TDP-treated composites, the stiffness showed higher

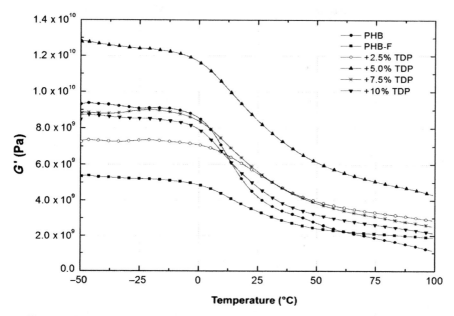

Fig. 13-25   Storage moduli of PHB composites with varying TDP concentrations.

values than the unmodified PHB composite, even after the glass–rubber transition. The stiffness of the composites increased as the concentration of TDP increased from 2.5% to 5.0%. Thereafter, the G' values were greater than for the unmodified PHB composite but lower than that with 5% TDP. Wong et al., attributed the phenomenon to the TDP migrating to the matrix when "excessive" amounts are present, which may cause the matrix to exhibit lower mechanical properties. A reduction in crystallinity, as determined by DSC, corresponded with a diminution of modulus.

The damping factor, tan δ was used to indicate the level of interfacial adhesion of the composites (Fig. 13-26). It can clearly be seen that the intensity of the damping factor was higher in the presence of the unmodified fibers, indicative of lower interfacial interactions. With an increase in TDP level up to 5%, the intensity of the damping factor diminished, implying that interfacial interaction is greater, most likely due to hydrogen bonding at the interface provided by TDP. The interaction parameter, A first introduced by Avella et al. (2007) was used to quantify the level of adhesion; the results are tabulated in Table 13-4. The calculated A values illustrated that the optimal TDP level for maximum adhesion was attained at 5%. Above and below this level, the adhesion was lower but was still better than that of the unmodified composite.

Further evidence to support the improved level of adhesion was observed from the cross-section of the fractured composites with SEM.

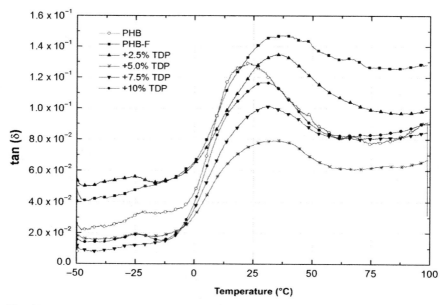

Fig. 13-26 Damping factor (tan δ) of PHB composites with varying TDP concentration.

All composites (Fig. 13-27) showed reasonable dispersion of the fibers within the PHB matrix as no fiber segregation was observed. However, it is evident that there is some limitation on their compatibility as cavities are seen surrounding the fibers in all composites. With TDP, the number of cavities as well as their size was reduced compared with the unmodified composite. Along with the closer contact between the fibers and the matrix with TDP, optical microscopy (Fig. 13-28) showed that the density of the transcrystalline regions was lower than in the unmodified composite. The effect of transcrystallinity on the performance of composites has been investigated extensively and the results are contradictory in the published literature. Some authors concluded that transcrystallinity was beneficial to the mechanical properties by preventing debonding at the fiber–matrix interface (Zhang et al., 1996).

TABLE 13-4 Adhesion Parameter *A* of PHB Composites with Varying TDP Levels

Composite	*A* Value at Temperature		
	0°C	25°C	50°C
PHB-F	1.85	1.15	2.04
+2.5% TDP	1.80	0.94	1.59
+5% TDP	0.35	0.19	0.48
+7.5% TDP	0.55	0.52	0.86
+10% TDP	0.71	0.77	0.93

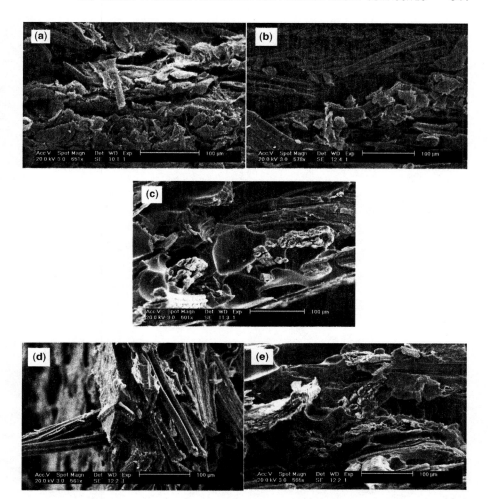

Fig. 13-27 SEM images of the cross-section of (a) unmodified PHB composite, and composites with (b) 2.5% TDP, (c) 5% TDP, (d) 7.5% TDP, and (e) 10% TDP. (Reprinted from Wong et al., 2004, Interfacial improvements in poly(3-hydroxybutyrate)–flax fiber composites with hydrogen bonding additives, *Composites Science and Technology* 64(9):1321–1330. Reproduced with permission from Elsevier.)

Others argue that a transcrystalline region caused a reduction in the mechanical properties by premature brittle failure occurring at the zone where the growing transcrystalline layers meet (Teishev and Maron, 1995). In this study, with the decrease in the density of the transcrystalline regions and induction of hydrogen bonding at the interface, higher G' values were attained compared with the unmodified composite. But at higher levels TDP migrated to the matrix, weakening the matrix by lowering the crystallinity, resulting in lower G' values than those at lower concentrations.

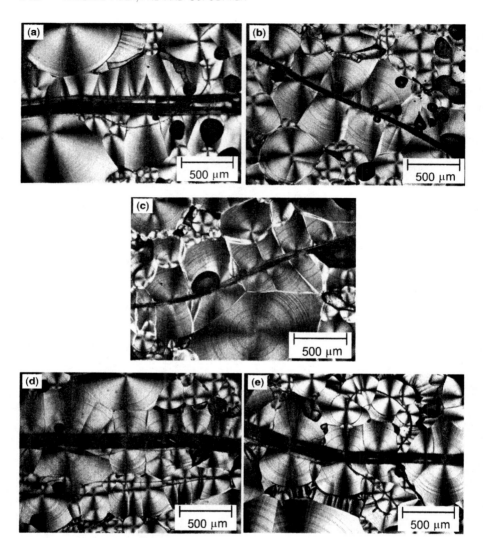

Fig. 13-28 Optical micrographs of (a) unmodified PHB composite, and composites with (b) 2.5% TDP, (c) 5% TDP, (d) 7.5% TDP, and (e) 10% TDP. (Reprinted from Wong et al., 2004, Interfacial improvements in poly(3-hydroxybutyrate)–flax fiber composites with hydrogen bonding additives, *Composites Science and Technology* 64(9): 1321–1330. Reproduced with permission from Elsevier.)

13.6 SUMMARY

Composites comprised of natural fibers and biodegradable thermoplastics can be used as alternative materials to traditional composites that are made of nonbiodegradable constituents such as glass or carbon fibers with polypropylene or epoxy resins.

The use of natural fibers offers the advantages of low density, good mechanical and manufacturing properties, and overall reduction in costs. However, even with these advantages, the use of natural fibers in composites is limited by the lower composite properties compared with those of traditional composites. The lower mechanical behavior was found to be due to two main problems that can be effectively resolved through modifications to the constituents.

The first problem is the incompatibility arising from the hydrophobic polymeric matrices and the hydrophilic nature of the natural fibers. The limited fiber–matrix interactions usually lead to reduced mechanical properties due to inefficient stress transfer from the matrix to the fibers. The second problem is the hydrophilic nature of the fibers as a result of which they absorb moisture. This not only contributes to lower mechanical properties but also affects the dimensional stability. The undesirable variability of moisture content in the composites under different environmental conditions can lead to erratic properties during their service life.

The use of additives and fiber surface treatments in the composites can address these problems and help improve the mechanical behavior of the composites. These effectively introduce a stronger fiber–matrix interface either by altering the fiber surface chemistry to reduce the incompatibility between the constituents or by forming chemical bonds between the components. For some fiber treatments and additives, the composites can be less moisture-sensitive and hence they possess improved dimensional stability. Since there are a range of treatments and additives available, this offers a window of flexibility for composite manufacturers to use appropriate methods without dramatic changes to processing equipment.

REFERENCES

Avella M, Bogoeva-Gaceva G, et al. 2007. Poly(3-hydroxybutyrate-co-3-hydroxyvalerate)-based biocomposites reinforced with kenaf fibres. *Journal of Applied Polymer Science* 104:3192–3200.

Baiardo MF, Scandola M, Licciardello A. 2002. Surface chemical modification of natural cellulose fibres. *Journal of Applied Polymer Science* 83:38–45.

Bauer H, Owen AJ. 1988. Some structural and mechanical properties of bacterially produced poly-beat-hydroxybutyrate-co-beta-hydroxyvalerate. *Colloid and Polymer Science* 266:241–247.

Bledzki AK, Gassan J. 1999. Composites reinforced with cellulose based fibres. *Progress in Polymer Science* 24:221–274.

Brouwer WD. 2000. Natural fibre composites: where can flax compete with glass? *SAMPE Journal* 36:18–23.

Chen C, Peng S, Fei B, et al. 2003. Synthesis and characterisation of maleated poly(3-hydroxybutyrate). *Journal of Applied Polymer Science* 88:659–668.

Corden TJ, Jone A, Rudd CD, Christian P, Downes S. 1999. Initial development into a novel technique for manufacturing a long fibre thermoplastic bioabsorbable composite: in-situ polymerisation of poly(epsilon-caprolactone). *Composites Part A: Applied Science and Manufacturing* 30:737–755.

Coutinho FMB, Costa THS, Carvalho DL, Gorelova MM, de Santa Maria LC. 1999. Thermal behaviour of modified wood fibres. *Polymer Testing* 17:299–310.

Das M, Chakraborty D. 2006. Influence of alkali treatment on the fine structure and morphology of bamboo fibres. *Journal of Applied Polymer Science* 102:5050–5056.

Drumright DE, Gruber PR, Heton DE. 2000. Polylactic acid technology. *Advanced Materials* 12:1841–1846.

El-Hadi A, Schnabel R, Straube E, Muller G, Henning S. 2002. Correlation between degree of crystallinity, morphology, glass temperature, mechanical properties and biodegradation of poly(3-hydroxyalkanoate) PHAs and their blends. *Polymer Testing* 21:665–674.

Fujimaki T. 1998. Processibility and properties of aliphatic polyesters, "Bionelle," synthesised by polycondensation reaction. *Polymer Degradation and Stability* 59:209–214.

Gao SL, Kim JK. 2000. Cooling rate influences in carbon fibre/PEEK composites. Part 1. Crystallinity and interface adhesion. *Composites Part A: Applied Science and Manufacturing* 31:517–530.

Gassan J, Bledzki AK. 1999. Possibilities for improving the mechanical properties of jute/epoxy composites by alkali treatment of fibres. *Composites Science and Technology* 59:1303–1309.

Gassan J, Bledzki AK. 2001. Thermal degradation of flax and jute fibres. *Journal of Applied Polymer Science* 82:1417–1422.

Ha CS, Cho WJ. 2002. Miscibility, properties, and biodegradability of microbial polyester containing blends. *Progress in Polymer Science* 27:759–809.

Herrmann AS, Nickel J, Riedel U. 1998. Construction materials based upon biologically renewable resources—from components to finished parts. *Polymer Degradation and Stability* 59:251–261.

Hill CAS, Abdul Khalil HPS. 2000. Effect of fibre treatments on the mechanical properties of coir or oil palm fibre reinforced polyester composites. *Journal of Applied Polymer Science* 78:1685–1697.

Huda M, Yasui M, Mohri N, Fujimura T, Kimura Y. 2002. Dynamic mechanical properties of solution-cast poly(L-lactide) films. *Materials Science and Engineering* 333:98–105.

Janigova I, Lacik I, Chodak I. 2002. Thermal degradation of plasticised poly(3-hydroxybutyrate) investigated by DSC. *Polymer Degradation and Stability* 77:35–41.

Joffe R, Janis A, Wallstron L. 2003. Strength and adhesion characteristics of elementary flax fibres with different surface treatments. *Composites Part A: Applied Science and Manufacturing* 34:603–612.

Karus MK, Kaup M. 1999. Use of Natural fibres in the German automotive industry. *Journal of the International Hemp Association* 6:72–74.

Kazayawoko M, Balatinecz JJ, Woodhams RT. 1997. Diffuse reflectance fourier transform infrared spectra of wood fibres treated with maleated polypropylenes. *Journal of Applied Polymer Science* 66:1163–1173.

Ke T, Sun X. 2001. Thermal and mechanical properties of poly(lactic acid) and starch blends with various plasticisers. *Transactions of the ASAE* 44:939–944.

Khalil AR, Mitra BC, Lawther M, Banerjee AN. 1997. Studies of acetylation of jute using simplified procedure and its characterisation. *Polymer Plastics Technology and Engineering* 39:757–781.

Kopinke F, MacKenzie K. 1997. Mechanistic aspects of the thermal degradation of poly(lactic acid) and poly(beta-hydroxybutyric acid). *Journal of Analytical and Applied Pyrolysis* 40–41:43–53.

Labrecque LV, Dave V, Gross RA, McCarthy SP. 1995. Citrate esters as biodegradable plasticisers for poly(lactic acid). ANTEC '95.

Labrecque LV, Dave V, Gross RA, McCarthy SP. 1997. Citrate esters as plasticisers for poly(lactic acid). *Journal of Applied Polymer Science* 66:1507–1513.

Li Y, Mai YW, Lin Y. 2000. Sisal fibre and its composites: a review of recent developments. *Composites Science and Technology* 60:2037–2055.

Ljungberg N, Wesslen B. 2002. The effects of plasticisers on the dynamic mechanical and thermal properties of poly(lactic acid). *Journal of Applied Polymer Science* 86:1227–1234.

Lu JZ, Wu Q, McNabb HS. 2000. Chemical coupling in wood fibre and *polymer composites*: a review of coupling agents and treatments. *Wood and Fibre Science* 32:88–104.

Lui S, Netravali AN. 1999. Mechanical and thermal properties of environment-friendly "green" composites made from pineapple leaf fibres and poly(hydroxybutyrate-co-valerate) resin. *Polymer Composites* 20:367–378.

Martin O, Averous L. 2001. Poly(lactic acid): plasticisation and properties of biodegradable multiphase systems. *Polymer* 42:6209–6219.

Mishra S, Misra M, Tripathy SS, Nayak SK, Mohanty AK. 2001. Graft copolymerisation of acrylonitrile of chemically modified sisal fibres. *Macromolecular Materials and Engineering* 286:107–113.

Mohanty A, Khan MA, Hinrichsen G. 2000a. Influence of chemical surface modification on the properties of biodegradable jute fabrics–polyester amide composites. *Composites Part A: Applied Science and Manufacturing* 31:143–150.

Mohanty A, Misra M, Hinrichsen G. 2000b. Biofibres, Biodegradable Polymers and Biocomposites: An Overview. *Macromolecular Materials and Engineering* 276/277:1–24.

Mohanty AK, Khan MA, Sahoo S, Hinrichsen G. 2000c. Effect of chemical modification on the performance of biodegradable jute yarn–biopol composites. *Journal of Materials Science* 35:2589–2595.

Mohanty A, Misra M, Drzal LT. 2001. Surface modifications of natural fibres and performance of the resulting biocomposites: an overview. *Composite Interfaces* 8:313–343.

Mwaikambo LY, Ansell MP. 1999. The effect of chemical treatment on the properties of hemp, sisal, jute and kapok for composite reinforcement. *Die Angewante Makromolekulare Chemie* 272:108–116.

Park JW, Im SS, Kim SH, Kim YH. 2000. Biodegradable polymer blends of poly(L-lactic acid) and gelatinized starch. *Polymer Engineering and Science* 40:2539–2550.

Perego G, Cella GD, Bastioli C. 1996. Effect of molecular weight and crystallinity on poly (lactic acid) mechancial properties. *Journal of Analytical and Applied Pyrolysis* 59:37–43.

Pou JB, Quintero QF, Lusquinos F, Soto R, Perez-Amor L. 2001. Comparative study of the cutting of car interior trim panels reinforced by natural fibres. *Journal of Laser Applications* 13:90–95.

Ray D, Sarkar BK. 2001. Characterisation of alkali-treated jute fibres for physical and mechanical properties. *Journal of Applied Polymer Science* 80:1013–1020.

Rong M, Zhang MQ, Lui Y, Yang GC, Zeng HM. 2001. The effect of fibre treatment on the mechanical properties of unidirectional sisal-reinforced epoxy composites. *Composites Science and Technology* 61:1437–1447.

Rouette HK 2001. Flax stem structure. In: *Encyclopedia of Textile Finishing*. Cambridge, UK: Woodhead Publishing. p. 74.

Rout JT, Misra M, Nayak SK, Mohanty AK. 2001. Scanning electron microscopy study of chemically modified coir fibres. *Journal of Applied Polymer Science* 79:1169–1177.

Rozman JT, Misra S, Nayak M, Mohanty AK. 2001. Preliminary studies on the use of modified ALCELL lignin as a coupling agent in the biofibre composites. *Journal of Applied Polymer Science* 81:1333–1340.

Saha AK, Das S, Basak K, Bhatta D, Mitra BC. 2000. Improvement of functional properties of jute-based composite by acrylonitrile pretreatment. *Journal of Applied Polymer Science* 78:495–506.

Samal RK. 1994. Chemical modification of lignocellulosic fibres I. Functionality changes and graft copolymerisation of acrylonitrile onto pineapple leaf fibres; their characterisation and behaviour. *Journal of Applied Polymer Science* 52:1675–1685.

Samal RK, Ray MC. 1997. Effect of chemical modifications on FTIR spectra II. Physicochemical behaviour of pineapple leaf fibre (PALF). *Journal of Applied Polymer Science* 64:2119–2125.

Samal RK, Mohanty AK, Ray MC. 2001. FTIR spectra and physico-chemical behaviour of vinyl ester participated transesterification and curing of jute. *Journal of Applied Polymer Science* 79:575–581.

Samal RK, Rout SK, Mohanty AK. 1995. Effect of chemical modification on FTIR spectra I. Physical and chemical behaviour of coir. *Journal of Applied Polymer Science* 58:745–752.

Shanks R, Hodzic A, Wong S. 2004. Thermoplastic biopolyester natural fibre composites. *Journal of Applied Polymer Science* 91:2114–2121.

Sharma HSS, Gaughey G, Lyons G. 1999. Comparsion of physical, chemical, and thermal chanracteristics of water-, dew-, and enzyme-retted flax fibres. *Journal of Applied Polymer Science* 74:139–143.

Sheth M, Kumar A, Dave V, Gross RA, McCarthy SP. 1997. Biodegradable polymer blends of poly(lactic acid) and poly(ethylene glycol). *Journal of Applied Polymer Science* 66:1495–1505.

Shibata M, Takachiyo KI, Ozawa K, Yosomiya R, Takeishi H. 2002. Biodegradable polyester composites reinforced with short abaca fibre. *Journal of Applied Polymer Science* 85:129–138.

Singh BS, Verma M, Tyagi OS. 2000. FTIR microscopic studies on coupling agents: treated natural fibres. *Polymer International* 49:1444–1451.

Singleton A, Ballie CA, Beaumont PWR, Peijs T. 2003. On the mechanical properties, deformation and fracture of a natural fibre/recycled polymer composites. *Composites Part B: Engineering* 34:519–526.

Stamboulis AB, Baillie CA, Garkhail SK, van Melick HBH, Peijs T. 2000. Environmental durability of flax fibres and their composites based on polypropylene matrix. *Applied Composite Materials* 7:273–294.

Stamboulis AB, Baillie CA, Peijs T. 2001. Effects of enviromental conditions on mechanical and physical properties of flax fibres. *Composites Part A: Applied Science and Manufacturing* 32:1105–1115.

Teishev A, Maron G. 1995. The effect of transcrystallinity on the transverse mechanical properties of single-polymer polyethylene composites. *Journal of Applied Polymer Science* 56:959–966.

Tserki V, Zafeiropoulos NE, Simon F, Panayiotou C. 2005. A study of the effect of acetylation and propionylation surface treatments on natural fibres. *Composites Part A: Applied Science and Manufacturing* 36:1110–1118.

Valadex-Gozalez A, Cervantes JM, Olayo R, Herrera-Franco PJ. 1999. Chemical modification of henequen fibres with an organosilane coupling agent. *Composites Part B: Engineering* 30:321–331.

Van de Velde K, Baetens E. 2001. Thermal and mechanical properties of flax fibres as potential composite reinforcement *Macromolecular Materials and Engineering* 286:342–349.

Van de Weyenberg I, Truong TC, Vangrimde B, Verpoest I. 2006. Improving the properties of UD flax fibre reinforced composites by applying an alkaline fibre treatment. *Composites Part A: Applied Science and Manufacturing* 37:1368–1376.

Van Den Oever MJA, Bos HL, Van Kemenade MJJM. 2000. Influence of the physical structure of flax fibres on the mechanical properties of flax fibre reinforced polypropylene composites. *Applied Composite Materials* 7:387–402.

Wong S, Shanks RA, Hodzic A. 2002. Properties of poly(3-hydroxybutyric acid) composites with flax fibres modified by plasticiser absorption. *Macromolecular Materials and Engineering* 287:647–655.

Wong S, Shanks RA, Hodzic A. 2004. Interfacial improvements in poly(3-hydroxybutyrate)–flax fibre composites with hydrogen bonding additives. *Composites Science and Technology* 64:1321–1330.

Zafeiropoulos NE, Williams DR, Baillie CA, Matthews FL. 2002. Engineering and characterisation of the interface in flax fibre/polypropylene composite materials. Part I. Development and investigation of surface treatments. *Composites Part A: Applied Science and Manufacturing* 33:1083–1093.

Zhang M, Xu J, Zeng H, Xiong X. 1996. Effect of transcrystallinity on tensile behaviour of discontinuous carbon fibre reinforced semicrystalline thermoplastic composites. *Polymer* 37:5151–5158.

Zini E, Focarete ML, Noda I, Scandola M. 2007. Bio-composite of bacterial poly(3-hydroxybutyrate-co-3-hydroxyhexanoate) reinforced with vegetable fibres. *Composites Science and Technology* 67:2085–2094.

■■■■ **CHAPTER 14**

Starch–Fiber Composites

MILFORD A. HANNA and YIXIANG XU

Industrial Agricultural Products Center, University of Nebraska, Lincoln, NE 68583–0730, USA

14.1   INTRODUCTION

Plastic products have occupied a dominant position since the 1950s when the petroleum industry grew rapidly (Xu and Hanna, 2005). Almost every product we buy and most of the food we eat comes encased in plastic, due to its easy processing, light weight, durability, and low cost (Australian Academy of Science, 2002). While we enjoy the convenience offered by plastic products, the side-effects of the excessive use of synthetic plastics have also become more obvious. Municipal solid wastes consist of 7.2% by weight, or 18% by volume, of plastics (Thiebaud et al., 1997). Serious environmental problems are associated with disposal of used plastic products, attributable to their nondegradability and long-term survival in landfills, resulting in overburdening of landfills and harming of wildlife. For example, it was found that 1 in 30 cetaceans had choked on plastic debris in the oceans (Demicheli, 1996).

Biodegradable Polymer Blends and Composites from Renewable Resources. Edited by Long Yu
Copyright © 2009 John Wiley & Sons, Inc.

In addition to the ecological problems, the pressure on fossil energy resources and growing awareness of their finiteness have triggered interest in reduced dependence on these petroleum-derived plastics. Accordingly, during the past two decades, many research efforts have been put forth to develop environmentally compatible biodegradable products. The potential advantages of such materials are their biodegradability and biocompatibility; they can be easily and naturally decomposed by microbial action after use rather than accumulating in landfills and waterways (Dufresne et al., 2000).

In general, biodegradable polymers can be classified into two categories: natural biopolymers such as starch and cellulose, and synthetic biopolymers including poly(lactic acid), polycaprolactone, poly(hydroxyalkanoate)s, and poly(vinyl alcohol) (Jang et al., 2001). Although these synthetic biopolymers possess excellent mechanical properties, they are expensive, which reduces their competitiveness with their traditional counterparts. More attention has therefore been paid to renewable natural biopolymers from agricultural sources, because of their low cost and availability, and their being totally biodegraded after usage. However, the intrinsic deficiencies of natural biopolymers, including poor mechanical properties, poor water resistance, and difficult processability, limit their wide use. Consequently, the development of biocomposites has become a subject of increasing research interest. Biocomposites, consisting of biodegradable polymers as the matrix material and biodegradable fillers, are expected to fully biodegradable since both components are biodegradable (Averous and Boquillon, 2004). In this chapter, we focus on the preparation and characterization of starch–natural fiber composites.

14.2   STARCH-BASED BIOPOLYMERS

14.2.1   Starch Composition, Structure, and Properties

Starch is a natural and renewable polysaccharide and exists in the form of fine white granules. Starch granules are composed of amylose and amylopectin with the basic composite unit of glucose (Fig. 14-1). Linear amylose, consisting of α-1,4 linked D-glucose, has an average molecular weight of 2×10^5 and is responsible for the amorphous region. Amylopectin, a branched chain with both α-1,4 and α-1,6-linked glucopyranose, has an average molecular weight of 2×10^8 and forms the double helix crystalline structure in starch molecules (Andersen et al., 1999; Xu et al., 2004). The discrete and partially crystalline microscopic granules are held together by intra- and inter-molecular hydrogen bonds (Dufresne and Vignon, 1998).

Although starch granules are insoluble in cold water, they form a high-viscosity paste when heated in the presence of water. This physical change of form of starch granules in hot water is termed gelatinization. Starch gelatinization is a process whereby the intermolecular bonds of starch molecules are broken down in the presence of water and heat, resulting in the collapse of the crystalline structures of starch granules. Starch gelatinization is a slow process. During initial heating of starch, the granules do not change their appearance. When a critical temperature is

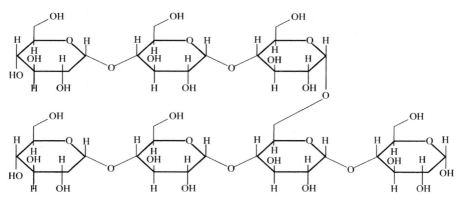

Fig. 14-1   Illustration of starch molecular structure.

reached and the granules swell up and absorb water, starch chains lose their ability to hold together. Increasing temperature further leads to all chains pulling out or away from each other. The gelatinized stage is reached with the formation of a viscous solution (Torres et al., 2007). Gelatinization plays an important role in industrial application of starch. However, gelatinized starch retrogrades with cooling or standing, caused by alignment of the molecules and recrystallization of amylose.

Starch granules are found in various sites of plants, such as roots (sweet potatoes, tapioca), tubers (potatoes), stems (sago palm), cereal grains (corn, rice, wheat, barley, oat, sorghum), and legume seeds (peas and beans) (Swinkels, 1985). Tapioca, potatoes, corn, rice, wheat, and peas are the most widely commercially available starch sources (Dufresne and Vignon, 1998). The shape, composition, and properties of the individual starches vary considerably, depending on the sources as shown in Table 14-1.

TABLE 14-1   Shape, Composition and Properties of Six Commercial Starches

Starch Type	Shape	Diameter (μm)	Amylose Content (%)	Amylopectin Content (%)	Pasting Temperature (°C)
Tapioca	Round, oval[a]	4–35[a]	16–17[b]	83–84[b]	65–70[a]
Potato	Oval, spherical[a]	5–100[a]	20–21[b]	79–80[b]	60–65[a]
Corn	Round, polygonal[a]	2–30[a]	25–28[b]	72–75[b]	75–80[a]
Rice	Polyhedral[e]	3–5[e]	17–30[b]	70–83[b]	79–86[b]
Wheat	Round, lenticular[a]	1–45[a]	25–30[b]	70–75[b]	80–85[a]
Pea[c]	Round, oval[d]	5–20[d]	33–49[d]	51–67[d]	60–67[d]

[a]Swinkels (1985).
[b]Gregorová et al. (2006).
[c]Refers to smooth pea.
[d]Ratnayake et al. (2002).
[e]Champagne (1996).

14.2.2 Starch-Based Biopolymers

There has been increasing interest in the use of starch as a biodegradable thermoplastic polymer and particulate filler because of its abundance, availability, and low cost. However, starch, by itself, is not a good alternative for synthetic polymers because of various inherent drawbacks. Starch is not truly thermoplastic and degrades, under high temperature, to form products with low molecular weight (Dufresne and Vignon, 1998). However, starch can be plasticized by disruption and plasticization of native starch in the presence of water and other plasticizers with high boiling-points, such as triethylene glycol, glycerol, and oleic acid, by injection, extrusion, or blow molding (Cinelli et al., 2003; Corradini et al., 2006). The resultant thermoplastic starch melts and flows under milder conditions. Thus, the use of starch to produce biodegradable plastics has become popular.

Starch-Based Biopolymers Type and Processing Method

Starch-Based Films   Starch-based films and coatings have been used mainly for protecting food and pharmaceuticals from oxygen and moisture, and for encapsulating active ingredients for controlled release (Liu and Han, 2005). The functional properties of starch films and coatings, such as tensile strength, elongation at break, water vapor permeability, and oxygen permeability, are highly dependent on their compositions and processing conditions. In general, starch films, in the absence of any plasticizer, are very brittle and are readily broken into fragments. They are good barriers to oxygen but have poor water resistance, owing to their hydrophilicity. The films are very sensitive to the humidity of environments where they are used and stored. Plasticizers, including glycerol, sorbitol, and poly(ethylene glycerol) (PEG), contain many hydroxyl groups and are compatible with starch film-forming solutions to give homogeneous mixtures without phase separation (Zhang and Han, 2006). Addition of plasticizers can enhance the flexibility and extensibility of the films by reducing intermolecular interaction between starch molecules and increasing the mobility of the chains. However, these improvements are related to the type of plasticizer. Glycerol generally gives the greatest effects on the functional properties of the films, resulting in significant decreases in water vapor permeability and tensile strength but increasing elongation. PEG had the most pronounced effect on oxygen permeability (Laohakunjit and Noomhorm, 2004). Furthermore, the functional properties also are dependent on the amylose content. Strong and flexible films were obtained from starch with high amylose content (Palviainen et al., 2001).

Generally, solvent casting and thermoplastic processing (extrusion and injection molding) have been used to prepare starch films. In the solvent casting method, starch and plasticizer are dissolved in water and heated until the starch gelatinizes. After cooling, the film-forming solution is poured onto a plate and dried at ambient conditions to obtain a film. The advantages of thermoplastic processing over solvent casting include that no preprocessing steps, such as gelatinization and destruction of granular starch, are required, and its environmentally benign character due to the absence of solvents other than water.

Starch-Based Foams   Starch is a good alternative for expanded polystyrene as loose-fill packaging materials to protect fragile products by absorbing or isolating impact energy during transportation and handling (Altieri and Lacourse, 1990; Wang and Shogren, 1997; Fang and Hanna, 2000a). A foam having a dense outer skin and a less dense rigid interior with large, mostly open cells is formed by thermal processing (extrusion), which consists of swelling, gelatinization if the starch and network building in the presence of a blowing agent, such as water, sodium bicarbonate, and citric acid and a nucleating agent such as talcum, calcium carbonate, barium sulfate, aluminum oxide, and silicon dioxide (Parra et al., 2006). Unfortunately, the low elastic modulus (strength) and high hydrophilicity of these native starch-based products do not meet the requirements of some applications. Strategies adopted to overcome these problems include chain modification and blending with other polymers.

Starch Modification   Chemical modification of starch involves replacing the starch's hydroxyl groups (OH) with ester or ether groups to produce hydrophobic thermoplastic materials (Sagar and Merrill, 1995; Shogren, 1996; Jantas, 1997; Bayazeed et al., 1998; Aburto et al., 1999; Miladinov and Hanna, 2000; Sitohy et al., 2000; Xu et al., 2004). After substitution of the hydrophilic hydroxyl groups of starch by ester and ether groups, starch's water resistance and miscibilities with other hydrophobic synthetic polymers are improved. In addition, these substituted groups also function as plasticizers, inhibiting the strong hydrogen-bonding networks between starch molecules, thus enhancing the toughness and thermal stability.

Currently, hydroxypropylated starch is being used in the preparation of commercial packaging material. At the same time, there is a great deal of interest in starch acetylation because of its relative ease. During acetylation, three free hydroxyl groups located on C-2, C-3, and C-6 of the starch molecule are substituted with acetyl groups, resulting in a theoretical maximum degree of substitution of 3. Although starch acetate has excellent functional properties, the biggest hurdle which limits its wide application is its production cost—on average 10 times higher than for native potato starch (Chen et al., 2006).

Mixtures with Other Polymers   Starch has been added, as a natural filler, to synthetic polymer matrixes to accelerate the deterioration of plastics under bioenvironmental conditions (Goheen and Wool, 1991; Matzions et al., 2001). It has also been blended with synthetic polymers such as polystyrene and poly(ethylene-vinyl alcohol) to strengthen the mechanical properties of starch biopolymers (Simmons and Thomas, 1995; Fang and Hanna, 2000b; Kalambur and Rizvi, 2006). However, whichever of the above methods was used, increasing the amount of starch in the matrix caused decreases in the mechanical and physical properties. This was attributed mainly to phase separation caused by immiscibility of starch and hydrophobic polymers at the molecular level.

Attempts have been made to improve the compatibility of starch and hydrophobic polymers, including grafting of functional groups such as carboxylic acid, anhydride, epoxy, urethane, or oxazoline on the polymers. These functional groups react with the

OH groups on starch to form hydrogen bonds, which result in a stable morphology (Shogren et al., 1991; Jang et al., 2001; Wu, 2003). However, some functional groups were found to inhibit the rate of starch biodegradation (Bikiaris and Panayiotou, 1998). Other methods to improve miscibility between starch and other hydrophobic polymers include chemical modifications of starch, as discussed above.

To preserve renewability and biodegradability, starch also has been blended with natural polymers including proteins, chitosan, and natural fibers (Dufresne and Vignon, 1998; Jagannath et al., 2003; Xu et al., 2005a). In this chapter, we focus on starch–natural fiber blends.

14.3 NATURAL FIBERS

14.3.1 Natural Fiber Composition, Structure, and Properties

Natural fibers originate mainly from plant materials. They are three-dimensional bio-polymers and are composed of a variety of chemical substances including cellulose, hemicellulose, lignin, pectin, and small amounts of waxes and fats (Parra et al., 2006). Among these components, cellulose, hemicellulose, and lignin can serve as reinforcement (Guan and Hanna, 2004). The chemical compositions and physical properties of fibers vary with different sources as summarized in Table 14-2.

Cellulose is the most abundant ingredient in all fibers and the content varies with different sources, ranging from 26–43% for bamboo to 87–91% for ramie. Cellulose is a linear polysaccharide polymer consisting of β-(1,4)-linked D-glucose units (Fig. 14-2), with an average molecular weight ranging from 10,000 to 150,000 (Rowell et al., 2000).

The formation of intra- and intermolecular hydrogen bonds results in a highly crystalline structure, with a 80% crystalline region for most plants (Rowell et al., 2000). At the same time, the large numbers of hydroxyl groups make cellulose hydrophilic. In contrast, hemicellulose usually contains more than one sugar. The reason the term "pentosan," instead of "hemicellulose," appears in Table 14-2 is that part of the hemicellulose fraction is composed of five-carbon sugars, namely D-xylose and L-arabinose (Rowell et al., 2000).

14.3.2 Natural Fiber Applications and Modifications

Natural fibers have been used primarily as animal feed. In recent years, interest in the use of natural fibers as reinforcement in both thermoplastic and thermoset polymers has grown. This is attributed to the intrinsic virtues of natural fibers including low cost, abundance and renewability, low density, good mechanical properties, and significant processing advantages such as being less abrasive to processing equipment. They also benefit our ecosystem since CO_2 is emitted (Romhány et al., 2003). Notwithstanding these attractive properties, the mechanical properties of a fiber-reinforced polymer composite depend on many factors, including fiber source, volume fraction, orientation, aspect ratio, and fiber–matrix adhesion (Parra et al., 2006).

TABLE 14-2 Chemical Composition and Physical Properties of Several Common Natural Fibers[a]

| Type of Fiber | Chemical Composition (%) | | | | Physical Properties | | | |
| | | | | | Fiber Length (mm) | | Fiber Width (mm) | |
	Cellulose	Lignin	Pentosan	Ash	Avg	Range	Avg	Range
k fiber								
Rice	28–48	12–16	23–28	15–20	1.4	0.4–3.4	8	4–16
Wheat	29–51	16–21	26–32	4.5–9	1.4	0.4–3.2	15	8–34
Cane fiber								
Bagasse	32–48	19–24	27–32	1.5–5	1.7	0.8–2.8	20	10–34
Bamboo	26–43	21–31	15–26	1.7–5	2.7	1.5–4.4	14	7–27
Grass fiber								
Esparto	33–38	17–19	27–32	6–8	1.2	0.2–3.3	13	6–22
Sabai	–	22	24	6	2.1	0.5–4.9	9	4–28
Reed fiber								
Communis	44–46	22–24	20	3	2.0	1.0–3.0	16	10–20
Bast fiber								
Seed flax	43–47	21–23	24–26	5	33	9–70	19	5–38
Kenaf	44–57	15–19	22–23	2–5	5	2–6	21	14–33
Jute	45–63	21–26	18–21	0.5–2	2	2–5	20	10–25
Hemp	57–77	9–13	14–17	0.8	25	5–55	25	10–51
Ramie	87–91	–	5–8	–	120	60–250	50	11–80
Leaf fiber								
Abaca	56–63	7–9	15–17	1–3	6	2–12	24	16–32
Sisal (agave)	43–62	7–9	21–24	0.6–1	3	1–8	20	8–41
Seed hull fiber								
Cotton	85–96	0.7–1.6	1–3	0.8–2	18	10–40	20	12–38
Wood fiber								
Coniferous	40–45	26–34	7–14	<1	–	–	–	–
Deciduous	38–49	23–30	19–26	<1	–	–	–	–

Note: – indicates no available data.
[a]From Rowell et al. (2000).

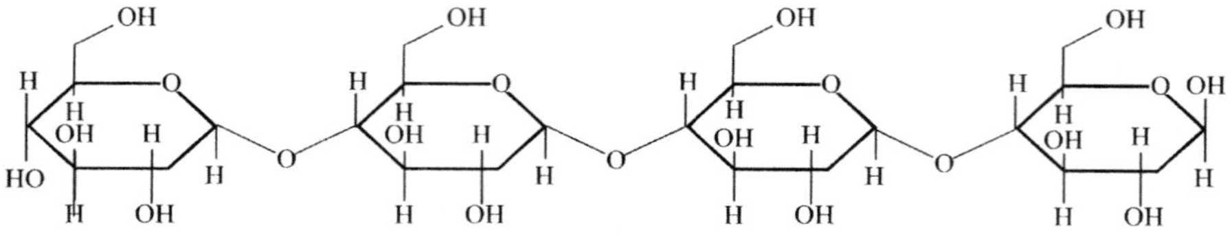

Fig. 14-2 Illustration of cellulose molecular structure.

The inherent polar and hydrophilic nature of cellulose makes it difficult to disperse in nonpolar and hydrophobic thermoplastic polymers, resulting in composites of poor performance (Dufresne et al., 2000).

Strategies to enhance the compatibility between hydrophobic thermoplastic polymers and hydrophilic cellulose fiber include modification of the polymeric matrices or fiber surfaces (Vallo et al., 2004). Alkaline treatment is the most commonly used chemical method to partially remove lignin and hemicellulose from natural fibers. The treatment is at high temperature and for long times. Consequently, the amount of crystalline cellulose increases and a rough surface topography is formed (Alvarez et al., 2003). FT-IR spectra showed that alkali treated fiber had an increased intensity of OH peak, along with disappearance of the C$=$O stretching of carboxylic group, compared to untreated group. The crystal structure of cellulose changes from the parallel polymer chains of cellulose I to aligned antiparallel cellulose II. This leads to higher exposure and concentration of OH groups to interact with groups on the polymer matrix (Vallo et al., 2004). Furthermore, the fiber density, elastic modulus, and fibrillation increased with treatment (Cyras et al., 2001). All of these factors improved the physical and mechanical properties of the composites.

14.4 STARCH–NATURAL FIBER BLENDS

The use of natural fibers as reinforcement opens a window of opportunity for starch-based composite developments because of the fibers' outstanding properties.

14.4.1 Preparation Methods

The preparation methods for starch and natural fiber blends include compound molding and solvent casting. Compound molding usually involves a premix process. Starch, fiber, and other additives are mixed vigorously at a high speed until no major fiber clumps are observed. This intensive premixing is not only efficient in separating the bundles of fibers to ensure a good dispersion, but it also results in a considerable reduction of fiber length and diameter and in a higher aspect ratio (Alvarez et al., 2004). The mixture is then transformed by different compounding techniques, including injection molding, compression, and extrusion. An injection mold typically consists of two parts, a core and a cavity, held together by a clamp (Fig. 14-3).

Starch, fiber, and additives

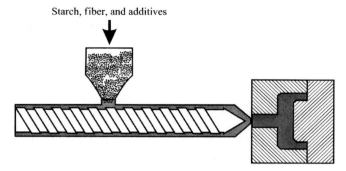

Fig. 14-3 A schematic representation of injection molding.

Compression molding is achieved by applying top pressure, usually from a hydraulic press, to force preheated material into a hot mold to cure (Fig. 14-4).

The effects of these two processing techniques on the properties of fiber composites were investigated by Alvarez et al. (2004). The results showed that complex viscosity varied as a function of frequency and temperature, and compression-molded samples had higher viscosity than injection-molded ones, implying that compression-molded material is more difficult to process than injected materials.

Extrusion is another popular composite polymer processing technique, and is used to produce expanded foams, films, and pellets. The premixed starch and fiber mixtures are compounded either with a single-screw or a twin-screw extruder (Fig. 14-5).

Selection of processing parameters for thermal compounding molding depends on the starch gelatinization temperature and thermal degradation temperatures of the starch and the fibers.

In addition, starch/fiber composites also can be prepared by the methods of filming stacking. In this method, no previous mixing is involved, and the thermoplastic starch film and a layer of fiber are placed on one another alternately, with fibers oriented either unidirectionally or cross-ply. Pressing is performed at a prescribed pressure and temperature in a matched tool (Romhány et al., 2003).

For solvent casting, a fiber suspension is mixed with a gelatinized starch solution. The mixture is homogenized and then cast into a Teflon mold and dried at ambient temperature (Dufresne and Vignon, 1998).

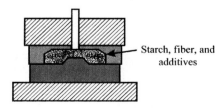

Starch, fiber, and additives

Fig. 14-4 A schematic representation of compression molding.

Starch, fiber, and additives

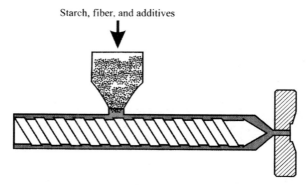

Fig. 14-5   A schematic representation of extrusion molding.

The maximum fiber content that can be introduced into the polymer depends on the selected processing method, with a maximum of 40 wt%. Above 40 wt%, fiber cannot be wetted totally by starch (Romhány et al., 2003). In thermal compounding, the fiber content is governed by viscosity. The viscosity increases with increase in fiber content and the great increase in viscosity at higher fiber contents makes processing difficult (Wollerdorfer and Bader, 1998).

14.4.2   Characterization of Starch−Fiber Blends

Morphology of Starch−Fiber Blends   The morphology of starch−fiber blends can be observed by scanning electron microscopy (SEM) to investigate fiber dispersion and interaction between the starch matrix and the fibers after processing. A typical micrograph of the fractured surface of a fiber-filled thermoplastic starch composite is shown in Fig. 14-6.

It can be seen clearly that there was good adhesion of fiber in the starch matrix. Also, the fiber's surface was covered by starch, further indicating a strong adhesion between the starch and fiber (Curvelo et al., 2001). This good adhesion between starch and fiber can be attributed to the chemical structural similarities of starch and fiber. Both of them are polysaccharides, polar, and hydrophilic; as a result, there is an interaction between these two components.

Transparency of Starch−Fiber Films   The influence of addition of fibers on the transparency of starch-based films was investigated by Ban et al. (2006). The decrease in film transparency with increases in fiber content was ascribed to the light−product interaction and to mixture structure. Film transparency is controlled by light diffusion and transmission: the higher the light diffusion at the interface, the lower the light transmission and transparency. In starch−fiber blends, although fibers can be dispersed in the starch matrix, they cannot be dissolved completely in a starch solution. Instead, the blends produce a discontinuous phase after drying.

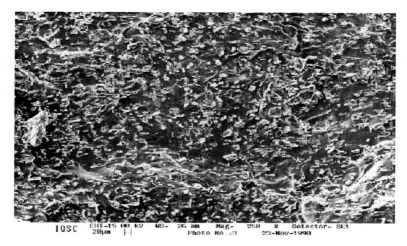

Fig. 14-6   SEM image of a fragile fracture surface of 16% fiber-filled thermoplastic starch (30% glycerin) composite (250× magnification). (From Curvelo et al., 2001.)

As a result, the increase in light diffusion at the starch–fiber interface reduces the light transmission, resulting in low transparency of the film.

Water Absorption   High water absorbency is one of the most limiting factors to many practical applications of pure starch products. The strong hydrophilicity of starch products is ascribed to its high number of hydroxyl groups in the molecules and amorphous structure. Many efforts, including blending of starch with protein and chitosan, have been proposed to improve the water resistance of starch products (Jagannath et al., 2003; Xu et al., 2005a). The water resistance of these products is also enhanced significantly by incorporation of small amounts of fiber (Funke et al., 1998). In fact, cellulose, like starch, is a hydrophilic polymer, owing to high numbers of hydroxyl groups in the molecules. However, differently from starch, high crystallinity and tight microfibril structure in the fiber render it more hydrophobic than starch (Ban et al., 2006). The improvement in water resistance of starch products with addition of fiber is controlled by various factors, including fiber content, fiber modification, the presence of glycerine, and environmental relative humidity. Water sensitivity is generally expressed by water uptake rate, and is measured as a function of time. Water uptake rate decreases dramatically with increasing fiber content. Ban et al. (2006) reported that a 6.7% fiber addition in the starch film resulted in approximately 40% reduction in total water absorbency, while the film water absorbency was reduced further to more than 50% when the cellulosic fiber content was 12.5%. No further improvement was observed at higher fiber contents. Alkaline treatment changes the structure of cellulose from type I to type II. Cellulose II is more hydrophilic than its type I counterpart. Therefore, a starch and alkaline-treated fiber blend has a higher water uptake rate than the blends with untreated fiber (Alvarez et al., 2003). Glycerin acts as a plasticizer in a starch film, and the plasticized starch is more sensitive to water uptake rate than an unplasticized one. Dufresne and Vignon (1998) found that addition of fiber resulted in a more pronounced reduction in water uptake for plasticized starch film than for unplasticized film. A 30% reduction was achieved with a glycerol-plasticized film compared with less than 20% for its unplasticized counterpart, when 40% fiber was added. Starch film is more hydrophilic at high relative humidity, irrespective of the fiber content.

Gas Permeation Gas permeability is an important functional property of starch-based films. Ban et al. (2006) found that incorporation of cellulose prohibited CO_2 permeation. The CO_2 concentration in a vial covered by a pure starch film reached almost the same level as that in air after 10 min. However, the CO_2 concentration increased 50% when using a film with 19% cellulosic fiber content. Further, the permeation rate of CO_2 decreased as the cellulosic fiber content of the starch-based film increased.

Mechanical Properties Mechanical properties, including Young's modulus, ultimate tensile strength, elongation at break, compression, and spring index, are important characteristics of starch-based films and foams. Of these properties, the first three are mainly film properties, while the last two are foam properties. Young's modulus, often referred to as tensile modulus, measures the stiffness of a material, and is expressed as the ratio of the change rate of stress with strain. Ultimate tensile strength is the maximum tensile stress a material can withstand before it breaks. Elongation at break is measured as percentage elongation of sample at the point of rupture to its initial length. Compressibility describes the cushioning ability of a foam material and is related to its relative softness or hardness. Spring index relates to resiliency and refers to the ability of a foam material to recover its original shape after it has been deformed. Poor mechanical properties of neat starch products limit their wide use in diverse commercial applications. In general, the addition of fiber into a starch matrix appreciably improves the mechanical properties. Ban et al. (2006) found that increases in film tensile strength of as much as fivefold were achieved by incorporating up to 22% fiber into these films. However, the extent of fiber enhancement of starch products is influenced by many factors.

Type of Fibers Chemical composition and physical properties of the fibers vary with type. Generally, fibers containing high cellulose contents offer larger enhancement in mechanical properties owing to the intrinsic characteristics of cellulose. Averous and Boquillon (2004) compared cellulose fiber and lignocellulose fiber. The surface properties of these two fibers were found to vary according to the fiber nature. Cellulose fibers had more polar components and greater surface tensions than lignocellulose fibers. The presence of less polar lignins on the surface of lignocellulose fibers decreases the adhesion between fiber and starch, thereby producing lower mechanical properties. Corradini et al. (2006) reported higher Young's modulus and ultimate tensile strength values for matrices with sisal fiber incorporated than for those with coconut and jute fibers, owing to sisal's high cellulose content (67–78% for sisal fiber, 61–71.5% for jute fiber, and 36–43% for coconut).

Torres et al. (2007) confirmed that sisal fibers gave the highest tensile strengths compared with jute and cabuya fibers. The tensile strength increased almost 100% for potato starch with a 10% sisal content with respect to the unreinforced matrix, while improvements were 54% and 15%, respectively, for matrices with additions of jute and cabuya fiber.

Chemical Treatment of Fiber   The goal of chemical treatment of fiber is to change the surface morphology to improve the compatibility between matrix and fiber. Corradini et al. (2006) reported higher Young's modulus and ultimate tensile strength values for composites reinforced with mercerized coconut and jute fibers than for those reinforced with nonmercerized fibers. Mohanty et al. (2000) observed that alkali-treated fabrics had better tensile strengths than dewaxed samples, although dewaxing of fiber also helps improve the fabric–matrix interaction. Alkali treatment of fiber not only produces a rough surface topography to improve the adhesive characteristics of fiber surface by removing natural and artificial impurities, but also fibrillates fiber bundles into smaller fibers, thus increasing the effective surface area available for contact with matrix polymers. The high values of the aspect ratio and rough surface have favorable effects on mechanical properties (Vallo et al., 2004).

Fiber Content   Fibers act as reinforcement for a starch matrix. In general, an increase in mechanical properties of starch/fiber composites is observed with fiber content, irrespective of fiber orientation and damage that may occur during mixing (Alvarez et al., 2003). The fiber content is often expressed by fiber volume fraction, and a composite's Young's modulus and ultimate tensile strength are functions of fiber volume fraction, having an increasing trend with increase in fiber volume fraction (Alvarez et al., 2006). Elongation decreases with fiber content (Averous and Boquillon, 2004). The improvement of the mechanical properties suggests that there is good fiber–starch matrix adhesion. Starch and natural fiber are polysaccharides and have the same unit structure. The intrinsic chemical similarities between these two components enhance compatibility. Torres et al. (2007) found that the ultimate tensile strength of the unreinforced potato starch matrix increased from 4.15 MPa to 6.5 MPa when 5 wt% of sisal fiber was added. However, when fiber content was above 10 wt%, fiber dispersion in the composites became difficult, resulting in fiber clumps and voids in the specimens, thereby decreasing the tensile strength of specimens and increasing the standard deviations. Lawton et al. (2004) showed that trays with a fiber content of 15–30% had the best performance at all humidities. Tray strength declined with fiber contents above 30%.

Fiber Length   Investigations of the effects of different fiber lengths on the strength of starch films indicated that short fiber reinforced the tensile properties of the film, since an effective load transfer from the matrix to the fiber provided a strong fiber–matrix interfacial bond (Alvarez et al., 2005). On the other hand, long fibers result in reduced film uniformity. Takagi and Ichihara (2004) indicated that 15 mm was a critical fiber length. The negative effect of long fibers on mechanical properties may be attributable to entanglement from long fiber–long fiber interactions, difficulty in uniformly dispersing such fibers in a starch matrix, and the occurrence of fiber pull-out (Ban et al., 2006).

Fiber Alignment   Fiber direction in a starch matrix plays a critical role in the reinforcement effect. Reinforcement is relatively low when fibers are laid transverse to the direction of loading. This can be explained by the fact that the fibers loaded in the transverse direction act as barriers that prevent the distribution of stress throughout the matrix (Alvarez et al., 2006). In contrast, when the fibers are aligned in the direction of loading, the strength of starch composites is double that of the neat matrix (Romhány et al., 2003).

Foam   Compression resistance and flexibility are the most commonly used parameters in evaluating the mechanical properties of starch-based foams. Parra et al. (2006) prepared cassava starch foam with addition of cassava and wheat fibers. Foam with 1% cassava fiber had the higher compression resistance, and compression resistance decreased with increasing fiber content up to 3%. The flexibility of the cassava starch increased with fiber content up to 2%, followed by a decrease with further increases in fiber content to 3%.

Thermal Behavior   Thermal characteristics, including phase transitions, melting, and thermal degradability play important roles in determining thermal processing conditions for starch-fiber composites.

Differential scanning calorimetry (DSC) is an accepted method for determining the thermal transition of starch polymer and its fiber blends. Samples sealed in stainless steel DSC pans are heated from 20 to 250°C at a heating rate of 10°C/min in a nitrogen atmosphere. The glass transition temperature and melting temperature are taken as the inflection point of the increment of specific heat capacity and as the peak value of the endothermal process in the DSC curves, respectively. Starch has a semicrystalline structure, showing a glass transition temperature and melting temperature on a DSC heating curve, with the presence of water or other plasticizers. Addition of fiber into the starch matrix increases its glass transition temperature (Curvelo et al., 2001). This is attributed to the interaction between the fiber and the plasticizer, leading to less-plasticized starch.

Measurements of the thermal stability of starch and its fiber composites are carried out by thermogravimetric analysis (TGA). Samples of 3–6 mg are placed in the balance system and heated from 50 to 650°C at a heating rate of 20°C/min in a nitrogen atmosphere. Degradation temperatures are determined as the peak maxima (Averous and Boquillon, 2004). Initial degradation of pure starch began at 297°C and ended at 326°C with a weight loss of 70.7% (Xu et al., 2005b). The decomposition of starch can be explained by inter- or intramolecular dehydration reactions of the starch molecules, with water as the main product of decomposition (Thiebaud et al., 1997). For fibers, two thermal decomposition temperatures are observed. The first at 300°C, corresponding to hemicellulose and glucosidic link depolymerization, overlaps with that of starch, while the second one at 360°C has been assigned to the thermal degradation of the α-cellulose in the fiber

(Alvarez and Vázquez, 2004). Addition of fiber into the starch matrix has a beneficial effect on the thermal stability of the starch matrix, with the maximum onset of degradation temperature shifting to a higher value and the mass loss decreasing. The improvement in the thermal stability of a starch–fiber blend, compared with an unfilled starch matrix, was attributed to overall low water content in the composite. In addition, diverse interactions between starch, fiber, and plasticizer reduce original water sites on thermoplastic starch (Averous and Boquillon, 2004).

14.5 SUMMARY

Starch-based films and foams have limited commercial applications because of their lack of flexibility and water resistance, among other characteristics. The addition of fiber into starch matrices enhances functional properties, that is, it decreases the water absorption and CO_2 permeability and increases Young's modulus and tensile strength. These improvements in functional properties extend the use of starch-based products in new applications that are moisture sensitive, including gas-selective membranes and ion channels. In addition, starches and natural fibers low cost and are natural and renewable polysaccharides from agricultural resources, which means that starch–natural fiber composites are fully biodegradable. The potential of these "green" composites to share the market with petroleum-based packaging becomes evident with stricter waste disposal regulations.

REFERENCES

Aburto J, Alric I, Thiebaud S, et al. 1999. Synthesis, characterization, and biodegradability of fatty-acid ester of amylose and starch. *J Appl Polym Sci* 74:1440–1451.

Altieri PA, Lacourse NL. 1990. Starch-based protective loose-fill material. *Proceedings of the Corn Utilization Conference III*, June 20–21, St. Louis, MO. National Corn Growers Association. Section 2, p. 1–4.

Alvarez VA, Vázquez A. 2004. Thermal degradation of cellulose derivatives/starch blends and sisal fiber biocomposites. *Polym Degrad Stab* 84:13–21.

Alvarez VA, Ruscekaite RA, Vázquez A. 2003. Mechanical properties and water adsorption behavior of composites made from a biodegradable matrix and alkaline-treated sisal fibers. *J Compos Mater* 37(17):1575–1588.

Alvarez VA, Rerenzi A, Kenny JM, Vázquez A. 2004. Melt rheological behavior of starch-based matrix composites reinforced with short sisal fibers. *Polym Eng Sci* 44(10):1907–1914.

Alvarez V, Vázquez A, Bernal C. 2005. Fracture behavior of sisal fiber-reinforced starch-based composites. *Polym Compos* 26:316–323.

Alvarez V, Vázquez A, Bernal C. 2006. Effect of microstructure on the tensile and fracture properties of sisal fiber/starch-based composites. *J Compos Mater* 40(1):21–35.

Andersen PJ, Kumar A, Hodson SK. 1999. Inorganically filled starch based fiber reinforced composite foam materials for food packaging. *Mater Res Innov* 3:2–8.

Australian Academy of Science. 2002. Making packaging greener—biodegradable plastics. http://www.science.org.au/nova/061/061key.htm (accessed August 28, 2007).

Averous L, Boquillon N. 2004. Biocomposites based on plasticized starch: thermal and mechanical behaviours. *Carbohydr Polym* 56:111–122.

Ban WP, Song JG, Argyropoulos DS, Lucia LA. 2006. Improved the physical and chemical functionality of starch-derived films with biopolymers. *J Appl Polym Sci* 100:2542–2548.

Bayazeed A, Farag S, Shaarawy S, Hebeish A. 1998. Chemical modification of starch via etherification with methyl methacrylate. *Starch/Staerke* 50:89–93.

Bikiaris D, Panayiotou C. 1998. LDPE/starch blends compatibilized with PE-g-MA copolymers. *J Appl Polym Sci* 70:1503–1521.

Champagne ET. 1996. Rice starch composition and characteristics. *Cereal Food World* 41(11):833–838.

Chen Y, Ishikawa Y, Maekawa T, Zhang Z. 2006. Preparation of acetylated starch/bagasse fiber composite by extrusion. *Trans ASAE* 49(1):85–90.

Cinelli P, Chiellini E, Gordon SH, Imam SH. 2003. Characteristics and degradation of hybrid composite films prepared from PVA, starch and lignocellulosics. *Macromol Symp* 197:143–155.

Corradini E, de Morais LC, Rosa MdF, Mazzetto SE, Mattoso LHC, Agnelli JAM. 2006. A preliminary study for the use of natural fibers as reinforcement in starch-gluten-glycerol matrix. *Macromol Symp* 246:558–564.

Curvelo AAS, de Carvalho AJF, Agnelli JAM. 2001. Thermoplastic starch–cellulosic fiber composites: preliminary result. *Carbohydr Polym* 45:183–188.

Cyras VP, Iannace S, Kenny JM, Vazquez A. 2001. Relationship between processing and properties of biodegradable composites based on PLC/starch matrix and sisal fibers. *Polym Compos* 22(1):104–110.

Demicheli M. 1996. Biodegradable plastics from renewable sources. http://www.jrc.es/iptsreport/vol10/english/Env1E106.htm#Contacts (accessed by Sep. 20, 2003).

Dufresne A, Vignon MR. 1998. Improvement of starch film performances using cellulose microfibrils. *Macromolecules* 31:2693–2696.

Dufresne A, Dupeyre D, Vignon MR. 2000. Cellulose microfibrils from potato tuber cells: processing and characterization of starch–cellulose microfibril composites. *J Appl Polym Sci* 76:2080–2092.

Fang Q, Hanna MA. 2000a. Mechanical properties of starch-based foams as affected by ingredient formulation and foam physical characteristics. *Trans ASAE* 43:1715–1723.

Fang Q, Hanna MA. 2000b. Functional properties of polylactic acid starch-based loose fill packaging foams. *Cereal Chem* 77(6):779–783.

Funke U, Bergthaller W, Lindhauer MG. 1998. Processing and characterization of biodegradable products based on starch. *Polym Degrad Stab* 59:293–296.

Goheen SM, Wool RP. 1991. Degradation of polyethylene-starch blends in soil. *J Appl Polym Sci* 42:2691–2701.

Gregorová E, Pabst W, Bohačenko I. 2006. Characterization of different starch types for their application in ceramic processing. *J Eur Ceram Soc* 26:1301–1309.

Guan JJ, Hanna MA. 2004. Functional properties of extruded foam composites of starch acetate and corn cob fiber. *Ind Crops Prod* 19:255–269.

Jagannath JH, Nanjappa C, Das Gupta DK, Bawa AS. 2003. Mechanical and barrier properties of edible starch-protein-based films. *J Appl Polym Sci* 88:64–71.

Jang BC, Huh SY, Jang JG, Bae YC. 2001. Mechanical properties and morphology of the modified HDPE/starch reactive blend. *J Appl Polym Sci* 82:3313–3320.

Jantas R. 1997. Synthesis and characterization of acryloyloxystarch. *J Appl Polym Sci* 65:2123–2129.

Kalambur S, Rizvi SSH. 2006. An overview of starch-based plastic blends from reactive extrusion. *J Plast Film Sheet* 22:39–57.

Laohakunjit N, Noomhorm A. 2004. Effects of plasticizers on mechanical and barrier properties of rice starch film. *Starch/Staerke* 56:348–356.

Lawton JW, Shogren RL, Tiefenbacher KF. 2004. Aspen fiber addition improves the mechanical properties of baked cornstarch foams. *Ind Crops Prod* 19:41–48.

Liu Z, Han JH. 2005. Film-forming characteristics of starches. *J Food Sci* 70(1):E31–36.

Matzions P, Bikiaris D, Kokkou S, Panayiotou C. 2001. Processing and characterization of LDPE/starch products. *J Appl Polym Sci* 79:2548–2557.

Miladinov VD, Hanna MA. 2000. Starch esterification by reactive extrusion. *Ind Crops Prod* 11:51–57.

Mohanty AK, Khan MA, Hinrichsen G. 2000. Surface modification of jute and its influence on performance of biodegradable jute-fabric/Biopol composites. *Compos Sci Technol* 60:1115–1124.

Palviainen P, Heinamaki J, Myllarinen P, Lahtinen R, Yliruusi J, Forssell P. 2001. Corn starches as film former in aqueous-based film coating. *Pharm Dev Technol* 6(3):353–361.

Parra DF, Carr LG, Ponce P, Tadini CC, Lugao AB. 2006. Biodegradable foams made of cassava starch and fibers: influence in the mechanical properties. *Proceedings of the 2nd CIGR Section VI International Symposium on Future of Food Engineering*, April 26–28, Warsaw, Poland.

Ratnayake WS, Hoover R, Warkentin T. 2002. Pea starch: composition, structure and properties—a review. *Starch/Staerke* 54:217–234.

Romhány G, Karger-Kocsis J, Czigány T. 2003. Tensile fracture and failure behavior of thermoplastic starch with unidirectional and cross-ply flax fiber reinforcements. *Macromol Mater Eng* 288:699–707.

Rowell RM, Han JS, Rowell JS. 2000. Characterization and factors effecting fiber properties. In: Frollini E, Leão AL, Mattoso LHC, editors. *Natural Polymers and Agrofibers Composites*. Brazil: Embrapa Instrumentação Agropecuária. p. 115–134.

Sagar AD, Merrill EW. 1995. Properties of fatty-acid esters of starch. *J Appl Polym Sci* 58:1647–1656.

Shogren RL. 1996. Preparation, thermal properties, and extrusion of high-amylose starch acetate. *Carbohydr Polym* 29:57–62.

Shogren RL, Thompson AR, Greene RV, Gordon SH, Cote G. 1991. Complexes of starch polysaccharides and poly (ethylene co-acrylic acid): structural characterization in the solid state. *J Appl Polym Sci* 47:2279–2286.

Simmnons S, Thomas EL. 1995. Structural characteristics of biodegradable thermoplastic starch/poly (ethylene-vinyl alcohol) blends. *J Appl Polym Sci* 58:2259–2285.

Sitohy MZ, Labib SM, El-Saadany SS, Ramadan MF. 2000. Optimizing the conditions for starch dry phosphorylation with sodium mono- and dihydrogen orthophosphate under heat and vacuum. *Starch/Staerke* 4:95–100.

Swinkels J.J.M. 1985. Composition and properties of commercial native starches. *Starch/ Staerke* 37:1–5.

Takagi H, Ichihara Y. 2004. Effect of fiber length on mechanical properties of "green" composites using a starch-based resin and short bamboo fibers. *JSME International Journal, Series A* 47(4):551–555.

Thiebaud S, Aburto J, Alric L, et al. 1997. Properties of fatty-acid esters of starch and their blends with LDPE. *J Appl Polym Sci* 65:705–721.

Torres FG, Arroyo OH, Gomez C. 2007. Processing and mechanical properties of natural fiber reinforced thermoplastic starch biocomposites. *J Thermoplast Compos Mater* 20:207–223.

Vallo C, Kenny JM, Vazquez A, Cyras VP. 2004. Effect of chemical treatment on the mechanical properties of starch-based blends reinforced with sisal fiber. *J Compos Mater* 38(16):1387–1399.

Wang L, Shogren RL. 1997. Preparation and properties of corn-based loose fill foams. In *Proceedings of the 6th Annual Meeting of the Bio/Environmentally Degradable Polymer Society*, St. Paul, MN.

Wollerdorfer M, Bader H. 1998. Influence of natural fibers on the mechanical properties of biodegradable polymers. *Ind Crops Prod* 8:105–112.

Wu CS. 2003. Physical properties and biodegradability of maleated-polycaprolactone/starch composite. *Polym Degrad Stab* 80:127–134.

Xu YX, Hanna MA. 2005. Physical, mechanical, and morphological characteristics of extruded starch acetate foams. *J Polym Environ* 13(3):221–230.

Xu YX, Miladinov V, Hanna MA. 2004. Synthesis and characterization of starch acetates with high degree of substitution. *Cereal Chem* 81(6):735–740.

Xu YX, Kim KM, Hanna MA, Nag D. 2005a. Chitosan-starch composite film: Preparation and characterization. *Ind Crops Prod* 21(2):185–192.

Xu YX, Dzenis YA, Hanna MA. 2005b. Water absorption, thermal, and biodegradability of starch acetates foams. *Ind Crops Prod* 21(3):361–368.

Zhang Y, Han JH. 2006. Mechanical and thermal characteristics of pea starch films plasticized with monosaccharides and polyols. *J Food Sci* 71(2):E109–118.

PART V

BIODEGRADABLE COMPOSITES

CHAPTER 15

Starch-Based Nanocomposites Using Layered Minerals

H. R. FISCHER and J. J. DE VLIEGER

TNO Science and Technology 5600 HE Eindhoven, The Netherlands

15.1 INTRODUCTION

The worldwide production of synthetic plastics is still growing. Out of the total plastics production approximately 40% is used in packaging. This short-term application will cause a global environmental problem because these plastics are not compostable and are difficult to recycle. The limited fossil resources require a renewable alternative that will not eventually contribute to the amount of greenhouse gases. Plastics from renewable resources are in most cases compostable, solving landfill and litter problems. The bioplastics in the market today are used in specific sectors where biodegradability is required such as packaging, agriculture, and hygiene. Bioplastics based on starch exploit the benefits of natural polymerization and the availability of the raw material. They offer a renewable resource for compostable plastics at potentially low and competitive prices.

Starch is nature's primary means of storing energy and is found in granule form in seeds, roots, and tuber cells as well as in stems, leaves, and fruits of plants. The two main components of starch are polymers of glucose: amylose (MW 10^5–10^6),

Biodegradable Polymer Blends and Composites from Renewable Resources. Edited by Long Yu
Copyright © 2009 John Wiley & Sons, Inc.

the major component and an essentially linear molecule; and amylopectin (MW $10^7 - 10^9$), a highly branched molecule. Starch granules are semi-crystalline with crystallinity varying from 15% to 45% depending on the source. The term "native starch" is mostly used for industrially extracted starch. It is an inexpensive (<0.5 euro/kg) and abundant product, available from potato, maize, wheat, and tapioca.

Thermoplastic starch (TPS) or destructurized starch is a homogeneous thermoplastic substance made from native starch by swelling in a solvent (plasticizer, mainly water and/or glycerol) and a consecutive "extrusion" treatment consisting of a combined kneading and heating process. Due to the destructurization treatment, the starch undergoes a thermomechanical transformation from the semicrystalline starch granules into a homogeneous amorphous polymeric material. One of the major problems connected with the use of most natural polymers, especially of carbohydrates, is their high water permeability and associated swelling behaviour in contact with water. All this contributes to a considerable loss of mechanical properties, which prohibits straightforward use in most applications. So far, special processing or after-treatment procedures are necessary to sustain an acceptable product quality. Presently applied methods for decreasing the hydrophilicity and increasing and stabilizing mechanical properties include:

- Application of hydrophobic coating(s) on the surface of bioplastic products
- Blending with different, hydrophobic, biodegradable synthetic polymers (polyesters)
- Reactive extrusion of natural polymers (graft- and co-polymerization, esterification during extrusion process), which diminishes the opportunities for hydrolysis.

The resulting products are, however, rather expensive compared with common alternative plastics. New concepts are required to solve the intrinsic problem of the hydrophilicity and mechanical instability of starch-based bioplastics without too much added cost. Application of the nanocomposite concept has proved to be a promising option in this respect. Moreover, the combination of two materials from a renewable/totally natural source into a new engineering material with enhanced properties, as in the case of combination of starch and natural clay as nanocomposite materials, offers a new concept in dealing with environmental problems caused by extensive use of disposable plastic products.

This chapter focuses on starch-based nanocomposites using layered minerals. An extensive review of developments in nanocomposites with many other biodegradable polymers as well has recently been published (Yang et al., 2007).

15.2   STARCH–MONTMORILLONITE NANOCOMPOSITES

For the preparation of nanocomposite materials consisting of a polymer and clay, it is necessary to use special compatibilizing agents (modifiers) between the two basic materials. However, since starch is plasticized mostly by compounds capable of hydrogen bonding and organized by the formation of (multiple) hydrogen bonds, such as water, glycerol, and amines, during processing at elevated temperatures, those plasticizers may also be used for distribution of the clay platelets. Amorphous starch recrystallizes upon storage, and the crystal form developed as well as the crystallinity depend on the water content. In order to achieve the final clay–starch nanocomposite material, "clay modification" and "extrusion" processing steps can be distinguished; these are described below.

The starch and the modified or preswelled clay can be mixed at temperatures above the softening point of the polymer by polymer melt processing (extrusion) to obtain exfoliated nanocomposite materials. Fully destructurization is needed for successfully polymer melt processing of starch.

The process of clay exfoliation requires essentially a modification of the surface such that the interlayer width exceeds some threshold value that depends on the difference of absorption energy in order to be thermodynamically favorable (Kudryavtsev et al., 2004). For exfoliation in the presence of starch, spontaneous mixing of (modified) clay and starch is required with enthalpic interactions between TPS and modifier units being favorable. Hydrophilic compounds such as the plasticizers mentioned earlier also interact strongly with the remaining ions found in the galleries of natural clay or with hydrophilic clay modifiers that might be used to meet these conditions and allow exfoliation to occur.

The following reports of studies present various attempts to obtain exfoliated structures in order to find a relation between improved properties and the nanocomposite structures.

True starch nanocomposites were first described in a study in which potato starch and modified clay were mixed above the softening point of starch by polymer melt processing (extrusion) (Fischer et al., 2001). The screw speeds were shown to be of great importance for the clay dispersion. The higher the screw speed, the better the dispersion of the clay particles within the starch matrix. TEM analysis showed an excellent and homogeneous distribution of the individual clay sheets (see Fig. 15-1). Sample preparations are not extensively described but no agglomerations of clay were found in X-ray diffraction (XRD) analysis. The resulting retrogradated materials displayed remarkable stability against cold water; no swelling occurred, and blown film bags filled with water retained their mechanical integrity and did not leak over a period of about 3 weeks. The clay in the starch–clay nanocomposite certainly affected the water stability properties (see Fig. 15-2).

TPS samples consisting of starch/water/glycerol in the proportions 5/2/3 were premixed at 110°C for 25 min in a Haake mixer, then cooled, cut and dried, and mixed with predried clays (three organically modified montmorillonite [MMT] with different ammonium cations, and one unmodified Na^+MMT [Cloisite Na^+] were used); the content of clay ranged from 2.5 to 10 wt% at 110°C (Park et al., 2002, 2003). The blends after preparation were placed in a sealed bag to prevent any moisture absorption. Only some partial intercalation was detected by XRD analysis; the samples containing the nonmodified clay showed hardly any intercalation with the starch. The modification of the clays does not match well thermodynamically

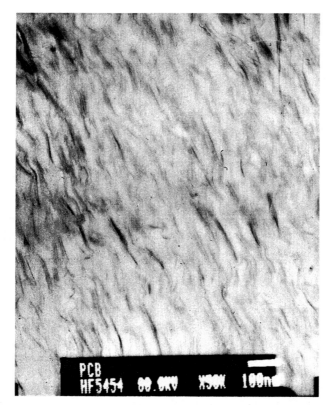

Fig. 15-1 TEM image of a TPS/5% montmorillonite nanocomposite plasticized with 30% glycerol/18% water, showing excellent exfoliation of the clay sheets and a homogeneous distribution and alignment after extrusion direction. (After Fischer et al., 2001.)

with the TPS matrix, and consequently the dispersion is poor. Although the tensile strength for all samples was almost the same, a slight increase was reported again for the samples containing the nonmodified clay (3.3 MPa instead of 2.6 MPa), together with slightly enhanced water vapor barrier properties.

A matrix of four different types of starches and three different clay samples have been studied for their ability to form nanocomposite compositions (Chiou et al., 2005). The samples were prepared by first adding the nanoclays (the hydrophilic Cloisite Na^+ clay as well as the more hydrophobic Cloisite 30B, 10A, and 15A clays) in concentrations of 2.5, 5.0, 7.5, and 10 wt% relative to starch. This starch–nanoclay powder was mixed manually in a plastic bag and the moisture content, up to 51 wt%, was then adjusted with deionized water. One particular sample also contained 15 wt% glycerol and only 36 wt% water. However, under such conditions it is not likely that exfoliation takes place. After mixing at room temperature, samples were analyzed in dynamic tests and creep tests using parallel plates and by XRD analysis.

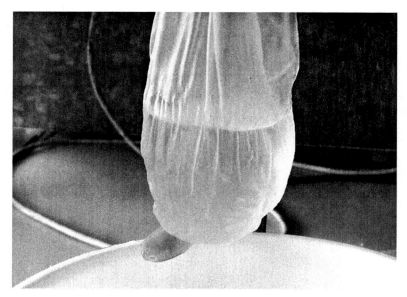

Fig. 15-2 A bag manufactured from a blown film of a TPS/5% montmorillonite nanocomposite plasticized with 30% glycerol/18% water. The bag was filled with water and stored for 20 days, showing excellent stability against cold water. (After Fischer et al., 2001.)

In short, at higher clay concentrations the Na-Cloisite samples had larger, frequency-independent elastic moduli and lower creep compliances than the other samples, indicating that these samples formed more gel-like materials. During analysis at higher temperatures (where gelatinization occurred), amylose that leached from the granules interacted more with the clay interlayers than was the case for the other clay samples. Potato starch samples showed a higher swelling capacity than other starches. This produced softer granules and led to lower elastic moduli in the clay–starch mixtures.

It is clear that major steps in improvement of mechanical properties can be achieved only if true nanocomposites are formed in the sample preparation steps.

Similar results were obtained when chitosan was used as a compatibilizing agent in order to disperse clay particles in a starch matrix (Kampeerapappun et al., 2007). Samples were prepared as follows: 100 parts of a mixture of starch with 0–15 wt% of chitosan and 1–15 wt% clay were mixed with 20 parts of glycerol. Distilled water was also added, followed by acidification with acetic acid to 1% (v/v) in order to dissolve the chitosan. The mixture was then heated to the gelatinization temperature of 70–80°C for 1 hour. The starch solution was cast onto an acrylic sheet mold with a wet thickness of 3 mm and dried overnight. For most analyses samples were conditioned at 50% RH at 23°C. XRD of samples of chitosan-treated MMT shows an increase in interlayer space from 14.78 to 15.80 Å. The starch composite films display a peak at 5.015° which corresponds to an interlayer space of 17.62 Å. This shift was due to the intercalation of glycerol in the clay. However, a completely exfoliated structure usually results in complete loss of registry between the clay layers. Consequently, SEM images show (large) aggregates of clay particles.

The tensile strength of the samples is remarkably high even for the free starch film; addition of MMT has hardly any effect and films containing chitosan show some increase in strength, from 21 MPa in starch/clay films to 25 MPa in starch/clay/ chitosan films. Also the Young's modulus was quite high for these type of samples (1000 MPa), but the elongation at break was very low (only 5%). Perhaps this can be attributed to a quite low content of plasticizer in the samples. The water vapor transmission rates decreased from 2000 to 1082 $g\,m^{-2}\,day^{-1}$ depending on the chitosan content. It was concluded that residual acetyl groups played an important role in hindering transport of water vapor due to reduced surface wettability. Similarly, the moisture absorption rates decreased significantly from 125% to 61% with respect to chitosan contents of 0 and 20 wt%, respectively.

The incorporation of nanoclay sheets into biopolymers can in general have a large positive effect on the water sensitivity and related stability problems of bioplastic products (Fischer et al., 2001). The nature of this positive effect lies in the fact that clay particles act as barrier elements since the highly crystalline silicate sheets are essentially impermeable even for small gas molecules such as oxygen or water, which has a large effect both on the permeation of incoming molecules (water or gases) and for molecules that tend to migrate out of the biopolymer, such as the water used as plasticizer in TPS. In other words; a nanocomposite material with well-dispersed nanoscaled barrier elements should show not only increased mechanical properties but also increased long-term stability of these properties and a related reduction of aging effects.

Mixing of wheat starch plasticized with 36 wt% glycerol and montmorillonites modified with aminododecanoic acid, stearyl dihydroxyethyl ammonium chloride, and distearyldimethyl ammonium chloride did not lead to true nanocomposite materials (Bagdi et al., 2006). The modified clays were swollen in glycerol for 1 day, homogenized with the starch powder, and processed in a mixer at 150°C for 10 min. Plates were compression-molded from the melt at 150°C in 5 min and stored under dry conditions. The swelling of the clay in glycerol resulted in some competitive dissolution adsorption of the plasticizer but not in exfoliation. The clay modification used in this study is certainly not optimal for generating nanocomposites and the mixing step might require more shear force for the intercalation and exfoliation process to reach favorable thermodynamic conditions. It was also found that the T_g of the composites is around 40°C, indicating that hardly any retrogradation can occur. The majority of the thermoplastic starch was found to be amorphous, but some crystallinity different from that of neat starch was found.

Thermodynamic conditions for preparing nanocomposites starting with organoclay are better when thermoplastic acetylated starch (TPAS) is used instead of neat TPS (Qiao et al., 2005). Compositions of acetylated starch/glycerol/clay with weight ratios 100/50/5 are described. The three components were mixed by hand and sealed for 12 h to effect sufficient swelling. The mixture was processed at 150°C for 10 min in a roller mixer. The composites obtained were hot pressed at 160°C to achieve samples of 1 mm thickness, the samples were stored in tightly sealed polyethylene bags to prevent moisture absorption. The intercalation of the TPS into the clay layers has been demonstrated by XRD; the tensile strength increased as expected from 5.5 to 10.3 MPa compared with TPAS and TPAS/OMMT; and the elongation at break decreased, also as expected. The storage modulus decreased rapidly above the T_g of 50°C due to the action of the layered silicates; however, the T_g was increased by the incorporation of clay.

Other approaches to obtain starch clay nanocomposites using montmorillonite with different modifiers and differently plasticized starches have been described in several studies from one academic group (Huang et al., 2004, 2005a,b, 2006; Huang and Yu, 2006). Here, conditions for deriving true nanocomposites were still not favorable, because a single-screw extruder was used for mixing and shear forces were perhaps not high enough.

Glycerol-plasticized thermoplastic starch was used to manufacture montmorillonite-reinforced starch composites (Huang et al., 2004). The samples were made by premixing glycerol and starch with a mixer and then stored in tightly sealed bags for 36 h, and the swelled mixtures were subsequently fed into a single-screw extruder and cut into small particles. These particles were again mixed with clay in the single-screw extruder. The blends were stored in sealed polyethylene bags. The clay was found to be uniformly dispersed in the starch on a submicrometer scale. The introduction of clay decreased the water absorption of the composites. The tensile strength of samples with a clay content of 30% stored at 39% RH for 2 weeks reached 27 MPa. Most importantly, the clay restrained crystallization of the starch for a long time (>90 days).

Similar behavior was found in formamide/ethanolamine-plasticized starch together with ethanolamine-activated clay (Huang et al., 2005a,b). Formamide was dissolved in the ethanolamine and the solution was premixed with the starch to form compounds with total 30% plasticizer calculated on weight starch. The preparation of nanocomposites was the same as described above. SEM and TEM showed that montmorillonite layers were expanded and uniformly dispersed in the nanometer range, and intercalated nanocomposites were formed. When the content of clay is in the range 2.5–10% the mechanical properties are clearly improved. Samples were stored at 25% RH before use for 2 weeks. The tensile strength increased from 6.5 to 9.7 MPa, the yield stress from 5.1 to 8.5 MPa, and the modulus from 68 to 184 MPa, while the elongation at break decreased from 88% to 74%.

The postcrystallization of starch was affected only by the incorporation of clays. This is not surprising since postcrystallization can only occur with sufficient moisture present. In general, starch crystallization cannot be avoided but is detrimental in maintaining good mechanical properties because the material becomes more brittle in time.

The nanoplastics prepared in the latter study (Huang et al., 2005a) showed restrained crystallization behavior when samples were stored at RH 50% for 30, 60, and 90 days. It was shown that the formamide and ethanolamine plasticizers did not impede the crystallization behavior of thermoplastic starch and it was concluded that the clay sheets must influence this noncrystallization behavior. Samples with clay were further found also to be more thermal stable. The water absorption at RH 75% and 100% was found to be lower than in the nanocomposites with the plasticized starch.

The conditions for nanocomposite formation were much better when starch was plasticized with urea and formamide and the clay was activated with citric acid. The $d001$ peak in the XRD pattern completely disappeared, indicating that the crystal lattice structure of the clay was totally dispersed and the slice layers exfoliated in the thermoplastic starch (Huang et al. 2006). The tensile stress with 10 wt% clay was highest, with a value of 24.9 MPa. The nanocomposite with 5 wt% clay had the maximal tensile strain 134%. The effect of water content on mechanical properties was also measured. Increasing of water content from 5% to 35% caused first an increase the tensile stress and then a decrease; the maximal tensile stress of nanocomposites reached 23 MPa at a water content of 17%. Intercalation of monmorrilonite (MMT) with sorbitol in a single screw extruder as a first step followed by mixing with starch as matrix, sorbitol and formamide leads to composite materials with an increase in tensile modulus by the factor of 4 at a load of 10% MMT together with an increase in strength (factor 3) and only a slight decrease in elongation at break (138 to 93%) (Ma et al., 2007).

Another study used two different clays, sodium montmorillonite (Na-MMT) and sodium fluorohectorite (Na-FHT) (Dean et al., 2007; Yu et al., 2007). It was shown that there was an optimum level of both plasticizer and nanoclay to produce a gelatinized starch film with the highest levels of exfoliation and consequent superior properties.

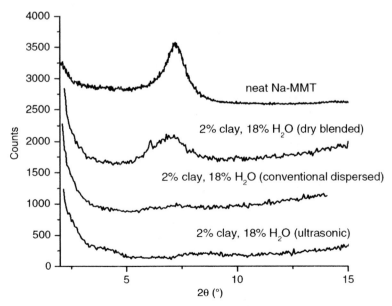

Fig. 15-3   XRD traces for a series of medium water content (18 wt%) Na-MMT/gelatinized starch nanocomposites with 2 wt% clay. (After Dean et al., 2007.)

Nanoclays were mixed with starch by dry blending (DB), or conventional dispersing in water using conventional mixers (CD), or dispersing in water using ultrasonics (U). The water (in the case of DB formulation) or water/clay dispersions were then added dropwise to the starch in a high-speed mixer prior to extrusion. A twin-screw extruder was used to process the starch nanocomposites using a profile producing a melt temperature of 110°C. The starch nanocomposite was extruded directly into sheet form via a 0.5 mm die. This study showed that there was an optimum level of both plasticizer and nanoclay for each type of clay under the conditions used, where exfoliated nancomposites are formed after extrusion only when hydrophilic clays are used and enough water is present.

X-ray analysis showed an increase in intergallery separation ranging from 15.5 Å to 62 Å, depending on water content during sonification prior to extrusion; the latter value was observed for a clay/water ratio of 1:10. Similar results were obtained for Na-MMT and Na-FHT. The extruded samples showed complete exfoliation only when enough water (18 wt%) was used in the case of the hydrophilic Na-MMT (with 2 wt% Na-MMT complete exfoliation was achieved, for 3.2 wt% clay some agglomerates are present) but not when using the more hydrophobic Na-FHT (Fig. 15-3).

The most significant improvement in moduli was observed for higher levels of clay (2.6–3.2 wt%), from 2500 MPa for native starch up to 5000 MPa for the nanocomposite. Similar improvements were observed for yield strength from 30 MPa up to 55 MPa (Fig. 15-4). The elongation at break decreases, not unexpectedly, comparing the nanocomposites with neat starch samples. This was explained as follows: Once melt extruded, the thermoplastic starch begins to undergo retrogradation, causing embrittlement and thus reducing the break elongation. On the other hand, clay

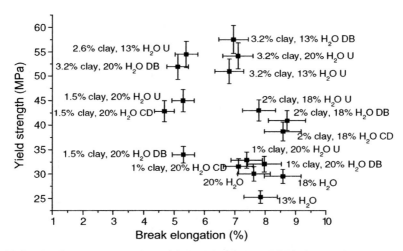

Fig. 15-4 Yield strength versus break elongation for Na-MMT/gelatinized starch nanocomposites for various clay loading, water loadings, and dispersion methods (DB, clay and starch dry blended; CD, clay dispersed using conventional mixing; U, clay dispersed using ultrasonics). (After Dean et al., 2007.)

retains moisture in the sample, leading to a more plasticized material with a greater elongation at break. Consequently a peak in the elongation at break at 2 wt% clay content was observed. For the Na-FHT samples, the yield elongation decreased with increasing clay content, indicating that the Na-FHT was not capable of disrupting recrystallization or binding water tightly into the matrix. Similar trends were observed for the yield strength. Here, the 2.6–3.2 wt% nanocomposites containing 13 wt% water showed the best ratio of strength to elongation.

The order of addition of components in preparing nanocomposites can be of great importance to the final properties of the product and Pandey and Singh 2005 investigated the effect of addition of components, in particular the plasticizer in relation to starch and clay. Glycerol and starch both enter the clay galleries, but it was found that glycerol is favored due to its smaller molecular size. The filler dispersion became highly heterogeneous and the product became more brittle if starch was plasticized before filling with clay. Better mechanical properties were obtained if the plasticizer was added after mixing the clay in the starch matrix. Starch/clay/glycerol was mixed in the four following ways:

1. Starch was gelatinized with water followed by plasticization and then clay slurry was added. This mixture was heated for 30 min to boiling (STN1).
2. A clay slurry in water was mixed with starch in water and heated to boiling, followed by addition of glycerol (STN2).
3. Starch, clay slurry, and glycerol were mixed and heated to boiling (STN3).
4. Glycerol was mixed with clay slurry, stirred for 5 hours, mixed with starch, and heated to boiling (STN4).

Solutions were poured into Petri dishes, water was evaporated, and films of 150–200 μm thickness were obtained and equilibrated at defined RH and temperature. XRD analysis showed the highest increase in *d* spacing for STN2 and the smallest for STN1. In the latter, probably large structures had formed which negatively affected global mobility. The gallery height was not increased after plasticization, suggesting that the plasticizer can be accommodated among the starch chains. The modulus increased slightly from 790 MPa for neat plasticized starch to 820 MPa for the best nanocomposite sample. The maximum strain was highest at 12% for STN2 and lowest at 6% for STN1, with neat starch having 10%.

In conclusion, better dispersion of clay can be achieved by mixing of starch and filler first, followed by plasticization.

15.3 STARCH-BASED NANOCOMPOSITES USING DIFFERENT LAYERED MATERIALS

Preliminary insight on composites of thermoplastic starch and kaolinite was reported by Carvalho et al. (2001) while studying the use of kaolinite as filler reinforcement for thermoplastic starch to improve its mechanical properties. The composites were prepared using regular cornstarch plasticized with 30% glycerin and hydrated

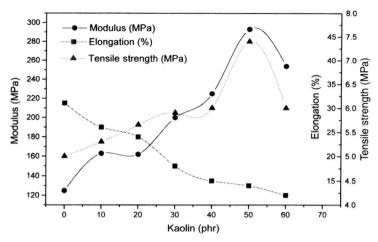

Fig. 15-5 Modulus, ultimate tensile strength, and elongation for thermoplastic starch and its composites with kaolinite. (After Carvalho et al., 2001.)

kaolinite with average particle size of 0.5 μm pre-mixed and processed in an intensive batch mixer at 170°C. Compounds with 0, 10, 20, 30, 40, 50, and 60 parts of kaolinite per hundred parts of thermoplastic starch were prepared. The composite filled with 50 parts per hundred (pph) kaolinite seemed to be the best composition. This showed an increase in the tensile strength from 5 to 7.5 MPa. The modulus of elasticity increased from 120 to 290 MPa and the tensile strain at break decreased from 30% to 14%. Beyond this value the amount of matrix was insufficient to wet kaolinite and the composite became fragile (Fig. 15-5).

Scanning electron microscopy of fractured surfaces revealed a very good dispersion of kaolinite, as there are no smooth areas or large agglomerates. The adhesion between the matrix and kaolinite is also very good and the particles of kaolinite are strongly bonded to the matrix, even after the fracture. Besides the good mechanical properties, the use of these clays also leads to an increase in the water resistance of thermoplastic starch compounds. The presence of kaolinite decreases the water uptake of the composites, this decrease being more pronounced starting at 20 pph (decrease in moisture uptake from 27% to 14% at 100% relative humidity). The observed decrease in glass transition temperature (T_g) is proportional to the amount of kaolinite present in the composites.

The water solubility of starch in kaolinite–starch composite prepared by adding clay to a 3–4% uncooked raw corn starch suspension and subsequent stirring and cooking at 95°C for 30 min was less than 3% at 50°C for 30 min. The clays were distributed well inside the composite and were perfectly coated by starch. When the starch was exposed to high temperature, the solubility increased to 5%. However, crosslinked starch could reduce the solubility even at high temperature. Although the solubility of crosslinked starch increased slightly as temperature increased, the result was only 1.25% at 90°C (Yoon and Deng, 2006).

The influence of different layered compounds (clays) on the properties of starch–clay composites has been described by Wilhelm et al. (2003a, b). Glycerol-plasticized starch films were modified by addition of various layered compounds as fillers, two being of natural origin (kaolinite as a neutral mineral clay, Ca-hectorite as a cationic exchanger mineral clay) and two synthetic (layered double hydroxide [LDH] as an anionic exchanger and brucite, which has a neutral structure).

The composite materials were prepared by dispersing the layered material in distilled water and adding the dispersion to an aqueous starch dispersion followed by degassing and heating to 100°C in a sealed tube to gelatinize the starch granules. Glycerol (20% w/w, relative to starch on a dry basis) was added to the heated solution and the mixtures were then poured into dishes, allowing solvent evaporation at 40–50°C.

The results obtained in the first investigation showed that the hectorite clay was intercalated in the unplasticized starch matrix, while glycerol–hectorite intercalation was observed in the plasticized films. The unplasticized composite films were very brittle, as expected. A film obtained with 30% w/w of clay showed an increase of more than 70% in the Young's modulus compared to nonreinforced plasticized starch. Both XRD and infrared spectroscopy showed that glycerol can be intercalated into the clay galleries and that there is a possible conformational change of starch in the plasticized starch/clay composite films.

In the subsequent study, it was found that no intercalation took place in presence of kaolinite, LDH, or brucite, while the starch/hectorite composite showed an increase of the interplanar basal distance from 14.4 to 18.3°Å. This basal spacing corresponds to the intercalation of glycerol molecules between the silicate layers, but nonplasticized starch and/or oxidized starch can also be intercalated into the hectorite galleries. The glycerol intercalation is minimized in plasticized–oxidized starch films where the oxidized starch chains are preferentially intercalated. The storage modulus of the composite materials was higher in the order kaolinite>brucite>hectorite than for LDH starch composites; only the hectorite composites showed an increased T_g.

The rheological properties of composite gels of starch extracted from yam roots and Ca-hectorite were studied in aqueous solutions as a function of hectorite concentration (Wilhelm et al., 2005). The elastic (G') and viscous (G'') moduli of the composite gels were dependent on the clay content. For all samples G' was larger than G''. Composite gels with clay contents of less than 30% exhibited higher G', G'', and viscosity $|\eta_0|$ values compared with pure starch gel, while those with a content of 50% showed lower values. The addition of hectorite significantly inhibited the creeping properties in relation to pure starch at 25°C. However, after heating–cooling cycles between 25 and 85°C, this effect was no longer observable and the composite gel displayed a similar behavior to that of pure starch.

Nanocomposites of glycerol-plasticized starch with untreated montmorillonite and hectorite were examined (Chen and Evans, 2005). The composites were prepared by mixing starch and glycerol at a ratio of 10 : 3 by mass and melt processing to thermoplastic starch on a two-roll mill at 120°C. The clay was added during the processing on the two-roll mill. Organo-treated hectorite and kaolinite were added to produce conventional composites within the same clay volume fraction range for comparison; 10 vol% untreated clay was approximately the maximum loading that this type of TPS could sustain. Unlike natural montmorillonite and hectorite, kaolinite is a nonswelling clay and should produce conventional composites with the starch. TPS/montmorillonite and TPS/hectorite nanocomposites showed an increase of $d001$ from 1.23 to 1.8 nm.

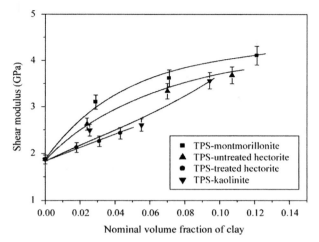

Fig. 15-6 Young's modulus and shear modulus of TPS/clay composites. (After Chen and Evans, 2005.)

The broad 001 peak suggests a wide range of interlayer spacings as a result of the coexistence of intercalation and exfoliation This is further supported by TEM images. However, no changes in the XRD pattern were recorded for the organo-treated hectorite/starch composites, indicating the formation of a conventional composite material.

In all composites, the presence of clay increased both Young's and shear modulus, and the montmorillonite and unmodified hectorite provided significantly more increase than kaolinite, which formed conventional composites with TPS (Fig. 15-6). That the treated-hectorite composites had similar elastic moduli to those prepared from kaolinite, which gives conventional composites, supports the previous observation that they also formed conventional composites. Hectorite particles have a lower aspect ratio and larger surface area than montmorillonite, the former factor tending to give a smaller modulus increase and the latter giving a greater increase. At low clay fractions, montmorillonite provided a higher Young's modulus than the hectorite composite but at higher clay fractions, reinforcement deteriorated. Nevertheless, it is clear that a nanocomposite provides a higher modulus than a conventional composite with the same clay content. Studies of the water absorption of TPS and the TPS/clay nanocomposite containing 6 wt% montmorillonite showed a dramatic softening, swelling, and partial disintegration of the unfilled TPS after immersing in distilled water for 2 hours. However, the nanocomposite remained generally intact after the same immersion conditions. The presence of clay obviously enhances the water resistance of TPS, although exact water absorption data were not obtained.

15.4 BIODEGRADABLE STARCH–POLYESTER NANOCOMPOSITE MATERIALS

There are many sources of biodegradable plastics, from synthetic polymers (polyesters, poly[lactic acid], poly[hydroxybutyrate]s) to natural polymers (starch, gelatin, chitosan). Synthetic polymers, although having excellent properties, are costly and typically manufactured using nonrenewable petroleum resources. Starch-based thermoplastics, on the other hand, are relatively cheap and, more importantly, are manufactured using renewable sources of raw materials. However, high viscosity and poor melt properties make them difficult to process, and products made from starch are often brittle and water sensitive. To alleviate this problem, starch polymers are often blended with synthetic polymers.

Recent research has indicated that organoclays in combination with biodegradable polyesters show much promise for starch-based polymer nanocomposites in terms of improving their mechanical properties. Organoclay using modifiers consisting of alcohols and hydrogenated tallow, which may be more thermodynamically compatible with the polyester, starch, and an unspecified polyester were blended simultaneously in the twin-screw extruder in order to minimize the time of processing of the starch (McGlashan and Halley, 2003) or in a Haake compounder (Perez et al., 2007). The addition of an organoclay (from 0 to 5 wt%) significantly improved both the processing and tensile properties over the original starch blends; the best results were obtained for 30 wt% starch blends. The increase in tensile strength and Young's modulus is mostly likely due to the enhanced interfacial area of the MMT, providing nanoscale reinforcement of the blend. The tensile strength and Young's modulus of the 30/70 blend nanocomposites were improved by approximately 50% and 200%, respectively. Improvements were also noted for the 50 and 70 wt% starch nanocomposite blends. The incremental improvement in mechanical properties was smaller as the content of starch was increased at clay loading of 1.5%. The same was observed for the higher loading (5 wt%) of MMT. The strain at break was dramatically improved upon the addition of the MMT to the 30 wt% starch (>1500%). The authors suggest that this improvement in elongation could be attributed to the addition of MMT and to better gelatinization and modified crystalline structure of the extrudate. The level of delamination depends on the ratio of starch to polyester and the amount of organoclay added. The crystallization temperature of the nanocomposite blends is significantly lower than that of the base blend. This is probably due to the platelets inhibiting order, and hence crystallization, of the starch and polyester. The effect of organoclay on the melting temperature for any given starch–polyester blend is not significant. The dispersed MMT should make a more tortuous path for the volatile starch plasticizers to escape, which has the twofold effect of lowering the gelatinization temperature of the starch and decreasing the viscosity of the starch–polyester blend. The more amorphous blend deforms plastically rather than failing through brittle fracture. However, the higher the starch loading, then the less the MMT seems to delaminate, and hence the advantage of the large interfacial area of MMT becomes less effective.

The nanocomposite blends were easier to process into films than the base blends using a film-blowing tower. Production of 10 μm-thick film blown on a laboratory film tower and of 30 μm thick films blown on an industrial-scale film blowing unit at production rates typical for low-density polyethylene has been demonstrated.

Starch–polycaprolactone blends filled with Nanomer I.30 E (Nanocor, IL, USA) with loads between 1 and 9 wt% were prepared (Kalambur et al., 2004, 2005). In order to obtain a more "hydrophilic" starch that would form stronger hydrogen bonds with the clay surface and improve the adhesion, carbonyl or carboxyl groups along the polysaccharide chains were introduced through peroxide-promoted oxidation reactions during a reactive extrusion step. The nanocomposites have much increased solvent resistance properties, most likely due to the increased resistance of the nanocomposite to diffusion. A moderate improvement in tensile strength and much improvement in elongation at break (from 200% to 900%) was achieved.

A more recent study focuses on the improvement of physical and dynamic properties of starch, including thermal plasticity, by blending with polycaprolactone (PCL) (Ikeo et al., 2006). Starch was blended in a first step with glycerin and water to achieve thermal plasticity and improve compatibility with PCL. To improve the compatibility between PCL and TPS, maleic anhydride (MAH) was added to denature the PCL. Using a single-screw extruder, 1 wt% maleic anhydride and 99 wt% PCL were kneaded with 0.3 pph of benzoyl peroxide (BPO) as the

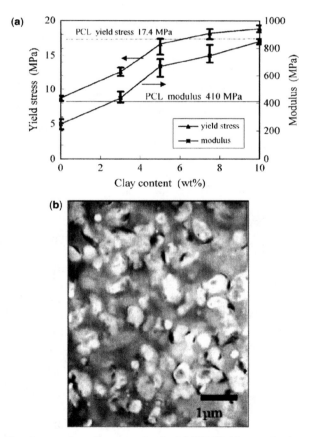

Fig. 15-7   (a) The change of tensile properties for MAHPCL/thermoplastic starch/clay from 50:50:0 to 50:40:10 wt%); (b) TEM image of a microtomed sample of MAHPCL/TPS/clay 50/47/3%. (After Ikeo et al., 2006.)

polymerization initiator. Finally, the addition of natural clay (Na-MMT) results in dispersed thermoplastic starch, which improves compatibility with PCL. The clay was dispersed in thermoplastic starch in advance, by kneading in an extruder, to prepare a nanocomposite so as to compensate for degradation of properties caused by adding thermoplastic starch to PCL. Two types of samples were prepared: one containing 10 wt% clay and the other 20 wt%. The clay content in the PCL/thermoplastic starch blends was 3–10 wt%.

According to the result of tensile tests, the strain of PCL/thermoplastic starch/clay blends is extremely low at around 3%, while that of MAHPCL/thermoplastic starch/clay exceeds 500% (Fig. 15-7). As can be seen from this result, both strain and yield strength were significantly improved after adding maleic anhydride to PCL. The blends with 5 wt% or higher clay content have modulus and yield strength equal to or higher than PCL alone. The blend with 10 wt% or higher clay content shows modulus two times or higher that of PCL alone; hence adding maleic anhydride and clay to the PCL/thermoplastic starch blend significantly improved its properties. The addition of clay is believed to have increased the viscosity of the thermoplastic starch phase close to that of PCL, which resulted in improved compatibility.

Novel biodegradable starch/clay/polyester (Ecoflex®, BASF) nanocomposite films, to be used as food packaging, were obtained by homogeneously dispersing native Na-MMT nanoparticles in different starch-based materials via polymer melt processing techniques (Avella et al., 2005). For starch/clay material, the results show good intercalation of the polymeric phase into clay interlayer galleries, together with an increase in mechanical parameters, such as modulus and tensile strength. In the case of starch/polyester/clay, a separation of the polyester phase was observed.

15.5 DISCUSSION AND CONCLUSION

Homogeneous incorporation of clay particles into a starch matrix on a true nano scale has proved possible. The stiffness, the strength, and the toughness of the nanocomposite materials are improved even at very low loadings of clay (2–4%) if the platelets are completely exfoliated and can be adjusted by varying the water content (Fischer et al., 2001; Dean et al., 2007). Starch/clay nanocomposites can be formed when complete TPS (verification of the absence of crystals is essential) is used and enough plasticizer is present in a combination of glycerol/sorbitol and water and when sufficient shear force is applied, preferably in a twin-screw extruder. In addition, knowledge of the degree of retrogradation is essential for eventually marketable products in consumer applications.

The addition of clay during processing supports and intensifies the destructurization of starch, providing a means of easier processing (lower extrusion temperature and less use of processing agents). The starch/clay nanocomposite films show a very strong decrease in hydrophilicity, no dramatic softening, swelling, and partial disintegration, unlike unfilled TPS after immersing in distilled water.

However, if no complete exfoliation has been achieved, it is very difficult to judge the effect of the reinforcing minerals. When (some) clusters of aggregated clay particles are present within the samples they can influence the overall mechanical properties and dominate the deformation behavior. In such cases only minor improvements can be expected and such are reported. Starch/clay nanocomposites can only be formed if the starch has been totally destructurized into a homogeneous thermoplastic material to ensure a homogeneous distribution of the nanoparticles. The destructurization always occurs above the glass transition temperature (T_g) and shifts the final T_g to lower temperatures. TPS with a plasticizer content below 30% has a T_g above room temperature and thus remains brittle. The modulus change is reversed proportionally with the quantity of plasticizer (or water) and therefore the mechanical properties will change when the relative humidity of the atmosphere changes. The modulus determined from test samples will be substantially lower on humid days compared with values obtained on dry days. Clay delays the brittleness by delaying the migration of plasticizer (moisture) to the surroundings but it cannot prevent the sample eventually becoming brittle in time.

Water stability can only be obtained in completely retrograded samples, conditioned at high relative humidities RH > 90%. It is known that samples conditioned at 50% RH will not be stable over time because of slow recrystallization/retrogradation.

Results from different research groups cannot always be compared due to the different sample preparation steps employed. It would be useful to agree some standard on how to prepare samples for analysis.

When comparing different clays it is seen that the actions of untreated montmorillonite and hectorite are very similar but are fundamentally different from that of fillers such as kaolinite and organo-treated hectorite and kaolinite. Both are nonswelling clays in aqueous and plasticizer environments and produce conventional (micro)composites with the starch. The adhesion between the matrix and filler particles is in general very good and the particles are strongly bonded to the matrix, even after fracture.

When combining starch with biodegradable polyesters a choice has to made of which phase should be reinforced by the clay particles and consequently the means of compatibilization to be used. Mostly, organoclay using modifiers consisting of amines substituted with alcohols and hydrogenated tallow are used, which are more thermodynamically compatible with the polyester, The addition of an organoclay (from 0 to 5 wt%) to starch/polyester blends may offer advantages with respect to the processing and tensile properties over the original starch blends; the increase in tensile strength and Young's modulus is most likely due to the enhanced interfacial area.

Because of water sensitivity, samples stored under normal ambient conditions will always eventually become brittle. Addition of biodegradable polyesters to starch–clay composites is always necessary for applications in packaging, whether in blown films or molded products. In these kinds of applications it then becomes questionable whether clay makes a large difference when polyesters are used in large quantities. More research is needed to discover whether the use of starch–clay nanocomposites might be effective in reducing the amount of polyester and thus help products made from these blends to become eventually more compatible with conventional plastics.

REFERENCES

Avella M, De Vlieger JJ, Errico ME, Fischer S, Vacca P, Volpe MG. 2005. Biodegradable starch/clay nanocomposite films for food packaging applications. *Food Chemistry* 93:467–474.

Bagdi K, Müller P, Pukánszky B. 2006. Thermoplastic starch/layered silicate composites: structure, interaction, properties. *Composites Interfaces* 13:1–17.

Carvalho AJF, Curvelo AAS, Agnelli JAM. 2001. A first insight on composites of thermoplastic starch and kaolinite. *Carbohydrate Polymers* 45:189–194.

Chen B, Evans JRG. 2005. Thermoplastic starch-clay nanocomposites and their characteristics. *Carbohydrate Polymers* 51:455–463.

Chiou BS, Lee E, Glenn GM, Orts WJ. 2005. Rheology of starch–clay nanocomposites. *Carbohydrate Polymers* 59:467–475.

Dean K, Yu L, Wu DY. 2007. Preparation and characterisation of melt-extruded thermoplastic starch/clay nanocomposites. *Composites Science and Technology* 67:413–421.

Fischer S, De Vlieger J, Kock T, Batenburg L, Fischer H. 2001. "Green" nano-composite materials—new possibilities for bioplastics. *Material Research Society Symposium Proceedings* 661:KK221–KK226.

Huang M, Yu J. 2006. Structure and properties of thermoplastic corn starch/montmorillonite biodegradable composites. *Journal of Applied Polymer Science* 99:170–176.

Huang MF, Yu JG, Ma XF. 2004. Studies on the properties of montmorillonite-reinforced thermoplastic starch composites. *Polymer* 45:7017–7023.

Huang MF, Yu JG, Ma XF, Jin P. 2005a. High performance biodegradable thermoplastic starch-ENMT thermoplastics. *Polymer* 46:3157–3162.

Huang MF, Yu JG, Ma XF. 2005b. Preparation of the thermoplastic starch/montmorillonite nanocomposites by melt-intercalation. *Chinese Chemical Letters* 16:561–564.

Huang M, Yu J, Ma X. 2006. High mechanical performance MMT-urea and formamide-plasticized thermoplastic cornstarch biodegradable nanocomposites. *Carbohydrate Polymers* 63:393–399.

Ikeo Y, Aoki K, Kishi H, Matsuda S, Murakami A. 2006. Nano clay reinforced biodegradable plastics of PCL starch blends. *Polymers for Advanced Technologies* 17:940–944.

Kalambur SB, Rizvi SSH. 2004. Starch-based nanocompositesby reactive extrusion processing. *Polymer International* 53:1413–1416.

Kalambur SB, Rizvi SSH. 2005. Biodegradable and functionally superior starch-polyester nanocomposites from reactive extrusion. *Journal of Applied Polymer Science* 96:1072–1082.

Kampeerapappun P, Aht-ong D, Pentrakoon D, Srikulkit K. 2007. Preparation of cassava starch/montmorillonite composite film. *Carbohydrate Polymers* 67:155–163.

Kudryavtsev YV, Govorun EN, Litmanovich AD, Fischer H. 2004. Polymer melt intercalation in clay modified by diblock copolymers. *Macromolecular Theory and Simulations* 13:392–399.

Ma X, Yu J, Wang N. 2007. Production of thermoplastic starch/MMt-sorbitol nanocomposites by dual-melt extrusion processing. *Macromolecular Material Engineering* 292:723–728

McGlashan SA, Halley PJ. 2003. Preparation and characterisation of biodegradable starch-based nanocomposite materials. *Polymer International* 52:1767–1773.

Pandey JK, Singh RP. 2005. Green nanocomposites from renewable resources: Effect of plasticizer on the structure and material properties of clay-filled starch. *Starch/Staerke* 57:8–15.

Park HM, Li X, Jin CZ, Park CY, Cho WJ, Ha CS. 2002. Preparation and properties of biodegradable thermoplastic starch/clay hybrids. *Macromolecular Material Engineering* 287:553–558.

Park HM, Lee WK, Park CY, Cho WJ, Ha CS. 2003. Environmentally friendly polymer hybrids, Part 1. Mechanical, thermal, and barrier properties of thermoplastic starch/clay nanocomposites. *Journal of Materials Science* 38:909–915.

Pérez, CJ, Alvarez, VA, Mondragón I, Vázquez A. 2007. Mechanical properties of layered silicate/starch polycaprolactone blend nanocomposites. *Polymer International* 56:686–693.

Qiao X, Jiang W, Sun K. 2005. Reinforced thermoplastic acetylated starch with layered silicates. *Starch/Staerke* 57:581–586.

Wilhelm HM, Sierakowski MR, Souza GP, Wypych F. 2003a. Starch films reinforced with mineral clay. *Carbohydrate Polymers* 52:101–110.

Wilhelm HM, Sierakowski MR, Souza GP, Wypych F. 2003b. The influence of layered compounds on the properties of starch/layered compound composites. *Polymer International* 52:1035–1044.

Wilhelm HM, Sierakowski MR, Reicher F, Wypych F, Souza GP. 2005. Dynamic rheological properties of yam starch/hectorite composite gels. *Polymer International* 54:814–822.

Yang KK, Wang XL, Wang YZ. 2007. Progress in nanocomposite biodegradable polymer. *Journal of Industrial Engineering Chemistry* 13:485–500.

Yoon SY, Deng Y. 2006. Clay–starch composites and their application in papermaking. *Journal of Applied Polymer Science* 100:1032–1038.

Yu L, Petinakis S, Dean K., Bilyk A., Dongyang W. 2007. Green polymer blends and composites from renewable resources. *Macromolecular Symposium* 249:535–539.

CHAPTER 16

Polylactide-Based Nanocomposites

SUPRAKAS SINHA RAY and JAMES RAMONTJA

National Centre for Nano-Structured Materials, Council for Scientific and Industrial Research, Pretoria 0001, Republic of South Africa

16.1 INTRODUCTION

In recent years, biodegradable polymers from renewable resources have attracted great research attention (Ikada and Tsuji, 2000). Biodegradable polymers are defined as those that undergo microbially induced chain scission leading to mineralization (Sinha Ray and Bousmina, 2005). Specific conditions in term of pH, humidity, oxygenation and the presence of some metals are required to ensure the biodegradation of such polymers. Renewable sources of polymeric materials offer an alternative for maintaining sustainable development of economically and ecologically attractive technology. Innovations in the development of materials from biodegradable

Biodegradable Polymer Blends and Composites from Renewable Resources. Edited by Long Yu
Copyright © 2009 John Wiley & Sons, Inc.

polymers, the preservation of fossil-based raw materials, complete biological degradability, reduction in the volume of garbage and compostability in the natural cycle, protection of the climate through the reduction of carbon dioxide released, as well as the possible application of agriculture resources for the production of green materials are some of the reasons why such materials have attracted academic and industrial interest.

One of the most promising polymers in this direction is polylactide (PLA) because it is made wholly from agricultural products and is readily biodegradable (Gruber and Brien, 2002). PLA is not a new polymer and has been the subject of many investigations for over a century. In 1845, Pelouze condensed lactic acid by distillation of water to form low-molecular-weight PLA and the cyclic dimer of lactic acid, known as lactide (Pelouze, 1845). About a half-century later, an attempt was made by Bischoff and Walden to polymerize lactide to PLA (Bischoff and Walden, 1894); however, the method was unsuitable for practical use (Watson, 1948). In 1948, Watson published a review on the possible uses of PLA for coatings and as a constituent in resins. Despite being known for over 100 years, PLA has not been commercially viable or practically useful, although it was described as having potential as a commodity plastic (Lipinsky and Sinclair, 1986). However, recent developments in the manufacturing of the monomer economically from agricultural products have placed this material at the forefront of the emerging biodegradable plastics industries (Vink et al., 2003).

In recent years, high-molecular-weight PLA has generally been produced by the ring-opening polymerization of the lactide monomer. The conversion of lactide to high-molecular-weight polylactide is achieved by two routes: (i) direct condensation, which involves solvents under high vacuum, and (ii) formation of the cyclic dimer intermediate (lactide), which is solvent free. Until 1990, the monomer lactide was produced commercially by fermentation of petrochemical feedstocks. The monomer produced by this route is an optically inactive racemic mixture of L- and D-enantiomers. Today the most popular route is fermentation in which corn starch is converted into lactide monomer by bacterial fermentation (Drumright and Gruber, 2000).

Recently, Cargill-Dow used a solvent-free process and a novel distillation process to produce a range of PLAs (Lunt, 1998; Drumright and Gruber, 2000). The essential novelty of this process lies in its ability to go from lactic acid to a low-molecular-weight poly(lactic acid), followed by controlled depolymerization to produce the cyclic dimer, commonly referred to as lactide. This lactide is maintained in liquid form and purified by distillation. Catalytic ring-opening polymerization of the lactide intermediate results in the production of PLAs with controlled molecular weights. The process is continuous with no need to separate the intermediate lactide.

In contrast, Mitsui Toatsu (now, Mitsui Chemicals) utilizes a solvent-based process in which a high-molecular-weight PLA is produced by the direct condensation using azeotropic distillation to continuously remove the condensation water (Enomoto et al., 1994). Commercially available PLA grades are copolymers of poly(L-lactide) with *meso*-lactide or D-lactide. The amount of the D enantiomer is

TABLE 16-1 Physical Properties of PLA

Property	Typical Value
Molecular weight (kg/mol)	100–300
Glass transition temperature, T_g (°C)	55–70
Melting temperature, T_m (°C)	130–215
Heat of melting, ΔH_m (J/g)	8.1–93.1
Degree of crystallinity, X (%)	10–40
Surface energy (dynes)	38
Solubility parameter, δ $(J/ml)^{1/2}$	19–20.5
Density, ρ (kg/m^3)	1.25
Melt flow rate, MRF (g/10 min)	2–20
Permeability of O_2 and CO_2 (fmol m^{-1} s^{-1} Pa^{-1})	4.25 and 23.2
Tensile modulus, E (GPa)	1.9–4.1
Yield strength (MPa)	70/53
Strength at break (MPa)	66/44
Flexural strength (MPa)	119/88
Elongation at break (%)	100–180
Notched Izod impact strength (J/m)	66/18
Decomposition temperature (K)	500–600

known to affect the properties of PLA—melting temperature, degree of crystallinity, and so on.

PLA has a balance of mechanical properties, thermal plasticity, and biodegradability, and is readily fabricated; it is thus a promising polymer for various end-uses (Fang and Hanna, 1999; Gu et al., 1992). Various properties of PLA are summarized in Table 16-1. Even when burned, it produces no nitrogen oxide gases and only one-third of the combustion heat generated by polyolefins; it does not damage the incinerator and provides significant energy savings.

The increasing appreciation of the various intrinsic properties of PLA, coupled with knowledge of how such properties can be improved to achieve compatibility with thermoplastics processing, manufacturing, and end-use requirements, has fuelled technological and commercial interest in PLA.

Over the last few years, a wealth of investigations have been undertaken to enhance the mechanical properties and the impact resistance of PLA. It can therefore compete with other low-cost biodegradable/biocompatible or commodity polymers. These efforts have made use of biodegradable and nonbiodegradable fillers and plasticizers or blending of PLA with other polymers (Martin and Averous, 2001).

In recent years the nanoscale, and the associated excitement surrounding nanoscience and technology, has afforded unique opportunities to create revolutionary material combinations. These new materials promise the circumvention of classical material performance trade-offs by accessing new properties and exploiting unique synergisms between constituents that occur only when the length scale of the morphology and the critical length associated with the fundamental physics of a given property coincide.

Nanostructured materials or nanocomposites based on polymers have been an area of intense industrial and academic research over the past one and a half decades (LeBaron et al., 1999; Alexander and Doubis, 2000; Zanetti et al., 2000; Biswas and Sinha Ray, 2001; Sinha Ray and Okamoto, 2003a). In principle, nanocomposites are an extreme case of composite materials in which interfacial interactions between two phases are maximized. In the literature, the term nanocomposite is generally used for polymers with submicrometer dispersions. In polymer-based nanocomposites, nanometer-sized particles of inorganic or organic-materials are homogeneously dispersed as separate particles in a polymer matrix. This is one way of characterizing this type of material. There is, in fact, a wide variety of nanoparticles and of ways to differentiate them and to classify them by the number of dimensions they possess. Their shape varies and includes: (i) needlelike or tubelike structures regarded as one-dimensional particles, for example, inorganic nanotubes, carbon nanotubes, or sepiolites; (ii) two-dimensional platelet structures, for example, layered silicates; and (iii) spheroidal three-dimensional structures, for example, silica or zinc oxide.

The main objective of this chapter is to provide a snapshot of the rapidly developing field of nanocomposite materials based on PLA. To date, various types of nanoreinforcements such as nanoclay, cellulose nanowhiskers, ultrafine layered titanate, nano-alumina, and carbon nanotubes (Mohanty et al., 2003; Nazhat et al., 2001; Hiroi et al., 2004; Nishida et al., 2005; Mark 2006; Kim et al., 2006). have been used for the preparation of nanocomposites with PLA. Progress in each particular system is discussed chronologically starting from the pioneering works. Various physicochemical characterizations and improved mechanical properties are summarized. Ongoing developments and promising avenues are also discussed. Finally, possible applications and prospect for nanocomposites based on PLA are mentioned.

16.2 PLA NANOCOMPOSITES BASED ON CLAY

Polymer nanocomposites based on clay minerals have attracted great interest from researchers in recent times, both in industry and in academia, because they often exhibit concurrent improvements in properties over the neat polymers. These improvements can include high moduli, increased strength and heat resistance, decreased gas permeability and flammability, and increased degradability of biodegradable polymers (LeBaron et al., 1999; Alexander and Doubis, 2000; Zanetti et al., 2000; Biswas and Sinha Ray, 2001; Sinha Ray and Okamoto, 2003a). On the other hand, these materials have also been demonstrated to be unique model systems for the study of the structure and dynamics of polymers in confined environments (Sinha Ray, 2006).

16.2.1 Structure and Properties of Clay

The commonly used clay minerals for the preparation of polymer/clay nanocomposites belong to the same general family of 2 : 1 layered silicates or phyllosilicates (Brindly and Brown, 1980; Sinha Ray and Okamoto, 2003a). Their crystal structure consists of layers made up of two tetrahedrally coordinated silicon atoms fused to an edge-shared octahedral sheet of either aluminum or magnesium hydroxide. The layer thickness is around 1 nm, and the lateral dimensions of these layers may vary from 30 nm to several micrometers or larger, depending on the particular clay minerals. Stacking of the layers leads to a regular van der Waals gap between the layers called the interlayer or gallery. Isomorphic substitution within the layers (for example, Al^{3+} replaced by Mg^{2+} or Fe^{2+}, or Mg^{2+} replaced by Li^+) generates negative charges that are counterbalanced by alkali and alkaline earth cations situated inside the galleries. This type of clay mineral is characterized by a moderate surface charge, known as the cation exchange capacity (CEC), and generally expressed as mequiv/100 g. This charge is not locally constant but varies from layer to layer, and must be considered as an average value over the whole crystal.

Montmorillonite (MMT), hectorite, and saponite are the most commonly used clay minerals for the preparation of nanocomposites. Clay minerals generally have two types of structure: tetrahedral-substituted and octahedral-substituted. In the case of tetrahedrally-substituted clay minerals, the negative charge is located on the surface of the silicate layers, and hence the polymer matrices can interact more readily with these than with octahedrally-substituted material.

Two particular characteristics of clay minerals are generally considered in the preparation of clay-containing polymer nanocomposites. The first is the ability of the silicate particles to disperse into individual layers. The second characteristic is the ability to modify their surface chemistry through ion-exchange reactions with organic and inorganic cations. These two characteristics are, of course, interrelated since the degree of dispersion of silicate layers in a particular polymer matrix depends on the interlayer cation.

Pristine layered silicates usually contain hydrated Na^+ or K^+ ions (Sinha Ray and Okamoto, 2003a; Sinha Ray and Bousmina, 2005). Obviously, in this pristine state, layered silicates are only miscible with hydrophilic polymers. To render clay particles miscible with PLA matrix, one must convert the normally hydrophilic silicate surface into an organophilic one, making possible the intercalation of PLA chains into the silicate galleries. Generally, this can be done by ion-exchange reactions with cationic surfactants including primary, secondary, tertiary, and quaternary alkylammonium or alkylphosphonium cations. Alkylammonium or alkylphosphonium cations in the organosilicates lower the surface energy of the inorganic host and improve the wetting characteristics of the polymer matrix, and the result is a larger interlayer spacing. Additionally, the alkylammonium or alkylphosphonium cations can provide functional groups that can react with the polymer matrix, or in some cases initiate the polymerization of monomers to improve the adhesion between the inorganic and the polymer matrix.

16.2.2 Preparation and Characterization of PLA/Clay Nanocomposites

Ogata and co-workers first reported the preparation of PLA/organoclay blends by dissolving the PLA in hot chloroform in the presence of dimethyldistearylammonium modified MMT (2C18MMT) (Ogata et al., 1997). In the case of PLA/MMT composites, wide-angle X-ray diffraction (WXRD) and small-angle X-ray scattering (SAXS) results showed that the silicate layers forming the clay could not be intercalated in the PLA/MMT blends prepared by the solvent-casting method. In other words, the clay existed in the form of tactoids, consisting of several stacked silicate monolayers. These tactoids are responsible for the formation of particular geometrical structures in the blends, which leads to the formation of superstructures in the thickness of the blended film. This kind of structural feature increases the Young's modulus of the hybrid. Then Bandyopadhyay et al. (1999) reported the preparation of intercalated PLA/organoclay nanocomposites with much improved mechanical and thermal properties. Two different kinds of clay, fluorohectorite (FH) and MMT, both modified with dioctadecyltrimethyl ammonium cation, were used for the preparation of nanocomposites with PLA.

Sinha Ray and co-workers used the same melt intercalation technique for the preparation of intercalated PLA/organoclay nanocomposites (Sinha Ray et al., 2002a,b). XRD patterns and TEM observations clearly established that the silicate layers of the clay were intercalated and randomly distributed in the PLA matrix (see Fig. 16-1). Incorporation of a very small amount of oligo-PCL as a compatibilizer in the nanocomposites led to a better parallel stacking of the silicate layers, and

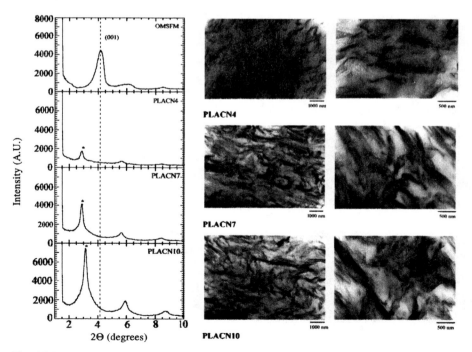

Fig. 16-1   XRD patterns and bright-field images of PLA nanocomposites prepared with organically modified synthetic fluorine mica. The number indicates amount of organoclay loading.

also to much stronger flocculation due to the hydroxylated edge–edge interaction of the silicate layers. Owing to the interaction between clay platelets and the PLA matrix in the presence of a very small amount of o-PCL, the strength of the disk–disk interaction plays an important role in determining the stability of the clay particles, and hence the enhancement of mechanical properties of such nanocomposites.

In subsequent research, Sinha Ray and co-workers (Sinha Ray et al., 2002c, 2003a,b,c,d) prepared PLA nanocomposites with organically modified synthetic fluorine mica (OMSFM). For the characterization of structure and morphology of prepared nanocomposites they first used XRD and conventional TEM (CTEM), and high resolution TEM (HRTEM), to examine the final structure of the PLACNs (PLACN is abbreviation of PLA Nanocomposites). In their further study Sinha Ray et al., 2003c) prepared a series of PLA-based nanostructured materials with various types of organoclays to investigate the effect of organic modification on the morphology, properties, and degradability of the final nanocomposites.

Maiti and co-workers prepared a series of PLA-based nanostructured materials with three different types of pristine clays—saponite (SAP), MMT, and synthetic mica (SM)—and each was modified with alkylphosphonium salts having different chain lengths. In their work, they first tried to determine the effect of varying the chain length of the alkylphosphonium modifier on the properties of organoclay, and how the various clays behave with the same organic modifier. They also studied the effects of dispersion, intercalation, and the aspect ratio of the clay particles on the properties of PLA (Maiti et al., 2002).

Paul et al. (2003) used an in-situ intercalative method for the preparation of exfoliated PLA/clay nanocomposites. They used two different kinds of organoclays (C30B and C25A, commercial names; Southern Clay Products) for the preparation of nanocomposites with PLA. In a typical synthetic procedure, the clay was first dried overnight at 70°C in a ventilated oven, and then, at the same temperature under reduced pressure, directly in the flame-dried polymerization vial for 3.5 h. A 0.025 molar solution of L-lactide in dried tetrahydrofuran (THF) was then transferred under nitrogen to the polymerization vial and the solvent was eliminated under reduced pressure. Polymerizations were conducted in bulk at 120°C for 48 h, after 1 h of clay swelling in the monomer melt. When C30B was used, the polymerization was co-initiated by a molar equivalent of $AlEt_3$, with respect to the hydroxyl groups borne by the ammonium cations of the filler, in order to form aluminum alkoxide active species, and was added before the L,L-lactide. $Sn(Oct)_2$ (monomer/$Sn(Oct)_2 = 300$) was used to catalyze the polymerization of L,L-lactide in the presence of C25A. The same group (Paul et al., 2003) also reported the preparation of plasticized PLA/MMT nanocomposites by melt intercalation technique. The organoclay used was MMT modified with bis-(2-hydroxyethyl) methyl (hydrogenated tallow alkyl) ammonium cations. XRD analyses confirmed the formation of intercalated nanocomposites.

Chang et al. (2003a) reported the preparation of PLA-based nanocomposites with three different kinds of organoclays a via solution intercalation method. They used N,N'-dimethylacetamide (DMA) for the preparation of nanocomposites. XRD patterns indicate the formation of intercalated nanocomposites whatever the organoclay. TEM images proved that most of the clay layers were dispersed homogeneously in the PLA matrix, although some clusters or agglomerated particles were also detected.

Krikorian and co-workers explored the effect of compatibility of different organic modifiers on the overall extent of dispersion of layered silicate layers in a PLA matrix (Krikorian and Pochan, 2003). Three different types of commercially available organoclays were used as a reinforcement phase. Nanocomposites were prepared using the solution-intercalation film-casting technique.

Lee et al. (2003) prepared PLA/MMT nanocomposite for the purpose of tailoring mechanical stiffness of PLA porous scaffold systems. They used a salt leaching/gas foaming method for the preparation of the nanocomposite scaffold. A viscous solution with a concentration of 0.1 g/ml was prepared by dissolving PLA polymer in chloroform. NH_4HCO_3/NaCl salt particles sieved in the range of 150–300 μm and dimethyl dehydrogenated tallow ammonium modified MMT (2M2HT-MMT) clays were added to the PLA solution and thoroughly mixed. The amount of the 2M2HT-MMT clay was 2.24, 3.58, and 5.79 vol% relative to PLA.

The paste mixture of polymer/salts/solvent was then cast into a special device equipped with a glass slide as a sheet model. The cast film was obtained after being air-dried under atmospheric pressure for 2 h. When the film became semisolid, a two-step salt leaching was performed. The film was first immersed in a 90°C hot water bath to leach out the NH_4HCO_3 particles, concomitantly generating gaseous ammonia and carbon dioxide in the polymer matrix. When no gas bubbles were generated, the film was subsequently immersed into another beaker containing hot water ($\sim$60°C), and kept there for 30 min to leach out the remaining NaCl particles. It was finally, freeze dried for two days.

From the XRD patterns, it was seen that pure 2M2HT-MMT demonstrated a sharp peak at $2\theta = 3.76°$ and this peak was not observed in the case of the nanocomposite, indicating the formation of exfoliated PLA/2M2HT-MMT nanocomposite, but the authors did not report any TEM observations. Recently, various authors have reported the preparation of PLA/clay nanocomposites using different synthetic routes (Kramschuster et al., 2004; Nam et al., 2004; Ninomiya et al., 2004).

16.3 PLA NANOCOMPOSITES BASED ON CARBON NANOTUBES

In the past decade, carbon nanotubes (CNTs) have attracted more and more interest from both scientists and engineers because of their extraordinarily high strength and high modulus (the strongest material known nowadays), their excellent electrical conductivity along with their important thermal conductivity and stability, and their low density associated with their high aspect ratio and one-dimensional tubular structure, and hence their tremendous potential in applications such as nanoengineering and bio-nanotechnology (Thostenson et al., 2001; Zhang et al., 2003; Andrews and Weisenberger, 2004; Xie et al., 2005; Awasthi et al., 2005; Sinha Ray et al., 2006; Banerjee et al., 2005; Miyagawa et al., 2005; Fernando et al., 2005; Yang et al., 2005; Kim et al., 2005; Vaudreuil et al., 2007; Shi et al. 2005, 2006; Baibarac and Gomez-Romero, 2006; Cui et al., 2006; Zhang et al., 2006).

CNTs are giant fullerenes. A fullerene, by definition, is a closed, convex cage molecule containing only hexagonal and pentagonal faces. CNTs have many structures, differencing in length, thickness, types of spiral, and number of layers, although they are formed from essentially the same graphite sheet. Graphite has three-coordinate sp^2 carbons forming planar sheets, whose motif is the flat six-membered benzene ring. In fullerene the three-coordinate carbon atoms tile the spherical or nearly spherical surfaces, the best known example being C_{60} with a truncated icosahedral structure formed by 12 pentagonal rings and 20 hexagonal rings. There are two main types of CNTs: single-walled CNTs (SWCNTs) and multi-walled CNTs (MWCNTs).

Theoretically, it is possible to construct an sp^2-hybrized carbon tubule by rolling up a hexagonal graphene sheet. This leads to two different types of arrangements such as "nonchiral" and "chiral". In the nonchiral arrangements, the honeycomb lattices, located at the top and bottom of the tube, are always parallel to the tube axis and these configurations are known as armchair and zig-zag. In the armchair structure, two C–C bonds on opposite sides of each hexagon are perpendicular to the tube axis, whereas in the zig-zag arrangement, these bonds are parallel to the tube axis. All other conformations in which the C–C bonds lie at an angle to the tube axis are known as chiral or helical structures.

Moon and co-workers prepared PLA/MWCNT nanocomposites by the solvent casting method (Moon et al., 2005). They used two different methods for the synthesis of nanocomposites: (i) A 10 wt% solution of PLA in chloroform was prepared, which was then combined with previously dispersed MWCNTs in chloroform and the whole mixture was sonicated for 6 h. Subsequently, the mixture was poured into Teflon dishes and dried at room temperature for one week and then sample was vacuum-dried at 80°C for 8 h. (ii) The As-cast composite films from method 1 were folded and broken into pieces of 0.5–1.0 cm^2 and stacked between two metal plates. This stack was then hot pressed at 200°C and 150 kgf/cm^2 for 15 min. As a result 100–200 μm thick films were obtained. Transmission electron microscopic analysis shows uniform dispersion of MWCNTs into the PLA matrix. This uniform dispersion of MWCNTs changes the physical properties of the pure polymer.

Chen et al. (2005) synthesized PLA/CNT nanocomposites using a "grafting to" technique. To elucidate the effect of molecular weight on the properties of final nanocomposites, they used PLA of three different molecular weights. They first oxidized MWCNTs by acid treatment. In a typical procedure, a 500 ml flask charged with 2.5 g of the crude MWCNTs and 200 ml of 60% HNO_3 aqueous solution was sonicated in a bath (28 kHz) for 30 min. The mixture was then stirred for 12 h under reflux. After the solution was cooled to room temperature, it was diluted with 400 ml of deionized water and vacuum-filtered through a 0.22 μm polycarbonate membrane. The solid was washed with deionized water until the pH of the filtrate reached approximately to 7. The solid was then dried under vacuum for 12 h at 60°C to yield 1.5 g of the carboxylic-acid-functionalized MWCNT (MWCNT-COOH).

For the preparation of nanocomposites, MWCNT-COOH was first reacted with excess thionyl chloride ($SOCl_2$) for 24 h under reflux, and then the residual $SOCl_2$ was removed by reduced-pressure distillation to yield the acyl-chloride-functionalized

MWCNT (MWCNT-COCL). The MWCNT-COCl was added to chloroform, and the reactor was then immersed in an oil bath at 70°C with methanol stirring for 1 h to remove the solvent. The reaction was allowed to proceed for 24 h at 180°C and 1 atm. The resulting reaction medium was dissolved in excess chloroform and vacuum-filtered three times through 0.22 μm polycarbonate membrane to yield the MWCNT-g-PLLA hybrid by filtering the chloroform-soluble substances such as the unbound PLLA. The MWCNT-g-L-PLA hybrids were prepared using the PLLAs with three different molecular weights of 1000 g/mol, 3000 g/mol, 11,000 g/mol, and 15,000 g/mol, which are represented as MWCNT-g-PLLA1, MWCNT-g-PLLA2, MWCNT-g-PLLA3, and MWCNT-g-PLLA4, respectively.

Various techniques were used for the characterization of the prepared nanocomposite samples. The composition of the resulting MWCNT-g-PLLA nanocomposites was confirmed by FTIR. The results showed the retention of PLLA even after extensive washing with a good solvent for the PLLA. Raman analysis revealed that the D- and G-bands of the MWCNT at 1287 and 1598 cm^{-1} for both MWCNT-COOH and MWCNT-g-PLLA2, which were attributed to the defects and disorder-induced peaks and tangential-mode peaks. The peak intensity of the MWCNT-g-PLLA2 was much weaker than that of the MWCNT-COOH, which means that the characteristic absorption peaks were strongly attenuated due to the grafted PLLA.

Thermogravimetric analysis (TGA) and TEM observations indicated that the amount of grafted PLLA and its morphology depended strongly on the molecular weight of the PLLA. As the molecular weight of the PLLA increased from 1000 to 3000, the PLLA coating on the MWCNT became thicker and more uniform. When the molecular weight of the PLLA was increased further to 11,000 and 15,000, the surface of the MWCNT was not covered wholly but was sparsely stained with the PLLA. The grafted PLLA formed bolds exhibiting a morphology that was like a squid leg. The MWCNT-g-PLLA prepared using PLLA with molecular weight of 3000 was more readily dispersed in inorganic solvent such as chloroform and N,N-dimethylformamide (DMF) than that obtained using other PLLAs. This was attributed to the higher PLLA content and the smaller area of the MWCNT surface. More recently, Song et al. (2007) reported the one-step synthesis of PLA-g-MWCNT nanocomposite.

16.4   PLA NANOCOMPOSITES BASED ON VARIOUS OTHER NANOFILLERS

Hiroi and co-workers prepared PLA/organically modified layered titanate nanocomposites by a melt-extrusion method (Hiroi et al., 2004). For the preparation of organically modified layered titanate, a blend of K_2CO_3 (304 g), Li_2CO_3 (54 g), TiO_2 (762 g), and KCl (136 g) was mixed intimately and heated at 1020°C for 4 h in an electric furnace. After cooling, the powder was dispersed to a 5 wt% water slurry and 10 wt% of H_2SO_4 water solution was added while agitating for 2 h until a pH of 7 was reached.

This slurry was filtered off and rinsed with water, and then dried at 110°C. After this, it was heated at 600°C for more than 3 h in an electric furnace. The white powder obtained was $K_{0.7}Ti_{1.73}Li_{0.27}O_{3.95}$ with an average particle size of 32 mm. $K_{0.7}Ti_{1.73}Li_{0.27}O_{3.95}$ (130 g) was stirred in 0.5 M HCl solution for 1.5 h at ambient temperature. After stirring, the product hydrated titanate (HTO) was collected by filtration and washed with water. The HTO was ion-exchanged, 72% of potassium ion and 99% or more lithium ion to protons. The chemical composition was determined by X-ray fluorescence analysis.

The recovered HTO and N-(cocoalkyl)-N,N-(bis(2-hydroxyethyl))-N-methyl-ammonium chloride (157.5 g) were stirred at 80°C for 1 h in double-distilled water. After 1 h, the product was filtered and washed at 80°C with water. The organo-HTO was first dried at 40°C for 1 day under air to prevent particle aggregation, and then at 160°C for 12 h under vacuum.

For nanocomposite preparation, the organo-HTO (OHTO) (dried at 120°C for 8 h) and PLA were first dry-mixed by shaking them in a bag. The mixture was then melt-extruded using a twin-screw extruder (KZW15-30TGN, Technovel Corp., Japan) operated at 195°C (screw speed 300 rpm, feed rate 22 g/min) to yield nanocomposite strands. XRD patterns and TEM observations show the formation of intercalated structured.

Nishida et al. (2005) reported the preparation of aluminum hydroxide (Al[OH]$_3$)-based nanocomposites of PLA to achieve the chemical recycling of flame-resistant properties of PLA. For the preparation of PLLA/Al(OH)$_3$ hybrids, PLLA was first synthesized by the ring-opening polymerization of L,L-lactide catalyzed by Sn(2-ethylhexanoate)$_2$. The polymerized PLLA was purified in a three-stage process: first extracting the catalyst and residues from the PLLA/chloroform solution with a 1 M HCl aqueous solution, then washing with distilled water until the aqueous phase became totally neutral, and finally precipitating the polymer with methanol before vacuum drying. The purified PLLA was then mixed with Al(OH)$_3$ in a prescribed weight ratio in a chloroform solution and vigorously stirred for 1 h to disperse the inorganic particles uniformly. The mixture was then cast on glass Petri dishes.

16.5 PROPERTIES OF PLA NANOCOMPOSITES

Nanocomposites consisting of PLA and nano-fillers frequently exhibit significant improvement in mechanical and various other properties compared with those of pure polymers. Improvements generally include a higher modulus both in solid and the melt states, increased strength and thermal stability, decreased gas permeability, and increased rate of degradability.

Dynamic mechanical analysis (DMA) has been used to study the temperature dependence of G' of PLA upon nanocomposite formation under different experimental conditions. Figure 16-2 shows the temperature dependence of G' for various PLA/clay nanocomposites and pristine PLA. For all PLACNs, the enhancement of G' can be seen in the investigated temperature range when compared with the neat PLA, indicating that organically modified clay particles have a strong effect on the elastic properties of virgin PLA. Below T_g, the enhancement of G' is clear for all nanocomposites. On the other hand, all PLACNs show a greater increase in G' at high temperature compared to that of the PLA matrix. This is due to both mechanical

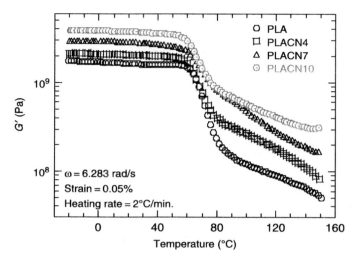

Fig. 16-2   Temperature dependence of storage modulus (G') of neat PLA and various PLACNs prepared with organically modified synthetic fluorine mica. The number indicates amount of organoclay loading.

reinforcement by the silicate layers and extended intercalation at high temperature (Sinha Ray and Okamoto, 2003b). Above T_g, when materials become soft, the reinforcement effect of the silicate layers becomes prominent due to the restricted movement of the polymer chains. This is accompanied by the observed enhancement of G'.

The tensile modulus of polymeric materials has been shown to be remarkably improved when nanocomposites are formed with various types of nano-fillers. In the case of PLA-based nanostructured materials, most studies report the tensile properties as a function of filler content. In most conventionally filled polymer systems, the modulus increases linearly with the filler volume fraction, whereas for these nanoparticles at much lower filler concentrations, the modulus increased sharply and to a much larger extent. The dramatic enhancement of the modulus for such extremely low clay concentrations cannot be attributed simply to the introduction of the higher modulus inorganic filler layers.

A theoretical approach assumes a layer of affected polymer on the filler surface, with a much higher modulus than the bulk equivalent polymer (Shia et al., 1998). This affected polymer can be thought of as the region of the polymer matrix that is physisorbed onto the silicate surface and is thus stiffened through its affinity for and adhesion to the filler surfaces (Shia et al., 1998). For such high-aspect-ratio fillers as our layered silicate layers, the surface area exposed to the polymer is huge and the significant increases in the modulus with very low filler content are not surprising. Furthermore, beyond the percolation limit the additional silicate layers are incorporated into polymer regions that are already affected by other silicate layers, and thus it is expected that the enhancement of modulus will be much less dramatic.

Lee et al. (2003) reported the MMT content dependence of tensile modulus of PLLA nanocomposites scaffolds. The modulus of the nanocomposites systematically increased with increasing MMT loading. According to these authors, the crystallinity and the glass transition temperature of PLLA nanocomposites were lower than those of neat PLLA, but the modulus of neat PLLA was significantly increased in the presence of a small amount of MMT loading. This observation suggests that the layered silicates of MMT act a mechanical reinforcement of polymer chains.

In the case of PLLA/MWCNT nanocomposites, it was observed that the Young's modulus increases with MWCNT loading in the nanocomposite films compare to pure PLLA films, although an increase in the MWCNT content did not cause a significant increase in Young's modulus. The average Young's modulus of the 5 wt% nanocomposite was approximately 2.5 GPa, which was approximately 150% higher than that of the pure PLLA film. It was also observed that increasing the loading of MWCNTs in these nanocomposites caused a significant increase in stiffness, which eventually led to brittle fracture, as indicated by the low elongation at break in the tensile test.

Sinha Ray and co-workers reported the detailed measurement of flexural properties of neat PLA and various nanocomposites prepared with organoclay (Sinha Ray et al., 2003a). They conducted flexural property measurements with injection-molded samples according to the ASTM D-790 method. There was a significant increase in flexural modulus for nanocomposite prepared with 4 wt% of organoclay when compared to that of neat PLA, followed by a much slower increase with increasing organoclay content, and a maximum of 50% for nanocomposite prepared with 10 wt% of organoclay. On the other hand, the flexural strength showed a remarkable increase with PLACN7 (nanocomposite prepared with 7 wt% clay), and then gradually decreased with organoclay loading. According to the authors, this behavior may be due to the high clay content, which leads to brittleness in the material. They also measured the flexural properties of PLA nanocomposites prepared with various kinds of organically modified MMT, and the results showed a similar trend. This means that there is an optimal amount of organoclay needed in a nanocomposite to achieve the greatest improvement in its properties.

Generally, the incorporation of nano-fillers into the polymer matrices was found to enhance the thermal stability by acting as a superior insulator and mass transport barrier to the volatile products generated during decomposition. Bandyopadhyay et al. (1999) reported the first improved thermal stability of nanostructured materials that combined PLA and organically modified fluorohectorite (FH) or MMT clay. These nanocomposites were prepared by melt intercalation. The authors showed that the PLA that was intercalated between the galleries of FH or MMT clay resisted the thermal degradation under conditions that would otherwise completely degrade pure PLA. They argue that the silicate layers act as a barrier for both the incoming gas and also the gaseous by-products, which increases the degradation onset temperature and also widens the degradation process. The addition of clay enhances the performance of the char formed, by acting as a superior insulator and mass transport barrier to the volatile products generated during decomposition. Recently, there have been many reports concerned with the improved thermal stability of PLA-based nanocomposites prepared with various kinds of organically modified layered silicates (OMLS) (Paul et al., 2003). Chang et al. (2003b) conducted detailed thermogravimetric analyses of PLA-based nanocomposites of three different OMLS. In the case of C16MMT- or C25A-based hybrids, the initial degradation temperatures (T_{iD}) of the nanocomposites decreased linearly with increasing amount of OMLS. On the other hand, in nanocomposite prepared with DTAMMT clay, the initial degradation temperature was nearly constant over code of clay loadings from 2 to 8 wt%. This observation indicates that the thermal stability of the nanocomposites is directly related to the stability of OMLS used for the preparation of nanocomposites.

Paul et al. (2003) prepared PLA layered silicate nanocomposites by melt intercalation in the presence of a stabilizer to decrease the possibility of host matrix degradation due to heating. The degradation of PLA during processing takes place even in the presence of antioxidant, and 41.2% decrease in number average molecular weight was observed compared with native PLA. An increase in the thermal stability under oxidative conditions was found and it was suggested that a physical barrier between the polymer medium and superficial zone of flame combustion may be generated due to the char formation. The use of PLA for automotive parts has been studied as a possible contribution to restraining the increase in CO_2 emissions. For this application, major improvements in heat and impact resistance are needed. It was found that in-mold crystallization of the PLA/clay nanocomposite led to a large suppression of the decrease in storage modulus at high temperature, which in turn improved the heat resistance of PLA.

16.6   BIODEGRADABILITY

The degradation process of aliphatic polyesters is a complex process, which can proceed via hydrolysis (most often catalyzed by enzymes) and/or oxidation (UV- or thermo-induced). The stereoconfiguration and possible crystallinity, relative hydrophobicity of the polymer matrix, presence of substituents, or even filler and conformational flexibility contribute to the biodegradability of synthetic polymers with hydrolysable and/or oxidizable linkages in their main chain. On the other hand, the morphology of the polymer samples or the material structure as recovered after loading with given filler; greatly affect their rate of biodegradation.

A major problem with the PLA matrix is the very slow rate of degradation. Despite the considerable number of reports concerning the enzymatic degradation of PLA (Gruber and Brien, 2002) and various PLA blends (Sinha Ray and Bousmina, 2005) very little is reported about the composting degradability of PLA, except in recent publications by the Sinha Ray and co-workers (Sinha Ray et al., 2003d; Sinha Ray and Bousmina, 2005). Figure 16-3 illustrates samples recovered from compost after various times.

Sinha Ray and his group also conducted a respirometric test to study the degradation of the PLA matrix in a compost environment (Sinha Ray et al., 2002c, 2003d). For this test the compost used was prepared from a mixture of bean-curd refuse, food waste, and cattle feces. Unlike weight loss or fragmentation, which

after 32 days after 50 days after 60 days

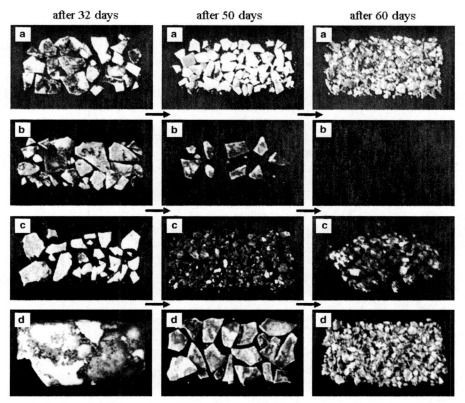

Fig. 16-3 Biodegradability of (a) neat PLA, (b) PLA/qC18-MMT4, (c) PLA/C18-MMT4, and (d) PLA/qC18-Mica4 recovered from compost after various times. The initial shape of the crystallized samples was 3 cm × 10 cm × 0.1 cm.

reflects the structural changes in the test sample, CO_2 evolution provides an indicator of the ultimate biodegradability, that is, mineralization, of the test samples. Experimental results clearly indicate that the biodegradability of the PLA component in PLA/qC13(OH)-mica4 or PLA/qC16-SAP4 was enhanced significantly. On the other hand, the PLA component in PLA/C18-MMT4 shows a slightly higher biodegradation rate, while the rates of degradation of pure PLA and PLA/qC18-MMT4 are almost the same. The compost degradation of PLA occurs by a two-step process. During the initial phases of degradation, the high-molecular-weight PLA chains hydrolyze to lower-molecular-weight oligomers. This reaction can be accelerated by acids or bases and is also affected by both temperature and moisture. Fragmentation of the plastic occurs during this step at a point where the M_n decreases to less than about 40,000. At about this same M_n value, microorganisms in the compost environment continue the degradation process by converting these lower-molecular-weight components to CO_2, water, and humus (Ikada and Tsuji, 2000; Gruber and Brien, 2002). Therefore, any factor that increases the tendency to hydrolysis of the PLA matrix ultimately controls the degradation of PLA.

The incorporation of OMLS fillers into the PLA matrix resulted in a small reduction in the molecular weight of the matrix. It is well known that PLA of relatively low molecular weight may show higher rates of enzymatic degradation because of, for example, the high concentration of accessible chain end groups (Sinha Ray et al., 2003d). However, in these cases the rate of molecular weight change of pure PLA and PLA in various nanocomposites is almost the same. Thus the initial molecular weight is not the main factor controlling the degradability of nanocomposites. Another factor that controls degradability of PLA in nanocomposite is the different degree of dispersion of silicate layers in the polymer matrix, which actually depends on the nature of the surfactant used to modify the clay surface. Therefore, we can control the degradability of PLA by judicious choice of OMLS.

To understand to what extent the incorporation of pristine and organically modified clay influences the degradation behavior of the PLA matrix, Paul et al. (2003) conducted the hydrolytic degradation of composites based on the same amounts (3 wt%) of Cloisite Na$^+$, Cloisite 25A, and Cloisite 30B for more than 5 months, directly compared with the pure PLA. In a typical experimental procedure, the samples were first shaped as films about 0.5 mm thick. In a second step, each film was cut into 1.3 cm × 3.0 cm rectangular specimens (three specimens per sample). Each specimen was then dipped in a flask containing 25 ml of 0.1 M phosphate buffer at pH 7.4. The flasks were immersed in a water bath at 37°C. At predetermined periods (1 week, 2 weeks, 1 month, 2.5 months, and 5.5 months) the specimen were picked out of the buffered solution and rinsed several times with distilled water. Finally, the residual water was wiped off from the sample surface before drying it by wrapping it in a small paper bag placed in a desiccator.

The results indicated that faster hydrolysis, leading to an increase of the crystallinity of the PLA matrix, is found for the Cloisite Na$^+$-based blend, that is, that it is due to the microcomposite structure. It was also concluded that both composite structure, either microcomposite or intercalated nanocomposite, and the relative hydrophilicity of the clay play determining roles in the hydrolytic degradation process. Indeed, the more hydrophilic the filler, the more pronounced is the degradation.

16.7   MELT RHEOLOGY

In the case of polymer nanocomposites, measurements of melt rheological properties are not only important for understand the processability of these materials but are also helpful in elucidating the strength of the polymer–filler and the structure–property relationships in nanocomposites. This is because rheological behavior is strongly influenced by nanoscale structure and interfacial characteristics.

From the master curves for G' and G'' of pure PLA and various nanocomposites with different weight percentages of C18MMT loading, at high frequencies ($a_T\omega > 10$ rad/s), the viscoelastic behavior was the same for all nanocomposites. In contrast, at low frequencies ($a_T\omega < 10$ rad/s), both moduli exhibited weak frequency dependence with increasing C18MMT content, such that there are gradual changes of behavior from liquidlike (G' and $G'' \propto \omega$) to solidlike with increasing C18MMT content (Sinha Ray and Okamoto, 2003b).

The slopes of G' and G'' in the terminal region of the master curves of the PLA matrix were 1.85 and 1, respectively, values in the range expected for polydisperse polymers (Hoffmann et al., 2000). On the other hand, the slopes of G' and G'' were considerably lower for all PLACNs compared to those of pure PLA. In fact, for PLACNs with high C18MMT content, G' becomes nearly independent at low $a_T\omega$ and exceeds G'', characteristic of materials exhibiting a pseudo solidlike behavior.

In the dynamic complex viscosity $|\eta^*|$ master curves for the pure PLA and nanocomposites, based on linear dynamic oscillatory shear measurements, in the low $a_T\omega$ region ($< 10 \, \text{rad/s}$), pure PLA exhibited almost newtonian behavior while all nanocomposites showed very strong shear-thinning tendency. On the other hand, M_w and PDI (Poly dispersity index) of pure PLA and various nanocomposites were almost the same; thus the high viscosity of PLACNs was explained by the flow restrictions of polymer chains in the molten state due to the presence of MMT particles.

The shear rate-dependent viscosity of pure PLA and various PLACNs at 175°C was also measured by Sinha Ray and Okamoto (2003b). The pure PLA exhibited almost newtonian behavior at all shear rates, whereas PLACNs exhibited nonnewtonian behavior. All PLACNs showed a very strong shear-thinning behavior at all measured shear rates and this behavior is analogous to the results obtained in the case of oscillatory shear measurements. Additionally, at very high shear rates, the steady shear viscosities of PLACNs were comparable to that of pure PLA. These observations suggest that the silicate layers are strongly oriented toward the flow direction (perhaps perpendicular alignment of the silicate layers toward the stretching direction) at high shear rates, whereas pure polymer dominates shear-thinning behavior at high shear rates.

Sinha Ray and Okamoto (2003b) first conducted elongation tests of PLA/clay nanocomposite prepared with 5 wt% of C18MMT in the molten state at constant Hencky strain rate, $\dot{\varepsilon}_0$ using elongation rheometry. On each run of the elongation test, samples of size $60 \, \text{mm} \times 7 \, \text{mm} \times 1 \, \text{mm}^3$ were annealed at a predetermined temperature for 3 min before starting the run in the rheometer, and uniaxial elongation experiments were conducted at various $\dot{\varepsilon}_0$ values. Figure 16-4 shows double-logarithmic plots of the transient elongation viscosity $\eta_E(\dot{\varepsilon}_0; t)$ versus time t, observed for PLACN5 (nanocomposite prepared with 5 wt% of clay) at 170°C with various $\dot{\varepsilon}_0$ values ranging from 0.01 to $1.0 \, \text{s}^{-1}$. The figure shows a strong *strain-induced hardening* behavior for PLACN5. In the early stage, η_E gradually increases with t but is almost independent of $\dot{\varepsilon}_0$. This is generally called the *linear region* of the viscosity curve. After a certain time, t_{η_E}, which is the *up-rising* time (marked with the upward arrows in the figure), η_E was strongly dependent on $\dot{\varepsilon}_0$, and a rapid upward deviation of η_E from the curves of the linear region was observed. The authors tried to measure the elongational viscosity of pure PLA but they were unable to do so accurately; the low viscosity of pure PLA may be the main reason. However, they confirmed that neither strain-induced hardening in elongation nor rheopexy in shear flow took place in the case of pure PLA having the same molecular weight and polydispersity as PLACN3.

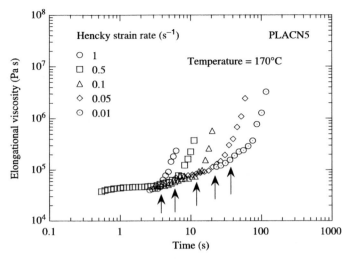

Fig. 16-4   Strain rate dependence of up-rising Hencky strain. Reproduced from Sinha Ray and Okamoto (2003b) by permission of Wiley-VCH Verlag GmbH, Germany.

As in polypropylene/OMLS systems, the extended Trouton rule, $3\eta_0(\dot{\gamma};t) \cong \eta_E(\dot{\varepsilon}_0;t)$, does not hold for PLACN5 melt, as opposed to the melts of pure polymers (Okamoto et al., 2001a; Nam et al., 2002). These results indicate that in the case of PLACN5, flow-induced internal structural changes also occurred in elongation flow, but the changes were quite different from those in shear flow. The strong rheopexy observed in shear measurements for PLACN5 at very slow shear rates reflected the fact that the shear-induced structural change involved a process with an extremely long relaxation time.

16.8   FOAM PROCESSING

Polymeric foams have become widely used as packing materials because they are lightweight, have a high strength/weight ratio, have superior insulating properties, and have high energy-absorbing performance. One of the most widely used polymers for the preparation of foam is PS (polystyrene). It is produced from fossil fuels, consumed, and discarded into the environment, ending up as waste that is not spontaneously degradable. Use of foams made from biodegradable polymeric materials would alleviate this problem.

Biodegradable polymers such as PLA have some limitations in foam processing because such polymers do not exhibit high strain-induced hardening, which is the primary requirement to withstand the stretching forces experienced during the latter stages of bubble growth. Branching of polymer chains, grafting with another copolymer, or blending of branched and linear polymers are the common methods used to improve the extensional viscosity of a polymer in order to make it suitable for foam formation. PLACNs have already been shown to exhibit a high modulus and, under uniaxial elongation, a tendency toward strong strain-induced hardening. On the basis of these results, Sinha Ray and his group (Sinha Ray and Okamoto, 2003b) first conducted foam processing of PLACNs with the expectation that they would provide advanced polymeric foams with many desirable properties. They used a physical foaming method, a batch process, for foam processing. This process consists of four stages: (a) saturation of CO_2 in the sample at the desired temperature; (b) cell nucleation when the release of CO_2 pressure begins (forming supersaturated CO_2); (c) cell growth to an equilibrium size during the release of CO_2; and (4) stabilization of the cell via cooling of the foamed sample.

Figure 16-5 shows typical SEM images of the freeze-fracture surfaces of neat PLA and two different nanocomposites foamed at 140°C for PLA and PLA/C18MMT5 and at 165°C for PLA/qC18MMT5. All foams exhibited a nice closed-cell structure with homogeneous cells in the case of nanocomposites, while the neat PLA foam showed nonuniform cell structure having large cell size ($\sim$230 μm). Also, the nanocomposite foams showed a smaller cell size (d) and larger cell density (N_c) than neat PLA foam, suggesting that the dispersed silicate particles act as nucleating sites for cell formation (Okamoto et al., 2001b). They calculated the distribution function of cell size from SEM images and showed that the nanocomposite foams nicely

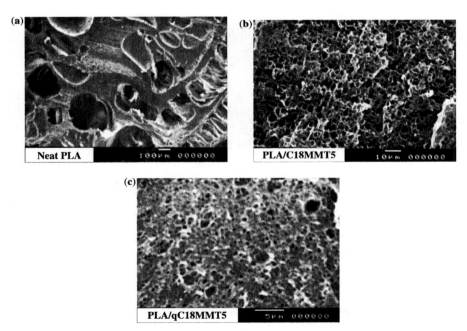

Fig. 16-5 SEM images of freeze-fracture surfaces of (a) neat PLA, (b) PLA/C18MMT5, and (c) PLA/qC18MMT5 foams. Reproduced from Fujimoto, Sinha Ray, Okamoto, Ogami and Ueda by permission of Wiley-VCH Verlag GmbH, Germany.

obeyed a Gaussian distribution. In the case of PLA/qC18MMT5 (Fig. 16-5b), the width of the distribution peaks, which indicates the dispersity for cell size, became narrow and were accompanied by a finer dispersion of silicate particles. The PLA/qC18MMT5 (nanocellular) foam has a smaller d value ($\sim$360 nm) and a huge N_c (1.2×10^{14} cells/cm^{-3}) compared with PLA/C18MMT5 (microcellular) foam ($d = 2.59$ μm and $N_c = 3.56 \times 10^{11}$ cells/cm^{-3}) (Sinha Ray and Okamoto, 2003b). These results indicate that the nature of the dispersion plays a vital role in controlling the size of the cell during foaming. On the other hand, the very high value of N_c for the PLA/qC18MMT5 foam indicates that the final ρ_f value is controlled by the competitive processes in cell nucleation, growth, and coalescence. The cell nucleation, in the case of the nanocomposite systems, took place in the boundary between the matrix polymer and the dispersed silicate particles. For this reason, the cell growth and coalescence were strongly affected by the characteristic parameter and the storage and loss modulus ($\approx$ viscosity component) of the materials during processing. This may create nanocellular foams without the loss of mechanical properties in the case of polymeric nanocomposites.

16.9 POTENTIAL FOR APPLICATIONS AND FUTURE PROSPECTS

New environmental polices, societal concerns, and growing environmental awareness have triggered the search for new products and processes that are benign to the environment. PLA is considered as an alternative to the existing petroleum-based plastic materials, but the intrinsic properties of unmodified PLA, such as its high crystallinity ($\sim$40%) and rigidity and slow degradation, put limits on its wide applicability (Leaversuch, 2002; Gupta, 2007). PLA-based nanostructured materials offer unique combinations of properties, including biodegradability and thermoplastic processibility, that offer potential applications as commodity plastics, such as in packaging, agricultural products, and disposable materials. The versatility of the nano-fillers used with respect to transformation into various shapes and morphologies along with good mechanical properties offers a wide range of applications.

The thermoplastic character of PLA is very useful in allowing transformation of the polymer materials into various shapes. The apparel sector is very promising. For example, the University of Tennessee (USA) has been active in work on spunlaid and melt-blown nonwovens based on PLA. Kanebo Ltd (Japan) has produced a PLA fiber under the brand name LACTRON^{TM}, which was exhibited in garments during the Nagano Olympics under the umbrella of "Fashion for the Earth" (Lunt and Shafer, 2000). The low modulus of the fiber has been exploited for better drape and feel of fabrics. A market for PLA and PLA-based nanostructured materials has also developed in sportswear, especially as an inner wicking layer.

Polylactide-based nanocomposites are an important part of the family of novel biomaterials though their history is no more than 10 years. From the considerations of versatile properties and wide range of possible applications, these materials have attracted a great deal of scientific and industrial interest and have led research in biomaterials science in new directions. This chapter has provided a snapshot of the synthesis and characterization procedures for PLA and various PLA-based nanostructured materials. Useful progress has been made to date toward improvement of the processability and potential for application of pure PLA to make it suitable for wide range of applications from commodity plastics to medical applications. By changing the nature of the nano-fillers and the processing conditions, some of the physical properties of pure PLA—mechanical, thermal, and electrical properties, biodegradability, etc.—have been improved significantly, which was not possible by simple mixing with conventional fillers. Considering the progress to date of research in this field, it can be concluded that these materials will be commercially available very soon and their appropriate utilization will open further research directions.

REFERENCES

Alexander M, Doubis P. 2000. Polymer-layered silicate nanocomposites, properties and uses of a new class of materials. *Mater Sci Eng* R28:1.

Andrews R, Weisenberger MC. 2004. Carbon nanotube polymer composites. *Curr Opin Solid State Mater Sci* 8:31.

Awasthi K, Srivastava A, Srivastava ON. 2005. Synthesis of carbon nanotubes. *J Nanosci Nanotechnol* 5:1616.

Baibarac M, Gomez-Romero P. 2006 Nanocomposites based on conducting polymers and carbon nanotubes: From fancy materials to functional applications. *J Nanosci Nanotechnol* 6:289–302.

Bandyopadhyay S, Chen R, Giannelis EP. 1999. Polylactic acid layered silicate nanocomposites. *Polym Mater Sci Eng* 81:159.

Banerjee S, Hemraj-Benny T, Wong SS. 2005. Routes towards separating metallic and semiconducting nanotubes. *J Nanosci Nanotechnol* 5:841.

Bischoff CA, Walden P. 1894. Synthesis of polyglycolic and polyglycolic acid. *Liebigs Ann Chem* 279:45–51.

Biswas M, Sinha Ray S. 2001. Recent progress in synthesis and evaluation of polymer-montmorillonite nanocomposites. *Adv Polym Sci* 155:167.

Brindly SW, Brown G. 1980. *Crystal Structure of Clay Minerals and their X-ray Diffraction.* London: The Mineralogical Society.

Chang JH, An YU, Sur GS. 2003a. Poly(lactic acid) nanocomposites with various organoclays. I. Thermomechanical properties, morphology, and gas permeability. *J Polym Sci Part B: Polym Phys* 41:94.

Chang JH, An YU, Cho D, Giannelis EP. 2003b. Poly(lactic acid) nanocomposites: comparison of their properties with montmorillonite and synthetic mica (II). *Polymer* 44:3715.

Chen GX, Kim HS, Park BH, Yoon JS. 2005. Controlled functionalization of multiwalled carbon nanotubes with various molecular-weight poly(L-lactic acid). *J Phys Chem B* 109:22237.

Cui X, Engelhard MH, Lin Y. 2006. Preparation, characterization and anion exchange properties of polypyrrole/carbon nanotube nanocomposites. *J Nanosci Nanotechnol* 6:547–553.

Drumright RE, Gruber PR. 2000. Polylactic acid technology. *Adv Mater* 23:1841.

Enomoto K, Ajioka M, Yamaguchi A. 1994. Polyhydroxycarboxylic acid and preparation process thereof. US Patent 5,310,865.

Fang Q, Hanna MA. 1999. Rheological properties of amorphous and semicrystalline polylactic acid polymers. *Ind Crops Prod* 10:47.

Fernando KAS, Lin Y, Zhou B, et al. 2005. Poly(ethylene-*co*-vinyl alcohol) functionalized single-walled carbon nanotubes and related nanocomposites. *J Nanosci Nanotechnol* 5:1050.

Fujimoto Y, Sinha Ray S, Okamoto M, Ogami A, Yamada K, Ueda K. 2003. Well-controlled biodegradable nanocomposite foams: from microcellular to nanocellular. *Macromol Rapid Commun* 24:457–461.

Gruber P, Brien MO. 2002. Polylactides NatureWorks[TM] PLA. In: Doi Y, Steinbuchel A, editors. *Biopolymers*, Volume 4, Polyesters III Applications and Commercial Products. Weinheim: Wiley-VCH. p. 235.

Gu JD, Gada M, Kharas G, Eberiel D, McCarthy SP, Gross RA. 1992. *Polym Mater Sci Eng* 67:351.

Gupta B, Revagade N, Hilborn J. 2007. Poly(lactic acid) fiber: an overview. *Prog Polym Sci* 32:455.

Hiroi R, Sinha Ray S, Okamoto M, Shiroi T. 2004. Organically modified layered titanate: a new nanofiller to improve the performance of biodegradable polylactide. *Macromol Rapid Commun* 25, 1359.

Hoffmann B, Kressler J, Stoppelmann G, Friedrich Chr, Kim GM. 2000. Rheology of nanocomposites based on layered silicates and polyamide-12. *Colloid Polym Sci* 278:629.

Ikada Y, Tsuji H. 2000. Biodegradable polyesters for medical and ecological applications. *Macromol Rapid Commun* 21:117.

Kim HW, Lee HH, Knowles CJ. 2006. Electrospinning biomedical nanocomposite fibers of hydroxyapatite/poly(lactic acid) for bone regeneration. *J Biomed Mater Res* 79A:643.

Kim K, Cho SJ, Kim ST, Chin IJ, Choi HJ. 2005. Formation of two-dimensional array of multiwalled carbon nanotubes in polystyrene/poly(methyl methacrylate) thin film. *Macromolecules* 38:10623.

Kramschuster A, Gong S, Turng LS, Li T. 2007. Injection-molded solid and microcellular polylactide and polylactide nanocomposites. *J Biobased Mater Bioenergy* 1:37.

Krikorian V, Pochan DJ. 2003. Poly (L-lactic acid)/layered silicate nanocomposite: fabrication, characterization, and properties. *Chem Mater* 15:4317.

LeBaron PC, Wang Z, Pinnavaia TJ. 1999. Polymer-layered silicate nanocomposites: an overview. *Appl Clay Sci* 15:11–35.

Lee JH, Park TG, Park HS, et al. 2003. Thermal and mechanical characteristics of poly(L-lactic acid) nanocomposite scaffold. *Biomaterials* 24:2773.

Leaversuch R. 2002. Renewable PLA polymer gets 'green light' for packaging uses. *Plastic Technology*. On-line article: www.plastictechnology.com/articles/200209fa3.html.

Lipinsky ES, Sinclair RG. 1986. Is lactic acid is a commodity chemical? *Chem Eng Prog* 82:26–32.

Lunt J. 1998. Large-scale production, properties and commercial applications of polylactic acid polymers. *Polym Degrad Stab* 59:149.

Lunt J, Shafer AL. 2000. Polylactic acid polymers from corn, Applications in the textiles industry. *J Ind Text* 29:191.

Maiti P, Yamada K, Okamoto M, Ueda K, Okamoto K. 2002. New polylactide/layered silicate nanocomposites: role of organoclays. *Chem Mater* 14:4654.

Mark JE. 2006. Some novel polymeric nanocomposites. *Acc Chem Res* 39, 881.

Martin O, Averous L. 2001. Poly(lactic acid): plasticization and properties of biodegradable multiphase systems. *Polymer* 42:6209.

Miyagawa H, Misra M, Mohanty AK. 2005. Mechanical properties of carbon nanotubes and their polymer nanocomposites. *J Nanosci Nanotechnol* 5:1593–1615.

Mohanty AK, Drzal LT, Misra M. 2003. Nano reinforcements of bio-based polymers—the hope and the reality. *Polym Mater Sci Eng* 88:60.

Moon S, Jin F, Lee CJ, Tsutsumi S, Hyon SH. 2005. Novel carbon nanotube/poly(L-lactic acid) nanocomposites; their modulus, thermal stability, and electrical conductivity. *Macromol Symp* 224:287.

Nam PH, Maiti P, Okamoto M, et al. 2002. Foam processing and cellular structure of polypropylene/clay nanocomposites. *Polym Eng Sci* 42:1907.

Nam PH, Fujimori A, Masuko T. 2004a. Flocculation characteristics of organo-modified clay particles in poly(L-lactide)/montmorillonite hybrid systems. *e-Polymers* 005. http://www.e-polymers.org/.

Nam PH, Fujimori A, Masuko T. 2004b. The dispersion behavior of clay particles in poly(L-lactide)/organo-modified montmorillonite hybrid systems. *J Appl Polym Sci* 93:2711.

Nam PH, Kaneko M, Ninomiya N, Fujimori A, Masuko T. 2004c. Melt intercalation of poly(L-lactide) chains into clay galleries. *Polymer* 46:7403.

Nazhat SN, Kellomaki M, Tormala P, Tanner KE, Bonfield W. 2001. Dynamic mechanical characterization of biodegradable composites of hydroxyapatite and polylactides. *J Biomed Mater Res* 58:335.

Ninomiya N, Nam PH, Fujimori A, Masuko T. 2004. Distribution of clay particles in the spherulitic texture of poly(L-lactide)/organo-modified montmorillonite hybrids. *e-Polymers* 41. http://www.e-polymers.org/.

Nishida H, Fan Y, Mori T, Oyagi N, Shirai Y, Endo T. 2005. Feedstock recycling of flame-resisting poly(lactic acid)/aluminum hydroxide composite to L,L-lactide. *Ind Eng Chem Res* 44:1433.

Ogata N, Jimenez G, Kawai H, Ogihara T. 1997. Structure and thermal/mechanical properties of poly(L-lactide)-clay blend. *J Polym Sci Part B: Polym Phys* 35:389.

Okamoto M, Nam PH, Maiti M, et al. 2001a. Biaxial flow-induced alignment of silicate layers in polypropylene/clay nanocomposite foam. *Nano Lett* 1:503.

Okamoto M, Nam PH, Maiti P, Kotaka T, Hasegawa N, Usuki A. 2001b. A house of cards structure in polypropylene/clay nanocomposites under elongational flow. *Nano Lett* 1:295.

Paul MA, Alexandre M, Degee P, Henrist C, Rulmont A, Dubois P. 2003. New nanocomposite materials based on plasticized poly(L-lactide) and organo-modified montmorillonites: thermal and morphological study. *Polymer* 44:443.

Paul MA, Delcourt C, Alexandre M, Degee Ph, Monteverde F, Dubois Ph. 2005. Polylactide/montmorillonite nanocomposites: study of the hydrolytic degradation. *Polym Degrad Stab* 87:535.

Pelouze J. 1845. Synthesis of low-molecular weight polylactic acid. *Ann Chimie* 13:257–262.

Shi X, Hudson JL, Spicer PP, Tour JM, Krishnamoorti R, Mikos AG. 2006. Injectable nanocomposites of single-walled carbon nanotubes and biodegradable polymers for bone tissue engineering. *Biomacromolecules* 7:2237.

Shi X, Hudson JL, Spicer PP, Tour JM, Krishnamoorti R, Mikos AG. 2005. Rheological behaviour and mechanical characterization of injectable poly(propylene fumarate)/single-walled carbon nanotube composites for bone tissue engineering. *Nanotechnology* 16:S531.

Shia D, Hui CY, Burnside SD, Giannelis EP. 1998. An interface model for the prediction of Young's modulus of layered silicate-elastomer nanocomposites. *Polym Compos* 19:608.

Sinha Ray S. 2006. Rheology of polymer/layered silicate nanocomposites. *J Ind Eng Chem* 12:811.

Sinha Ray S. 2007. In: Nalwa HS, editor. Clay containing polymer nanocomposites. *Polymeric Nanostructures and Their Applications*, Volume 1, Chapter 1. Los Angeles: American Scientific Publishers.

Sinha Ray S, Bousmina M. 2005. Biodegradable polymer and their layered silicate nanocomposite: In greening the 21st century materials science. *Prog Mater Sci* 50:962–1079.

Sinha Ray S, Okamoto M. 2003a. Polymer/layered silicate nanocomposites: a review from preparation to processing. *Prog Polym Sci* 28:1539.

Sinha Ray S, Okamoto M. 2003b. Polylactide/layered silicate nanocomposites.6. Melt rheology and foam processing. *Macromol Mater Eng* 288:936.

Sinha Ray S, Yamada K, Ogami A, Okamoto M, Ueda K. 2002a. New polylactide/layered silicate nanocomposite: Nanoscale control over multiple properties. *Macromol Rapid Commun* 23:493.

Sinha Ray S, Maiti P, Okamoto M, Yamada K, Ueda K. 2002b. Polylactide/layered silicate nanocomposites. 1. Preparation, characterization and properties. *Macromolecules* 35:3104.

Sinha Ray S, Yamada K, Okamoto M, Ueda K. 2002c. Polylactide/layered silicate nanocomposite: A novel biodegradable material. *Nano Lett* 2:1093.

Sinha Ray S, Okamoto K, Okamoto M. 2003a. Structure–property relationship in biodegradable poly(butylene succinate)/layered silicate nanocomposites. *Macromolecules* 36:2355.

Sinha Ray S, Yamada K, Okamoto M, Ogami A, Ueda K. 2003b. Polylactide/layered silicate nanocomposites. Part 3. High performance biodegradable materials. *Chem Mater* 15:1456.

Sinha Ray S, Yamada K, Okamoto M, Fujimoto Y, Ogami A, Ueda K. 2003c. Polylactide/layered silicate nanocomposites. 5. Designing of materials with desired properties. *Polymer* 44:6633–6646.

Sinha Ray S, Yamada K, Okamoto M, Ueda K. 2003d. Polylactide/layered silicate nanocomposite. 2. Concurrent improvements of materials properties, biodegradability and melt rheology. *Polymer* 44:857.

Sinha Ray S, Yamada K, Okamoto M, Ueda K. 2003e. Biodegradable polylactide/montmorillonite nanocomposites. *J Nanosci Nanotechnol* 3:503.

Sinha Ray S, Yamada K, Okamoto M, Ogami A, Ueda K. 2003f. New polylactide/layered silicate nanocomposites, 4. Structure, properties and biodegradability. *Compos Interfaces* 10:435.

Sinha Ray S, Vaudreuil S, Maazouz A, Bousmina M. 2006. Dispersion of multi-walled carbon nanotubes in biodegradable poly(butylene succinate) matrix. *J Nanosci Nanotechnol* 6:2191–2195.

Song W, Zheng Z, Tang W, Wang W. 2007. A facile approach to covalently functionalized carbon nanotubes with biocompatible polymer. *Polymer* 48:3658–3663.

Thostenson ET, Ren ZF, Chou TW. 2001. Advances in the science and technology of carbon nanotubes and their composites: a review. *Compos Sci Technol* 61:1899–1912.

Vaudreuil S, Labzour A, Sinha Ray S, Maazouz A, Bousmina M. 2007. Dispersion characteristics and properties of poly(methyl methacrylate)/multi-walled carbon nanotubes Nanocomposites. *J Nanosci Nanotechnol* 7:2349.

Vink ETH, Rabago KR, Glassner DA, Gruber PR. 2003. Applications of life cycle assessment to NatureWorks™ polylactide (PLA) production. *Polym Degrad Stab* 80:403.

Watson PD. 1948. Lactic acid polymers as constituents of synthetic resins and coatings. *Ind Eng Chem* 40:1393.

Yang Y, Gupta MC, Dudley KL, Lawrence RW. 2005. A comparative study of EMI shielding properties of carbon nanofiber and multi-walled carbon nanotube filled polymer composites. *J Nanosci Nanotechnol* 5:927.

Zanetti M, Lomakin S, Camino G. 2000. Polymer layered silicate nanocomposites. *Macromol Mater Eng* 279:1.

Zhang W, Suhr J, Koratkar N. 2006. Carbon nanotube/polycarbonate composites as multifunctional strain sensors. *J Nanosci Nanotechnol* 6:960.

Zhang XF, Liu T, Sreekumar TV, et al. 2003. Poly(vinyl alcohol)/SWNT composite film. *Nano Lett* 3:1285.

CHAPTER 17

Advances in Natural Rubber/ Montmorillonite Nanocomposites

DEMIN JIA, LAN LIU, XIAOPING WANG, BAOCHUN GUO, and YUANFANG LUO

College of Materials Science and Engineering, South China University of Technology Guangzhou, China

Biodegradable Polymer Blends and Composites from Renewable Resources. Edited by Long Yu
Copyright © 2009 John Wiley & Sons, Inc.

17.1 INTRODUCTION

Natural rubber (NR) is a natural polymeric material collected from *Hevea brasiliensis* or Guayule. Since the first vulcanized natural rubber factory was established in 1839, NR has been used extensively in tires, hoses, belts, shoes, sealing products, vibration insulators, sports and medical products, and adhesives, as well as in military and high-technology fields.

The production of synthetic rubbers in 1930s ended the single contribution of NR in the rubber industry. With the rapid development of the petrochemical industry from the 1950s, the yield of synthesized rubbers exceeded that of natural rubber. Currently, the relative yields of natural rubber and synthesized rubber are approximately 40% and 60%, respectively. However, synthetic rubbers come from nonrenewable petroleum resources, and the production process is energy-intensive and polluting. With the increasing depletion of petroleum resources, the importance and developable potential of NR as a green, renewable, and degradable elastomer material for environment concerns and sustainable development will increase (Anil and Howard, 2001).

Natural rubber has the molecule structure of *cis*-1,4-polyisoprene. Generally, its molecular weight is in range $10^4 - 10^7$, and the index of molecular weight distribution is from 2.5 to 10 (Anil and Howard, 2001). With the high flexibility of NR molecular chains, the material exhibits excellent elasticity, high fatigue resistance, and low hysteresis loss. At the same time, the high stereoregularity of the NR molecular structure affords tensile crystallization and orientation, which leads to a self-reinforcing action and hence lead to high tensile strength, tear strength, abrasive resistance, and so on. In addition, NR has excellent processability.

Ordinarily, the properties of NR pure gum vulcanizate cannot satisfy the requirements of rubber products. Reinforcing agents have to be added to increase the hardness, modulus, tear strength, tensile strength, abrasive resistance, fatigue resistance, etc. The traditional reinforcing agent for NR and other rubbers is carbon black. The dosage of carbon black can reach 50–60% of the amount of rubber. As well as carbon black, silica and certain resins can be used as reinforcing agents. Some inert fillers, such as clay, calcium carbonate, and talcum, are also usually added to rubber compounds. These do not play a role in the reinforcement of the rubber and are applied only to increase the volume and decrease cost (Dick, 2001).

Since the latter part of the twentieth century, along with the development of nanotechnology around the world, research on rubber-based nanocomposites reinforced with nanometer particles such as montmorillonite, kaolin, nano-calcium carbonate, nanosilica, nano-magnesium hydroxide, attapulgite clay, and halloysite, has attracted great interest in rubber science and technology. The common features of these nanocomposites are that they do not depend on petroleum, and that most of the raw materials are of natural occurrence or are prepared from natural resources that are of low cost. The price of NR has soared because of the effects of natural disasters and the like; and the price of carbon black, which is mostly made from petroleum, has also increased significantly. Therefore, research on NR-based nanocomposites reinforced with cheap natural nano-fillers is important for the manufacture of rubber products that have low cost and high performance as well as being environmentally friendly and offering sustainable development.

Polymer layered silicate (PLS) nanocomposites are hybrids consisting of an organic phase (the polymer) and an inorganic phase (the silicate). The choice of the silicate determines the nanoscopic dispersion of the nanocomposites. The silicates employed belong to the family of layered silicates known as phyllosilicates, such as mica, talc, montmorillonite, vermiculite, hectorite, and saponite. They belong to the general family of so-called 2 : 1 layered silicates. Their crystal structure consists of layers made up of two silica tetrahedra fused to an edge-shared octahedral sheet of either alumina or magnesia. Stacking of the layers leads to a regular van der Waals gap between the layers called the interlayer or gallery. Isomorphic substitution within the layers generates negative charges that are normally counterbalanced by cations residing in the interlayer (Utracki et al., 2007).

Pristine layered silicates usually contain hydrated sodium or potassium ions. Ion-exchange reactions with cationic surfactants render the normally hydrophilic silicate surface organophilic, which makes possible the intercalation of many engineering polymers. The role of alkylammonium cations in the organosilicates is to lower the surface energy of the inorganic component and improve the wetting characteristics with the polymer. Additionally, the alkylammonium cations can provide functional groups that can react with the polymer or initiate polymerization of monomers to improve the strength of the interface between the inorganic component and the polymer (Utracki et al., 2007; Okada and Usuki, 2006; Alexandre and Dubois, 2000; LeBaron et al., 1999; Giannelis, 1998).

In general, two types of hybrid structures are possible: intercalated, in which a single, extended polymer chain is intercalated between the silicate layers, resulting in a well-ordered multilayer with alternating polymer/inorganic layers; and disordered or delaminated, in which the silicate layers (1 nm thick) are exfoliated and dispersed in a continuous polymer matrix. It was not until 1988 that the first industrial application was provided by Okada and co-workers at Toyota's central research laboratories in Japan. In this case a Nylon 6 nanocomposite was formed by polymerization in the presence of the inserted monomer. It is currently used to make the timing belt cover of Toyota's car engines and for the production of packaging film (Kojima et al., 1993b; Okada et al., 1988; Pinnavaia et al., 2001). Since then, much work has been done on polymer/silicate nanocomposites worldwide.

Since the 1980s, with the development of the technology of intercalating nanocomposites polymers and inorganics, clay fillers with nanolayered structure, for example, montmorillonite, allowed dispersion on the nanometer scale in rubber matrix and hence led to the formation of rubber/clay nanocomposites, in which clay has remarkable reinforcing effects on the rubber matrix. The barrier properties, thermal stability, aging resistance, and flammability resistance of rubber were also improved. Okada and co-workers (Okada et al., 1995; Kojima et al., 1993a) reported that only 10 phr of organoclay was required in Nitrile Butadiene Rubber (NBR) to achieve tensile strength comparable to that of the compound containing 40 phr of carbon black. The permeability of hydrogen and water decreased to 70% with addition of 3.9% w/w montmorillonite. These rubber-based nanocomposites can be used as tire inner-liners.

The preparation of the rubber/clay nanocomposites typically involved the intercalation of rubber in the organoclay during mixing, in the latex or in solution state (Okada et al., 1991; Burnside and Giannelis, 1995; Laus et al., 1997; Wang and Pinnavaia, 1998; Wang et al., 1998; Zhang et al., 2000a,b; Arroyo et al., 2003; Jeon et al., 2003; Gatos et al., 2004; López-Manchado et al., 2004; Zheng et al., 2004). In these conditions, the surfaces of clay layers are sufficiently compatible with the rubber that the rubber molecular chains can infiltrate into the interlayer space of the clay and form either intercalated or exfoliated nanocomposites. However, the layers of clay dispersing in rubber/clay nanocomposites always form aggregates with an average thickness in the range of about 10–200 nm because the poor interfacial adhesion between the rubber and the layers of clay causes ten or more layers to congregate together. It is therefore important to enhance the interfacial adhesion between the inorganic layers and the rubber matrix.

In this chapter, the results of the authors' research on nanocomposites of NR and a typical clay filler, montmorillonite, are reviewed.

17.2 MATERIALS AND PROCESS

17.2.1 Materials

Natural rubber latex with 60 wt% of solid content was produced in Hainan province, China; Natural Rubber with trade name ISNR-3 was supplied from Thailand. Na-montmorillonite (MMT), with a cationic exchange capacity of 90 meq/100 g, was supplied by Nanhai Inorganic Materials Factory, China. Modified organomontmorillonites USMMT, HMMT, GMMT, AMMT were prepared by a solid-phase method in our laboratory (Jia et al., 2002; Wang and Jia, 2004) (See later for the specification of these compounds); resorcinol and hexamethylenetetramine complex (RH) was supplied by Zhujiang Tire Corp., China. Butyl acrylate (BA) monomer and other reagents were of pure chemicals grade.

17.2.2 Processing and Procedures

Preparation of NR/BA/USMMT Nanocomposites by Grafting and Intercalating Method in Latex   NR latex, USMMT aqueous suspension, BA monomer, the initiator potassium persulfate, and the surfactant sodium dodecyl sulfate were mixed and stirred for 2 h at room temperature, and then the mixture was reacted for 6 h at 80°C. Electrolyte was added to co-coagulate the mixture. The coagulum was washed several times with water and dried in an oven at 60°C, and finally the NR/BA/USMMT compound was obtained.

The coagulated NR/BA/USMMT compound and other ingredients were mixed on a 160 mm-diameter open roll mill by a common procedure and then the compound was cured in a compression mold at 143°C. The optimum cure time was determined by RPA 2000 Rubber Process Analyzer (Alpha Technologies, USA).

Preparation of NR/HMMT/RH Nanocomposite by Grafting and Intercalating Method in Mixing and Curing Process NR, organomodified montmorillonite HMMT, RH complex, and other rubber ingredients were mixed on a 160 mm-diameter two-roll mill by a common procedure. The compound was vulcanized in an electrically heated press at 143°C. The optimum cure time t_{90} was determined by RPA 2000 Rubber Process Analyzer.

Preparation of NR/GMMT Nanocomposite by Reacting and Intercalating Method in Mixing and Curing Process NR and organomodified montmorillonite GMMT with multiple sulfur bonds as well as other ingredients were mixed on a 160 mm-diameter two-roll mill by a common procedure and the compound was vulcanized in an electrically heated press at 143°C for the optimum cure time t_{90}.

17.3 CHARACTERIZATION

17.3.1 X-Ray Diffraction

X-ray diffraction analysis was performed with the D/MAX-III power diffractometer using CuKα radiation ($\lambda = 1.54$ Å) and a curved crystal graphite monochromator.

17.3.2 Dynamic Mechanical Analysis

A TA Instruments Universal V1.7F DMA2980 instrument was used for dynamic mechanical analysis (DMA) in the tension mode on a sample of approximately 6 mm width and 1.5 mm height. Temperature scans from −120°C to 200°C were carried out at a heating rate of 3°C/min and frequency of 10 Hz.

17.3.3 Transmission Electron Microscopy

Transmission electron microscopy observation was performed with a JEOL JEM-100SX microscope using an acceleration voltage of 80 kV. The samples were prepared using an ultramicrotome in a liquid-nitrogen trap.

17.3.4 Mechanical Properties and Aging Resistance Testing

Mechanical properties testing and aging resistance testing were performed according to the relevant Chinese Standards (GB). Tensile tests were measured using a Shimadzu AG-I tensile machine with an extension rate of 500 mm/min at room temperature.

17.3.5 Rubber Process Analyzer

A Rubber Process Analyzer RPA 2000 (Alpha Technologies) was used to determine the curing characteristics of unvulcanized rubber and measure the mechanical loss factor of vulcanized rubber at frequencies from 0.5 to 25 Hz with 0.5 degree strain at 50–100°C.

17.4 RESULTS AND DISCUSSION

17.4.1 Solid-Phase Method for Modification of Montmorillonite

The organic modification of montmorillonite is usually achieved in aqueous suspension system, and requires complex processes such as decentralization, swelling, heat, reaction, filtering, washing, drying, grinding, and so on. The complexity of this method leads to high prices and pollutes the environment and also hinders the application of polymer/montmorillonite nanocomposites. Accordingly, the authors have successfully developed a new technique, intercalation of montmorillonite in the solid state, in which the montmorillonite in the solid state is mixed with organic modifiers, stirred, and heated. This greatly simplifies the process of montmorillonite intercalation, significantly reduces the cost, and is nonpolluting, which is conducive to industrialization and wider utilization.

Six kinds of solid-phase intercalated montmorillonites have been developed: (1) USMMT, which was intercalated with unsaturated chemicals with C=C double bonds; (2) UAMMT, which was intercalated with long-chain nonpolar organics; (3) EMMT, which was intercalated with chemicals with epoxy groups; (4) AMMT, which was intercalated with chemicals with amido groups; (5) TMMT, which was intercalated with chemicals with phosphate groups; (6) GMMT, which was intercalated with chemicals with multiple sulfur bonds (Liu et al., 2002a,b, 2004, 2006a,b; Wang et al., 2004).

17.4.2 Natural Rubber/Montmorillonite Nanocomposites Prepared by Grafting and Intercalating Method in Latex

Rubber/clay nanocomposites have attracted considerable research attention due to their outstanding mechanical properties, thermal stability, and barrier properties. The

TABLE 17-1 d_{001} **Values of Solid-Phase-Modified Montmorillonite Determined by XRD (Liu et al., 2004, 2006a,b; Wang et al., 2004)**

	2θ (°)	d_{001} (nm)
MMT	6.90	1.25
USMMT	4.90	1.76
UAMMT	2.48	3.80
AMMT	6.90	1.25
	3.31	2.62
EMMT	6.90	1.25
	2.55	3.37
TMMT	5.76	1.53
GMMT	1.52	5.81

preparation of the rubber/clay nanocomposites typically involved the intercalation of rubber in the organoclay in mixing, in the latex or in solution state, which requires that the surfaces of clay layers are sufficiently compatible with the rubber that the rubber molecular chains can infiltrate into the interlayer space of the clay and form either intercalated or exfoliated nanocomposites. Several studies have been undertaken to prepare rubber/clay nanocomposites in latex by co-coagulating rubber latex and clay aqueous suspension (Wu et al., 2001; Jia et al., 2002; Liu et al., 2002).

In this work, NR/montmorillonite nanocomposites were prepared in latex by a grafting and intercalating method (Liu et al., 2006), in which an organomontmorillonite with unsaturated carbon–carbon double bonds on the interlayer surfaces (USMMT) and a monomer, butyl acrylate (BA), were added to NR latex and reacted for some time under appropriate conditions; then the latex mixture was co-coagulated and dried. The grafting and intercalating method in latex is shown schematically in Fig. 17-1. It is expected that the monomers can intercalate into the interlayer galleries of USMMT and polymerize in situ together with C$=$C bonds on the interlayer surfaces and, at the same time, the monomers can undergo grafting copolymerization onto NR molecular chains and form a conjugated three-component interpenetrating network structure (Jia et al., 1991, 1994). Consequently the interfacial combination between the layered silicate and rubber should be strengthened and the mechanical properties of the nanocomposites should be improved. This is a novel way of preparing rubber/organoclay nanocomposites.

XRD was used to measure the change of the interlayer spacing of the silicate before and after introducing the monomer BA, and the results are shown in Fig. 17-2. The results show that the XRD patterns of USMMT contain a peak at $2\theta = 4.9°$, while a strong peak at $2\theta = 2.18°$ is observed in the NR/BA/USMMT (100/10/5) nanocomposite, which is a basal reflection from the silicate layers. The d_{001} values of USMMT and NR/BA/USMMT (100/10/5) are 1.8 nm and 4.0 nm, respectively, indicating that some polymer chains intercalate into the galleries of USMMT; the monomer BA was added to participate in the polymerization in situ in the rubber matrix. The d_{001} value of NR/BA/USMMT (100/10/10) reaches 4.2 nm, which means that more macromolecular chains could intercalate into the

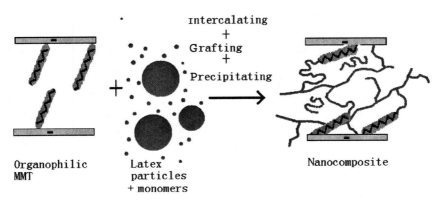

Fig. 17-1 Schematic of the method of grafting and intercalating in latex.

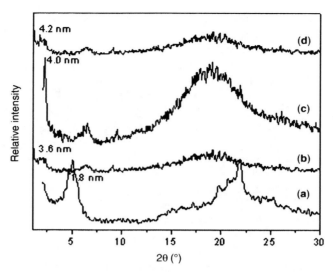

Fig. 17-2 XRD curves of NR/MMT nanocomposites by grafting and intercalating method in latex. (a) USMMT ($d_{001}=1.8$ nm); (b) NR/USMMT(100/10) ($d_{001}=3.6$nm); (c) NR/BA/USMMT (100/10/5) ($d_{001}=4.0$nm); (d) NR/BA/USMMT(100/10/10) ($d_{001}=4.2$nm).

galleries. The XRD results show that monomer BA can intercalate into the interlayer galleries of modified clay and polymerize in situ, and at the same time graft onto NR through emulsion polymerization.

The mechanical properties of the nanocomposites are summarized in Table 17-2. Compared with the NR, NR/BA and NR/USMMT nanocomposites, the mechanical properties of NR/BA/USMMT nanocomposites are clearly improved in 300% and 500% modulus, tensile strength, and tear strength while good elasticity is still retained. As discussed above, the chains of NR can intercalate into modified clay galleries in the latex state and monomer BA can connect the rubber chains and layers of

TABLE 17-2 Mechanical Properties of the Vulcanizates of NR, NR/USMMT, and NR/BA/USMMT Composites

	NR	NR/BA	NR/USMMT	NR/BA/USMMT	NR/BA/USMMT
NR	100	100	100	100	100
USMMT	–	–	5	5	5
BA	–	10		5	10
300% modulus (MPa)	2.78	2.72	2.90	3.89	3.94
500% modulus (MPa)	4.85	5.63	5.04	7.54	7.62
Tensile strength (MPa)	17.56	22.38	21.43	24.38	26.41
Elongation at break (%)	800	800	750	900	800
Tear strength (kN/m)	24.62	27.14	25.71	30.45	29.61
Hardness (Shore A)	37	40	44	40	44

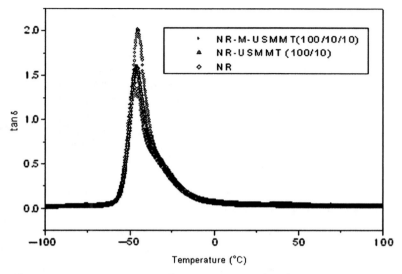

Fig. 17-3 DMA spectra of NR, NR/USMMT, and NR/BA/USMMT nanocomposites.

clay through polymerization in situ. As a result, the interfacial combination between rubber and clay layers is enhanced and the mechanical properties of NR are improved significantly.

Dynamic losses are usually associated with hysteresis and some mechanism of molecular or structural motion in rubber materials. The damping characteristics are extensively measured as the tangent of the phase angle (tan δ). Figure 17-3 shows the tan δ curves of NR, NR/USMMT, and NR/BA/USMMT nanocomposites and the T_g and tan δ values at 0°C and 60°C of NR/BA/USMMT system are shown in Table 17-3. NR exhibits T_g at −46.9°C. The T_g of NR/BA/USMMT nanocomposite increases to −45.3°C due to the interaction between the layers and the chains of rubber through the grafting and intercalation of BA. It is known that for tire applications, high tan δ of the vulcanizate at 0°C is favorable for wet grip, and a low tan δ at 60°C is favorable for reducing rolling resistance. From Table 17-3, the tan δ value of NR/BA/USMMT nanocomposites at 60°C is lower than that of NR, which means the NR/BA/USMMT nanocomposite has a lower rolling loss.

TABLE 17-3   The T_g Values of the NR/USM/BA System from DMA

	T_g (°C)	tan δ (0°C)	tan δ (60°C)
NR	−46.9	0.073	0.036
NR/USM (100/10)	−46.1	0.078	0.034
NBR/USM/BA (100/10/10)	−45.3	0.048	0.013

17.4.3 Natural Rubber/Montmorillonite Nanocomposites Prepared by Grafting and Intercalating Method in Mixing and Curing Process

The solution intercalation method needs a compatible rubber-solvent system and modified organoclay. Its disadvantages are the environment pollution due to the solvents and the difficulty of removing the solvents. In latex intercalation it is necessary to co-coagulate the rubber latex and clay suspension and then dry the mixture. The procedure is complicated. Compared with the latex and solution intercalation method, the method of mechanical mixing intercalation is a convenient and easy way to realize industrialization. The approach is suitable to produce rubber-based nanocomposites and it can be applied to most rubbers.

In general, however, only some segments of the rubber macromolecules can intercalate into the interlayer galleries of the clay in the mixing intercalation method since rubber is a high-molecular-weight polymer with very high viscosity in the processing state and has very poor interfacial adhesion with the layers of clay.

In our previous work (Liu et al., 2002a,b; Jia et al., 2002), a new method for preparing rubber/clay nanocomposites by introducing some monomers into the NR latex/organoclay system was investigated. This method can enhance the interfacial adhesion between the inorganic layers and the rubber matrix so that the mechanical properties of NR/organoclay nanocomposites are greatly improved. However, this latex intercalation needs to polymerize the monomer and then co-coagulate the rubber latex and organoclay. The procedure is still complicated.

In this section, the authors develop a new preparation method for rubber/clay nanocomposites, which we term the grafting and intercalating method in mixing and curing process, by introducing some reactive monomers during the rubber mixing and vulcanization process. The organomontmorillonite (HMMT) was prepared by ion exchanging Na^+-MMT with hexadecyltrimethylammonium bromide. The monomer adopted was resorcinol and hexamethylenetetramine complex (RH; the structural formula is shown in Scheme 17-1). RH is an usual adhesive in the rubber industry; it can be decomposed into resorcinol and formaldehyde at temperatures above $110°C$ and can react with NR in the process of vulcanization (Jain and Nando, 1988). It was expected that RH would intercalate into the interlayer galleries of HMMT and polymerize in situ while RH and NR are reacting; consequently the interfacial combination between rubber and the layered silicate should be strengthened

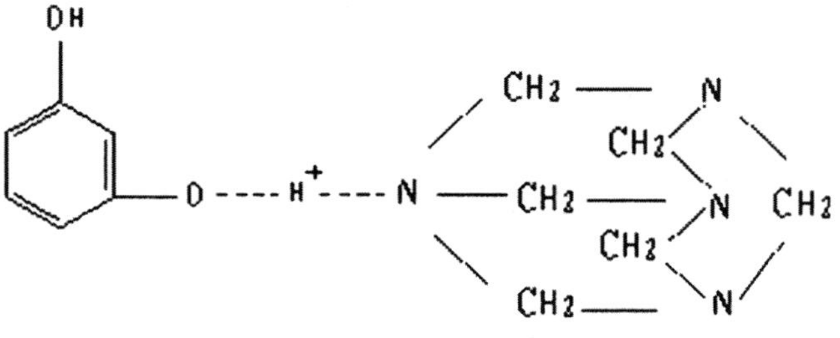

Scheme 17-1 Structural formula of RH.

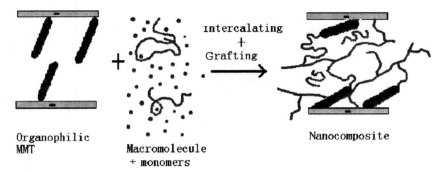

Fig. 17-4 Schematic of the reactive mixing intercalation method.

and the mechanical properties of the nanocomposites should be improved. The reactive mixing intercalation method is shown schematically in Fig. 17-4.

Table 17-4 lists the mechanical properties of the NR vulcanizate and its composites. Compared with neat NR, the mechanical properties of NR/HMMT nanocomposites increase dramatically. With the addition of RH, the mechanical properties of NR/HMMT/RH nanocomposites are further improved for 300% modulus, tensile strength, tear strength, and elongation at break. Furthermore, the results of aging tests demonstrate that the mechanical properties of NR/HMMT/RH nanocomposites after aging in air for 48 h at $100 \pm 1°C$ are better than those of NR/HMMT nanocomposites. The results show that the reaction and interactions of RH with NR and HMMT can improve the interfacial adhesion between NR and silicate layers and facilitate dispersion of the silicate layers in the rubber matrix at nanometer level. The high aspect ratio characteristic of silicate layers in nanocomposites can not only reinforce the rubber but also strongly reduce the hot air permeability, which can protect the macromolecules from thermal and oxidative aging.

The $\tan \delta$ values at $0°C$ and $60°C$ are also summarized in Table 17-4. The $\tan \delta$ value of NR/HMMT/RH nanocomposites is higher at $0°C$ and lower at $60°C$,

TABLE 17-4 Mechanical Properties of NR/NR/HMMT and NR/HMMT/RH Vulcanizates (Liu et al., 2004, 2006a,b)

	NR	NR/HMMT (100/10)	NR/HMMT/RH (100/10/8)
d_{001} of MMT/nm	–	2.6	5.5
Modulus at 300% (MPa)	1.36	1.62	2.11
Tensile strength (MPa)	17.60	19.44	26.34
Elongation at break (%)	800	750	900
Tear strength (kN/m)	28.72	26.74	35.72
Hardness (Shore A)	35	38	44
$\tan \delta$ at $60°C$	0.036	0.037	0.026
$\tan \delta$ at $0°C$	0.075	0.110	0.089
Property descend ratio after aging in air (%) ($100 \pm 1°C$; 48 h)	46.0	40.7	30.9

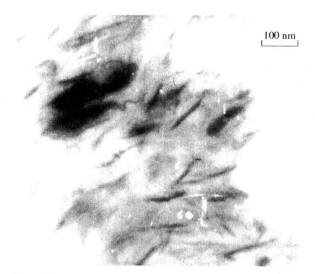

Fig. 17-5 TEM image of NR/RH/HMMT nanocomposite.

indicating that the nanocomposites have better damping properties around room temperature and lower heat build-up at higher temperature.

Figure 17-5 shows a TEM image of NR/HMMT/RH nanocomposite. The dark lines in the micrographs correspond to the intersections of the silicate layer. It is apparent that the layered silicate is further divided into thinner bundles with thicknesses of 5–20 nm and lengths of about 100 nm dispersed uniformly in rubber matrix, and there are still individual larger aggregates with sizes of more than 50 nm. The preparation of NR/HMMT/RH nanocomposites may be considered successful due to the addition of RH. RH improved the dispersion of HMMT in the rubber matrix and the interfacial adhesion between rubber macromolecule chains and silicate layers, and hence increased the intercalation efficiency.

17.4.4 Natural Rubber/Montmorillonite Nanocomposites Prepared by Reacting and Intercalating Method in Mixing and Curing Process

Compared with ordinary mixing intercalation, the intercalating effect of the mechanical mixing and grafting intercalation method is a clear improvement; however, the adding of suitable monomers is inconvenient for processing. Accordingly, a new technique termed the mechanical mixing and reaction intercalation method was developed by us (Wang et al., 2004). In this technique, first, in the course of the organic modification of montmorillonite by the solid-state method, some reactive groups are introduced onto the interlayer surfaces of montmorillonite; and then in the course of mechanical mixing and vulcanization of the rubber compound, NR macromolecules intercalate into the interlayers of montmorillonite by mechanical

TABLE 17-5   The Effect of Organomontmorillonite Amount on the Mechanical Properties of NR/GMMT Composities (Wang, 2004)

NR (pph)	100	100	100	100	100	100
GMMT (pph)	0	2	4	6	8	10
100% modulus (MPa)	0.82	0.77	0.89	1.01	1.17	1.44
300% modulus (MPa)	1.83	2.21	2.39	3.15	3.61	5.53
Tensile strength (MPa)	20.3	25.3	27.1	28.9	26.3	26.0
Elongation at break (%)	610	708	698	682	678	610
Permament set (%)	12	18	22	25	30	32
Tear strength (kN/m)	22.4	25.2	30.1	30.5	31.7	30.9
Shore A hardness (degrees)	35	37	38	40	42	45
Descend ratio of mechanical properties after aging (100°C; 48 h)	44.32	40.99	37.64	25.76	30.65	32.64

shear, compression, and thermochemical effects and react in situ with the reactive groups. The reactive intercalation leads to fine dispersion of montmorillonite in the rubber matrix at nanometer scale and enhances the interfacial combination between rubber and montmorillonite; hence, NR/montmorillonite nanocomposites with high performance are obtained. This new technique has good applicability prospects in industry with advantages of simple technology, low cost, and high efficiency of intercalation.

The mechanical properties and aging resistance of NR/GMMT nanocomposites prepared by the mechanical mixing and reaction intercalation method are listed in Table 17-5. The results show that the modulus, tensile strength, elongation at break, and tear strength of the nanocomposites are evidently improved; at the same time, the aging resistance is distinctly enhanced.

The TEM image of an NR/GMMT nanocomposite in Fig. 17-6 shows the modified montmorillonite dispersed in the rubber matrix at several to tens of nanometers,

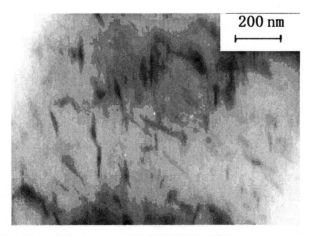

Fig. 17-6   TEM image of NR/GMMT (100/10) nanocomposite.

mostly less than 10 nm. A typical sandwich structure consisting of parallel alternating organoclay layers and rubber matrix indicates the coexistence of intercalated/exfoliated clay structure.

Figure 17-7 summarizes the change in properties of NR/AMMT nanocomposite after aging in air at 100°C. It is clear that the addition of modified montmorillonite significantly improves the mechanical properties in aging. NR/AMMT nanocomposite maintains about 67% of tensile strength and 68% of elongation at break after aging for 7 days. However, for the blank sample without montmorillonite the values are 26% and 27%, respectively.

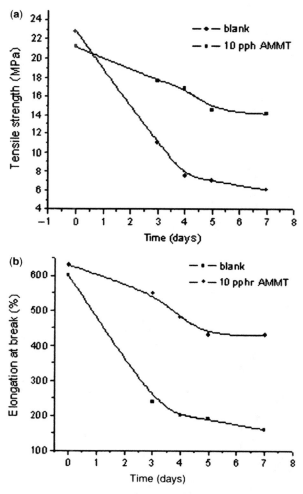

Fig. 17-7   The aging behavior of NR/AMMT (100/10) nanocomposites (100°C, in air). (a) Tensile strength and (b) elongation at break against aging time.

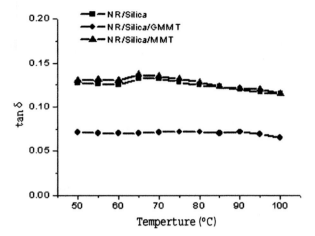

Fig. 17-8 The temperature sweep curves of NR/silica/GMMT nanocomposite and related materials.

The RPA temperature sweep curves of NR/silica/GMMT, NR/silica/MMT, and NR/silica composites shown in Fig. 17-8 show that the damping coefficient of NR/silica/GMMT nanocomposites in range 50–100°C is lower than that of NR/silica and NR/silica/MMT composites. This indicates that the heat-build and rolling resistance of NR/silica/GMMT nanocomposites is the lowest of the three materials. This behavior is quite valuable for rubber products working in dynamic conditions, such as tire and vibration-damping products.

Much work has been done to study the flammability properties of different kinds of polymer/clay nanocomposites using the cone calorimeter, and it is found that clay fillers could become a popular "green" alternative to current flame-retardant additives for polymers. Cone calorimeter data showed that both the peak and average heat release rate (HRR), were reduced significantly for intercalated and delaminated nanocomposites with low silicate mass fraction of 2–5% where no additional flame retardant is used (Gilman et al., 1998). The flammability properties of rubber/layered silicate nanocomposites systems have been much less reported in the literature (Song et al., 2005).

NR/organic-modified layered silicate TMMT nanocomposites prepared by the mixing intercalation method and the structure and flame-retardant properties of the nanocomposites were studied. The samples were analyzed by cone calorimeter to compare the effects of TMMT on polymer flammability, and the results are summarized in Table 17-6. Compared with the NR control, all the samples show reduced HRR and peak HRR values. However, the NR-TMMT nanocomposites show larger reduction in HRR than do NR-MMT microcomposites. The results in Table 17-6 show that when the mass fraction of TMMT is 10 pph, the peak HRR of NR-TMMT nanocomposites is reduced by 44%; the peak HRRs are further reduced as the mass fraction of TMMT increases. When the mass of TMMT is

**TABLE 17-6   Cone Calorimeter Results for NR, NR/MMT Composites
and NR/TMMT Nanocomposites**

	NR	NR-MMT Microcomposites	NR-TMMT Nanocomposites		
MMT	–	10	–	–	–
TMMT	–	–	10	20	30
Tig (s)	29	26	25	32	27
Peak HRR (kW/m^2)	1222	732	679	577	644
Average HRR (kW/m^2)	630	508	472	382	365

increased to 20 wt%, the peak HRR of the NR-TMMT nanocomposites is 53% lower than that of pure NR (Liu et al., 2008).

The flame retardancy mechanism can be deduced from the cone calorimeter data of NR-TMMT nanocomposites, combined with previous literature reports (Zanetti et al., 2002; Song et al., 2005). Once the nanocomposite is ignited, it burns slowly but does not self-extinguish until most of the fuel has been combusted. The nanodispersed layers of clay collapse down to form a clay-rich barrier that slows the pyrolysis of decomposing polymer and reduces the smoke produced, and at the same time, the tributyl phosphate situated in interlayers of TMMT reacts with decomposition products of NR and layers of clay to form a glassy coat and stable carbonaceous-silicious-phosphorated charred layers. This stable physical protective barrier on the surface of the polymer materials may insulate the underlying polymeric substrate from the heat source and retard heat and mass transfer between the gaseous and condensed phases.

17.5 SUMMARY

Natural rubber is an important renewable and degradable natural elastomer material. Montmorillonite is a cheap natural nanometer material. NR/montmorillonite nanocomposites as complete natural nanocomposites are important for the application of nanometer materials and for environmentally friendly and sustainable development.

Organic modification of montmorillonite in the solid phase is an effective new technique. It can greatly simplify the process of organic modification of montmorillonite, significantly reduce the cost, and reduce pollution, all of which favor the promotion of industrialization and utilization.

In the NR/montmorillonite nanocomposites prepared by grafting and intercalating method in latex, the monomer intercalated into the interlayer galleries of montmorillonite and polymerized in situ on the interlayer surfaces and, at the same time, the monomer underwent grafting copolymerization onto NR molecular chains; consequently the interfacial combination between the layered silicate and NR was strengthened and the mechanical properties of the nanocomposites were remarkably improved.

The grafting and intercalating method in mixing and curing process for preparation of rubber/montmorillonite nanocomposites is a convenient and easy way to realize industrialization. The reactive monomer intercalated into the interlayer galleries of montmorillonite and polymerized in situ while the monomer and NR was reacting and grafting; consequently the interfacial combination between rubber and the layered silicate was again strengthened and the mechanical properties of the nanocomposites were improved. The mechanical mixing and grafting intercalation method is applicable to the production of rubber-based nanocomposites.

The reacting and intercalating method in mixing and curing process for preparing NR/MMT nanocomposites is a novel technique. In this process, NR macromolecules intercalate into the interlayers of montmorillonite and react in situ with the reactive groups in the interlayers, leading to fine dispersion of montmorillonite in the rubber matrix at nanometer scale and enhancing the interfacial combination between rubber and montmorillonite; hence NR/montmorillonite nanocomposites with high performance are obtained. This new technique has good application prospects in industry for its advantages of simple technology, low cost, and high efficiency of intercalation.

Thus, due to enhanced the interfacial adhesion between the inorganic layers and the rubber matrix, the rubber/clay nanocomposites show improved mechanical properties, thermal properties, and gas barrier performance, and reduced flammability when compared with conventional filled polymers.

ACKNOWLEDGMENTS

This work was supported by the National Natural Science Foundation of China (Grants No. 59933060 and No. 50573021).

REFERENCES

Alexandre M, Dubois P. 2000. Polymer-layered silicate nanocomposites: preparation, properties and uses of a new class of materials. *Mater Sci Eng* 28:1.

Anil KB, Howard LS. 2001. *Handbook of Elastomers*. New York: Marcel Dekker.

Arroyo M, López-Manchado MA, Herrero B. 2003. Organo-montmorillonite as substitute of carbon black in natural rubber compounds. *Polymer* 44(8):2447.

Burnside SD, Giannelis EP. 1995. Synthesis and properties of new poly(dimethylsiloxane) nanocomposites. *Chem Mater* 7:1597.

Dick JS. 2001. *Rubber Technology: Compounding and Testing for Performance*. Munich: Hanser.

Gatos KG, Thomann R, Karger-Kocsis J. 2004. Characteristics of ethylene propylene diene monomer rubber/organoclay nanocomposites resulting from different processing conditions and formulations. *Polym Int* 53(8):1191.

Giannelis EP. 1996. Polymer layered silicate nanocomposites. *Adv Mater* 8(1):29–35.

Giannelis EP. 1998. *Appl Organometal Chem* 12:675–680.

Gilman JW, Kashiwagi T, Lomakin S, et al. 1998. *Fire Retardancy of Polymers: the Use of Intumescence.* Cambridge: The Royal Society of Chemistry.

Jain R, Nando GB. 1988. A novel system for nylon-6 tire cord-to-natural rubber adhesion. *Rubber World* 199(2):40–43.

Jeon HS, Rameshwaram JK. 2003. Characterization of polyisoprene-clay nanocomposites prepared by solution blending. *Polymer* 44(19):5749.

Jia DM, Pang YX, Dai ZS. 1991. Conjugate three-component interpenetrating polymer networks and their applications. *Polym Mater Sci Eng* 65:167.

Jia DM, Pang YX, Liang X. 1994. Mechanism of adhesion of polyurethane/polymethacrylate simultaneous interpenetrating networks adhesives to polymer substrates. *J Polym Sci: Part B Polym Phys* 32:817.

Jia DM, Luo YF, Liu L, Zheng ZY. 2002. Rubber-layered silicate nanocomposites and their preparation method. China Patent ZL02134581.3.

Kojima Y, Fukumori K, Okada A, Kurachi T. 1993a. Gas permeabilities in rubber-clay hybrid. *J Mater Sci Lett* 12(12):889–890.

Kojima Y, Usuki A, Kawasumi M, et al. 1993b. Mechanical properties of nylon-6-clay hybrid. *J Mater Res* 6:1185.

Laus M, Francesangeli O, Sandrolini F. 1997. New hybrid nanocomposites based on an organophilic clay and poly(styrene-*b*-butadiene) copolymers. *J Mater Res* 12:3134.

LeBaron PC, Wany Z, Pinnavaia TJ, 1999. Polymer-layered silicate nanocomposites: an overview. *Applied Clay Science* 15:11–29.

Lei Song, Yuan Hu, Yong Tang, Rui Zhang, Zuyao Chen, Weicheng Fan. 2005. Study on the properties of flame retardant polyurethane/organoclay nanocomposites. *Polym Degrad Stab* 87:111–116.

Liu L, Luo YF, Jia DM. 2008. Flammability properties of NR-organoclay nanocomposites. *Polymer Composites* (in press).

Liu L, Luo YF, Zhang F, Huang MY, Jia DM. 2002. Properties of NR/HMMT nanocomposites prepared by intercalation in latex. *China Synthetic Rubber Industry* 25(4):262.

Liu L, Luo YF, Jia DM, Fu WW, Guo BC. 2004. Studies on NBR-ZDMA-OMMT nanocomposites prepared by reactive mixing intercalation method. *Int Polym Process* 12:374.

Liu L, Luo YF, Jia DM, Fu WW, Guo BC. 2006a. Preparation, structure and properties of nitrile-butadiene rubber-organoclay nanocomposites by reactive mixing intercalation method. *J Appl Polym Sci* 100(3):1905.

Liu L, Luo YF, Jia DM, Fu WW, Guo BC. 2006b. Structure and properties of natural rubber-organoclay nanocomposites prepared by grafting and intercalating method in latex. *J Elastomers Plast* 38(2):147–161.

López-Manchado MA, Herrero B, Arroyo M. 2004. Organoclay-natural rubber nanocomposites synthesized by mechanical and solution mixing methods. *Polym Int.* 53(11):1766.

Okada A, Usuki A. 2006. Twenty years of polymer-clay nanocomposites. *Macromol Mater Eng* 291(12):1449–1476.

Okada A, Fukumori K, Usuki A, Kojima Y, Kurauchi T, Kamigaito O. 1991. Rubber-clay hybrid: synthesis and properties. *Polym Prep* 32:540.

Okada A, Usuki A, Kurauchi T, et al. 1995. *Hybrid Organic-Inorganic Composites.* ACS Symposium Series. p. 55–65.

Okada A, Fukoshima Y, Inagaki S, et al. 1988. US Patent 4,739,007.

Pinnavaia TJ, Beall GW. 2001. *Polymer-Clay Nanocomposites*. Chichester: Wiley.

Utracki LA, Sepehr M, Boccaleri E. 2007. Review of synthetic layered nanoparticles for polymeric nanocomposites. *Polym Adv Technol* 18:1–37.

Wang SJ, Long CF Long, et al. 1998. Synthesis and properties of silicone rubber/organomontmorillonite hybrid nanocomposites. *J Appl Polym Sci* 69(8):1557–1561.

Wang XP, Jia DM. 2004. China Patent 200410051956.0.

Wang, Pinnavaia TJ. 1998. Nanolayer reinforcement of elastomeric polyurethane, *Chem Mater* 10(12):3769–3771.

Wu YP, Zhang LQ, et al. 2001. Structure of CNBR-clay nanocomposites by co-coagulating rubber latex and clay aqueous suspension. *J Appl Polym Sci* 82(11):2842–2848.

Wang YZ, Zhang LQ, et al. 2000. Preparation and characterization of rubber-clay nanocomposites. *J Appl Polym Sci* 78(11):1879–1883.

Zanetti M, Kashiwagi T, Falqui L, Camino G. 2002. Cone calorimeter combustion and gasification studies of polymer layered silicate nanocomposites. *Chem Mater* 14:881–887.

Zheng H, Zhang Y, Peng Z, Zhang Y. 2004. Influence of clay modification on the structure and mechanical properties of EPDM/montmorillonite nanocomposites. *Polym Test* 23(2):217.

Zhang LQ, Wang YZ, et al. 2000a. Morphology and mechanical properties of Clay/SBR nanocomposites. *J Appl Polym Sci* 78(11):1873–1878.

Zhang L, Wang Y, et al. 2000b. Morphology and mechanical properties of clay/styrene-butadiene rubber nanocomposites. *J Appl Polym Sci* 78(11):1873.

PART VI

MULTILAYER DESIGNED MATERIALS

■■■■ CHAPTER 18

Multilayer Coextrusion of Starch/Biopolyester

L. AVÉROUS

ECPM-LIPHT (UMR CNRS 7165), University Louis Pasteur, Strasbourg France

18.1 INTRODUCTION

Advanced technology in petrochemical polymers has brought many benefits to mankind. However, it is becoming increasingly obvious that the ecosystem is considerably disturbed and damaged as a result of the use of nondegradable materials for disposable items. The environmental impact of persistent plastic wastes is becoming a global concern, and alternative disposal methods are not unlimited. Besides this, petroleum resources are finite and approaching depletion. It is becoming important to find plastic substitutes based on sustainability, especially for short-term packaging and disposable applications. Starch may offer a substitute for petroleum-based plastics. Starch is a renewable and degradable carbohydrate and can be obtained from various botanical sources (wheat, maize, potato, and others). Starch, by itself, has severe limitations due to its water sensitivity (Rouilly and Rigal, 2002). Articles made from starch swell and deform upon exposure to moisture. In recent decades, several authors (Tomka, 1991; Swanson, 1993; Avérous, 2004) have shown the possibility of transforming native starch into thermoplastic resinlike products under destructuring and plasticizing conditions. Unfortunately, plasticized starch, also called "thermoplastic starch," is a very hydrophilic material with limited performance. One way to overcome these difficulties and to maintain its biodegradability consists of associating plasticized starch with another biodegradable polymer (Avérous, 2004). Biodegradable polymers show a large range of properties and they can now compete with nonbiodegradable thermoplastics in various fields (packaging, textiles, biomedical, among others) (Van de Velde and Kiekens, 2002; Steinbuchel, 2003). Avérous has proposed a tentative classification of these biodegradable polymers in two main groups and four different families (Avérous, 2004). The main groups are (i) agropolymers (polysaccharides, proteins, etc.) and (ii) biopolyesters (biodegradable polyesters) such as poly(ester amide) (PEA), poly(lactic acid) (PLA), poly(ε-caprolactone) (PCL), poly(hydroxyalkanoate) (PHA), and aromatic and aliphatic copolyesters (Avérous, 2004).

Melt blending is one established method for associating different polymers. Plasticized starch (PLS) has been blended with various biodegradable polyesters, such as PCL (Huang et al., 1993; Bastioli et al., 1995; Koenig and Huang, 1995; Narayan and Krishnan, 1995; Pranamuda et al., 1996; Amass et al., 1998; Myllymäki et al., 1998; Vikman et al., 1999; Avérous et al., 2000b; Matzinos et al., 2002; Schwach and Avérous, 2004), PLA (Martin and Avérous, 2001; Schwach and Avérous, 2004; Huneault and Li, 2007), PHA (Huang et al., 1993; Ramsay et al., 1993; Koenig and Huang, 1995; Verhoogt et al., 1995), poly(butylene succinate-*co*-butylene adipate) (PBSA) (Ratto et al., 1999; Avérous and Fringant, 2001; Schwach and Avérous, 2004), poly(butylene adipate-*co*-butylene terephthalate) (PBAT) (Avérous and Fringant, 2001), and PEA (Avérous et al., 2000b; Schwach and Avérous, 2004). Blending PLS with these biopolyesters resulted in a significant improvement of the properties of plasticized starch. However, although a "protective" polyester skin layer was formed at the surface of some blends under certain conditions, for instance during injection molding or extrusion (Belard et al., 2005), the moisture sensitivity of PLS was not fully addressed. In an effort to develop starch-based applications, coating the starchy material with hydrophobic (compared to PLS) and biodegradable polyester layers would be preferred.

Multilayer coextrusion has been widely used in the past decades to combine the properties of two or more polymers into a single multilayered structure (Schrenk and Alfrey, 1978; Han, 1981). Some studies have reported the use of plasticized starch and biopolyesters in coextrusion (Avérous et al., 1999; Wang et al., 2000; Martin et al., 2001; Martin and Avérous, 2002; Schwach and Avérous, 2004).

Other techniques may be used for the preparation of starch-based multilayers, such as compression molding (Van Soest et al., 1996; Hulleman et al., 1998; Hulleman et al., 1999; Martin et al., 2001; Schwach and Avérous, 2004). Coating is also mentioned in the literature; coating has been achieved by spraying (Shogren and Lawton, 1998) or painting (Lawton, 1997) different dilute liquids such as biopolyester solutions onto the starch-based material. Nevertheless, coextrusion appears to be the best option since it offers the advantages of being a one-step, continuous, and versatile process. Realistic development of moisture-resistant starch-based products is attempted through multilayer coextrusion, allowing the preparation of sandwich-type structures with PLS as the central layer and the hydrophobic component as the surface outer layers.

However, there are some inherent problems with the multiphasic nature of the flow during coextrusion operations, such as nonuniform layer distribution, encapsulation, and interfacial instabilities, which are critical since they directly affect the quality and functionality of the multilayer products. There has been extensive experimental and theoretical investigation of these phenomena (Dooley, 2005). The layer encapsulation phenomenon corresponds to the surrounding of the more viscous polymer by the less viscous one (Lee and White, 1974). Experimental investigations on the shape of the interface have been reported (Southern and Ballman, 1973; Khan and Han, 1976) showing that viscosity differentials between respective layers dominate over elasticity ratios. Conversely, White et al. (1972) reported the influence of normal stress differences on the shape of the interface. In more recent experimental studies, Dooley and co-workers (Dooley and Stout, 1991; Dooley et al., 1998) investigated the layer rearrangement during coextrusion, and the importance of the channel geometry. They indicated that coextrusion of identical polymers (i.e., with matched viscosities) can lead to layer rearrangement, due to the die geometry.

Yih and Hickox (Yih, 1967; Hickox, 1971) pioneered studies on interfacial instabilities, suggesting that viscosity differences may cause instabilities of stratified flow. Schrenk and co-workers (Schrenk and Alfrey, 1978; Schrenck et al., 1978) investigated the factors responsible for the onset of instabilities, and suggested the existence of a critical shear stress value beyond which interfacial instabilities are likely to occur. Han and Shetty (1976, 1978) described in detail the factors responsible for the occurrence of instabilities, such as critical shear stress at the interface, viscosity and elasticity ratio, and layer thickness ratio. In addition, many authors (Sornberger et al., 1986a,b; Karagiannis et al., 1988; Khomani, 1990a,b; Su and Khomani, 1992a,b; Wilson and Khomani, 1992; Wilson Khomani, 1993; Gifford, 1997; Tzoganakis and Perdikoulias, 2000; Dooley, 2005) have modeled multilayer flows by computer simulation, and attempted to elucidate the influence of viscoelasticity, layer thickness, and die geometry on layer rearrangements and onset of instabilities. Karagiannis et al. (1988) modeled encapsulation phenomenon, and Sornberger and co-workers (Sornberger et al., 1986b) studied the interface position in two-layer flat film coextrusion. In a later work, Gifford (1997) attempted to account for the effects of viscoelasticity on the layer deformation. Most numerical investigations only partly addressed the layer uniformity problem, due to the complexity of the stratified flow systems. Khomani and co-workers (Khomani, 1990a; Su and Khomani,

1992a,b; Wilson and Khomani, 1992; Wilson and Khomani, 1993) contributed significantly to the understanding of the onset and propagation of interfacial instabilities and of interface deformations, thanks to a specially designed optical interface monitoring system allowing visualization, after image reconstruction, of the multilayer flow in the die. Khomani and Su (Khomani, 1990a; Su and Khomani, 1992a,b) examined the effects of elasticity and viscosity on the interfacial stability, according to the die geometry and layer depth ratio. They determined the role of elasticity in the mechanism of instabilities. Wilson and Komani (1992, 1993) studied experimentally and numerically the propagation of periodic flow disturbances, and determined the stable and unstable flow conditions, with a good agreement with models. In a more recent study, Tzoganakis and Perdikoulias (2000) studied experimentally the effects of material properties and flow geometry on the appearance of interfacial instability.

Despite the number and diversity of studies on multilayer flows and stability, only few articles (Avérous et al., 1999; Martin and Avérous, 2002; Wang et al., 2000; Martin et al., 2001; Schwach and Avérous, 2004) have reported the use of plasticized starch and polyester in coextrusion processes. Different stratified structures were processed by coextrusion and studied with PCL (Wang et al., 2000; Martin et al., 2001; Schwach and Avérous, 2004), PBSA (Wang et al., 2000; Martin et al., 2001; Schwach and Avérous, 2004), PEA (Wang et al., 2000; Martin et al., 2001; Martin and Avérous, 2002; Schwach and Avérous, 2004), PLA (Wang et al., 2000; Martin et al., 2001; Schwach and Avérous, 2004), polybutylene adipate-*co*-butylene terephthalate (PBAT) (Wang et al., 2000; Schwach and Avérous, 2004), or polyhydroxybutyrate-*co*-hydroxyvalerate (PHBV) (Avérous et al., 1999), a biopolyester which belong to the PHA family.

18.2 MATERIALS AND PROCESS

18.2.1 Materials

Native wheat starch was purchased from Chamtor (France). The starch contains 74% amylopectin and 26% amylose, with residual protein and lipid contents less than 0.2% and 0.7%, respectively. Glycerol of 99.5 % purity was used as a nonvolatile plasticizer. Various types of PLS (PLS1, PLS2, and PLS3) were prepared, differing in the glycerol:starch ratio and the water content. Table 18-1 shows the different PLS formulations prepared (with starch, water, and glycerol contents before and after extrusion) and some resulting properties. Both powdery and extruded products were used in the study for the coextrusion process. Most PLS properties and mechanical behavior have been given in previous publications (Avérous, 2004).

The aliphatic poly(ester amide) (PEA) LP BAK 404 was kindly supplied by Bayer AG (Germany). The poly(ε-caprolactone) (PCL) CAPA 680 was purchased from Solvay (Belgium). The poly(lactic acid) (PLA), 92% L-lactide and 8% *meso*-lactide, was obtained from Cargill (USA). The poly(butylene succinate-*co*-butylene adipate) (PBSA) Bionolle 3000 was obtained from Showa Highpolymer Co. (Japan). Finally, PHBV (Biopol D600G, 12% HV) was kindly supplied by

TABLE 18-1   Characteristics of Different Biopolyesters

	PLA Cargill Nature Works	PHBV Monsanto Biopol D600G	PBSA Showa Bionolle 3000	PEA Bayer BAK 1095	PCL Solway CAPA 680
MW (g/mol)	150,000[a]	430,000[a]	234,000[b]	37,000[c]	80,000[a]
Density	125	125	123	107	111
Melting point (°C DSC)	158	144	114	112	65
Glass transition (°C DSC)	58	−1	−45	−29	−61
Modulus (MPa)	2050	500	249	262	190
Elongation at break (%)	9	20–30	>500	420	>500
Water permeability[d] (g/m^2/day) at 25°C	172	21	330 with 7% HV	680	177

[a]Data obtained from the provider.
[b]Source: He et al. (2000).
[c]Source: Krook et al. (2005).
[d]Source: Shogren (1997).

Monsanto (USA). For some coextrusion experiments, LDPE (Lacqtene 1070MG24, Arkema, France) was used, as a model.

18.2.2   Processing and Procedures

Plasticized Starch Preparation   The first step of the process consists of preparing plasticized wheat starch (PLS). Native wheat starch was introduced into a turbomixer. Glycerol and water were then added slowly under stirring according to the formulation (Table 18-2). After complete glycerol addition, the mixture was mixed at high speed to obtain a homogeneous dispersion. The mixture was placed in a vented oven at 160°C for 20 min and stirred to allow vaporization of water and diffusion of glycerol into the starch granules, and finally this dry blend was placed again in the oven for 20 min.

TABLE 18-2   Plasticized Starch Formulations

Formulation	Starch Content (wt%)	Glycerol Content[a] (wt%) GC	Glycerol/ Starch Ratio[a] G/S	Moisture Content[a] (wt%) MC	Density	Glass Transition (°C) T_g
PLS1	74	10 (11)	0.14 (0.14)	16 (8.5)	1.39	43
PLS2	70	18 (18)	0.26 (0.25)	12 (8.7)	1.37	8
PLS3	65	35 (30)	0.54 (0.50)	0 (12.6)	1.34	−20

[a]Compositions are given in wt% total wet basis, and values in parentheses are the glycerol or moisture contents, or glycerol/starch ratio after extrusion at 105°C and equilibration at 23°C, 50% RH.

For coextrusion process, starch can be introduced as a powder (dry blend) or as granules. In the latter case, the dry blend is extruded, granulated after air-cooling, and then equilibrated at a given relative humidity and temperature.

Coextrusion Procedure   Figure 18-1 presents a schematic view and photograph of the experimental set-up for the coextrusion, which consists of two single-screw extruders, a feedblock attached to a wide, flat die, and a three-roll calendering system. The PLS extrusion step was based on a single-screw extruder (SCAMIA S 2032, France), equipped with a conical-shaped element to provide high shearing with a 20 mm screw diameter and an L/D ratio of 11. A 30 mm diameter, 26 : 1 L/D single screw extruder was used for the cap layer. Figure 18-2 presents a

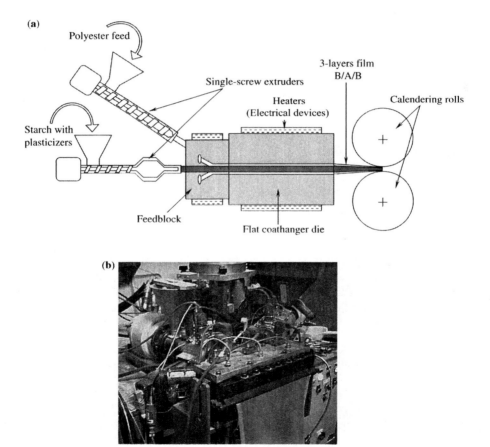

Fig. 18-1   (a) Schematic illustration of the coextrusion system; (b) photograph of the experimental set-up without the three-roll calendaring system on the right side; the mono-extruder is for biopolyester feed.

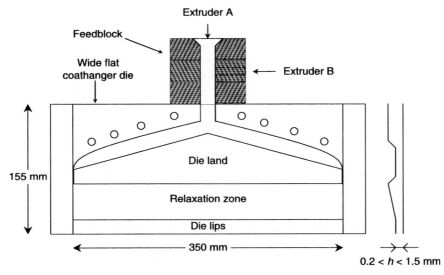

Fig. 18-2 Schematic view of the feedblock and flat coathanger die. The feedblock consists of three different modules: (i) the manifold module, (ii) the feed-port module, and (iii) the transition module.

schematic view of a flat coathanger die with the feedblock attachment. The die is constituted by a rectangular entry channel ($L \times W \times h = 50\,\text{mm} \times 30\,\text{mm} \times 4\,\text{mm}$), a coathanger of decreasing cross-section, a die land area of adjustable height, a relaxation area of decreasing thickness (from 4.5 to 1 mm), and finally the die lips ($L = 350\,\text{mm}$). Figure 18-3 shows a schematic of the feedblock, allowing

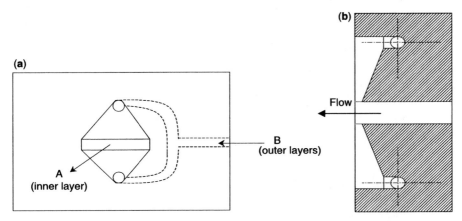

Fig. 18-3 Front view (a) and transverse cross-section (b) of the middle block of the feedblock. The rectangular flow channel of the feedblock is $W \times L \times h = 30\,\text{mm} \times 120\,\text{mm} \times 4\,\text{mm}$ and the side stream inlet tube diameter is 8 mm.

the polymer melt from each extruder to be combined into a stratified three-layer melt stream. The feedblock was designed to receive two feed streams, and the slit section has the dimensions length 120 mm, width 30 mm, and thickness 4 mm. As may be seen from Fig. 18-2, the three blocks are put together to form the feedblock. The melt stream forming the outer layers (from the polyester extruder) is split in half along two flow paths (Fig. 18-3a) and then merges with the main stream (inner layer, from the PLS extruder) in the slit section, 30 mm before the flat die entry. From the die inlet, the multilayer melt streams may flow through the die inlet and spread uniformly across the entire width of the die. The spreading of the three-layer flow across the die is due to the coathanger geometry, that is, restriction of the channel section along the flow direction. The die land area (Fig. 18-2), whose section is determined by a restriction bar, is adjustable in height by changing the thickness of the movable restriction bar. The die land channel height could be varied from 1.5 mm to 4 mm. Consequently, the flow path profile along the axial direction can be modulated from distorted to relatively smooth.

In routine experiments, the coextrusion line was run for at least 30 min to ensure steady-state operation. Initially, the flow rates of all polymers used were measured independently for each extruder, at the corresponding screw speeds, so that the A : B flow ratio could be controlled. In all experiments, the outer layer was always the minor component. After exiting the die, the coextruded film was cooled with the chill roll, and collected to analyze the wavy instabilities (magnitude and periodicity). A cooling air jet was applied to the film exiting the die in some cases, especially when the higher moisture content PLS was used, to prevent material expansion by water vaporization at the exit.

The flow rates of the polymers used were measured independently for each extruder, at the corresponding screw speeds, so that the PLS/polyester flow ratio could be controlled. In all experiments, the outer layers constituted the minor component, for cost efficiency reasons. PLS was processed within a temperature range of 100–130°C, and the outer layer polyesters were processed at 110°C, 120°C, 140°C, 160°C, and 180°C for PCL, PBSA, PEA, PLA, and PHBV, respectively. Because melt temperature differentials between respective layers should not be too great (Mitsoulis, 1988), the PLA had to be plasticized to lower its processing temperature. The effect of various plasticizers on the thermal properties of PLA has been reported previously (Martin and Avérous, 2001). In addition, a low-melting-temperature PHBV grade was used with Biopol D600G, which has a high HV content. Once the steady-state conditions were reached, the multilayer films were collected and set apart for further analyses.

18.3 CHARACTERIZATION

18.3.1 Peel Test

Peel tests were carried out on a Thwing-Albert peel tester (model 225-100) at a rate of 50 mm/min. The test specimens were conditioned at 23°C and 50% RH, and cut from the multilayer films into 100 mm × 20 mm strips prior testing. The outer polyester layer was delaminated manually and gripped onto the load cell, while the film was secured on a sliding plate, so that a constant 90° angle between the polyester layer and PLS was maintained during the test. Load data collection started after 3 seconds of pre-peel. A mean value was determined from each test, and each sample was tested 5 times.

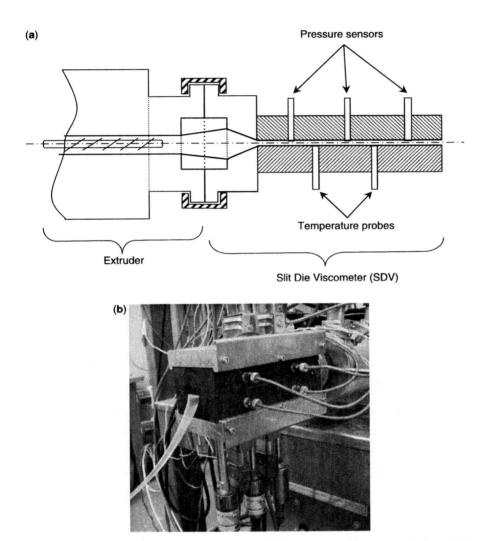

Fig. 18-4   (a) Schematic illustration of the in-line SDV rheometer; (b) photograph of the SDV system.

18.3.2 Surface Tension

A Krüss G23 goniometer (Germany) was used to measure the contact angle and to capture the kinetics of sorption (slope of the curve contact angle as a function of time) in relation to the material's hydrophobic character.

18.3.3 Optical Microscopy

A Leica transmission optical microscope was used to capture the shape of the interface between the respective layers of coextruded structures. Each film was either freeze-fractured, or cut with a diamond-like precision saw. The PLS phase was tinted with iodine vapor to better differentiate layers.

18.3.4 Melt Rheology

The viscosity of all products used in coextrusion experiments was measured using a specially designed slit die viscometer (SDV) as illustrated in Fig. 18-4. The use of this system was fully described in a previous publication (Martin et al., 2003). Shear rates in the range of $1-1000\,s^{-1}$ were obtained. PLS materials were tested in both powdery and pellet forms. To analyze the behavior of the multilayer materials, the SDV was associated with the feedblock, at the exit.

18.4 RESULTS AND DISCUSSION

18.4.1 Effect of Process Parameters: Melt-State Behavior

This part is more particularly focused on the analysis of the coextrusion of poly(ester amide) and plasticized starch. Because of the complexity of multilayer flows, attention will be paid to the rheological behavior of each component, the effects of operating conditions and viscosity ratios on the layer uniformity, and the mechanisms giving rise to instabilities at the interface.

Rheological Behavior of Neat Components   Table 18-3 presents the physical characteristics and rheological parameters of the different materials used in

TABLE 18-3   **Rheological Parameters**

Material	E/R (K)	Ko Consistency[a] (Pa s)^m
PEA	3360	9,920
LDPE	1960	9,000
PLS1	4500	19,300
PLS2	5860	12,600
PLS3	5860	10,350

[a]m = pseudoplasticity index.

this section. The viscosities were measured using the specially designed slit die viscometer at the exit of the extruder (see Fig. 18-4). The rheological properties of PEA and PLS3 products are presented in Fig. 18-5. The viscosity values of all PLS products (both powder and pellet types), determined using the SDV, are fully presented in a previous article (Martin et al., 2003). Depending on the temperature range, melt PEA viscosities lay around that of molten PLS. Melt viscosity ratios between PLS and PEA ranging from 1 to 5 could thus be obtained, by control of the process temperatures of the respective polymers. It should be noted that PEA melt temperature below 150°C could not be obtained; otherwise, some fluctuations of the extruder torque occur. Unsteady torque at given screw speed is known to cause flow disturbances due to pressure variations and to promote interfacial instabilities (Khomani, 1990a; Su and Khomani, 1992a,b; Wilson and Khomani, 1992) (see Fig. 18-6).

Interface Deformation and Encapsulation Experiments were carried out without any die to check whether the three layers were evenly distributed and were uniform in thickness at the feedblock exit. Extrusion grade LDPE was used in both extruders to perform these trials. To distinguish the interface between respective layers, the central layer and the cap layers were pigmented in black and white, respectively. Uniform layer distribution and flat interfaces are obtained when the LDPE streams have similar temperatures. However, when a melt differential as low as 20°C was imposed, the outer layers started to surround the central layer, along the 30 mm flow length (from the merging point to the feedblock exit). This configuration, consisting of a higher melt temperature and lower viscosity polymer as the cap layers, was chosen to closely reflect the conditions of PLS/polyester coextrusion. In effect, the temperature processing range of PLS is limited and starch melt viscosity is generally higher than that of PEA. The layer rearrangement observed, where the less viscous outer layers start to surround the other layer, is called "encapsulation." The encapsulation of the more viscous layer by the less viscous one is a well-known phenomenon, and has been extensively dealt with in the literature (Dooley, 2005). Other authors (Southern and Ballman, 1973; Lee and White, 1974) have shown that when the viscosity ratio increases, the degree of encapsulation increases accordingly. The interface shape change is explained by viscous encapsulation, in which layers rearrange themselves to minimize the total energy. The encapsulation effects may be decreased by reducing the viscosity difference between the layers.

However, the viscosity difference is not the only factor inducing encapsulation. The shape and length of the flow channel also have a nonnegligible influence. When the above three-layer systsem was passed through the feedblock associated with the SDV, some degree of encapsulation was found at the exit of the SDV with a viscosity ratio close to unity. The cap layers in contact with the die wall are exposed to higher shear stresses and greater heat dissipation than the central layer, resulting in a progressive viscosity evolution. A specific geometric design of the flow channel should compensate for these effects (Dooley, 2005).

When PLS and PEA were coextruded together, the PEA cap layers almost encapsulated the central layer, whatever the flow conditions. The cap component commonly encapsulated the central layer at the side wall of the die. The degree of

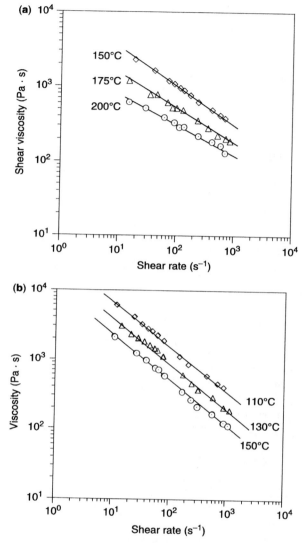

Fig. 18-5 Rheological properties of (a) PLS3 and (b) PEA at various temperatures. Source: Martin and Avérous (2002).

encapsulation, referred to the percentage area surrounded by the outer layers, ranged from partial (0–50%) to total (100%) in the film-forming die. Decreasing the cap layer flow rate usually resulted in a decrease of the extent of encapsulation on the edges of film, from 10–15 mm down to 2–4 mm. But the occurrence of encapsulation was not thought to be very critical in this case since it commonly affects the edges of the coextruded structures only, which are usually discarded.

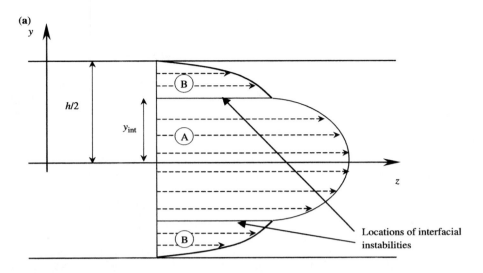

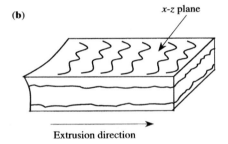

Fig. 18-6 (a) Velocity profile of a three-layer B/A/B through parallel plates; (b) schematic of a multilayer film exhibiting wavy instabilities at the interfaces; (c) photograph of a coextruded and delaminated film exhibiting wavelike instabilities.

Interfacial Instability   Unlike encapsulation effects, the occurrence of interfacial instabilities is very critical because it directly affects the quality and functionality of the coextruded films (Dooley, 2005). Figure 18-6a is an illustration of the velocity profile of a three-layer flow through a slit die of gap $y = h$. Owing to the symmetry, only one half of the channel height is considered. Under stable flow conditions, the interfaces of the final multilayer film may be flat and smooth. Conversely, interfacial instabilities manifest themselves as wavelike distortions of the interface between two polymers across the width of the film, as shown by the schematic (Fig. 18-6b) and the photograph (Fig. 18-6c).

In the system studied, consisting of shear-thinning polymers in simple shear flow, several mechanisms can give rise to interfacial instabilities. According to our experimental results, interfacial instability set in within the die. The variables known to play a important role in the occurrence of instabilities are the skin-layer viscosity, the layer thickness ratio, the total extrusion rate, and the die gap. Moreover, interfacial instabilities are known to result when a certain shear stress value at the interface of two polymers is exceeded. The onset of interfacial instabilities can thus be characterized by the critical interfacial shear stress (CISS). The interfacial shear stress τ_{int} can be calculated from the known coextrusion conditions and the flow characteristics by equation (18.1) where τ_w is the shear stress at the die wall (at $y = h/2$), y_{int} is the position of the interface, and h is the slit thickness.

$$\tau_{int} = \frac{\tau_w \cdot y_{int}}{(h/2)} \qquad (18.1)$$

The shear stress at the wall is determined by equation (18.2) where $(-dp/dz)$ is the axial pressure gradient measured by the wall normal stress measurement.

$$\tau_w = \frac{(dp/dz)}{(h/2)} \qquad (18.2)$$

According to equation (18.1), determining the interface position for a given flow ratio allows one to determine the shear stress at the die wall. To measure the interface positions inside the die, the solidified sprues obtained from different experiments were cut into slices along the flow direction, as presented in Fig. 18-7. In the absence of a direct visualization inside the flow channel, the cross-sections of the sprue allow us to determine the repartition of the layers in the distribution channel, the flow restriction zone, the relaxation zone, and the die lips, respectively. The exact interface position was measured with a binocular. The influence of main variables were investigated, while keeping other conditions constant, and flow conditions from stable to unstable were obtained. At the point of incipient instabilities, the CISS can be calculated thanks to the known flow conditions and equations (18.1) and (18.2).

Effect of Viscosity Ratio and Layer Thickness Ratio   PLS (dry-blend) and PEA were coextruded together through the feedblock and wide, flat film die. The conditions for the coextrusion of the PEA/PLS3/PEA three-layer system are reported in

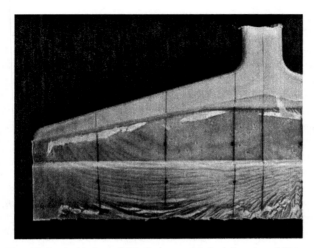

Fig. 18-7 Half-die solidified sprue with incipient instabilities.

Table 18-4. Note that only the total flow rate was measured, and that the individual flow rates were derived from the total flow rate and those determined from corresponding extruder speeds. The total flow rate was maintained in the same range in all experiments, to study the effects of viscosity and thickness ratio only. The flat die temperature was set close to the higher melt temperature layer, so that the

TABLE 18-4   Experimental Coextrusion Conditions of "PEA/PLS3/PEA" 3-Layer System[a]

Test No.	PLS3–Melt Temperature T_{PLS} (°C)	PEA–Melt Temperature T_{PEA} (°C)	Flow-Rate Q_T (g/min)	Thickness Ratio h_{PEA}/h_{PLS}	Interfacial Shear Stress τ (kPa)	Level of Stability
1	110	150	880	0.24	65	Unstable
2	110	150	864	0.38	52	Small instabilities
3	110	150	982	0.58	51	Stable
4	110	150	904	0.79	42	Stable
5	130	150	819	0.26	86	Unstable
6	130	150	870	0.36	90	Unstable
7	130	150	915	0.64	48	Small instabilities
8	130	150	959	0.84	56	Unstable
9	150	150	873	0.27	73	Unstable
10	150	150	902	0.35	64	Unstable
11	150	150	944	0.55	55	Stable
12	150	150	928	0.81	68	Unstable

[a]*Source*: Martin and Avérous (2002).

combined stream did not solidify at the die wall. The interface positions across the entire width of the film at the die exit corresponding to experiments are observed. It may be shown that the interface is not flat, due to some irregular layer distribution. The encapsulation that begins in the feedblock was enhanced by the flow through the die since about 10–15 mm of each edge of film was encapsulated by the cap layers. The coextrusion of the PLS/PEA three-layer system through the wide film die mostly led to unstable flow conditions, even when the viscosity of the respective layers was matched (experiments 5 to 8), and whatever the layer thickness ratio. Note that these trials were carried out with the narrow die gap geometry, that is, a channel height in the die land area as low as 1.5 mm.

Because changing variables like the viscosity of products and the layer thickness ratio did not yield satisfactory results, the combined streams were tested under similar conditions through the SDV instrumented flow channel. We assumed that the locations of interfaces inside the SDV die were equivalent to those at the wide, flat die entry channel, measured from the solidified sprue. As stated previously, the pressure gradient and interface positions allow us to calculate the wall shear stress and the interfacial shear stress. Figure 18-8 shows a plot of the shear stress as a

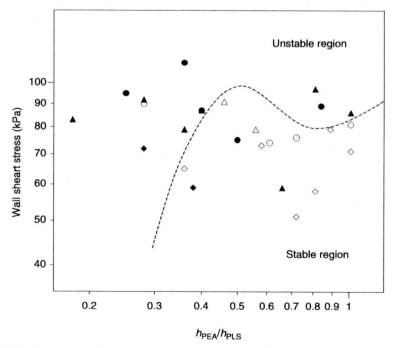

Fig. 18-8 Plot of the wall shear stress versus the layer thickness ratio of "PEA/PLS/PEA" coextruded films: $\circ$, $\bullet$, $\eta_B/\eta_A = 1$; $\triangle$, $\blacktriangle$ $\eta_B/\eta_A = 4$; $\diamond$, $\blacklozenge$ $\eta_B/\eta_A = 0.3$. Filled shapes denote unstable flow regions, whereas white ones represent stable flow conditions. Source: Martin and Avérous (2002).

function of the layer thickness ratio. Despite the scattering of data points, there seems to be a physical limit between stable and unstable regions. In addition, Fig. 18-8 shows that some extent of flow stability may be obtained when the low-viscosity component is the cap layer, with layer ratio between 0.5 and 1 and at moderate wall shear stress. Conversely, low layer thickness ratios or high shear stresses almost exclusively induced instabilities, whatever the viscosity ratio. From these results and the known interface position inside the die, the interfacial shear stress was calculated (Table 18-4). It is clear that higher shear stresses at the interface, due to higher flow rates or lower cap layer thickness, are responsible for the onset of instabilities in the multilayer flow. There seems to be a critical interfacial shear stress (CISS) value above which interfacial instabilities set in, depending on the layer thickness ratio. However, previous studies (Han and Shetty, 1978; Karagiannis et al., 1988; Mavradis and Schroff, 1994) on three-layer films demonstrated that the CISS, for a given polymer system, is independent on the process parameters varied, including layer thickness ratio. For our three-layer system, no such trend of CISS exist, since values at incipient instabilities ranged from 52 to 64 kPa, and because instability-free flows were observed beyond 55 kPa. These results are in apparent contradiction with published ones, probably due to the complexity of the system studied and specificity of the rheological behavior of PLS. It may be concluded from this section that low layer thickness ratio and viscosity differences with the cap layer as the more viscous component promote the onset of instabilities.

Effect of the Extrusion Rate and Die Geometry   The extrusion rate refers to the magnitude of the total volumetric flow rate measured at the exit of the die. Enhancing extrusion rates result in increasing pressure gradients, and thus increasing shear stress at the die wall. As shown by Han and Shetty (1978) in coextrusion of films, the shear stress is continuous across the interface. Therefore, higher interfacial shear stresses may be reached at high extrusion rates. Moreover, the height of the slit section of the die land may be varied between 1.5 and 4 mm. A change in the slit section height, before the relaxation zone, is expected to cause a diminution of the global pressure drop across the die, thus reducing the shear stress level.

Experiments were aimed at reducing the shear rate in the die. The flow rates of individual melt streams were accordingly varied to increase the extrusion rate across the flat film die of decreasing slit height. The resulting flow rate Q_T was measured at the die exit and the global pressure drop was recorded. The corresponding data are shown in Table 18-5. The most remarkable result is that no flow stability can be obtained with the narrow die gap geometry (G_1 with $h = 1.5$ mm), whereas some instability-free three-layer films were obtained through the larger die land slit section, principally with the G_3 geometry. Increasing the die temperature to $170°C$ did not seem to help. In fact, increasing the die land slit section induces significant decrease in the shear stress at the wall and at the interface. For instance, in condition 5, the shear stress ranged from 10^5 to 8×10^3 between the G_1 and the G_3 geometries. However, poor layer distribution resulted from the increase of the slit height. The elimination of interfacial instabilities was achieved at the expense of the uniformity of the product. The magnitude of shear stress change through modulation of the

TABLE 18-5 Effect of Extrusion Rate and Die Geometry on the Stability of Coextrusion Flows, Die Temperature $T_{\text{Die}} = 150°C^a$

		Global Pressure Drop $(MPa)^b$			Flow Stability^c		
Test No.	Total Flow Rate Q_T (g/min)	G_1	G_2	G_3	1	2	3
1	48	4.6	3.1	2.5	s	s	s, bld
2	56	5.1	3.9	3.1	u	s	s, bld
3	65	5.8	5.2	4.0	u	s, bld	s, bld
4	78	6.8	5.8	4.5	u	s, bld	u, bld
5	91	7.8	6.4	4.9	u	u, bld	u, bld
6	112	8.6	7.1	5.5	u	u, bld	u, bld

^{a}*Source*: Martin and Avérous (2002).
^{b}Subscripts 1, 2, and 3 designate the height of the slit section in die land area of 1.5, 2.5 and 3.5 mm, respectively.
^{c}s designates a stable flow; u designates an unstable flow; and bld means "bad layer distribution."

geometry and extrusion rate is significant. Increasing extrusion rates and reducing the flow section both favor the onset of instabilities. These results are in good agreement with previously published ones. One may conclude that the film coextrusion of polymers is a compromise between many factors, and that the influence of each parameter needs to be fully ascertained.

Control of the Instability Amplitude with the Shear Rate Ranjbaran and Khomani (1996) showed that a controlled amount of instability can significantly increase interfacial strength of multilayers, thanks to mechanical interlocking. Wang et al. (2000) also described the possible advantages of instabilities in terms of interfacial bonding increase.

We saw that the occurrence of interfacial instabilities is closely related to the shear stress at the interface. As a result, there may be several ways to increase the instabilities in coextruded film, such as increasing the extrusion rates, lowering the outer layer thickness, or reducing the gap in the die land area. Figure 18-9a shows that the amplitude of wave-instabilities is closely dependent on the shear rate. Figure 18-9b shows a micrograph of the cross-sections of PLS-PEA sprues, giving evidence of the penetration of the PEA layer into the PLS phase. In the micrograph, the dark phase is the PLS (tinted with iodine vapor) and the lighter outer phase is the PEA. The amplitude gradually increased from 10 to 170 μm as the shear rate increased. However, that clear trend was not observed for all coextrusion experiments, in particular when the mechanism giving rise to instabilities was not controlled, such as elasticity difference between layers, die exit phenomenon, or velocity profile difference. Finally, it should be added that the extent and, to a certain degree, the amplitude of interfacial instabilities could be readily controlled, and that the main conditions likely to generate instabilities are known. Under control, these instabilities can generate mechanical interlocking and so increase the interfacial adhesion strength on the coextruded material (Schwach and Avérous, 2004).

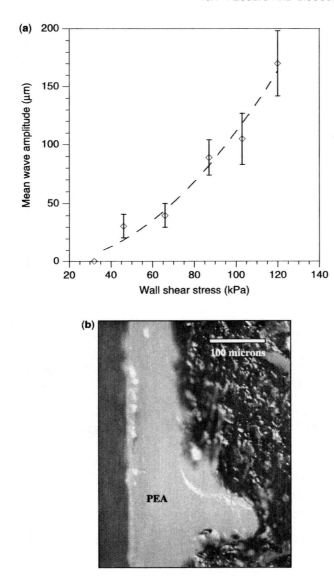

Fig. 18-9   (a) Mean wave amplitude as a function of shear stress; (b) illustration of the mechanical interlocking and wave amplitude for PEA/PLS2/PEA films. Source: Martin and Avérous (2002).

18.4.2 Coextrusion Analysis of the Solid-State Behavior

This section is more particularly focused on the analysis of the coextruded materials after cooling. The behavior of several polyesters has been tested in association with different PLS formulations.

Adhesion Strength Between PLS and Biopolyester Layers   Figure 18-10 shows peel strength results of coextruded films, based on PLS3 and different polyesters. The polyester amide (PEA) is seen to have the highest peel strength, which is linked to its polar nature; PBSA and PCL have medium values; and PLA and PHBV have the lowest ones. Although peel strength does not constitute an absolute measure of adhesion, it reflects the compatibility or affinity between PLS and the respective polyesters. This result confirms well some previous findings on the compatibility of plasticized starch with polyesters (Avérous, 2004) through the study of blended systems. These results are in good agreement with those of Biresaw and Carriere (2000) or, more recently, with those of Schwach and Avérous (2004).

It is well known that plasticizers migrating toward the surface of polymers can significantly affect the adhesion. The amount of plasticizer (water, glycerol) added to starch for processing purposes is likely to affect the adhesion strength between the layers of plasticized starch–polyester structures. We have previously reported the tendency of highly plasticized starch to exude toward the surface with time (Avérous et al., 2000a). Table 18-6 presents the peel strength data of PLS/polyester coextruded structures, with different PLS formulations as the central layer, and with various biopolyesters as the outer layers. From Table 18-6 it is obvious that the peel strength decreases as the glycerol/starch (G/S) ratio increases. For instance, the peel strength of PLS/PLA laminates ranges from 0.12 to 0.05 N/mm for G/S ratio comprised between 0.14 and 0.54. These results seem to show that polyol migration to the interface occurs readily, whatever the polyester skin layer. This is in agreement with the results of Wang et al. (2000). However, the multilayers prepared with PLS pellets (conditions 7 to 10, Table 18-6) did not show the same trend, in that the peel strength data lie around 0.2 N/mm while the G/S ratio varied from 0.14 to 0.50. This result could be interpreted in terms of the diminution of the water content after extrusion and equilibration of the PLS granules (Table 18-1). We obtained higher polysaccharide–glycerol interactions which are linked to exchanges of

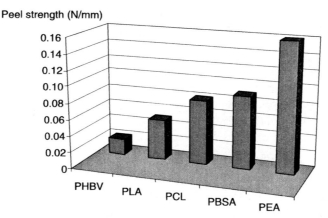

Fig. 18-10   Effect of the type of biopolyester on the peel strength of PLS3-based films.

TABLE 18-6   Effect of PLS Composition on the Peel Strength of Starch–Biopolyester Multilayers[a]

Test No.	Polyester	PLS Type	Film Thickness[b] (μm)	Polyester Layer Thickness[b] (μm)	Peel Strength (N/mm)
1	PEA[c]	PLS1	920	120	0.22 (0.05)[e]
2	PEA	PLS2	910	130	0.18 (0.04)
3	PEA	PLS3	830	120	0.16 (0.03)
4	PLA	PLS1	890	140	0.12 (0.02)
5	PLA[c]	PLS2	900	160	0.11 (0.01)
6	PLA	PLS3	850	140	0.05 (0.01)
7	PEA	PLS1[d]	900	135	0.22 (0.06)
8	PEA[c]	PLS2[d]	880	120	0.24 (0.05)
9	PLA	PLS2[d]	860	115	0.12 (0.04)
10	PEA[c]	PLS3[d]	820	110	0.19 (0.04)

[a]*Source*: Martin et al. (2001).
[b]The thickness was measured at the center, according to the film width.
[c]The multilayer film contained some defects, i.e., instabilities at the interface.
[d]The PLS had previously been extruded and pelletized.
[e]Values between brackets are the standard deviations.

occupation of some strong sites from water to glycerol molecules. This phenomenon induces a decrease of glycerol migrations.

The aging of films is known to influence the peel strength data (Wang et al., 2000). All our tests were performed on 1-week equilibrated (at 23°C, 50% RH) specimens, unless specified otherwise. It was noticed that the adhesion strength varied with time. For instance, the peel strength of PLS/PHBV ranged from slightly adhesive (0.02 N/mm, Fig. 18-10) to nonadhesive, since the PHBV layers spontaneously peeled off the starch layer after 50 weeks. In that case, it was not clearly identified whether the aging, the glycerol migration, or the low compatibility was responsible for the loss of adhesion.

Strategy to Improve Interfacial Adhesion   A strategy was tested to improve the adhesion strength between plasticized starch and some biopolyesters. The blended composition was used in the inner starch layer, to enhance the ties between starch and polyester layers. The opposite strategy, adding PLS into the cap layers, must be avoided since the skin layers act as water barriers.

The polyesters constituting the minor phase of blend (i.e., 5 wt% and 10 wt%) were blended with PLS. The influence of a blended PLS inner layer was tested with a selection of three polyesters: PEA, PLA, and PCL. The effects on the peel strength are presented in Fig. 18-11. The dark bars correspond to the peel strength results of coextruded starch-polyester films from Fig. 18-10. The gray and white bars represent the peel strength values of multilayers where the central layers are starch–polyester blends comprising 5% and 10% of polyester, respectively. An increase of 19% to 31% was observed for PEA-based films and 37% to 50% for

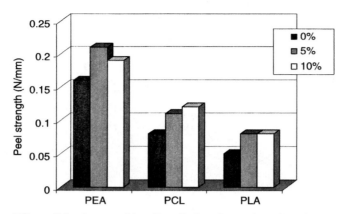

Fig. 18-11 Effect of blend composition (0 wt%, 5 wt%, or 10 wt%) and the type of biopolyester on the peel strength of PLS3/polyester coextruded films. Source: Martin et al. (2001).

PCL-based films, and an increase in peel strength of 40% was observed for PLA-based films. This strategy yielded satisfactory results, although the presence of some defects on the films, referred to as interfacial instabilities (Wang et al., 2000), may partially alter the results. Moreover, these improvements were obtained at the cost of a supplementary operation and use of more polyester, which is not fully desirable from an economic point of view.

Moisture Resistance Properties   Figure 18-12 shows the results of contact angle measurements performed on starch–polyester systems, as well as on neat polyesters and plasticized starch. PLS3-blends based on PEA, PLA, and PCL were chosen. Low values of the slope (in absolute value) indicate stronger hydrophobic character of the products. For all blends, it was observed that the hydrophilic character decreased rapidly from 0% to 10% polyester content and kept decreasing until the value of neat biopolyester. PLS/PLA blends, whatever the composition, showed the best characteristics. As expected, better barrier properties can be obtained with starch products laminated with moisture-resistant polyesters, compared with blends. Immersion tests of starch–polyester multilayers in water were also undertaken to check their ability to resist water penetration. Water penetration is known to have detrimental effects on the multilayers, such as the swelling of PLS layers, resulting in the loss of elastic modulus and spontaneous delamination of the cap layers. Over a few days, no significant product swell or delamination was observed. That result is satisfactory since it shows that multilayers have good resistance to moisture. This confirms results of Shogren (1997), who indicated that, although biopolyesters have lower water barrier properties than polyethylene, they can provide sufficient protection to starch products for short-terms applications. Finally, it is interesting to note that the moisture resistance of polyesters (PLA > PCL > PEA) is in inverse order to that of adhesion.

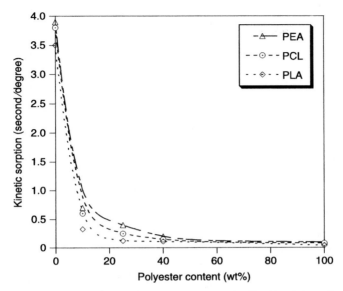

Fig. 18-12   Effect of the polyester content (PEA, PLA, and PCL, % w/w) on the hydrophilic character of the product. Source: Martin et al. (2001).

18.5   CONCLUSION

The development of compostable and low-cost multilayer materials based on plasticized starch and biodegradable polyesters is interesting in more than one sense. One requisite in the preparation of multilayered products based on starch is to obtain sufficient adhesion between layers, moisture barrier properties, and uniform layer thickness distribution.

Experimental results on melt-state during the process show that the key parameters are the skin-layer viscosity and thickness, the global extrusion rate, and the die geometry. Under certain conditions, reasonable layer uniformity was obtained across the wide die width, and encapsulation of the central starch layer at both edges of the die could not be avoided, even when the cap layers were the highest viscosity component. However, side wall encapsulation is not thought to be very critical. Closer attention should be given to the interfacial instabilities since they may be more detrimental to the coextruded material and final product. The occurrence of instabilities is strongly related to the shear stress at the interface. Flow conditions ranging from stable to unstable can be obtained. Investigation of the flow behavior of starch/ PEA multilayers allows us to classify the instabilities according to the dominating factors, such as shear stress-, viscosity difference-, or melt stream confluence point-driven instabilities. The interfacial instability amplitude of multilayer films can be controlled through the wall shear stress. Under control, these instabilities can generate mechanical interlocking and thus increase the interfacial adhesion strength of the coextruded material.

Experimental results on coextruded materials show that different levels of peel strength can be found, depending on the compatibility of plasticized starch with the respective biopolyesters. In particular, poly(ester amide) (PEA) presents the highest adhesion with the PLS layer, probably due to its polar amide groups. PCL and PBSA showed medium adhesion values, and PLA or PHBV were the less-compatible polyesters. However, it was possible to increase the adhesion properties of the film by up to 50%, by introducing polyester blends in the central layer. We show that the different multilayers exhibit satisfactory water resistance properties.

The use of the coextrusion technique is validated for preparation of compostable multilayer films based on plasticized starch for such uses as short-term packaging. Nevertheless, the scope of these investigations is limited on one hand by the narrow processing range of PLS products and, on the other, by the lack of viscoelastic data for low-moisture starch melts. In many cases, elastic properties of polymer melts are equally important in the mechanism of interfacial instabilities.

There are some inherent problems due to the multilayer flow conditions encountered in coextrusion, such as the phenomena of encapsulation and interfacial instabilities. It is crucial to address these problems because they can be detrimental to the product, affecting quality and functionality.

ACKNOWLEDGMENTS

This work was funded by Europol'Agro (Reims, France) through a research program devoted to the study and development of new packaging materials based on renewable resources. The authors thank Dr. Olivier Martin and Dr. Emmanuelle Schwach for their inputs into these studies.

REFERENCES

Amass W, Amass A, Tighe B. 1998. A review of biodegradable polymers: uses, current developments in the synthesis and characterization of biodegradable polyesters, blends of biodegradable polymers and recent advances in biodegradation studies. *Polym Int* 47:89–144.

Avérous L. 2004. Biodegradable multiphase systems based on plasticized starch: a review. *J Macomol Sci Part C Polym Rev* C4:231–274.

Avérous L, Fringant C. 2001. Association between plasticized starch and polyesters: processing and performances of injected biodegradable systems. *Polym Eng Sci* 41:727–734.

Avérous L, Fringant C, Martin O. 1999. Coextrusion of biodegradable starch-based materials. In: *Biopolymer Science: Food and Non-Food Applications*. Colonna P, Guilbert S, editors. Paris: INRA. p. 207–212.

Avérous L, Fauconnier N, Moro L, Fringant C. 2000a. Blends of thermoplastic starch and polyesteramide: processing and properties. *J Appl Polym Sci* 76:1117–1128.

Avérous L, Moro L, Dole P, Fringant C. 2000b. Properties of thermoplastic blends: starch-polycaprolactone. *Polymer* 41:4157–4167.

Bastioli C, Cerutti A, Guanella I, Romano GC, Tosin M. 1995. Physical state and biodegradation behavior of starch–polycaprolactone systems. *J Environ Polym Degrad* 3:81–95.

Belard L, Dole P, Avérous L. 2005. Current progress on biodegradable materials, based on plasticized starch. *Aust J Chem* 58:457–460.

Biresaw G, Carriere CJ. 2000. Interfacial properties of starch/biodegradable esters blends. *Polymer preprints* 41:64–65.

Dooley J. 2005. Co-extrusion instabilities. In: *Polymer Processing Instabilities: Control and Understanding*. Hatzikiriakos S, Migler K, editors. New York: CRC Press. p. 383–426.

Dooley J, Stout B. 1991. An experimental study of the factors affecting the layer thickness uniformity of coextruded structures. *Soc Plast Eng Annu Tech Meet* 37:62–65.

Dooley J, Hyun KS, Hugues K. 1998. An experimental study on the effect of polymer viscoelasticity on layer rearangement in coextruded structures. *Polym Eng Sci* 38:1060–1071.

Gifford WA. 1997. A three dimensional analysis of coextrusion. *Polym Eng Sci* 37:315–320.

Han CD. 1981. *Multiphase Flow in Polymer Processing*. New York: Academic Press.

Han CD, Shetty R. 1976. Studies on multilayer film coextrusion I The rheology of flat film coextrusion. *Polym Eng Sci* 16:697–705.

Han CD, Shetty R. 1978. Studies on multilayer film coextrusion II. Interfacial instability in flat film coextrusion. *Polym Eng Sci* 18:180–186.

He Y, Asakawa N, Masuda T, Cao A, Yoshie N, Inoue Y. 2000. The miscibility and biodegradability of poly3-hydroxybutyrate blends with polybutylene succinate-*co*-butylene adipate and polybutylene succinate-*co*-caprolactone. *Eur Polym J* 36:2221–2229.

Hickox CE. 1971. Instability due to viscosity stratification in axisymmetric pipe flow. *Phys Fluids* 14:251–262.

Huang SJ, Koening MF, Huang M. 1993. Design, synthesis, and properties of biodegradable composites. In: *Biodegradable Polymers and Packaging*. Ching C, Kaplan DL, Thomas EL, editors. Basel: Technomic Publication. p. 97–110.

Hulleman SHD, Janssen FHP, Feil H. 1998. The role of water during plasticization of native starches. *Polymer* 39:2043–2048.

Hulleman SHD, Kalisvaart MG, Janssen FHP, Feil H, Vliegenthart JFG. 1999. Origins of B-type crystallinity in glycerol-plasticised compression-moulded potato starches. *Carbohydr Polym* 39:351–360.

Huneault M, Li H, 2007. Morphology and properties of compatibilized polylactide/thermoplastic starch blends. *Polymer* 48:270–280.

Karagiannis A, Mavridis H, Hrymak AN, Vlachopoulos J. 1988. Interface determination in bicomponent extrusion. *Polym Eng Sci* 28:982–988.

Khan AA, Han CD. 1976. On the interface deformation in the stratified two-phase flow of viscoelastic fluids. *Trans Soc Rheol* 20:595–621.

Khomani B. 1990a. Interfacial instability and deformation of two stratified power law fluids in plane poiseuille flow 1 Stability analysis. *J Non-Newton Fluid Mech* 36:289–303.

Khomani B. 1990b. Interfacial instability and deformation of two stratified power law fluids in plane poiseuille flow 2 Interface deformation. *J Non-Newton Fluid Mech* 37:19–36.

Koenig MF, Huang SJ. 1995. Biodegradable blends and composites of polycaprolactone and starch derivatives. *Polymer* 36:1877–1882.

Krook M, Morgan G, Hedenqvist MS. 2005. Barrier and mechanical properties of injection molded montmorillonite/polyesteramide nanocomposites. *Polym Eng Sci* 45:135–141.

Lawton JW. 1997. Biodegradable coatings for thermoplastics starch. In: *Cereals: Novel Uses and Process*. Campbell GM, Webb C, McKee SL, editors. New York: Campbell Plenum Press. p. 43–47.

Lee BL, White JL. 1974. An experimental study of rheological properties of polymer melts in laminar shear flow and of interface deformation and its mechanisms in two-phase stratified flow. *Trans Soc Rheol* 18:467–492.

Martin O, Avérous L. 2001. Polylactic acid: plasticization and properties of biodegradable multiphase systems. *Polymer* 42:6209–6219.

Martin O, Avérous L. 2002. Comprehensive experimental study of a starch/polyesteramide coextrusion. *J Appl Polym Sci* 86:2586–2600.

Martin O, Schwach E, Avérous L, Couturier Y. 2001. Properties of biodegradable multilayer films based on plasticized wheat starch. *Starch/Staerke* 37:372–380.

Martin O, Avérous L, Della Valle G. 2003. Inline determination of plasticized wheat starch viscous behaviour: Impact of processing. *Carbohydr Polym* 53:169–182.

Matzinos P, Tserki V, Kontoyiannis A, Panayiotou C. 2002. Processing and characterization of starch/polycaprolactone products. *Polym Degrad Stab* 77:17–24.

Mavradis H, Schroff R. 1994. Multilayer extrusion: Experiments and computer simulation. *Polym Eng Sci* 34:559–569.

Mitsoulis E. 1988. Multilayer sheet coextrusion: analysis and design. *Adv Polym Tech* 8:225–242.

Myllymäki O, Myllärinen P, Forssell P, et al. 1998. Mechanical and permeability properties of biodegradable extruded starch/polycaprolactone films. *Pack Technol Sci* 11:265–274.

Narayan R, Krishnan M. 1995. Biodegradable composites and blends of starch with polycaprolactone. *Polym Mater Sci Eng* 72:186–187.

Perdikoulias J, Richard C, Vlcek J, Vlachopoulos J. 1991. A study of coextrusion flows in polymer processing. In: ANTEC Proceedings, SPE (Society of Plastics Engineers), Montreal, p. 2461–2464.

Pranamuda H, Tokiwa Y, Tanaka H. 1996. Physical properties and biodegradability of blends containing polycaprolactone and tropical starches, *J Environ Polym Degrad* 4:1–7.

Ramsay BA, Langlade V, Carreau PJ, Ramsay JA. 1993. Biodegradability and mechanical properties of polyhydroxybutyrate-*co*-hydroxyvalerate–starch blends. *Appl Environ Microbiol* 59:242–1246.

Ranjbaran MM, Khomani B. 1996. The effect of interfacial instability on the strength of the interface in two-layer plastic structures. *Polym Eng Sci* 36:1875–1885.

Ratto JA, Stenhouse PJ, Auerbach M, Mitchell J, Farrell R. 1999. Processing, performance and biodegradability of a thermoplastic aliphatic polyester/starch system. *Polymer* 40:6777–6788.

Rouilly A, Rigal L. 2002. Agro-materials: a bibliographic review. *J Macromol Sci Part C Polym Rev* 4:441–479.

Schrenck WJ, Alfrey T Jr. 1978. Coextruded multilayer films and sheets, In: *Polymer Blends*, Volume 2. Paul DR, Newman S, editors. London: Academic Press. p. 129–165.

Schrenck WJ, Bradley NL, Alfrey T Jr, Maack H. 1978. Interfacial flow instability in multilayer coextrusion. *Polym Eng Sci* 18:620–623.

Schwach E, Avérous L. 2004 Starch-based biodegradable blends: morphology and interface properties. *Polym Int* 53:2115–2124.

Shogren RL. 1997. Water vapor permeability of biodegradable polymers. *J Environ Polym Degrad* 5:91–95.

Shogren RL, Lawton JW. 1998. Enhanced water resistance of starch-based materials. US Patent 5,756,194.

Sornberger G, Vergnes B, Agassant JF. 1986a. Coextrusion of two molten polymers between parallel plates: non-isothermal computation and experimental study. *Polym Eng Sci* 26:682–689.

Sornberger G, Vergnes B, Agassant JF. 1986b. Two directional coextrusion flow of two molten polymers in flat dies. *Polym Eng Sci* 26:455–460.

Southern JH, Ballman RL. 1973. Stratified bicomponent flow of polymer melts in a tube. *J Appl Polym Sci* 20:175–189.

Steinbuchel A. 2003. *Biopolymers, General Aspects and Special Applications.* Weinheim: Wiley-VCH.

Su YY, Khomani B. 1992a. Purely elastic interfacial instabilities in superposed flow of polymeric fluids. *Rheol Acta* 31:413–420.

Su YY, Khomani B. 1992b. Interfacial stability of multilayer viscoelastic fluids in slit and converging channel die geometries *J Rheol* 36:357–387.

Swanson CL, Shogren RL, Fanta GF, Imam SH. 1993. Starch–plastic materials—preparation, physical properties, and biodegradability. *J Environ Polym Degrad* 1:155–166.

Tomka I. 1991. Thermoplastic starch. *Adv Exp Med Biol* 302:627–637.

Tzoganakis C, Perdikoulias J. 2000. Interfacial instabilites in coextrusion flows of low-density polyethylenes: Experimental studies. *Polym Eng Sci* 40:1056–1064.

Van de Velde K, Kiekens P. 2002. Biopolymers: overview of several properties and consequences on their applications. *Polym Test* 21:433–442.

Van Soest JJG, Hulleman SHD, De Wit D, Vliegenthart JFG. 1996. Crystallinity in starch plastics. *Ind Crops Prod* 5:11–22.

Van Tuil R, Schennink G, De Beukelaer H, Van Heemst J, Jaeger R. 2000. Converting biobased polymers into food packaging. In: Claus J, Weber, editor. *The Food Biopack Conference.* Proceedings of the meeting held in Copenhagen, Denmark, August 2000. The Royal Veterinary and Agricultural University, Denmark. p. 28–30.

Verhoogt H, St-Pierre N, Truchon FS, Ramsay BA, Favis BD, Ramsay JA. 1995. Blends containing polyhydroxybutyrate-*co*-12%-hydroxyvalerate and thermoplastic starch. *Can J Microbiol* 41:323–328.

Vikman M, Hulleman SHD, Van Der Zee M, Myllarinen P, Feil H. 1999. Morphology and enzymatic degradation of thermoplastic starch–polycaprolactone blends. *J Appl Polym Sci* 74:2494–2604.

Wang L, Shogren RL, Carriere C. 2000. Preparation and properties of thermoplastic starch-polyester laminate sheets by coextrusion. *Polym Eng Sci* 40:499–506.

White JL, Ufford RC, Dharod KR, Price RL. 1972. Experimental and theoritical study of the coextrusion of two-phase molten polymer systems. *J Appl Polym Sci* 16:1313–1330.

Wilson GM, Khomani B. 1992. An experimental investigation of interfacial instabilities in multilayer flow of viscoelastic fluids 1 Incompatible polymer systems. *J Non-Newton Fluid Mech* 45:355–384.

Wilson GM, Khomani B. 1993. An experimental investigation of interfacial instabilities in multilayer flow of viscoelastic fluids. 2. Elastic and non-linear effects in incompatible polymer systems. *J Rheol* 37:315–339.

Yih CS. 1967. Instability due to viscosity stratification. *J Fluid Mech* 27:337–352.

INDEX

AA. *See* Acetic anhydride (AA)
Abaca
    chemical composition and physical
      properties, 355
    description, 288
    surface modifications, 289
Abaca composites
    flexural properties, 289
    with PLA, 289
Acetic acid, 116
Acetic anhydride (AA), 328
Acetylated chitosan, 108
Acetylated starch, 41
Acetylation, 274, 326
    dew-retted flax, 316
    fiber composite, 329
Adsorption equilibrium, 138
Agar-agar, 116
Agave. *See* Sisal
AGE. *See* Allylglycidyl-ether (AGE)
    modified potato starch
Alginate
    with cellulose, 139–142
    miscibility and interactions, 140
Alginic acid, 139
Aliphatic polyester, 316
    degradation, 402
Aliphatic polyester amide, 440
Alkaline temperature-treated corn (ATS),
    23, 58
Alkali treatment, 274
    composite tensile strength, 326
    fiber, 276
    FTIR spectra, 356
    jute fibers, 325
Alkalization, 323

Alkyl succinic anhydride (ASA), 328
Allylglycidyl-ether (AGE) modified potato
    starch, 255
Allyl-3-methylimidazolium chloride
    (AMIMCl), 134
AMF. *See* Anhydrous milk fat (AMF)
Amide bond, 7
Amido montmorillonite (AMMT)
    nanocomposites, 428
AMIMCl. *See* Allyl-3-methylimidazolium
    chloride (AMIMCl)
Amine, 96
Amino acids, 7
Amipol, 259
AMMT. *See* Amido montmorillonite
    (AMMT) nanocomposites
Amylopectin, 87
    chemical structure, 88, 212
    structure, 4, 243
Amylose, 87
    chemical structure, 88, 212
    structure, 4, 243
Anhydride
    grafting, 330, 331
    PHB-HV, 331
    polymers, 330
    treatment, 328
Anhydrous milk fat (AMF), 63
Anisotropic composites, 333
Annealed polylactide composites
    SEM, 297
    tensile properties, 296
Antibacterial materials, 136
Aqueous metal-based solvent system,
    133–134
Aramide, 252

Biodegradable Polymer Blends and Composites from Renewable Resources. Edited by Long Yu
Copyright © 2009 John Wiley & Sons, Inc.